The Subnuclear Series • Volume 51

Proceedings of the International School of Subnuclear Physics

REFLECTIONS ON THE NEXT STEP FOR LHC

THE SUBNUCLEAR SERIES*

Series Editor: ANTONINO ZICHICHI, *European Physical Society, Geneva, Switzerland*

18. 1980 THE HIGH-ENERGY LIMIT
19. 1981 THE UNITY OF THE FUNDAMENTAL INTERACTIONS
20. 1982 GAUGE INTERACTIONS: Theory and Experiment
21. 1983 HOW FAR ARE WE FROM THE GAUGE FORCES?
22. 1984 QUARKS, LEPTONS, AND THEIR CONSTITUENTS
23. 1985 OLD AND NEW FORCES OF NATURE
24. 1986 THE SUPERWORLD I
25. 1987 THE SUPERWORLD II
26. 1988 THE SUPERWORLD III
27. 1989 THE CHALLENGING QUESTIONS
28. 1990 PHYSICS UP TO 200 TeV
29. 1991 PHYSICS AT THE HIGHEST ENERGY AND LUMINOSITY: To Understand the Origin of Mass
30. 1992 FROM SUPERSTRINGS TO THE REAL SUPERWORLD
31. 1993 FROM SUPERSYMMETRY TO THE ORIGIN OF SPACE–TIME
32. 1994 FROM SUPERSTRING TO PRESENT-DAY PHYSICS
33. 1995 VACUUM AND VACUA: The Physics of Nothing
34. 1996 EFFECTIVE THEORIES AND FUNDAMENTAL INTERACTIONS
35. 1997 HIGHLIGHTS OF SUBNUCLEAR PHYSICS: 50 Years Later
36. 1998 FROM THE PLANCK LENGTH TO THE HUBBLE RADIUS
37. 1999 BASICS AND HIGHLIGHTS IN FUNDAMENTAL PHYSICS
38. 2000 THEORY AND EXPERIMENT HEADING FOR NEW PHYSICS
39. 2001 NEW FIELDS AND STRINGS IN SUBNUCLEAR PHYSICS
40. 2002 FROM QUARKS AND GLUONS TO QUANTUM GRAVITY
41. 2003 FROM QUARKS TO BLACK HOLES: PROGRESS IN UNDERSTANDING THE LOGIC OF NATURE
42. 2004 HOW AND WHERE TO GO BEYOND THE STANDARD MODEL
43. 2005 TOWARDS NEW MILESTONES IN OUR QUEST TO GO BEYOND THE STANDARD MODEL
44. 2006 THE LOGIC OF NATURE, COMPLEXITY AND NEW PHYSICS: FROM QUARK-GLUON PLASMA TO SUPERSTRINGS, QUANTUM GRAVITY AND BEYOND
45. 2007 SEARCH FOR THE "TOTALLY UNEXPECTED" IN THE LHC ERA
46. 2008 PREDICTED AND TOTALLY UNEXPECTED IN THE ENERGY FRONTIER OPENED BY LHC
47. 2009 THE MOST UNEXPECTED AT LHC AND THE STATUS OF HIGH ENERGY FRONTIER
48. 2010 WHAT IS KNOWN AND UNEXPECTED AT LHC
49. 2011 SEARCHING FOR THE UNEXPECTED AT LHC AND THE STATUS OF OUR KNOWLEDGE
50. 2012 WHAT WE WOULD LIKE LHC TO GIVE US
51. 2013 REFLECTIONS ON THE NEXT STEP FOR LHC

Volume 1 was published by W. A. Benjamin, Inc., New York; 2–8 and 11–12 by Academic Press, New York and London; 9–10 by Editrice Compositori, Bologna; 13–29 by Plenum Press, New York and London; 30–51 by World Scientific, Singapore.

*For the complete list of titles in this series, please go to

http://www.worldscientific.com/series/ss

The Subnuclear Series • Volume 51

Proceedings of the International School of Subnuclear Physics

REFLECTIONS ON THE NEXT STEP FOR LHC

Edited by

Antonino Zichichi

European Physical Society
Geneva, Switzerland

World Scientific

NEW JERSEY • LONDON • SINGAPORE • BEIJING • SHANGHAI • HONG KONG • TAIPEI • CHENNAI

Published by

World Scientific Publishing Co. Pte. Ltd.
5 Toh Tuck Link, Singapore 596224
USA office: 27 Warren Street, Suite 401-402, Hackensack, NJ 07601
UK office: 57 Shelton Street, Covent Garden, London WC2H 9HE

British Library Cataloguing-in-Publication Data
A catalogue record for this book is available from the British Library.

ISBN 978-981-4678-10-0

In-house Editor: Rhaimie Wahap

Printed in Singapore

PREFACE

In June/July 2013, a group of 154 physicists from 22 countries met in Erice to participate in the 51st Course of the International School of Subnuclear Physics. The countries represented by the participants were: Brazil, China, Colombia, Denmark, Egypt, France, Germany, Greece, Hungary, India, Italy, Japan, Mexico, the Netherlands, Romania, Russia, Spain, Switzerland, Ukraine, United Kingdom, United States of America, and Uzbekistan.

The School was sponsored by the Academies of Sciences of Estonia, Georgia, Lithuania, Russia and Ukraine; the Chinese Academy of Sciences; the Commission of the European Communities; the European Physical Society; the Italian National Institute for Nuclear Physics; the Italian Ministry of Education, University and Scientific Research; the Sicilian Regional Government; the Weizmann Institute of Science; the World Federation of Scientists and the World Laboratory.

This year's Course was the third of the three dedicated to the celebrations of the 50th Anniversary of the School. Why three? Because Subnuclear Physics School needed three years for being established. In 1961 we started to discuss the bridge between University teaching and International Research Centres like CERN. In 1962 a formal paper for its existence was signed at CERN by Blackett, Rabi, Weisskopf and two young fellows, J.S. Bell and me. The first Course in Erice was held in 1963. The three years celebrations are coupled with big discoveries. In 2011, the negative sign of the β–function and asymptotic freedom; in 2012, QCD and in 2013, SSB and Instantons.

As usual, the lectures dealt with the latest and most significant achievements in both theoretical and experimental physics. Two special sessions were dedicated to celebrate SSB and Instantons. Highlights from the most important Laboratories and Astroparticle experimental facilities were presented, while the theoretical lectures were focused on fascinating topics such as Supersymmetry, Supergravity and Black Holes Physics. Several lectures have been devoted to reflections about the next steps for LHC, not forgetting that, if Complexity exists at the fundamental level, a totally unexpected discovery should be given to us in the forthcoming years.

An original feature of the School, introduced in 1996, is a series of special sessions devoted to "New Talents", with the aim of encouraging and promoting young physicists to achieve recognition at an international level. This is a serious problem in Experimental Physics, where collaborations count several hundreds of participants and it is almost impossible for young fellows to be known. Even if with much less emphasis, the problem exists also in Theoretical Physics. So it was decided to offer 51 young fellows the possibility to give an open presentation of the results of their studies, followed by a discussion. Among them 26 students were awarded a diploma; it was also decided to award a special diploma in honour of Bruno Zumino's 90th anniversary. One award was also assigned for the best student.

During the organization and the running of the Course, I enjoyed the collaboration of my colleague and friend, Gerardus 't Hooft, who shared with me the Directorship of the School. I would like to thank him, together with the group of invited scientists and all the people who have contributed to the success of the Course.

I hope the reader will enjoy the book as much as the students attending the lectures and the discussion sessions, which, as every year, have been the focal point of this School activity. Thanks to the work of the Scientific Secretaries, the discussions have been reproduced as faithfully as possible. At various stages of my work I have enjoyed the collaboration of many friends whose contributions have been extremely important for the School and are highly appreciated. I thank them most warmly. A final acknowledgement to all those in Erice, Bologna and Geneva who have helped me in so many occasions and to whom I feel very indebted.

Antonino Zichichi

Geneva, October 2013

CONTENTS

«ETTORE MAJORANA» FOUNDATION AND CENTRE FOR SCIENTIFIC CULTURE
TO PAY A PERMANENT TRIBUTE TO GALILEO GALILEI, FOUNDER OF MODERN SCIENCE AND TO ENRICO FERMI, THE "ITALIAN NAVIGATOR", FATHER OF THE WEAK FORCES

INTERNATIONAL SCHOOL OF SUBNUCLEAR PHYSICS

50TH ANNIVERSARY CELEBRATIONS 2011-2012-2013

51st Course: ***REFLECTIONS ON THE NEXT STEP FOR LHC***

ERICE-SICILY: 24 June – 3 July 2013

OPENING REMARKS

Antonino Zichichi
INFN and University of Bologna, Italy
CERN, Geneva, Switzerland
Enrico Fermi Centre, Rome, Italy
World Federation of Scientists, Beijing, Geneva, Moscow, New York

Paul A.M. Dirac Lecture Hall – Patrick M.S. Blackett Institute
Erice, 24 June 2013

OPENING REMARKS

Antonino Zichichi
INFN and University of Bologna, Italy
CERN, Geneva, Switzerland
Enrico Fermi Centre, Rome, Italy
World Federation of Scientists, Beijing, Geneva, Moscow, New York

OPENING REMARKS

PROGRAMME AND LECTURERS

SSB CELEBRATION

My Life as Boson
• P.W. HIGGS, University of Edinburgh, UK

INSTANTON CELEBRATION

Origin and Status of Instantons
• G. 't HOOFT, Utrecht University, NL

OPENING REMARKS AND REFLECTIONS

• A. ZICHICHI, CERN, Geneva, CH; INFN & University of Bologna, IT

HOT THEORETICAL TOPICS

Mass Hierarchy and Physics Beyond the Standard Model
• I. ANTONIADIS, Ecole Polytechnique, Palaiseau, FR

Magic Supergravity from Squaring Yang Mills
• M.J. DUFF, Imperial College London, London, UK

Gauge Forces: From QCD to Quantum Gravity
• L.N. LIPATOV, Saint-Petersburg State University, RU

The Internal Structure of the Weak Bosons
• H. FRITZSCH, Ludwig-Maximilians-Universitaet, Muenchen, DE

Embedding Oscillatory Modes of Quarks for Baryons in QCD
• P. MINKOWSKI, University of Bern, CH

No-scale Supergravity in the Light of LHC and Planck
• D.V. NANOPOULOS, Texas A&M University, College Station, TX, US

Inflation and Quantum Origin of Structure in the Universe
• V. MUKHANOV, Ludwig Maximilian University of Munich, DE

The Pedagogic Higgs – or Somebody's – Boson
• F. CLOSE, University of Oxford, UK

New Symmetries of N=4 Supersymmetric Gauge Theories
• E. SOKATCHEV, LAPTH, Annecy-le-Vieux Cedex, FR

Dualities in Supergravity
• S. FERRARA, CERN, Geneva, CH; LNF - INFN, Frascati, IT

Status of the Perturbative Approach to Supergravity
• Z. BERN, UCLA, Los Angeles, CA, US

Discreteness and Determinism in Superstring Theory
• G. 't HOOFT, Utrecht University, NL

HOT EXPERIMENTAL TOPICS

Highlights from ATLAS – ALICE – CMS
• S. BERTOLUCCI, CERN, CH

Status of the Gran Sasso Lab
• F. FERRONI, INFN, Rome, IT

Latest Results from BNL and RHIC
• M.J. TANNENBAUM, Brookhaven National Laboratory, Upton, NY, US

Highlights from ALICE
• P. GIUBELLINO, CERN, CH

Highlights from the Planck Satellite
• A. RIAZUELO, Institute d'Astrophysique de Paris, FR

SEMINARS ON SPECILIZED TOPICS

Roadmap at the LHC to the Higgs Boson and Beyond
• P. JENNI, CERN, Geneva, CH

Status of the Three Neutrinos
• A. BETTINI, INFN & Padoa University, IT; Canfranc Underground Lab, ES

Present status of the EMC effect
• K. RITH, DESY, Hamburg, DE; FAU Erlangen-Nuremberg, DE

Technology and Computing in relation to "Reflections on the Next Step for LHC"
• H. WENNINGER, CERN, Geneva, CH

What is the Ontological Status of the Higgs Particle?
• T.Y. CAO, Boston University, MA, US

New Spectroscopy with Charm and Beauty Multiquark States
• L. MAIANI, Sapienza University of Rome, IT

The Time of Flight (TOF) World Record
• A. ZICHICHI, CERN, Geneva, CH; INFN & University of Bologna, IT

CLOSING REMARKS

• G. 't HOOFT, Utrecht University, NL

«ETTORE MAJORANA» FOUNDATION AND CENTRE FOR SCIENTIFIC CULTURE
INTERNATIONAL SCHOOL OF SUBNUCLEAR PHYSICS – *50TH ANNIVERSARY CELEBRATIONS 2011-2012-2013*
51st Course: REFLECTIONS ON THE NEXT STEP FOR LHC

ERICE: 24 JUNE – 3 JULY 2013 **TIMETABLE** **Page 1**

Mon 24 Jun	Morning: arrival and registration at the Secretariat of the School							
	LECTURES						**DISCUSSION SESSIONS**	
	09.30 10.15	**10.15 11.00**		**11.15 12.00**	**12.00 12.45**		**18.00 – 20.00 Each Discussion Session lasts 30 mn**	
Tue 25 Jun	*OPENING REMARKS AND REFLECTIONS* **A. ZICHICHI**	**H. WENNINGER** *Technology and Computing in relation to "Reflections on the Next Step for LHC"*	**BREAK**	**P. JENNI** *Roadmap at the LHC to the Higgs Boson and Beyond*	**M.J. DUFF** *Magic Supergravity from Squaring Yang Mills*		ALL MORNING LECTURERS	**DINNER**
Wed 26 Jun	**G. 't HOOFT** *Origin and Status of Instantons*	**V. MUKHANOV** *Inflation and Quantum Origin of Structure in the Universe*		**P. JENNI** *Roadmap at the LHC to the Higgs Boson and Beyond*	**M.J. DUFF** *Magic Supergravity from Squaring Yang Mills*		ALL MORNING LECTURERS	
Thur 27 Jun	**L.N. LIPATOV** *Gauge Forces: From QCD to Quantum Gravity*	**H. FRITZSCH** *The Internal Structure of the Weak Bosons*		**D.V. NANOPOULOS** *No-scale Supergravity in the Light of LHC and Planck*	**A. RIAZUELO** *Highlights from the Planck Satellite*		ALL MORNING LECTURERS	
Fri 28 Jun	**A. ZICHICHI** *The Time of Flight (TOF) World Record*	**I. ANTONIADIS** *Mass Hierarchy and Physics Beyond the Standard Model*		**E. SOKATCHEV** *New Symmetries of N=4 Supersymmetric Gauge Theories*	**F. FERRONI** *Status of the Gran Sasso Lab*		ALL MORNING LECTURERS	
Sat 29 Jun	**Z. BERN** *Status of the Perturbative Approach to Supergravity*	**G. 't HOOFT** *Discreteness and Determinism in Superstring Theory*		**T.Y. CAO** *What is the Ontological Status of the Higgs Particle?*	**F. CLOSE** *The Pedagogic Higgs – or Somebody's – Boson*		**FREE FOR EXCURSION**	
Sun 30 Jun	**P.W. HIGGS** *My Life as Boson*	**L. MAIANI** *New Spectroscopy with Charm and Beauty Multiquark States*		**M.J. TANNENBAUM** *Latest Results from BNL and RHIC*	**K. RITH** *Present status of the EMC effect*		ALL MORNING LECTURERS	
Mon 1 Jul	**S. BERTOLUCCI** *Highlights from ATLAS – ALICE – CMS*	**S. FERRARA** *Dualities in Supergravity*		**P. MINKOWSKI** *Embedding Oscillatory Modes of Quarks for Baryons in QCD*	**A. BETTINI** *Status of the Three Neutrinos*		ALL MORNING LECTURERS	
Tue 2 Jul	**9.30 – 10.15** **P. GIUBELLINO** *Highlights from ALICE*	**10.15 – 10.30** **S. DAVIES** *Understanding the Ultraviolet Properties of Supergravity Theories*		**10.30 – 11.15** *CLOSING REMARKS* **G. 't HOOFT**	**11.15 – 12.15** *DISCUSSIONS* **Z. BERN and G. 't HOOFT**		***NEW TALENTS CEREMONY***	
Wed 3 Jul	Departures							

Discussion and New Talent's Sessions Organizers: Antonino ZICHICHI – Gerardus 't HOOFT –– Richard ARNOWITT – Michael J. DUFF – Harald FRITZSCH – Dieter HAIDT – – Christian KORTHALS-ALTES – Peter JENNI – Michael J. TANNENBAUM – Tom TAYLOR

Poster Sessions Organizer: Francesco NOFERINI – Barbara GUERZONI

SPECIAL SESSIONS FOR NEW TALENTS

«ETTORE MAJORANA» FOUNDATION AND CENTRE FOR SCIENTIFIC CULTURE
INTERNATIONAL SCHOOL OF SUBNUCLEAR PHYSICS – *50TH ANNIVERSARY CELEBRATIONS 2011-2012-2013*
51st Course: REFLECTIONS ON THE NEXT STEP FOR LHC

ERICE: 24 JUNE – 3 JULY 2013 **TIMETABLE** **Page 2**

Mon 24 Jun	Morning: arrival and registration at the Secretariat of the School									
	SPECIAL SESSIONS FOR NEW TALENTS									
	12.45 13.00	13.00 13.15		16.15 16.30	16.30 16.45	16.45 17.00	17.00 17.15	17.15 17.30	17.30 17.45	
Tue 25 Jun	**G. KARANANAS** Higgs-dilaton cosmology	**J. LANGE** Charge multiplication – a solution towards radiation-hard silicon detectors?	LUNCH	**A. TOKAREVA** Broken scale invariance and inflation	**J. SEELE** The PHENIX W program	**J. ZENNAMO** Identification of heavy flavour jets at DØ	**A.K. ROTHKOPF** From complex to stochastic potential: heavy quarkonia in the QGP	**M.V. CARLUCCI** Flavour as a spontaneously broken gauge symmetry	**R. LOPES DE SÀ** Precision calibration of the D0 liquid argon calorimeter	BREAK
Wed 26 Jun	**M. VIGNATI** Searching for Majorana neutrinos with cryogenic detectors	**M. SCHROEDER** Search for new physics in the multijet and missing transverse momentum $\sqrt{s} = 7$ TeV		**K. FUKUSHIMA** Real-time Yang-Mills simulation and thermalization problem at RHIC and LHC	**N. RUTHMANN** Search for the SM Higgs boson in the channel $H \to \tau\tau \to l_\nu l\nu_\tau$ had ν_τ with the ATLAS detector at the LHC	**O. GOGOTA** Double J/ψ production at DØ experiment	**C. HARTMANN** Neutrino masses and mixing	**K. DOROUD** Future of fast Timings	**P. SATUNIN** The influence of Lorentz violation on Ultra-high-energy photon detection	
Thur 27 Jun	**P. BIANCOFIORE** Rare heavy meson decays as a probe of new physics	**M.R. ALAM** Strange particle production off the nucleon		**K. ASTAPOV** Impact of sgoldstino on Higgs sector physics	**O.I. IVANYTSKYI** Non classical scaling laws within exactly solvable models	**R. DE PALO** Underground study of the ^{22}Ne+p reaction and its role in stellar nucleosynthesis	**A. ESPOSITO** *A Mechanism For Hadron Molecule Production in pp(pbar) collisions*	**E. SICKING** Physics and detectors at CLIC	**D. BRITZGER** Multi-jet cross sections in deep-inelastic scattering at HERA (H1) and extraction of alpha_s(M_Z)	
Fri 28 Jun	**M. ZOCCALI** A charming opportunity for T-violation	**M. SEIDEL** Precision measurement of the top-quark mass with the CMS detector		**J. NOHLE** Color-kinematics duality for pure Yang-Mills and gravity at one and two loops	**C. BACKHOUSE** Status and sensitivity of the NOνA experiment	**M. MOJAZA** Systematic all-orders method to eliminate the renormalization-scale and scheme ambiguities in pQCD	**C. WEVER** NLL soft and Coulomb resum-mation for squark and gluino produ-ction at the LHC	**A. SASTRE BRUNO** Chiral behaviour of pion properties from lattice QCD	**S. LEONTSINIS** Associated production of heavy quarkonia and electroweak bosons at LHC	
Sat 29 Jun	**S. TROFYMENKO** Transition radiation and ionization energy losses of ultra high-energy 'half-bare' electron	**I. TIMIRYASOV** Heavy-meson physics and flavour violation with a single generation		**FREE FOR EXCURSION**						
Sun 30 Jun	**M. KURKOV** *Universal Landau pole*	**M. KIEHN** The Mu3e experiment: searching for the lepton flavour changing decay $\mu^+ \to e^+e^-e^+$		**L. BORSTEN** *Superqubits*	**O. MOROZ** *Analytical formulas, general properties and calculation of transport coefficients in the hadron gas: shear and bulk viscosities*	**S. DOWNES** Hints of universality in small-field inflation	**A. PIKELNER** Standard Models three-loop beta-functions	**A. SEMENOVA** Composite super-conformal string model as a new class of consistent string models	**C. WYMANT** Making the most of MET: mass reconstruction from collimated decays	BREAK
Mon 1 Jul	**S. NAGY** A Magic Square from Yang-Mills Squared	**A. LÀSZLO** Understanding A+A collisions with h+A data at SPS energies		**M. CSANÀD** Hydrodynamic description of the sQGP at RHIC and LHC	**H.M.I.M. BAZA** The equation of state at LHC energy	**C.-T. YU** The naturally custodial Higgs triplet model	**W.J. TORRES BOBADILLA** Full one loop QCD scattering amplitudes	**Y. WANG** Study of the $a_0^0(980)$–$f_0(980)$ mixing	**A. MOROZOV** *Cabling procedure for HOMFLY polynomials*	
Tue 2 Jul										
Wed 3 Jul	Departures									

Discussion and New Talent's Sessions Organizers: Antonino ZICHICHI – Gerardus 't HOOFT –- Richard ARNOWITT – Michael J. DUFF – Harald FRITZSCH - Dieter HAIDT – – Christian KORTHALS-ALTES – Peter JENNI – Michael J. TANNENBAUM – Tom TAYLOR
Poster Sessions Organizer: Francesco NOFERINI – Barbara GUERZONI

We want to encourage New Talents the world over to demonstrate that New Talents really exist in the Queen of all Sciences, Physics, as defined by Enrico Fermi.

The proof that New Talents exist is the time devoted to their presentations.

Those who think that something is missing are invited to let us know in written form with no more than one page.

We want to go on with our group of people not very young (now) but giving the cpportunity to the New Talents to continue along our tracks.

CHAIRMAN: A. ZICHICHI

Scientific Secretaries: M. Mojaza, O. Gogota

DISCUSSION

- *Nohle:*

Why should we exclude chaos from the definition of complexity?

- *Zichichi:*

Since Professor 't Hooft is with us let me give the floor to him who, as I said, is the author of this definition of Complexity, where chaos is excluded.

- *'t Hooft:*

What usually is called "chaos" would be a very poor definition of "complexity". It is not sufficiently complex to my taste. There is a high amount of regularity in there. I don't think that is a sufficiently good representative of the really complex phenomena that can take place. Life can be a lot more complex that what we normally call chaos. Chaos often refers to the behaviour of the three body system, how perturbed and connected pendula behave as a function of time, etc. What you should really call complexity is the things that run out of control in such a way that your sloppy thinking is unable to resolve and follow them.

- *Wymant:*

Could you clarify the problem of transition between inert matter and living matter?

- *Zichichi:*

The problem is how many pieces of inert matter you need to build up the most simple example of living matter; the number which is not unanimously agreed is 10^{11}. The maximum number you can put in, in order to have the simplest example of living matter, cannot exceed 10^{11}. Why? Because living objects must be interacting objects in such a way to create something, which is identified as living system. For example, you, me, Gerardus 't Hooft, etc. are complex systems, which need billions of billions of billions of molecules in interactions. If we know the weight we can say how many protons, neutrons and electrons are in the body. But if you have this number of protons, neutrons and electrons you cannot get Gerardus 't Hooft. This is called the principle of transition from inert matter to living matter. If the famous evolution theory would be a scientifically rigorous theory, it should be able to describe this transition. No scientist knows how to do it. This problem is called "Minimal Life"; what are the laws and regularities, which govern the minimal life is the post. Laws and regularities which cannot be deduced from quantum electrodynamics (QED), despite the fact that you are studying the result of QED interactions among an enormous number of elements. People claim that they can establish quantum entanglement, provided the number is not greater than 10^{11}.

- *Wymant:*

Can you clarify the relationship between the so-called Anderson-Feynman-Beethoven principle and complexity?

- *Zichichi:*

The fundamental point is: in every field where people speak of complexity, the regularities, the laws of this field apparently could not care less about the fundamental interactions which are needed in order to produce these phenomena. This is the Anderson-Feynman-Beethoven phenomenology. So, no matter what field you are choosing which is presented to you as a complex field of activity, no matter what this field is; the regularities and laws that govern this field apparently have nothing to do with the fundamental forces without which these same phenomena could not exist. The most famous example of complexity is history. There is no limit to what extent complexity can be associated with history, but the two basic elements are there. The Anderson-Feynman-Beethoven phenomenology is obvious: in fact, without the fundamental forces, which we study, without our fundamental logic, history could not exist. This is not all. History has also UEEC (Unexpected Events with Enormous Consequences). The UEEC has not been invented by us, but by historians, and they called UEEC, Sarajevo type events. In Sarajevo a man was killed. How many people are killed in the world? But this fellow being killed has produced the First World War. According to historians, the Second World War was the consequence of the First World War. So, one man was assassinated in Sarajevo and tremendous effects came in. This is UEEC. If we now look at our physics: had anybody predicted the existence of the nucleus? Zero. Had anybody predicted the fact that space and time cannot be both real, but that space-time must be a complex quantity? This was discovered by Lorentz, by studying the properties of the Maxwell equations. Maxwell himself did not realize this, and Lorentz could not believe in what he was finding. The man that took this very seriously was Albert Einstein who extended the validity of Lorentz invariance from the electromagnetic phenomena to all physics phenomena, thus discovering $E=mc^2$. Peter Bergmann, the youngest assistant of Albert Einstein, told me that Einstein told him that he could not sleep for three months after $E=mc^2$. You know why? Because “m” was the symbol for matter and mass. If “E” is Energy, c^2 is the constant, and “m” is “matter” - we could not exist, because any form of matter could become Energy. And this is the reason why Einstein could not sleep. After three months he realized, that in 1897, J.J. Thomson discovered the electron whose charge is opposite to that of the proton. So he said that “m” is not **matter**, it is **mass**. After three months he started to sleep again. According to some historians, if J.J. Thomson had not discovered the electron, Einstein had never published his paper, because he was terrified by people saying “What are you talking about, if we are here, how can be that $E=mc^2$?”. If “m” is matter, if matter can become Energy, the stability of matter does not exist. He presented this strong argument associating stability of matter with the conservation of electric charge: the proton cannot go into an electron because of conservation of charge. Then Dirac came predicting the existence of the anti-electron; so the trouble came back again. All this had not been predicted, these are UEEC events. Two hundred years of totally unexpected discoveries in electricity, magnetism and optics produced the super-synthesis of James Clark Maxwell’s 4 equations. Give me only one quantity, the electric charge, in addition to Space and Time, and I produce all electromagnetic phenomena, including light. QED is the result of many UEEC events. No matter, what you can take as an example of complexity, the experimental evidence for the existence of complexity is AFB and UEEC. The biggest trouble I had was with nuclear physics, because everybody including ourselves make statements, which are not correct; no one knows how to go from QCD to Nuclear Physics. Nuclear Physics does not care about QCD; laws and regularities discovered in Nuclear Physics, should be derived from QCD. Summing up, all examples of complexity have their experimental evidence in UEEC and AFB.

- Haidt:

You stated that complexity exists at the fundamental level. In our field this is the microscopic level. I would like to know how entropy can be defined microscopically.

- Zichichi:

The answer is given by Constantino Tsallis: How you define entropy. If we admit that biological or linguistic evolution are complex phenomena, then the typical dynamics is somehow between strong chaos, which means a positive Lyapunov exponent and simple orbits i.e. negative Lyapunov exponents. In other words, complexity is deeply related to the age of chaos, i.e. maximum Lyapunov exponents. Since chaos appears to be related parametrically to an entropy index q, different from unity, there must be some deep connection between complexity and general entropy. These points are discussed by Constantino Tsallis and Murray Gell-Mann in a book.

- *'t Hooft:*

Something that is coming before the question of what is complexity, is why is there complexity? That has to do with the hierarchy problem. If all constants of Nature were of order one at the Planck scale, our Universe would be as tiny as the Planck scale. It would have been very simple. The fact is that there are constants that are much bigger than others. And that means that we have an enormously big variation in scales. Our Universe is very large and there is a lot of room for tremendously complex things happening there.

- *Zichichi:*

This is the problem of big numbers. There is a current of thought that wants to associate complexity with big numbers with the hierarchy problem. This is indeed a very active field. For example Donald Glazer, who was here, is working on the mathematical modeling of the brain. Our brain cannot be described as a system of electronic devices. The mathematical experts on the simulation of a brain claim that the activity of our brain is a system of antennas. How many antennas do you need in order to produce an original idea? Contrary to my hope, the number that the experts quote is bigger than 10^{123}. Why do I quote this number? The biggest number we have in physics is the ratio between the action of the Universe divided by the Planck action. This ratio is 10^{123}. Unfortunately for me, these experts claim that the number of antennas needed to get all of them in resonance in order to produce an original idea must be greater than 10^{123}.

«ETTORE MAJORANA» FOUNDATION AND CENTRE FOR SCIENTIFIC CULTURE
TO PAY A PERMANENT TRIBUTE TO GALILEO GALILEI, FOUNDER OF MODERN SCIENCE AND TO ENRICO FERMI, THE "ITALIAN NAVIGATOR", FATHER OF THE WEAK FORCES

INTERNATIONAL SCHOOL OF SUBNUCLEAR PHYSICS
50TH ANNIVERSARY CELEBRATIONS 2011-2012-2013
51st Course: ***REFLECTIONS ON THE NEXT STEP FOR LHC***
ERICE-SICILY: 24 June – 3 July 2013

FROM PLANCK TO COMPLEXITY

Antonino Zichichi
INFN and University of Bologna, Italy
CERN, Geneva, Switzerland
Enrico Fermi Centre, Rome, Italy
World Federation of Scientists, Beijing, Geneva, Moscow, New York

Paul A.M. Dirac Lecture Hall – Patrick M.S. Blackett Institute
Erice, 24 June 2013

FROM PLANCK TO COMPLEXITY

Antonino Zichichi
INFN and University of Bologna, Italy
CERN, Geneva, Switzerland
Enrico Fermi Centre, Rome, Italy
World Federation of Scientists, Beijing, Geneva, Moscow, New York

Forty years ago a great scientific novelty came; all experimental discoveries obtained with our powerful accelerators were to be considered only matters of **extremely low energy**. The novelty of our days is the so called "Science of Complexity". Figure 1 shows these two points which produce the trend from "Reductionism" to "Holism". In this lecture we prove that **Complexity** exists at the fundamental level and **Reductionism** is the only way to go on.

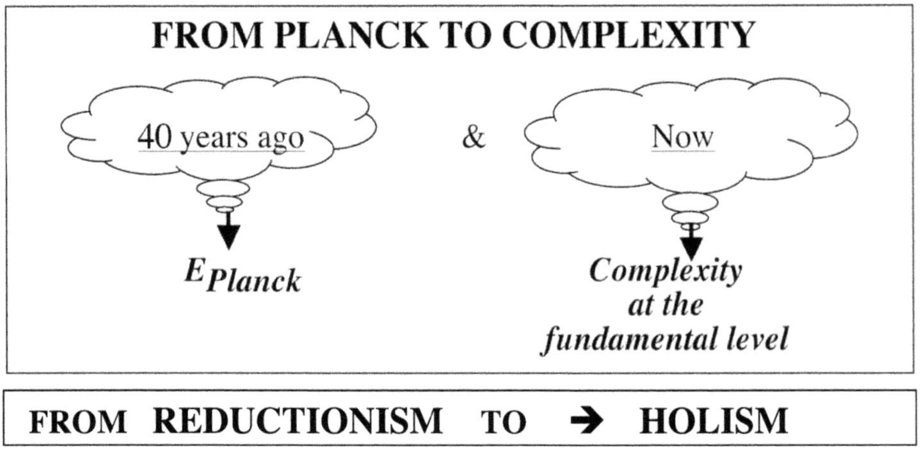

Figure 1

Let me start with the scale of energy on which to direct the attention in order to understand the Logic that rules the world, from the tiniest structures to the galactic ones. Forty years ago this scale of energy had to be shifted from the Fermi scale, $\mathbf{E_{Fermi}}$ that already seemed to be an extremely high level of energy,

$$\mathbf{E_{Fermi} \simeq 3 \cdot 10^2\ GeV}$$

to $\mathbf{E_{Planck}}$, something like seventeen powers of ten above.

Let me quote the first example of this new trend [1] where the energy scale was at 10^{15} GeV, i.e. something four powers of 10 below the Planck scale.

This is how the "**Desert**" came in our Physics [2].

At the Geneva Conference (**1979**) in my Opening Lecture [2] it was pointed out that, if the **Desert** was there, we better stop our efforts to encourage our young fellows to work in Subnuclear Physics.

On the other hand we cannot be so pessimistic, and probably the **Desert** is in our **Imagination**. In fact the title of my Opening Lecture was **Deserts are of the imagination** [2] (1979).

Four centuries of Galilean research work based on **Reductionism**, i.e. on the identification of the simplest elements in the study of Nature, has allowed us to get the greatest achievement of Science, i.e. the so called **Standard Model** and its extension (**SM&B**) as shown in Figure 2.

This extension predicts GUT (the Grand Unification Theory), the existence of the **Superworld** and the resolution of the quantum-gravity problem via the powerful theoretical structure of **RQST** (Relativistic Quantum String Theory).

Where we are now with the "Desert" is shown in Figure 3.

SM&B

THE STANDARD MODEL AND BEYOND

① RGEs ($\alpha_i\,(i \equiv 1, 2, 3)$; $m_j\ (j \equiv q,\ l,\ G,\ H)$) : $f\,(k^2)$.

- GUT ($\alpha_{GUT} \cong 1/24$) & GAP ($10^{16} - 10^{18}$) GeV.
- SUSY (to stabilize $m_F/m_P \cong 10^{-17}$).
- RQST (to quantize Gravity).

② Gauge Principle (hidden and expanded dimensions).

— How a Fundamental Force is generated: SU(3); SU(2); U(1) and Gravity.

③ The Physics of Imaginary Masses: SSB.

— The Imaginary Mass in SU(2)×U(1) produces masses ($m_{W^\pm}$; m_{Z^0} ; m_q; m_l), including $m_\gamma = 0$.

— The Imaginary Mass in SU(5)⇒SU(3)×SU(2)×U(1) or in any higher (not containing U(1)) Symmetry Group ⇒ SU(3)×SU(2)×U(1) produces Monopoles.

— The Imaginary Mass in $SU(3)_c$ generates Confinement.

④ Flavour Mixings & CP ≠ , T ≠ .

— No need for it but it is there.

⑤ Anomalies & Instantons.

— Basic Features of all Non-Abelian Forces.

Note:

q	≡	quark and squark;	m_F	≡	Fermi mass scale;
l	≡	lepton and slepton;	m_P	≡	Planck mass scale;
G	≡	Gauge boson and Gaugino;	k	≡	quadrimomentum;
H	≡	Higgs and Shiggs;	C	≡	Charge Conjugation;
RGEs	≡	Renormalization Group Equations;	P	≡	Parity;
GUT	≡	Grand Unified Theory;	T	≡	Time Reversal;
SUSY	≡	Supersymmetry;	≠	≡	Breakdown of Symmetry Operators.
RQST	≡	Relativistic Quantum String Theory;			
SSB	≡	Spontaneous Symmetry Breaking.			

The five basic steps in our understanding of nature. ① The renormalization group equations (RGEs) imply that the gauge couplings (α_i) and the masses (m_j) all run with k^2. It is this running which allows GUT, suggests SUSY and produces the need for a non point-like description (RQST) of physics processes, thus opening the way to quantize gravity. ② All forces originate in the same way: the gauge principle. ③ Imaginary masses play a central role in describing nature. ④ The mass-eigenstates are mixed when the Fermi forces come in. ⑤ The Abelian force QED has lost its role of being the guide for all fundamental forces. The non-Abelian gauge forces dominate and have features which are not present in QED.

Figure 2

WHERE WE ARE NOW WITH THE DESERT ?

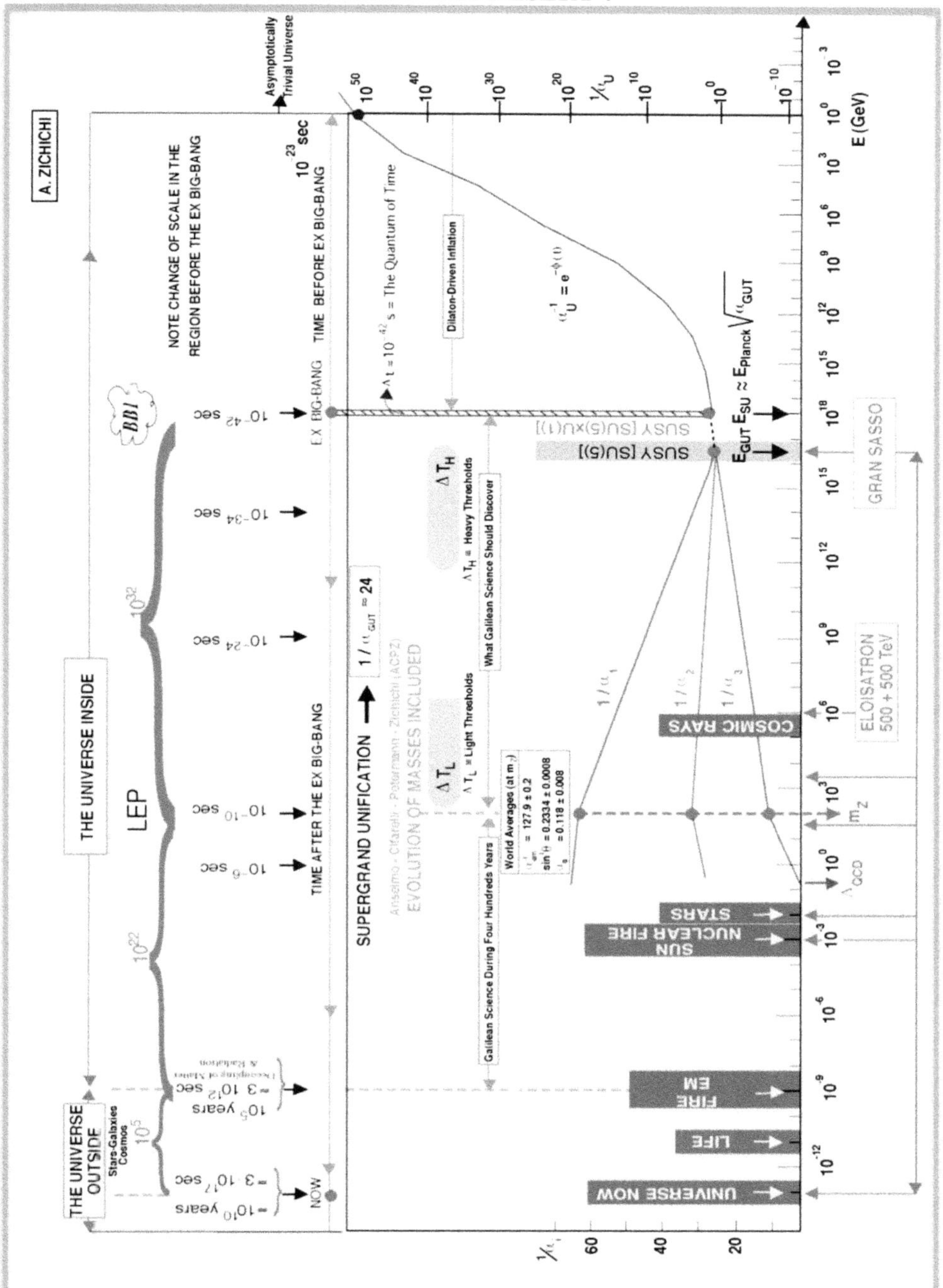

Figure 3

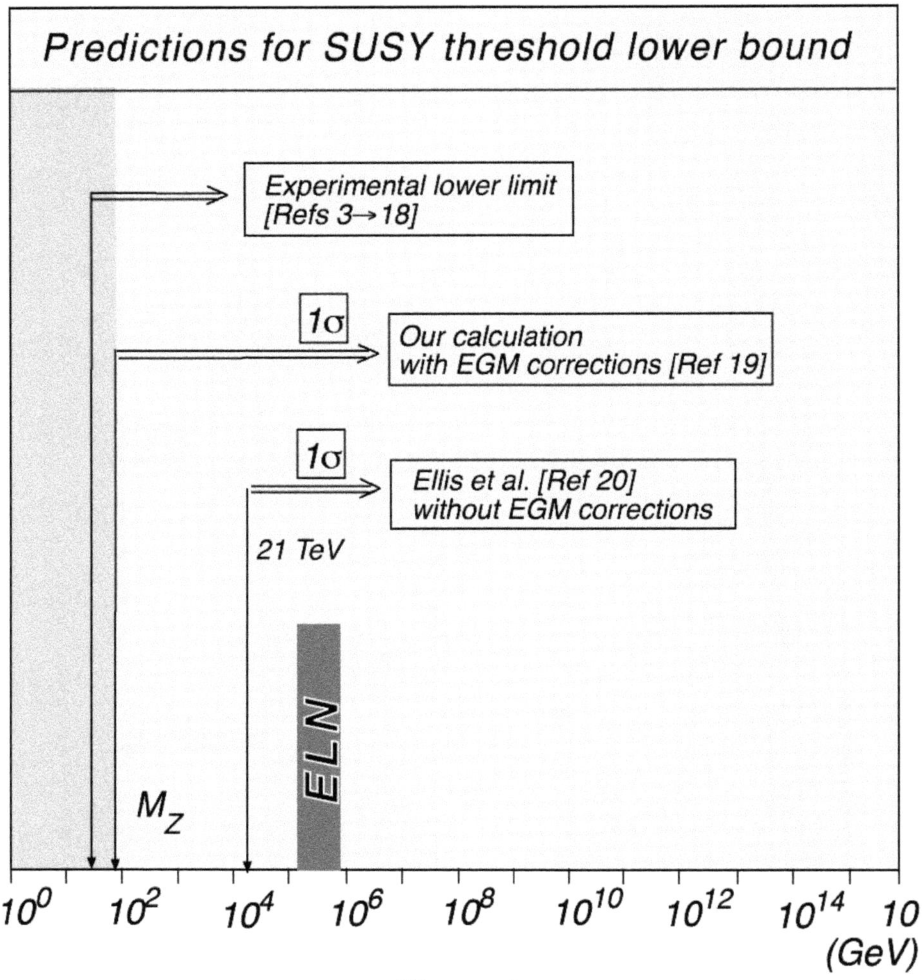

Figure 4

Figure 4 illustrates the EGM effect which lowers by a factor 700 the threshold for the production of the lightest superparticle.

The mathematical formalism used to obtain the results shown in Figures 3 and 4 is a system of three differential non-linear equations (shown in Figure 5) describing how the gauge couplings

$$\alpha_i\,,\ \alpha_j\ (\text{with } i = 1, 2, 3;\ \text{and } J = 1, 2, 3\ \text{ but } i \neq j),$$

vary with "μ", the basic parameter which depends on the energy of a given elementary process.

THE UNIFICATION OF ALL FUNDAMENTAL FORCES AND THE EGM EFFECT

In Figure 3 the three lines (α_1^{-1}, α_2^{-1}, α_3^{-1}) are the result from calculations executed with a supercomputer using the following system of equations:

$$\mu \frac{d\alpha_i}{d\mu} = \frac{b_i}{2\pi} \alpha_i^2 + \sum_j \frac{b_{ij}}{8\pi^2} \alpha_i \alpha_j$$

α_i, α_j *(with $i = 1, 2, 3$; and $J = 1, 2, 3$ but $i \neq j$).*

This is a system of three coupled non-linear differential equations where **the existence of the Superworld is included**. This system describes how the gauge couplings ($\alpha_1, \alpha_2, \alpha_3$) vary with "μ", the basic parameter which depends on the energy of the elementary process, from the maximum level of Energy (Planck Scale) to the energy level of our world. **During more than ten years (from 1979 to 1991), no one had realized that the energy threshold for the existence of the Superworld was strongly dependent on the "running" of the masses.**

This is now called: **the EGM effect** (from the initials of Evolution of Gaugino Masses). This effect produces a factor 700 in the threshold for the lightest supersymmetry particle. No one knows the value of the threshold Energy for the production of the lightest supersymmetry particle. The EGM effect lowers this value by a factor 700. Suppose that somebody would say that the threshold is at 700 TeV, thanks to the EGM effect this value is going to be 1 TeV.

Figure 5

During more than ten years (from 1979 to 1991), no one had realized that the energy threshold for the existence of the Superworld was strongly dependent on the "running" of the masses [21].

This is now called: the EGM effect (from the initials of Evolution of Gaugino Masses).

To compute the energy threshold using only the "running" of the gauge couplings (α_1, α_2, α_3) corresponds to neglecting nearly three orders of magnitude in the energy threshold for the discovery of the first particle (the lightest) of the Superworld, as illustrated in Figure 4.

A different way to describe how the gauge couplings α_1, α_2, α_3 vary with energy is reported in Figure 6.

PLATONIC VERSUS REAL GUT

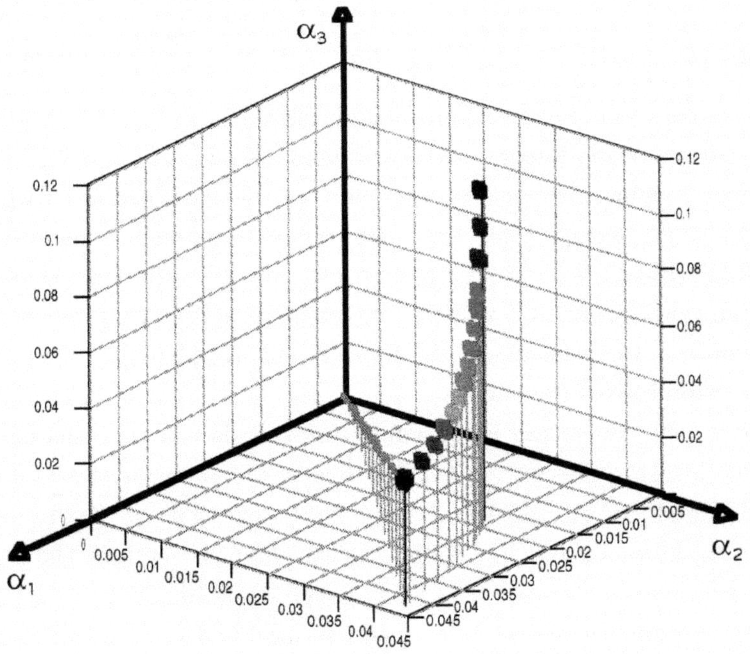

The "big" red points **represents the real GUT**. They have a sequence of 100 GeV in energy. The last point where the "ideal" platonic straight line intercepts the theoretical prediction is at the energy of the Grand Unification. This corresponds to $E_{GU} = 10^{16.2}$ GeV. Other detailed information on the theoretical inputs: the number of fermionic families, N_F, is 3; the number of Higgs particles, N_H, is 2. The input values of the gauge couplings at the Z^0-mass is $\alpha_3\,(M_Z) = 0.118 \pm 0.008$; the other input is the ratio of weak and electromagnetic couplings also measured at the Z^0-mass value: $\sin^2\theta_W\,(M_Z) = 0.2334 \pm 0.0008$.

The Platonic GUT is the straight line of the "dotted" blue points.

Figure 6

The simplest way to get GUT (the point where all fundamental forces are together: Grand Unification Theory) would be the straight line. But the real world does not follow this "platonic" straight line. The way nature goes is reported by the sequence of the big red points which are the result of the mathematics reported in Figure 5.

The sequence of points (the big red points), in steps of 100 GeV, is very different from the Platonic line (dotted blue points).

The first time where it was pointed out that Supersymmetry is needed for the convergence of the gauge couplings,

$$\alpha_1^{-1};\ \alpha_2^{-1};\ \alpha_3^{-1}\ ,$$

is at the EPS Conference, Geneva 1979, Proceedings; and Nuovo Cimento Rivista 2 (1979) 1 (A. Zichichi) [22].

We have been experimentally engaged to look for the **lightest Supersymmetric particle (LEP)**. Unfortunately no evidence from LHC (so far).

The apparent "novelty" is **Holism.** If this was a correct solution **Reductionism** which has produced **SM&B should be over**.

In order to follow the "novelty" we should introduce **Holism** in our research work.

How the Logic of Nature goes from the Basic Quantities to History is illustrated in Figure 7. The reason for introducing "Holism" is the interaction with the environment. But the interaction with the environment starts at the level of Proteins. Nevertheless the new guiding line should be, according to many fellows, Complexity.

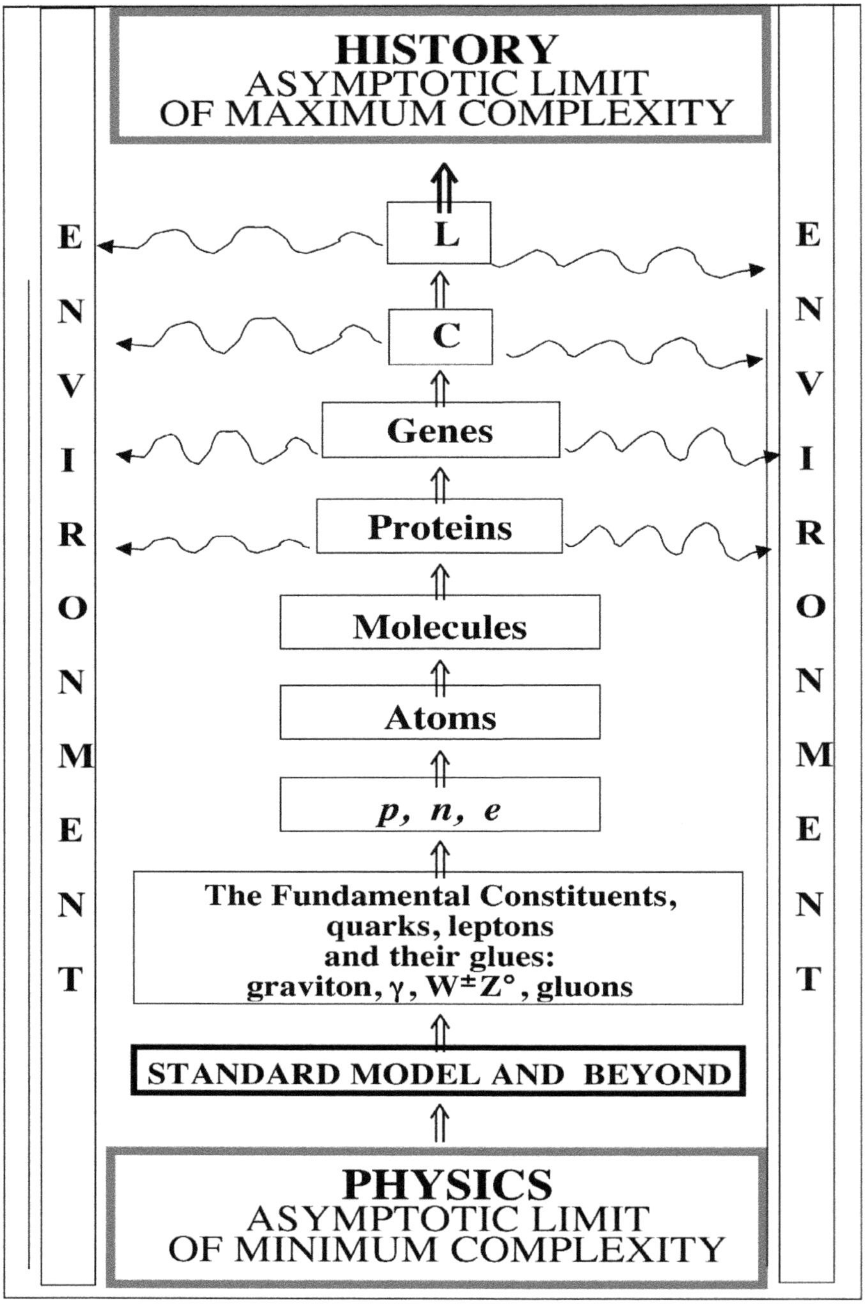
HISTORY
ASYMPTOTIC LIMIT
OF MAXIMUM COMPLEXITY
ENVIRONMENT
L
C
Genes
Proteins
Molecules
Atoms
p, n, e
The Fundamental Constituents,
quarks, leptons
and their glues:
graviton, γ, W±Z°, gluons
STANDARD MODEL AND BEYOND
PHYSICS
ASYMPTOTIC LIMIT
OF MINIMUM COMPLEXITY
ENVIRONMENT

Figure 7

Why? Because according to the latest "novelty" **Science of Complexity** is the **Real Science**.

People speak of '**Complexity**' as a source of new insights in physics, biology, geology, cosmology, social sciences, traffic flux, internet, and in all intellectual activities, as shown in Figure 8.

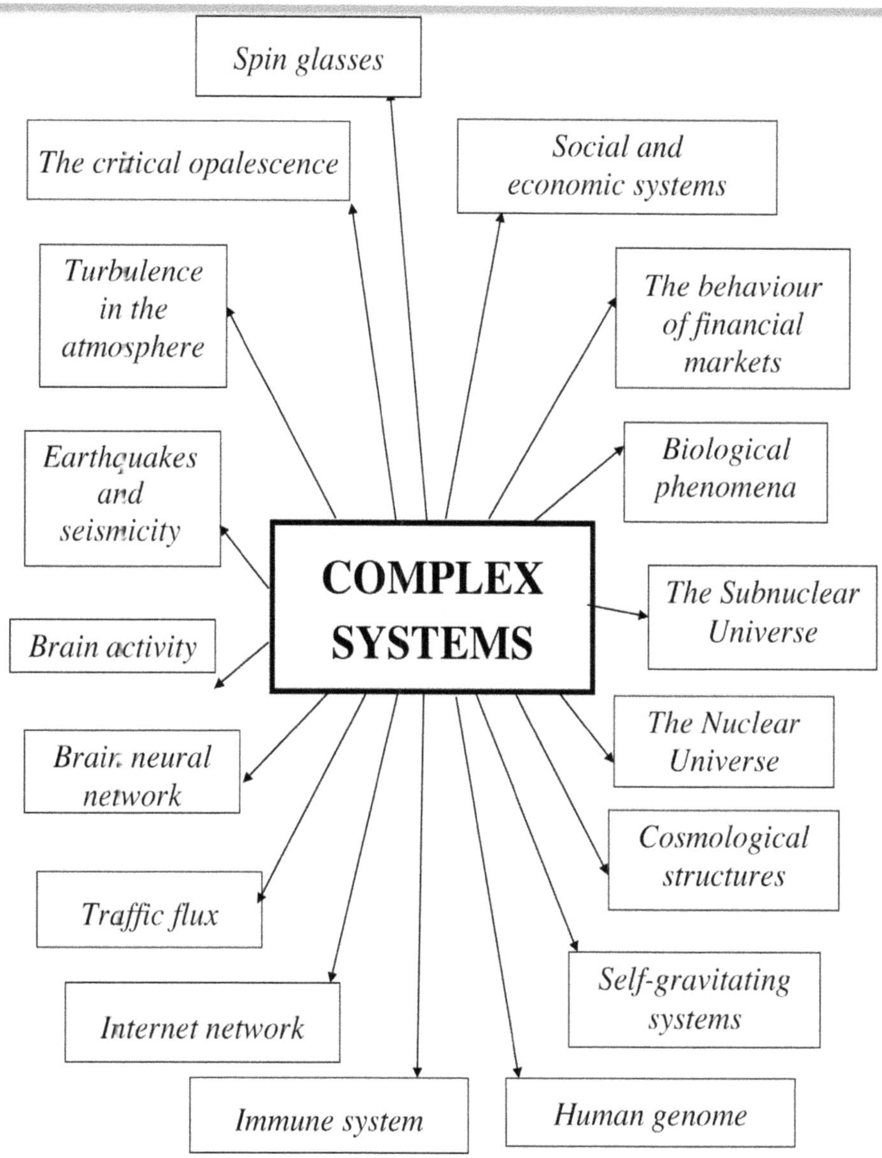

Figure 8

As you can see, we go from traffic flux, to the internet network, to earthquakes and seismicity, to social and economic systems, to the behaviour of financial markets, to the study of cosmological structures, and so on.

And where we are with Complexity?

Complexity has given rise to the Science of Complexity which dominates all activities from finance, traffic flux, internet, etc. ...

We cannot ignore the existence of *"Science of Complexity"*.

But 'Complexity' is ill-defined, as shown by the existence of very many definitions of **Complexity**.

Definitions

$$\gtrsim \mathbf{60}$$

I have selected Seven Definitions of **Complexity**. Let me point out that the **experimental evidences for the existence of Complexity** are only 2 (see later).

THE SEVEN SELECTED DEFINITIONS OF COMPLEXITY

DEFINITION NUMBER 1

Complexity is a property of systems that are somewhere in between a completely random and a completely regular state, often described by a highly non linear set of equations but sometimes not describable by equations at all.

DEFINITION NUMBER 2 (Gerardus 't Hooft)

Bad ones:

1) Chaos.
2) The need for lengthy calculations.
3) The need for many distinct variables.

Better ones:

4) Unexpected difficulty when attempting to describe something in a precisely formulated theory.
5) What is left over after all systematic approaches failed.

But it could also be that: Complexity is an excuse for sloppy thinking.

Definition Number 3 (Richard Kenway)

The Complexity of a theory (problem) is the minimum amount of computer time and storage required to simulate (solve) it to a specified level of precision.

Definition Number 4 (Constantino Tsallis)

If we admit that biological or linguistic evolution, or financial dynamics are complex phenomena, then their typical dynamics is somehow between strong chaos (i.e. positive Lyapunov exponents) and simple orbits (i.e. negative Lyapunov exponents). In other words, Complexity (or at least some form of it) is deeply related to the edge of chaos (i.e. vanishing maximal Lyapunov exponent). Since the edge of chaos appears to be related paradigmatically to an entropy index 'q' different from unity, there must be some deep connection between Complexity and generalized entropies such as 'S_q'.

Definition Number 5 (Leonard Susskind)

From the mathematical point of view:

- A problem can be polynomial, which means that it is not to hard to predict surprises.
- A problem can be NP or NP-complete, which represent different degrees of difficulty in predicting surprises.
- Surprises means: UEEC event (see later).
- That degree of difficulty can be associated with the level of Complexity.

Definition Number 6

A system is 'complex' when it is no longer useful to describe it in terms of its fundamental constituents.

Definition Number 7

The simplest definition of Complexity: '*Complexity is the opposite of Simplicity*'.

This definition comes from people who look at the world through the lens of a standard analysis in terms of either **Simplicity** or **Complexity.**

This is why we have studied the **Platonic Simplicity** for the **Grand Unification** and for **Supersymmetry**.

The seven definitions of Complexity must be compared with the whole of our knowledge (see later) in order to focus our attention on the key features needed to study our real world. Let me start with "Simplicity".

PLATONIC SIMPLICITY: two examples

- **The Platonic Grand Unification** (already discussed in Figure 6)
- **The Platonic Supersymmetry** (Figure 9)

THE PLATONIC SUPERSYMMETRY

The **Gauge Principle** should generate a
Gauge Force ⟶ **Gauge Bosons**
If *NATURE* was platonically *SUPERSYMMETRIC*
Supersymmetry Transformation should generate **Gauginos**

1st **DEVIATION FROM PLATONIC SIMPLICITY**
OUR FERMIONS ARE NOT THE GAUGINOS

2nd **DEVIATION FROM PLATONIC SIMPLICITY**
THE FUNDAMENTAL FERMIONS ARE OF TWO DIFFERENT CLASSES: LEPTONS AND QUARKS

3rd **DEVIATION FROM PLATONIC SIMPLICITY**
THERE IS NOT ONLY ONE BUT THREE FAMILIES OF FUNDAMENTAL FERMIONS

4th **DEVIATION FROM PLATONIC SIMPLICITY**
THE FUNDAMENTAL FERMIONS BECOME MIXED WHEN THE WEAK FORCES ARE SWITCHED ON: MIXINGS EXIST

5th **DEVIATION FROM PLATONIC SIMPLICITY**
THERE ARE DIFFERENT MIXINGS

Figure 9

THE PLATONIC SUPERSYMMETRY

The Gauge Principle should generate a

Gauge Force ⟶ Gauge Bosons.

If Nature was platonically supersymmetric Supersymmetry transformation should generate Gauginos.

1st deviation from platonic simplicity. Our fermions are not the Gauginos.

2nd deviation from platonic simplicity. The fundamental fermions are of two different classes: leptons and quarks.

3rd deviation from platonic simplicity. There is not only one but three families of fundamental fermions.

4th deviation from platonic simplicity. The fundamental fermions become mixed when the weak forces are switched on: mixings exist.

5th deviation from platonic simplicity. There are different mixings for quarks & for leptons.

The **seven definitions** of **Complexity** is a very strict selection done by **me**. The number of **definitions**, as mentioned before, exceeds **60**.

The crucial point is that the **number** of **experimental evidences** for the **existence of Complexity is two**.

THE EXPERIMENTAL EVIDENCES FOR THE EXISTENCE OF COMPLEXITY

The experimental evidences for the **existence** of **Complexity are two**:

1 — The Anderson-Feynman-Beethoven-type phenomena (**AFB**) i.e. phenomena whose laws and regularities ignore the existence of the Fundamental Laws of Nature from which they originate.

2 — The Sarajevo-type effects, i.e. Unexpected Events of quasi irrelevant magnitude which produce Enormous Consequences (**UEEC**).

The only certainty about Complexity is the existence of these experimentally observable effects.

Nature tell us that **AFB** and **UEEC** exist at all scales, and therefore **Complexity** exists at all scales, as illustrated in Figure 8.

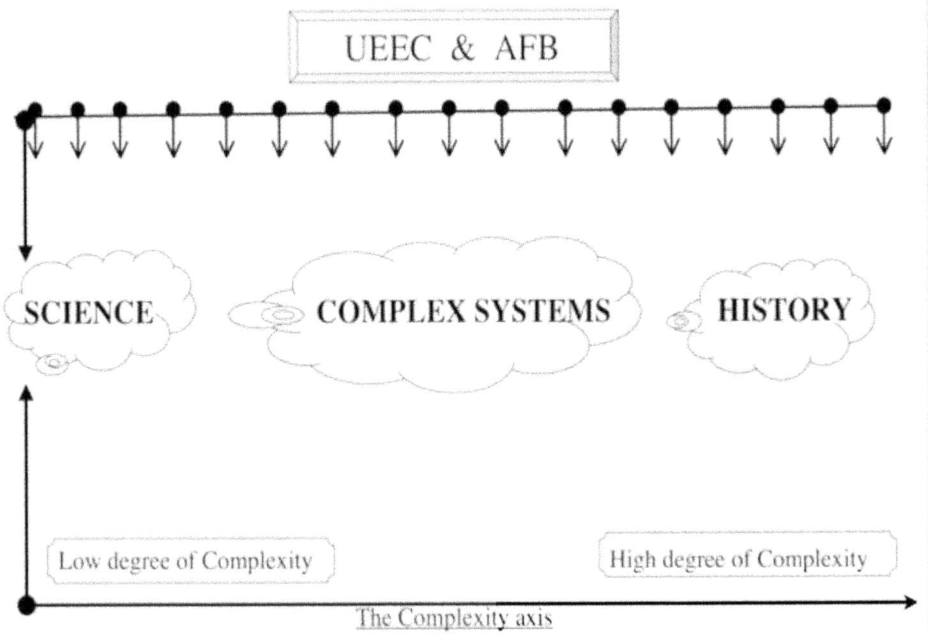

Figure 10

COMPLEXITY EXISTS AT ALL SCALES

The Logic of Nature allows the existence of a large variety of structures with their regularities and laws which appear to be independent from the basic constituents and Fundamental Laws of Nature which govern their interactions.

But, without these laws it would be impossible to have the real world which is in front of us and of which we are part of. The series of complex systems has been shown in Figure 8.

Now we will see that **Complexity** exists at the Fundamental Level. The proof is given by the fact that **AFB & UEEC** exist in **Physics**.

AFB Phenomena from Beethoven to the Superworld

Beethoven and the laws of acoustics.

Beethoven could compose superb masterpieces of music without any knowledge of the laws governing acoustic phenomena.

But these masterpieces could not exist if the laws of acoustics were not there.

The living cell and QED.

To study the mechanisms governing a living cell, people working in the field do not need to know the laws of electromagnetic phenomena whose advanced formulation is QED (Quantum ElectroDynamics).

All mechanisms needed for life are, to a great extent, examples of electromagnetic processes. If QED was not there, Life could not exist.

Nuclear physics and QCD.

Proton and neutron interactions appear as if a fundamental force of nature is at work: the nuclear force, with its rules and its regularities.

These interactions ignore that protons and neutrons are made with quarks and gluons.

Nuclear physics does not appear to care about the existence of Quantum ChromoDynamics (QCD), the fundamental force acting between quarks and gluons at the heart of the subnuclear world.

Nuclear physics ignores QCD but all phenomena occurring in nuclear physics have their roots in the interactions of quarks and gluons.

In other words, protons and neutrons behave like Beethoven: they interact and build up nuclear physics without any clear link with the laws governing QCD.

The most recent example of AFB type phenomenon comes from the frontier of our scientific knowledge: the Superworld. What we call the **World could *apparently* not care less about the existence of the Superworld**, whose foundation is not in the four dimensional space-time but in a superspace with 43 dimensions.

UEEC events and Nuclear Physics

Let me dedicate some attention to discuss UEEC events in nuclear physics.

It is considered standard wisdom the fact that nuclear physics is based on perfectly sound theoretical predictions. People forget the impressive series of UEEC events discovered in what I have decided to call the 'Yukawa goldmine' [23] on the occasion of the centennial celebration of the birth of Hideki Yukawa in Japan.

The opening Ceremony had the great honour of the presence of the Emperor and the Empress.

Let me quote just three of them:

1 The first experimental evidence for a cosmic ray particle believed to be the Yukawa meson was a lepton: the muon.

2 The decay-chain:

$$\pi \rightarrow \mu \rightarrow e$$

was found to break the symmetry laws of Parity and Charge Conjugation.

3 The intrinsic structure of the Yukawa particle was found to be governed by a new Fundamental Force of Nature, Quantum ChromoDynamics: QCD.

None of these three UEEC events was predicted.

UEEC EVENTS
IN THE CONSTRUCTION OF THE
SM&B ≡ MY PERSONAL EXPERIENCE

① ***The 3rd lepton***, **HL** (now called τ) with its own neutrino, ν_{HL} (now called ν_τ),
despite the abundance of neutrinos: ν_e and ν_μ.

② ***Antimatter***
despite S-matrix and C, P, CP, T breakings.

③ ***Nucleon Time-like*** *EM structure*
despite S-matrix.

④ ***No quarks*** *in violent (pp) collisions*
despite scaling.

⑤ ***Meson mixings***
$\theta_V \neq \theta_{PS}$: $(51^\circ) \neq (10^\circ) \neq 0$ **despite** $SU(3)_{uds}$.

⑥ ***Effective energy:*** the QCD-light
despite QCD.

⑦ ***The running*** *of* α_1 α_2 α_3 *versus* ***energy*** *at* **a point** E_{GU}.
(1979) (1991) **despite** straight line convergence.
EGM

Figure 11

CONCLUSION ABOUT AFB AND UEEC

We have proved that AFB and UEEC – which are at the origin of Complexity, with its consequences permeating all our existence, from molecular biology to life in all its innumerable forms up to our own, including History – do exist at the fundamental level [24–27] and [28].

It turns out that Complexity in the real world exists, no matter the mass-energy and space-time scales considered. We have discovered this, thanks to Reductionism.

Therefore we must go on with Reductionism.

*** *** *** ***

- **Totally Unexpected Effects** should **show up**.
- **Effects**, which are impossible to be predicted on the basis of our **present knowledge**.

*** *** *** ***

THE LESSON NEEDED FOR THE FUTURE

We should be prepared with powerful experimental instruments, technologically at the frontier of our knowledge, to discover Totally Unexpected Events in all laboratories, the world over (including CERN in Europe, Gran Sasso in Italy, and other facilities in Japan, USA, China and Russia).

All the pieces of the Yukawa gold mine [23] could not have been discovered if the experimental technology was not at the frontier of our knowledge.

Example: the **cloud-chambers** (Blackett, Anderson, Neddermeyer), the **high power magnetic fields** (Conversi, Pancini, Piccioni), the **photographic emulsions** (Lattes, Muirhead, Occhialini, Powell) and the **powerful particle accelerators and associated detectors for the discovery** – the world over – of the **SM&B** as synthetically reported in Figure 2.

This means that we must be prepared with **the most advanced technology** for the discovery of totally unexpected events like the ones found in the Yukawa gold mine [23].

NOTE CONCERNING THE CONFUSION ABOUT PREDICTIONS AND UEEC

The mathematical descriptions, and therefore the predictions – for new phenomena to be discovered in the field opened by an UEEC event – come after the UEEC event, never before.

<u>**Recall**</u>:

- The **discoveries in Electricity**, **Magnetism** & **Optics** (UEEC) produced the 4 Maxwell Equations & **QED** (UEEC).
- **Radioactivity** (UEEC)⇒**Weak Forces** (UEEC).
- **Cosmic Rays** (UEEC)⇒**Nuclear Physics** (UEEC).
- **Strange Particles** (UEEC) & Δ^{++} (UEEC)⇒**Subnuclear Physics** (UEEC).
- **3 Columns** (UEEC).
- **3 Fundamental Forces** (UEEC).

The present status of Science is reported in Figure 12, in order to show that **Holism is 10^4 years old**. It cannot be the **new** basis for a **new** Science (the **Science of Complexity**).

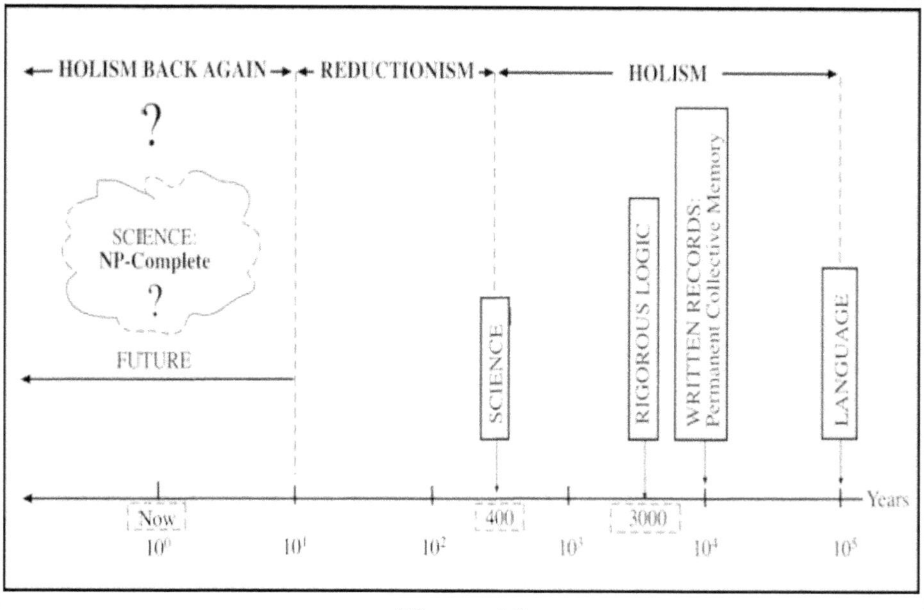

Figure 12

It could be that Science will be mathematically proved to be 'NP-complete'. This is the big question for the immediate future [29].

It is therefore instructive to see how Science fits in the whole of our knowledge as reported in Figure 13.

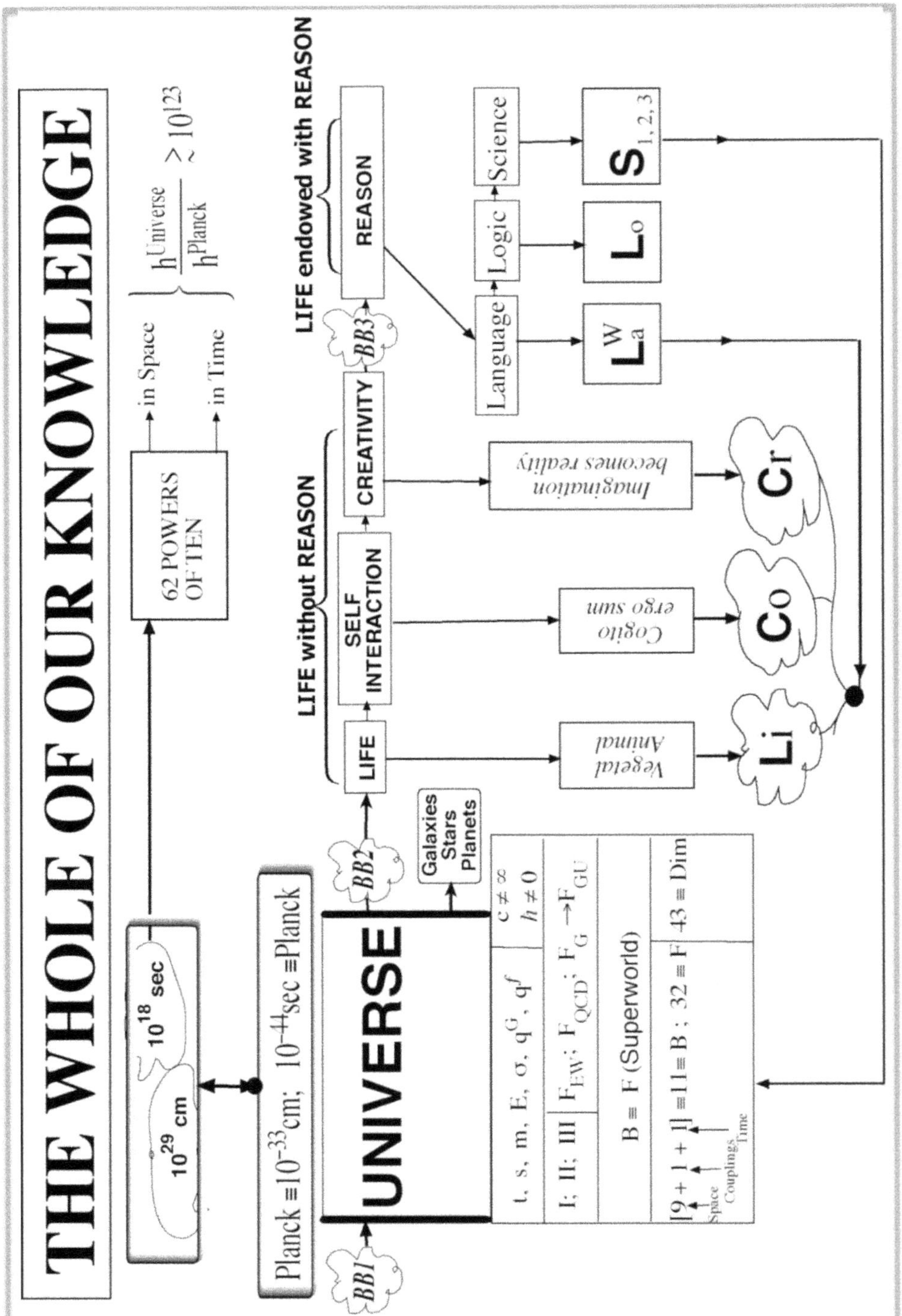
THE WHOLE OF OUR KNOWLEDGE
10^29 cm
10^18 sec
62 POWERS OF TEN
in Space
in Time
h^Universe / h^Planck ≳ 10^123
Planck ≡ 10^−33 cm; 10^−44 sec ≡ Planck
BB1
UNIVERSE
BB2
Galaxies Stars Planets
BB3
LIFE without REASON
LIFE endowed with REASON
LIFE
SELF INTERACTION
CREATIVITY
REASON
Vegetal Animal
Cogito ergo sum
Imagination becomes reality
Language
Logic
Science
Li
Co
Cr
L^W_a
L_o
S_1, 2, 3
t, s, m, E, σ, q^G, q^f
c ≠ ∞
h ≠ 0
I; II; III
F_EW; F_QCD; F_G → F_GU
B ≡ F (Superworld)
[9 + 1 + 1] ≡ 11 ≡ B ; 32 ≡ F
43 ≡ Dim
Space
Couplings
Time

Figure 13

The experimental evidence is that UEEC events dominate our life as the **E**volution of the **W**orld in its **R**eal **L**ife (EWRL ≡ **History**) and the **E**volution of our **B**asic **U**nderstanding of the laws governing the world in its **S**tructure (EBUS ≡ **Science**) do show (see Table 1).

UEEC EVENTS IN HISTORY AND IN SCIENCE

	In History ≡ EWRL		In Science ≡ EBUS
I	What if Julius Caesar had been assassinated many years before?	*I*	What if Galileo Galilei had not discovered that $F = mg$?
II	What if Charles VII had not been able to win the 100 years war?	*II*	What if Newton had not discovered that $F = G\frac{m_1 \cdot m_2}{R_{12}^2}$?
III	What if America had been discovered a few centuries later?	*III*	What if Maxwell had not discovered the unification of electricity, magnetism and optical phenomena, which allowed him to conclude that light is a vibration of the EM field?
IV	What if Napoleon had not been born?	*IV*	What if Becquerell had not discovered radioactivity?
V	What if Louis XVI had been able to win against the 'Storming of the Bastille'?	*V*	What if Planck had not discovered that $h \neq 0$?
VI	What if the 1908 Tunguska Comet had fallen somewhere in Europe instead of Tunguska in Siberia?	*VI*	What if Lorentz had not discovered that space and time cannot both be real?
VII	What if the killer of the Austrian Archduke Francisco Ferdinand had been arrested the day before the Sarajevo event?	*VII*	What if Einstein had not discovered the existence of time-like and space-like real worlds? Only in the time-like world, simultaneity does not change, with changing observer.
VIII	What if Lenin had been killed during his travelling through Germany?	*VIII*	What if Rutherford had not discovered the nucleus?
IX	What if Hitler had not been appointed Chancellor by the President of the Republic of Weimar Paul von Hindenburg?	*IX*	What if Hess had not discovered cosmic rays?
X	What if the first nuclear weapon had been built either by Japan before Pearl Harbour (1941) or by Hitler in 1942 or by Stalin in 1943?	*X*	What if Dirac had not discovered his equation, which opens new horizons, including the existence of the antiworld?
XI	What if Nazi Germany had defeated the Soviet Union?	*XI*	What if Fermi had not discovered the weak forces?
XII	What if Karol Wojtyla had not been elected Pope, thus becoming John Paul II?	*XII*	What if Fermi and Dirac had not discovered the Fermi–Dirac statistics?
XIII	What if Gorbachev had not been defeated by Yeltsin?	*XIII*	What if Yukawa had not proposed the existence of a "meson" in order to have the nuclear glue?
XIV	What if the USSR had not collapsed?	*XIV*	What if the 'strange particles' had not been discovered in the Blackett Lab?

Table 1

*** *** *** ***

Reductionisms is the only way we have to investigate the **Logic of Nature.**

*** *** *** ***

UEEC EVENTS, FROM GALILEI UP TO SM&B

In Figure 14 there is a sequence of **UEEC** events from Galilei to Fermi-Dirac and the "strange particles".

In Figures 15, 16, 17 from Fermi-Dirac to the construction of the Standard Model. The Standard Model and Beyond (**SM&B**) has already been discussed. Details on the first fifty years of Subnuclear Physics can be found in my book [30].

"UEEC"
TOTALLY UNEXPECTED DISCOVERIES
FROM GALILEI TO THE "STRANGE" PARTICLES

I	Galileo Galilei discovery of $F = mg$.
II	Newton discovery of $F = G\frac{m_1 \cdot m_2}{R_{12}^2}$.
III	Maxwell discovers the unification of electricity, magnetism and optical phenomena, which allows him to conclude that light is a vibration of the EM field.
IV	Planck discovery of $h \neq 0$.
V	Lorentz discovers that Space and Time cannot be both real.
VI	Einstein discovers the existence of Time-like and Space-like worlds. Only in the Time-like world, simultaneously does not change, with changing observer.
VII	Rutherford discovers the nucleus.
VIII	Hess discovers the cosmic rays.
IX	Dirac discovers his equation, which opens new horizons, including the existence of the antiworld.
X	Blackett discovers the "vacuum polarizations" (the first example of Virtual Phenomena).
XI	Fermi discovers the weak forces.
XII	Fermi–Dirac and Bose–Einstein discover two totally different statistical laws.
XIII	The "strange particles" are discovered in the Blackett Lab.
XIV	Heavy mesons with positive and negative strangeness are discovered in the Blackett Lab.

Figure 14

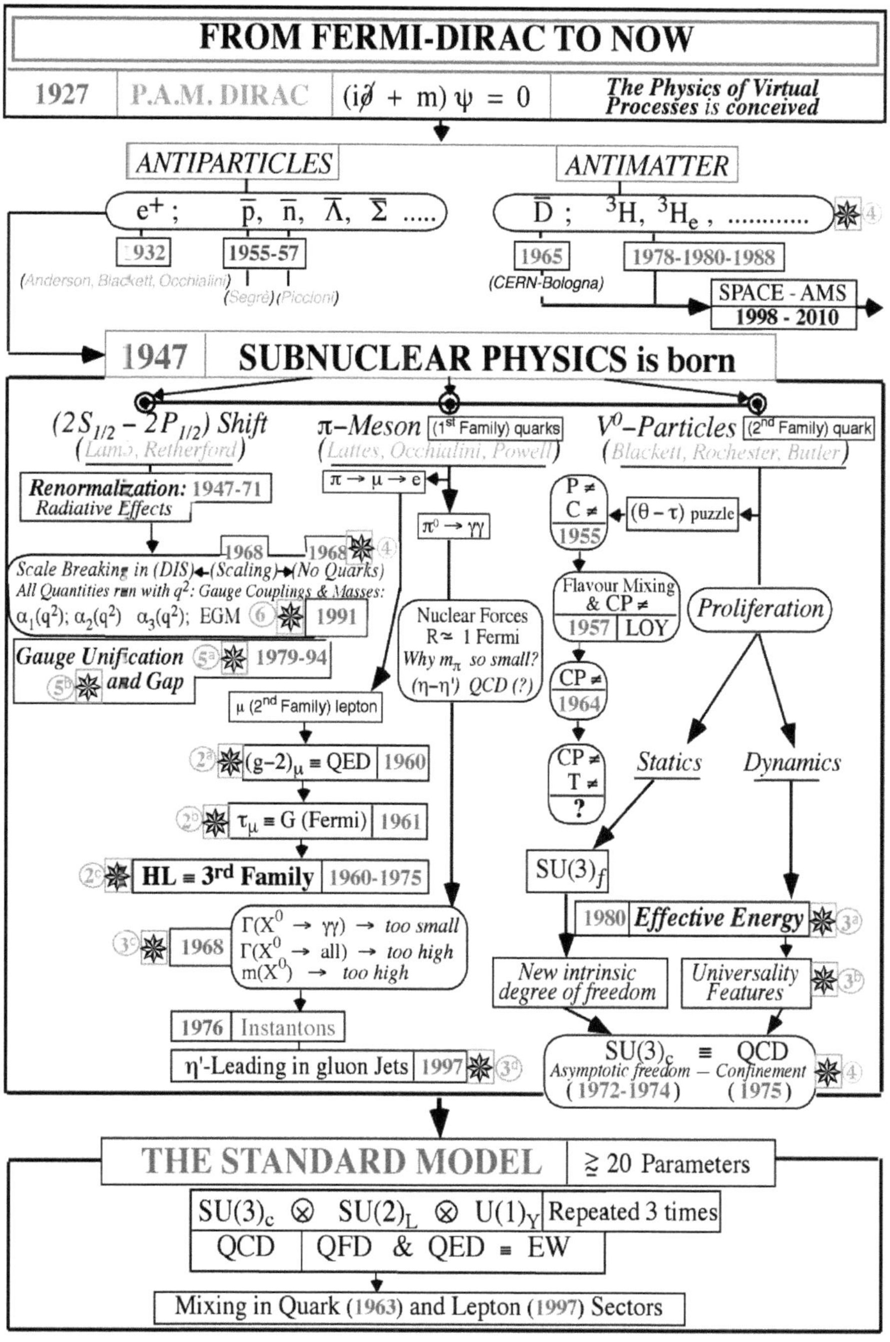

FROM FERMI-DIRAC TO NOW
1927
P.A.M. DIRAC
(i∂̸ + m) ψ = 0
The Physics of Virtual Processes is conceived
ANTIPARTICLES
ANTIMATTER
e+ ; p̄, n̄, Λ̄, Σ̄ .....
D̄ ; ³H, ³He , ...
1932
1955-57
(Anderson, Blackett, Occhialini)
(Segrè) (Piccioni)
1965
(CERN-Bologna)
1978-1980-1988
SPACE - AMS
1998 - 2010
1947
SUBNUCLEAR PHYSICS is born
(2S1/2 – 2P1/2) Shift
(Lamb, Retherford)
π–Meson
(1st Family) quarks
(Lattes, Occhialini, Powell)
V0–Particles
(2nd Family) quark
(Blackett, Rochester, Butler)
Renormalization: 1947-71
Radiative Effects
π → μ → e
π0 → γγ
P ≠
C ≠
1955
(θ – τ) puzzle
1968
1968
Scale Breaking in (DIS)←(Scaling)→(No Quarks)
All Quantities run with q²: Gauge Couplings & Masses:
α1(q²); α2(q²) α3(q²); EGM
1991
Flavour Mixing & CP ≠
1957 LOY
Proliferation
Nuclear Forces
R ≃ 1 Fermi
Why mπ so small?
(η–η') QCD (?)
Gauge Unification 1979-94
and Gap
CP ≠
1964
μ (2nd Family) lepton
Statics
Dynamics
(g–2)μ ≡ QED 1960
CP ≠
T ≠
?
τμ ≡ G (Fermi) 1961
HL ≡ 3rd Family 1960-1975
SU(3)f
1980 Effective Energy
1968
Γ(X0 → γγ) → too small
Γ(X0 → all) → too high
m(X0) → too high
New intrinsic degree of freedom
Universality Features
1976 Instantons
η'-Leading in gluon Jets 1997
SU(3)c ≡ QCD
Asymptotic freedom – Confinement
(1972-1974) (1975)
THE STANDARD MODEL
≳ 20 Parameters
SU(3)c ⊗ SU(2)L ⊗ U(1)Y
Repeated 3 times
QCD
QFD & QED ≡ EW
Mixing in Quark (1963) and Lepton (1997) Sectors

Figure 15

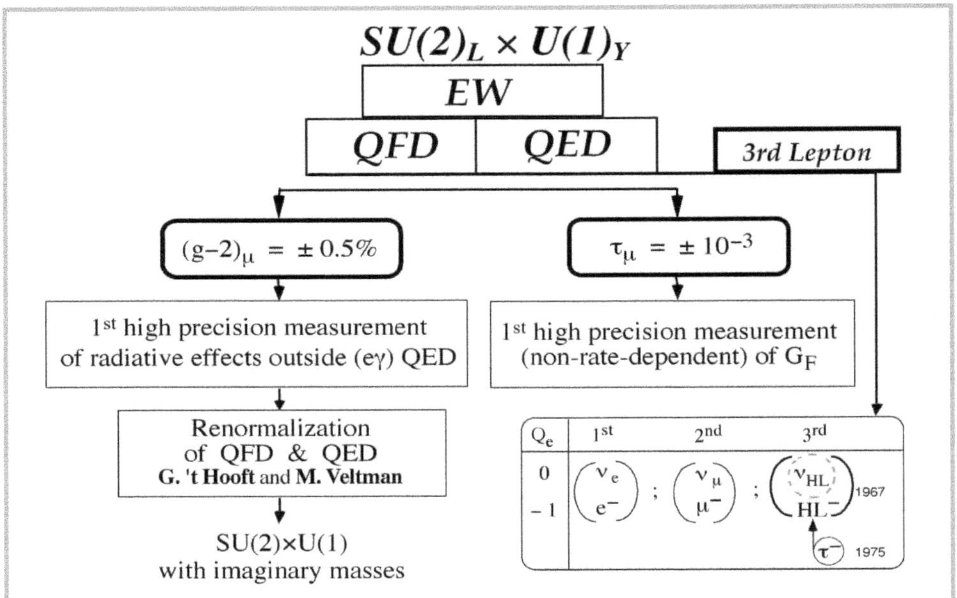

Figure 16: Details from Figure 15, concerning $SU(2)_L$ and $U(1)_Y$.

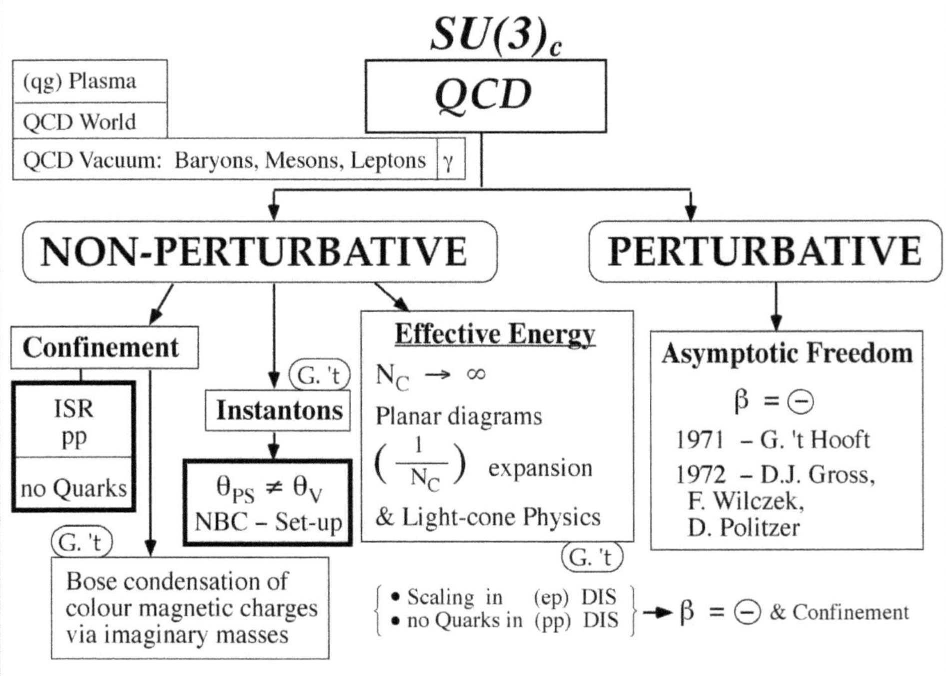

Figure 17: Details from Figure 15, concerning $SU(3)_c$.

Let me close with a list of very well known examples of UEEC Events in Physics, from Lorentz Invariance to the sequence of Fermi discoveries.

UEEC EVENTS IN PHYSICS

1 *Lorentz Invariance (Lorentz).*
2 *The quantum nature of the World (Planck).*
3 *Two worlds: "Time-like" and "Space-like" (Einstein).*
4 *The photon (Einstein).*
5 *Gauge Invariance (Weyl).*
6 *The Space-Time curvature (Einstein).*
7 *Bose-Einstein statistics (Bose-Einstein).*
8 *The structure of the atom (Bohr).*
9 *The wave nature of particles (de Broglie).*
10 *The wave function (Schrödinger) and its probabilistic interpretation (Born).*
11 *Nuclear Forces (Rutherford) and their "glue" (Yukawa).*
12 *The proof that Quantum Mechanics is self consistent (no contradictions) (Von Neumann).*
13 *The Weak Forces (Fermi).*
14 *The Exclusion Principle (Pauli).*
15 *The Uncertainty Principle (Heisenberg).*
16 *Fermi statistics (Fermi).*
17 *The antiparticles (Dirac).*
18 *The neutron (Chadwick).*
19 *The strange particles (Blackett).*
20 *Time Reversal Invariance (Wigner).*
21 *Other Invariance Laws (Wigner, Parity; Dirac-Weyl, charge conjugation; Pauli, CPT).*
22 *The neutrino (Pauli-Fermi).*
23 *The Stars are "nuclear-fusion" candles (Fermi-Bethe).*
24 *Electronic computing (Von Neumann).*
25 ***The sequence of unexpected Fermi discoveries**: Fermi-coupling, Fermi-gas, Fermi-momentum, Fermi-temperature, Fermi-surface, Fermi-statistics, Fermi-transition, Fermi-length (plus the other three quoted above).*

REFERENCES

[1] *Unity of All Elementary-Particle Forces*
H. Georgi and S.L. Glashow, *Phys. Rev. Letters* 32, 438 (1974).

[2] *Deserts are of the Imagination*
A. Zichichi, Opening Lecture in Proceedings of the EPS International Conference on "High-Energy Physics", Geneva, Switzerland, 27 June-4 July 1979 (CERN, Geneva, 1980), 3.

[3] *Search for Supersymmetric Particles using Acoplanar Charged Particle Pairs from Z^0 decays*
ALEPH Collab., D. Decamp et al., *Phys. Lett.* B236, 86 (1990).

[4] *Search for Neutral Higgs Bosons from Supersymmetry in Z decays*
ALEPH Collab., D. Decamp et al., *Phys. Lett.* B237, 291 (1990).

[5] *Search for Neutralino Production in Z decays*
ALEPH Collab., D. Decamp et al., *Phys. Lett.* B244, 541 (1990).

[6] *Search for the Neutral Higgs Bosons of the MSSM and other two Doublet Models*
ALEPH Collab., D. Decamp et al., *Phys. Lett.* B265, 475 (1991).

[7] *Search for Heavy Charged Scalars in Z^0 decays*
DELPHI Collab., P. Abreu et al., *Phys. Lett.* B241, 449 (1990).

[8] *Search for Pair Production of Neutral Higgs Bosons in Z^0 decays*
DELPHI Collab., P. Abreu et al., *Phys. Lett.* B245, 276 (1990).

[9] *Search for Scalar Quarks in Z^0 decays*
DELPHI Collab., P. Abreu et al., *Phys. Lett.* B247, 148 (1990).

[10] *A Search for Sleptons and Gauginos in Z^0 Decays*
DELPHI Collab., P. Abreu et al., *Phys. Lett.* B247, 157 (1990).

[11] *Mass Limits for Scalar Muons, Scalar Electrons and Winos from e^+e^- Collisions near S**(1/2) = 91–GeV*
L3 Collab., B. Adeva et al., *Phys. Lett.* B233, 530 (1989).

[12] *Search for the Neutral Higgs Bosons of the Minimal Supersymmetric Standard Model from Z^0 Decays*
L3 Collab., B. Adeva et al., *Phys. Lett.* B251, 311 (1990).

[13] *Search for the Charged Higgs Boson in Z^0 decay*
L3 Collab., B. Adeva et al., *Phys. Lett.* B252, 511 (1990).

[14] *A Search for Acoplanar Pairs of Leptons or Jets in Z^0 decays: Mass Limits on Supersymmetric Particles*
OPAL Collab., M.Z. Akrawy et al., *Phys. Lett.* B240, 261 (1990).

[15] *A Search for Technipions and Charged Higgs Bosons at LEP*
OPAL Collab., M.Z. Akrawy et al., *Phys. Lett.* B242, 299 (1990).

[16] *A Direct Search for Neutralino Production at LEP*
OPAL Collab., M.Z. Akrawy et al., *Phys. Lett.* B248, 211 (1990); P.D. Acton et al., preprint CERN-PPE/91-115, 22 July 1991.

[17] *Searches for Supersymmetric Particles Produced in Z Boson decay*
MARK II Collab., T. Barklow et al., *Phys. Rev. Lett.* 64, 2984 (1990).

[18] *Searches for New Particles at LEP*
M. Davier, LP-HEP 91 Conference, Geneva, CH, Preprint LAL 91-48, December 1991.

[19] *The Evolution of Gaugino Masses and the SUSY Threshold*
F. Anselmo, L. Cifarelli, A. Peterman and A. Zichichi, *Nuovo Cimento* 105A, 581 (1992).

[20] *A Detailed Comparison of LEP Data with the Predictions of the Minimal Supersymmetric SU(5) GUT*
J.R. Ellis, S. Kelley, D.V. Nanopoulos, preprint CERN-TH/6140-91, *Nucl. Phys.* B373, 55 (1992).

[21] *The Evolution of Gaugino Masses and the SUSY Threshold*
F. Anselmo, L. Cifarelli, A. Peterman and A. Zichichi, *Nuovo Cimento* 105A, 581 (1992).

[22] A. Zichichi, EPS Conference, Geneva 1979, Proceedings; and Nuovo Cimento Rivista 2 (1979) 1.

[23] *From the Yukawa Particle to the QGCW*
A. Zichichi, in Proceedings of the *'Symposium for the Centennial Celebration of Hideki Yukawa'*, International Nuclear Physics Conference, Tokyo, Japan, June 3-8 (2007).

The References [24–27] refer to the various occasions where I have presented papers on highly specialized topics and discussed the connection of these topics with Complexity. Each title on the upper part refers to the connection with Complexity while the specialized topic is reported in the detailed references.

[24] *Complexity at the Fundamental Level*
A. Zichichi
presented at:

- International Conference on 'Quantum [un]speakables' in Commemoration of John S. Bell, International Erwin Schrödinger Institut (ESI), Universität Wien (Austria), November 2000, *'John Bell and the Ten Challenges of Subnuclear Physics'*.
- 40th Course of the International School of Subnuclear Physics, Erice (Italy), September 2002, *'Language Logic and Science'*.
- 31st, 32nd and 33th Course of the International School of Solid State Physics, Erice (Italy), July 2004, *'Complexity at the Elementary Level'*.
- 42nd International School of Subnuclear Physics, Erice (Italy), August - September 2004, *'Complexity at the Elementary Level'*.
- Trinity College, Dublin (Ireland), February 2005, *'Complexity at the Elementary Level'*.
- Department of Physics, University of Padova (Italy), March 2005, *'Complexity at the Elementary Level'*.
- 43th Course of the International School of Subnuclear Physics, Erice (Italy), September 2005, *'Complexity at the Elementary Level'*.
- Italian Physics Society (SIF) XCI Annual National Congress, University of Catania (Italy), September 2005, *'Complexity at the Elementary Level'*.
- Desy, Hamburg, November 2005, *'Complexity at the Fundamental Level'*.
- 44th Course of the International School of Subnuclear Physics, Erice (Italy), September 2006, *'Complexity at the Fundamental Level'*.

[25] *The Logic of Nature and Complexity*
A. Zichichi
presented at:

- Pontificia Academia Scientiarum, The Vatican, Rome (Italy), November 2002, '*Scientific Culture and the Ten Statements of John Paul II*'; '*Elements of Rigour in the Theory of Evolution*'.
- The joint Session of:
6th Course of the International School of Biological Magnetic Resonance; Erice (Italy), July 2003, '*Language Logic and Science*'.
- 2nd Workshop on Science and Religion of the Advanced School of History of Physics; Erice (Italy), July 2003, '*Language Logic and Science*'.
- 10th Workshop of the International School of Liquid Crystals; Erice (Italy), July 2003, '*Language Logic and Science*'.
- International School on Complexity, 1st Workshop on Minimal Life, Erice (Italy), December 2004, '*Evolution and Complexity at the Elementary Level*'.

[26] *Complexity and New Physics*
A. Zichichi
presented at:

- INFN-Alice Meeting, University of Catania (Italy), January 2005, '*Complexity at the Elementary Level*'.
- INFN Eloisatron Project 'The 1st Physics ALICE Week', Erice (Italy), December 2005, '*Complexity and New Physics with ALICE*'.
- 50th Anniversary of INFN Bologna - ALICE Week, Bologna (Italy), June 2006, '*Complexity at the Fundamental Level*'.

[27] *Complexity and Planetary Emergencies*
A. Zichichi
presented at:

- 27th Sessions of the International Seminars on Planetary Emergencies, Erice (Italy), August 2002, '*Language, Logic and Science*'.
- 28th Sessions of the International Seminars on Planetary Emergencies, Erice (Italy), August 2003, '*Language Logic and Science, Evolution and Planetary Emergencies*'.
- 36th Sessions of the International Seminars on Planetary Emergencies, Erice (Italy), August 2006, '*Complexity and Planetary Emergencies*'.

[28] *Complexity exists at the Fundamental Level*
A. Zichichi in Proceedings of the 2004–Erice Subnuclear Physics School '*How and Where to Go Beyond the Standard Model*', World Scientific (2007) pages 251-333.

[29] *The Logic of Nature, Complexity and New Physics: From Quark-Gluon Plasma to Superstrings, Quantum Gravity and Beyond*
Proceedings of the 2006–Erice Subnuclear Physics School, World Scientific (2008).

[30] A. Zichichi in '*Subnuclear Physics - The first fifty years*', O. Barnabei, P. Pupillo and F. Roversi Monaco (eds), a joint publication by University and Academy of Sciences of Bologna, Italy (1998); World Scientific Series in 20th Century Physics, Vol. 24, two editions (2000–2001).

Mass Hierarchy and Physics Beyond the Standard Model

I. Antoniadis*
Department of Physics,
CERN - Theory Division
CH-1211 Geneva 23, Switzerland

I will discuss the status of the mass hierarchy problem and prospects for beyond the Standard Model physics in the light of the Higgs scalar discovery at the LHC and the experimental searches for new physics. In particular, I will discuss in this context low energy supersymmetry and large extra dimensions with low string scale.

1. INTRODUCTION

During the last few decades, physics beyond the Standard Model (SM) was guided from the problem of mass hierarchy. This can be formulated as the question of why gravity appears to us so weak compared to the other three known fundamental interactions corresponding to the electromagnetic, weak and strong nuclear forces. Indeed, gravitational interactions are suppressed by a very high energy scale, the Planck mass $M_P \sim 10^{19}$ GeV, associated to a length $l_P \sim 10^{-35}$ m, where they are expected to become important. In a quantum theory, the hierarchy implies a severe fine tuning of the fundamental parameters in more than 30 decimal places in order to keep the masses of elementary particles at their observed values. The reason is that quantum radiative corrections to all masses generated by the Higgs vacuum expectation value (VEV) are proportional to the ultraviolet cutoff which in the presence of gravity is fixed by the Planck mass. As a result, all masses are "attracted" to about 10^{16} times heavier than their observed values.

Besides compositeness, there are two main theories that have been proposed and studied extensively during the last years, corresponding to different approaches of dealing with the mass hierarchy problem. (1) Low energy supersymmetry with all superparticle masses in the TeV region. Indeed, in the limit of exact supersymmetry, quadratically divergent corrections to the Higgs self-energy are exactly cancelled, while in the softly broken case, they are cutoff by the supersymmetry breaking mass splittings. (2) TeV scale strings, in which quadratic divergences are cutoff by the string scale and low energy supersymmetry is not needed. Both ideas are experimentally testable at high-energy particle colliders and in particular at LHC. Another option is to live with the hierarchy, addressing differently questions such as gauge coupling unification, dark matter candidate and electroweak (EW) vacuum stability. Split supersymmetry is an interesting example of such a possibility. All these ideas are experimentally testable at high-energy particle colliders and in particular at the LHC.

On the other hand, the recent major discovery of the Higgs boson at the LHC with a mass

---

*Electronic address: ignatios.antoniadis@cern.ch - *On leave from CPHT (UMR CNRS 7644) Ecole Polytechnique, F-91128 Palaiseau*

around 126 GeV is so far compatible with the Standard Model within 2σ and its precision tests. It is also compatible with low energy supersymmetry, although with some degree of fine-tuning in its minimal version. Indeed, in the minimal supersymmetric Standard Model (MSSM), the lightest Higgs scalar mass m_h satisfies the following inequality:

$$m_h^2 \lesssim m_Z^2 \cos^2 2\beta + \frac{3}{(4\pi)^2}\frac{m_t^4}{v^2}\left[\ln\frac{m_{\tilde{t}}^2}{m_t^2} + \frac{A_t^2}{m_{\tilde{t}}^2}\left(1 - \frac{A_t^2}{12m_{\tilde{t}}^2}\right)\right] \lesssim (130GeV)^2\,, \tag{1}$$

where the first term in the r.h.s. corresponds to the tree-level prediction and the second term includes the one loop corrections due to the top and stop loops. Here, $m_Z, m_t, m_{\tilde{t}}$ are the Z-boson, the top and stop quark masses, respectively, $v = \sqrt{v_1^2 + v_2^2}$ with v_i the VEVs of the two higgses, $\tan\beta = v_2/v_1$, and A_t the trilinear stop scalar coupling. Thus, a Higgs mass around 126 GeV requires a heavy stop $m_{\tilde{t}} \simeq 3$ TeV for vanishing A_t, or $A_t \simeq 3m_{\tilde{t}} \simeq 1.5$ TeV in the 'best' case. These values are obviously consistent with the present LHC bounds on supersymmetry searches, but they will certainly be probed in the next run at double energy. Theoretically, they imply a fine-tuning of the EW scale at the percent to per mille level. This fine-tuning can be alleviated in supersymmetric models beyond the MSSM, as discussed in the next session.

2. MSSM HIGGS SECTOR WITH DIMENSION-FIVE AND DIMENSION-SIX OPERATORS

Although extremely successful, the Standard Model or its supersymmetric version (MSSM) is not a fundamental theory, and this motivated the theoretical efforts to understand the nature of new physics beyond it. This search can be done using an effective field theory approach, in which the "new physics" is parametrised by effective operators. The power of this approach resides in arranging these operators in powers of $1/M_*$ where M_* is the scale of new physics that generated them. To improve the predictive power, one considers additional organising principles, such as: **(i)** symmetry constraints that these operators should respect, often inspired by phenomenology (for example: R-parity, lepton or baryon number conservation, etc). **(ii)** a truncation of the series of operators to a given order in the power of the inverse scale $1/M_*$. The effective low-energy Lagrangian then takes the form

$$\mathcal{L} = \mathcal{L}_0 + \sum_{i,n} \frac{c_{n,i}}{M_*^n}\mathcal{O}_{n,i} \tag{2}$$

where $\mathcal{L}_0$ is the SM or the MSSM Lagrangian; $\mathcal{O}_{n,i}$ is an operator of dimension $d = n+4$ with the index i running over the set of operators of a given dimension; $c_{n,i}$ are some coefficients of order $\mathcal{O}(1)$. This description is appropriate for scales E which satisfy $E \ll M_*$. Constraints from phenomenology can then be used to set bounds on the scale of new physics M_*.

Regarding the origin of operators $\mathcal{O}_{n,i}$, they can be generated classically or at the quantum level. At the classical level, this can happen by integration of some new massive states, via the equations of motion and one then generates an infinite series. This can happen even in 4D renormalisable theories; indeed, even though the low energy interaction looks nonrenormalisable, it may actually point to a renormalisable theory valid up to a much higher scale (a familiar example is the Fermi interaction). Such effective operators are also generated at the quantum level, for example following compactification of a higher dimensional theory, by the radiative corrections associated with momentum and winding modes of the compactification [1–5].

The effects of these operators on the low energy observables can be comparable to the radiative effects of light states in the SM/MSSM [6] and this shows the importance of their study. In the following we shall study these effects to the case of the MSSM Higgs sector with additional operators of dimensions $d = 5$ and $d = 6$ [6, 7]. In particular we show that the mass of lightest SM-like Higgs can easily be increased close to the observed value by new physics in the region of few TeV.

2.1. MSSM with d=5 operators.

As an application, consider the MSSM extended by all possible $d = 5$ operators that respect R-parity, baryon and lepton number symmetry. The Lagrangian is $\mathcal{L} = \mathcal{L}_0 + \mathcal{L}^{(5)}$ where

$$\begin{aligned}\mathcal{L}_0 &= \int d^4\theta \sum_{i=1,2} Z_i(S,S^\dagger)\, H_i^\dagger\, e^{V_i}\, H_i + \Big\{ \int d^2\theta\, \mu_0\, (1 + B_0\, S)\, H_1.H_2 + h.c. \Big\} + \cdots \\ &+ \int d^2\theta \Big[Q\,\lambda_U(S)\, U^c\, H_2 - Q\,\lambda_D(S)\, D^c\, H_1 - L\,\lambda_E(S)\, E^c\, H_1 \Big] + h.c. \qquad (3)\end{aligned}$$

The dots stand for Higgs-independent terms and $Z_i(S,S^\dagger) = 1 - c_i\, S^\dagger S$, $c_i \sim \mathcal{O}(1)$. Further

$$\begin{aligned}\mathcal{L}^{(5)} &= \frac{1}{M_*} \int d^4\theta \Big[H_1^\dagger\, e^{V_1} Q\, Y_U U^c + H_2^\dagger\, e^{V_2} Q Y_D D^c + H_2{}^\dagger\, e^{V_2} L\, Y_E E^c \qquad (4) \\ &+ a\, D^\alpha \big[b\, H_2 e^{-V_1} \big] D_\alpha \big[c\, e^{V_1} H_1 \big] + \delta(\overline{\theta}^2) \big[Q U^c\, T_Q\, Q D^c + Q U^c\, T_L\, L E^c + \lambda_H (H_1 H_2)^2 \big] + h.c. \Big]\end{aligned}$$

with a standard notation. We introduced here some spurion dependent functions: $a, b, c, Z_{1,2}, Y_F$ $F = U, D, E$, which are general functions of $(S, S^\dagger)$ while T_Q, T_L, λ_H, are holomorphic functions of S. Here $S = m_0 \theta^2$ is the spurion superfield and m_0 the supersymmetry breaking scale. Any supersymmetry breaking associated with the presence of the above interactions is included using the spurion field technique. Not all operators in (4) are independent [5]. To remove the redundant operators we use field re-definitions:

$$\begin{aligned} H_1 &\to H_1 - \frac{1}{M_*} \overline{D}^2 \Big[\Delta_1\, H_2^\dagger\, e^{V_2}\, (i\,\sigma_2) \Big]^T + \frac{1}{M_*}\, Q\, \rho_U\, U^c \\ H_2 &\to H_2 + \frac{1}{M_*} \overline{D}^2 \Big[\Delta_2\, H_1^\dagger\, e^{V_1}\, (i\sigma_2) \Big]^T + \frac{1}{M_*}\, Q\, \rho_D\, D^c + \frac{1}{M_*}\, L\, \rho_E\, E^c \qquad (5)\end{aligned}$$

where $\rho_F = \rho_F(S)$, $F = U, D, E$, $\Delta_i = \Delta_i(S, S^\dagger)$, $i = 1, 2$, can be chosen arbitrarily. To avoid the presence of flavour changing neutral currents, the following simple ansatz can be made:

$$\begin{aligned} T_Q(S) &= c_Q(S)\, \lambda_U(0) \otimes \lambda_D(0), \qquad T_L(S) = c_L(S)\, \lambda_U(0) \otimes \lambda_E(0), \\ \rho_F(S) &= c_F(S)\, \lambda_F(0), \qquad Y_F(S, S^\dagger) = y_F(S, S^\dagger)\, \lambda_F(0), \quad F = U, D, E. \qquad (6)\end{aligned}$$

and, as usual $\lambda_F(S) = \lambda_F(0)\,(1+A_F\,S)$. Using a suitable choice for the (otherwise arbitrary) coefficients of the spurion entering in $\Delta_{1,2}$, one can set $T_Q = T_L = 0$ also $a = b = c = 0$ and $Y_F \to y_F(S^\dagger)\,\lambda_F(0)$. Then one finds [6]

$$\mathcal{L}^{(5)} = \frac{1}{M_*}\int d^4\theta \left[H_1^\dagger e^{V_1}\,Q\,Y_U'(S^\dagger)\,U^c + H_2^\dagger e^{V_2} Q\,Y_D'(S^\dagger)\,D^c + H_2^\dagger e^{V_2} L\,Y_E'(S^\dagger)\,E^c + h.c.\right]$$
$$+ \frac{1}{M_*}\int d^2\theta\;\lambda_H'(S)\,(H_1\,H_2)^2 + h.c. \qquad (7)$$

Detailed calculations show[6] that the new Yukawa couplings $Y_F'(S^\dagger)$ now depend on $S^\dagger$ only, $Y_F'(S^\dagger) = \lambda_F(0)\,(x_0^F + x_2^F\,S^\dagger)$. After (5), the couplings of $\mathcal{L}_0$ also acquired, at tree level, threshold corrections which depend on M_* [6]. The new form of $\mathcal{L}^{(5)}$ in (7) gives the minimal irreducible set of R-parity, B, L conserving $d = 5$ operators that can be present beyond MSSM.

A consequence of this analysis is the generation of new couplings, beyond those in the MSSM at the tree level. For example there is a "wrong"-Higgs Yukawa coupling, that exchanges the holomorphic dependence on one Higgs by that on the hermitian conjugate of the other [8, 9]. Such couplings do arise in the MSSM at one-loop, after integrating out massive squarks and are suppressed by $m_0^2/M_*^2\times$(loop-factor). Here they are suppressed by m_0/M_* only, as seen below:

$$\frac{M_s}{M_*}\Big[x_2^U\,[\lambda_U(0)]_{ij}\,(h_1^\dagger\,q_{Li})\;u_{Rj}^c + x_2^D\,[\lambda_D(0)]_{ij}\,(h_2^\dagger\,q_{Li})\;d_{Rj}^c + x_2^E\,[\lambda_E(0)]_{ij}\,(h_2^\dagger\,l_{Li})\;e_{Rj}^c + h.c.\Big] \qquad (8)$$

These couplings bring a $\tan\beta$ enhancement of a prediction for a physical observable, such as the bottom quark mass, relative to bottom quark Yukawa coupling:

$$m_b = (1/\sqrt{2})\,v\cos\beta\,\left(\lambda_b + \delta\lambda_b + \Delta\lambda_b\tan\beta\right) \qquad (9)$$

Here λ_b is the usual bottom quark Yukawa coupling, $\delta\lambda_b$ is its one-loop correction in MSSM and $\Delta\lambda_b$ is a "wrong"-Higgs coupling' corrections, obtained after integrating out at one-loop massive squarks in MSSM; in our case $\Delta\lambda_b$ receives an extra correction from (8), which can actually be larger than its one-loop-generated MSSM counterpart [8, 10–12]. This can bring a $\tan\beta$ enhancement of the Higgs decay rate into bottom quarks pairs (for further details see [6]).

2.2. MSSM Higgs sector with d=5 and d=6 operators

In the leading order, new physics beyond the MSSM Higgs sector can manifest itself as operators of either $d = 5$ [6, 13–15] or $d = 6$ [7, 16] or both. If generated by the same new physics, by comparing $\mathcal{O}(1/M^*)$ and $\mathcal{O}(1/M_*^2)$ terms one can estimate when the series expansion in $1/M_*$ breaks down. There is only one operator in the Higgs sector of dimension $d = 5$:

$$\mathcal{L}_1 = \frac{1}{M_*}\int d^2\theta\;\lambda_H'(S)\,(H_2.H_1)^2 + h.c. = 2\,\zeta_{10}\,(h_2.h_1)(h_2.F_1 + F_2.h_1) + \zeta_{11}\,m_0\,(h_2.h_1)^2 + h.c. \qquad (10)$$

where

$$\lambda'_H(S)/M_* = \zeta_{10} + \zeta_{11}\, m_0\, \theta\theta, \qquad \zeta_{10}, \zeta_{11} \sim 1/M_*, \tag{11}$$

$\mathcal{L}_1$ can be generated by integrating out a massive gauge singlet or $SU(2)$ triplet. Indeed, in the MSSM with a massive gauge singlet, with an F-term of type $M_*\Sigma^2 + \Sigma\, H_1.H_2$, when integrating out Σ generates $\mathcal{L}_1$.

We assume that $m_0 \ll M_*$, so that the effective approach is reliable. If this is not respected and the "new physics" is represented by "light" states (like the MSSM states), the $1/M_*$ expansion is not reliable and one should work in a setup where these are not integrated out.

The list of $d = 6$ operators is longer [17]:

$$\begin{aligned}
\mathcal{O}_1 &= \frac{1}{M_*^2}\int d^4\theta\; \mathcal{Z}_1\, (H_1^\dagger e^{V_1} H_1)^2, & \mathcal{O}_5 &= \frac{1}{M_*^2}\int d^4\theta\; \mathcal{Z}_5\, (H_1^\dagger e^{V_1} H_1)\, H_2.\, H_1 + h.c.\\
\mathcal{O}_2 &= \frac{1}{M_*^2}\int d^4\theta\; \mathcal{Z}_2\, (H_2^\dagger e^{V_2} H_2)^2, & \mathcal{O}_6 &= \frac{1}{M_*^2}\int d^4\theta\; \mathcal{Z}_6\, (H_2^\dagger e^{V_2} H_2)\, H_2.\, H_1 + h.c.\\
\mathcal{O}_3 &= \frac{1}{M_*^2}\int d^4\theta\; \mathcal{Z}_3\, (H_1^\dagger e^{V_1} H_1)\,(H_2^\dagger e^{V_2} H_2), & \mathcal{O}_7 &= \frac{1}{M_*^2}\int d^2\theta\; \mathcal{Z}_7\, \mathrm{Tr}\, W^\alpha\, W_\alpha\, (H_2\, H_1) + h.c.\\
\mathcal{O}_4 &= \frac{1}{M_*^2}\int d^4\theta\; \mathcal{Z}_4\, (H_2.\, H_1)\,(H_2.\, H_1)^\dagger, & \mathcal{O}_8 &= \frac{1}{M_*^2}\int d^4\theta\; \mathcal{Z}_8\, (H_2\, H_1)^2 + h.c.
\end{aligned} \tag{12}$$

where $W^\alpha = (-1/4)\,\overline{D}^2 e^{-V} D^\alpha e^V$ is the chiral field strength of $SU(2)_L$ or $U(1)_Y$ vector superfields V_w and V_Y respectively. Also $V_{1,2} = V_w^a(\sigma^a/2) + (\mp 1/2)\, V_Y$ with the upper (minus) sign for V_1. Finally, the wavefunction coefficients are spurion dependent and have the structure

$$(1/M_*^2)\, \mathcal{Z}_i(S, S^\dagger) = \alpha_{i0} + \alpha_{i1}\, m_0\, \theta\theta + \alpha_{i1}^*\, m_0\, \overline{\theta\theta} + \alpha_{i2}\, m_0^2\, \theta\theta\overline{\theta\theta}, \qquad \alpha_{ij} \sim 1/M_*^2. \tag{13}$$

Regarding the origin of these operators: $\mathcal{O}_{1,2,3}$ can be generated in the MSSM by an additional, massive $U(1)'$ gauge boson or $SU(2)$ triplets, when integrated out [13]. $\mathcal{O}_4$ can be generated by a massive gauge singlet or $SU(2)$ triplet, while $\mathcal{O}_{5,6}$ can be generated by a combination of $SU(2)$ doublets and massive gauge singlet. $\mathcal{O}_7$ is essentially a threshold correction to the gauge coupling, with a moduli field replaced by the Higgs. $\mathcal{O}_8$ exists only in non-susy case, but is generated when removing the $d = 5$ derivative operator $\mathcal{L}_2$ by field redefinitions [6], so we keep it.

2.3. Higgs mass corrections from $d = 5$ and $d = 6$ operators.

With the remaining set of independent, effective operators $\mathcal{L}_1$, $\mathcal{O}_{1,\ldots,8}$ of dimensions $d = 5$ and $d = 6$, one finds the scalar potential V and its EW minimum; this is perturbed by $\mathcal{O}(1/M_*^2)$ corrections from that of the MSSM. The expression of V is long and it is not given here (see [7] for its form). From V one computes the mass of CP-odd/even Higgs fields. One has:

$$m_A^2 = (m_A^2)_{\mathrm{MSSM}} - \frac{2\,\zeta_{10}\,\mu_0\, v^2}{\sin 2\beta} + 2\, m_0\, \zeta_{11}\, v^2 + \delta m_A^2, \quad \delta m_A^2 = \mathcal{O}(1/M_*^2) \tag{14}$$

for the pseudoscalar Higgs, with $(m_A^2)_{\rm MSSM}$ the MSSM value, with δm_A^2 due to $\mathcal{O}(1/M_*^2)$ corrections from $d=5$ and $d=6$ operators. For the CP-even Higgs one has [6, 13, 15]

$$\begin{aligned} m_{h,H}^2 &= (m_{h,H}^2)_{\rm MSSM} \\ &+ (2\,\zeta_{10}\,\mu_0)\,v^2\,\sin 2\beta\left[1\pm\frac{m_A^2+m_Z^2}{\sqrt{\tilde w}}\right]+\frac{(-2\,\zeta_{11}\,m_0)\,v^2}{2}\left[1\mp\frac{(m_A^2-m_Z^2)\,\cos^2 2\beta}{\sqrt{\tilde w}}\right] \\ &+ \delta m_{h,H}^2, \qquad \text{where} \qquad \delta m_{h,H}^2=\mathcal{O}(1/M_*^2) \end{aligned} \tag{15}$$

The upper (lower) signs correspond to h (H), and $\tilde w\equiv(m_A^2+m_Z^2)^2-4\,m_A^2\,m_Z^2\cos^2 2\beta$. With this result one can show that the mass m_h can be increased near the observed value, also with the help of quantum corrections [6, 13–15].

Regarding the $\mathcal{O}(1/M_*^2)$ corrections of $\delta m_{h,H}^2$, δm_A^2 and $\delta m_{h,H}^2$ of (14), (15), in the general case of including all operators and their associated supersymmetry breaking, they have a complicated form. Exact expressions can be found in [7, 16]. For most purposes, an expansion of these in $1/\tan\beta$ is accurate enough. At large $\tan\beta$, $d=6$ operators bring corrections comparable to those of $d=5$ operators. The relative $\tan\beta$ enhancement of $\mathcal{O}(1/M_*^2)$ corrections compensates for the extra suppression that these have relative to $\mathcal{O}(1/M_*)$ operators (which involve both h_1, h_2 and are not enhanced in this limit). Note however that in some models only $d=6$ operators may be present, depending on the details of the "new physics" generating the effective operators.

Let us present the correction $\mathcal{O}(1/M^2)$ to m_h^2 for the case m_A is kept fixed to an appropriate value. The result is, assuming $m_A>m_Z$, (otherwise δm_h^2 and δm_H^2 are exchanged):

$$\begin{aligned} \delta m_h^2 = &-2\,v^2\left[\alpha_{22}\,m_0^2+(\alpha_{30}+\alpha_{40})\mu_0^2+2\alpha_{61}\,m_0\,\mu_0-\alpha_{20}\,m_Z^2\right]-(2\,\zeta_{10}\,\mu_0)^2\,v^4\,(m_A^2-m_Z^2)^{-1} \\ &+v^2\cot\beta\Big[(m_A^2-m_Z^2)^{-1}\Big(4\,m_A^2\,\big((2\alpha_{21}+\alpha_{31}+\alpha_{41}+2\alpha_{81})\,m_0\,\mu_0+(2\alpha_{50}+\alpha_{60})\,\mu_0^2+\alpha_{62}\,m_0^2\big) \\ &-(2\alpha_{60}-3\alpha_{70})\,m_A^2\,m_Z^2-(2\alpha_{60}+\alpha_{70})\,m_Z^4\Big)+8\,(m_A^2+m_Z^2)\,(\mu_0\,m_0\,\zeta_{10}\,\zeta_{11})\,v^2/(m_A^2-m_Z^2)^2\Big] \\ &+\mathcal{O}(1/\tan^2\beta) \end{aligned} \tag{16}$$

The mass corrections in (16) must be added to the rhs of eqs.(15) to obtain the full value of m_h^2. Together with (12), (13), these corrections identify the operators of $d=6$ with the largest contributions, which is important for model building beyond the MSSM Higgs sector. These operators are $\mathcal{O}_{2,3,4}$ in the absence of supersymmetry breaking and $\mathcal{O}_{2,6}$ when this is broken. It is preferable, however, to increase m_h^2 by supersymmetric rather than supersymmetry-breaking effects of the effective operators, because the latter are less under control in the effective approach; also, one would favour a supersymmetric solution to the fine-tuning problem associated with increasing the MSSM Higgs mass. Therefore $\mathcal{O}_{2,3,4}$ are the leading operators, with the remark that $\mathcal{O}_2$ has a smaller effect, of order $(m_Z/\mu_0)^2$ relative to $\mathcal{O}_{3,4}$ (for similar α_{j0}, $j=2,3,4$). At smaller $\tan\beta$, $\mathcal{O}_{5,6}$ can also give significant contributions, while $\mathcal{O}_7$ has a relative suppression factor $(m_Z/\mu_0)^2$. Note that we kept all operators $\mathcal{O}_i$ independent. By doing so, one can easily single out the individual contribution of each operator, which helps in model building, since not all operators are present in a specific model.

One limit to consider is that where the operators of $d = 6$ have coefficients such that their contributions add up to maximise δm_h^2. Since α_{ij} are not known, one can choose:

$$-\alpha_{22} = -\alpha_{61} = -\alpha_{30} = -\alpha_{40} = \alpha_{20} > 0 \tag{17}$$

In this case, at large $\tan\beta$:

$$\delta m_h^2 \approx 2\, v^2 \alpha_{20} \big[m_0^2 + 2\, m_0 \mu_0 + 2\, \mu_0^2 + m_Z^2\big] \tag{18}$$

A simple numerical example is illustrative. For $m_0 = 1$ TeV, $\mu_0 = 350$ GeV, and with $v \approx 246$ GeV, one has $\delta m_h^2 \approx 2.36\, \alpha_{20} \times 10^{11}$ (GeV)^2. Assuming $M_* = 10$ TeV and ignoring $d = 5$ operators, with $\alpha_{20} \sim 1/M_*^2$ and the MSSM value of m_h taken to be its upper classical limit m_Z (reached for large $\tan\beta$), we obtain an increase of m_h from $d = 6$ operators alone of about $\Delta m_h = 12.15$ GeV to $m_h \approx 103$ GeV. An increase of α_{20} by a factor of 2.5 to $\alpha_{20} \sim 2.5/M_*^2$ would give $\Delta m_h \approx 28$ GeV to $m_h \approx 119.2$ GeV, which is already above the LEP bound. Note that this increase is realised even for a scale M_* of "new physics" beyond the LHC reach.

The above choice of $M_* = 10$ TeV was partly motivated by the fine-tuning results [14] (for $d = 5$ operators) and on convergence grounds: the expansion parameter of our effective analysis is m_q/M_* where m_q is any scale of the theory, in particular it can be m_0. For a susy breaking scale $m_0 \sim \mathcal{O}(1)$ TeV (say $m_0 = 3$ TeV) and $c_{1,2}$ or α_{ij} of $\mathcal{Z}_i(S, S^\dagger)$ of order unity (say $c_{1,2} = 2.5$) one has for $M_* = 10$ TeV that $c_{1,2}\, m_0/M_* = 0.75$ which is already close to unity, and at the limit of validity of the effective expansion in powers of $1/M_*$. To conclude, even for a scale of "new physics" above the LHC reach, one can still classically increase m_h to near the LHC measured value.

2.4. Final remarks

The final step is to identify the nature of "new physics" that generated the operators with the largest correction to m_h, ideally from a renormalisable model. At the level of dimension $d = 5$ operators, this is clear from previous discussion: a massive gauge singlet can generate operator $\mathcal{L}_1$ of (10) and the needed increase of m_h, for a scale $M_* \sim 5 - 10$ TeV [14]; this can provide a solution to the little hierarchy problem, provided that one can fix dynamically the scale M_*.

For dimension-six operators, from the above discussion one finds that to increase m_h it is needed that one or more of the following conditions are satisfied:

$$\alpha_{20} > 0,\ \alpha_{30} < 0,\ \alpha_{40} < 0 \tag{19}$$

First recall that $\mathcal{O}_{1,2,3}$ can be most easily generated by integrating out a massive gauge boson $U(1)'$ or $SU(2)$ triplets [13], while $\mathcal{O}_4$ can be generated by a massive gauge singlet or $SU(2)$ triplets. Let us discuss the signs of the operators when they are generated as above: **(a):** Integrating out a massive vector superfield $U(1)'$ under which Higgs fields have opposite charges (to avoid a Fayet-Iliopoulos term), one finds $\alpha_{20} < 0$ and $\alpha_{30} > 0$ (also $\alpha_{10} < 0$) [13], which is opposite to what we need. This can be changed, if for example there are additional pairs of massive Higgs doublets also charged under new $U(1)'$; then $\mathcal{O}_3$ could be generated with $\alpha_{30} < 0$.

(b): Integrating massive $SU(2)$ triplets that couple to the MSSM Higgs sector would bring $\alpha_{20} > 0$, $\alpha_{40} < 0$, $\alpha_{30} > 0$, so the first two relations agree with what we need.

(c): Integrating a massive gauge singlet would bring $\alpha_{40} > 0$, which would instead decrease m_h.

Finally, at large $\tan\beta$, due to additional corrections that effective operators bring to the ρ parameter [18], it turns out that α_{40} and α_{30} can have the largest correction to m_h^2, while avoiding ρ-parameter constraints. The case of a massive gauge singlet or additional $U(1)'$ vector superfield (giving $\mathcal{O}_{3,4}$) have the advantage of preserving gauge couplings unification at one-loop. Following the above information, one can proceed to construct explicit models with additional states that can generate these effective operators.

Let us mention that the method provided here to reduce the fine tuning in the MSSM for m_h, relies on introducing an additional scale in the visible sector, due to "new physics" in this sector. Other solutions to this problem may exist, which essentially rely on a low scale in the hidden sector of supersymmetry breaking [19]. In this case the quartic coupling and the mass of the SM-like Higgs are increased by a factor proportional to $(\mu^2/f)^2$, where f is the hidden sector supersymmetry breaking scale. While not without problems, the advantage of this latter method is that it does not pay the "cost" of an additional parameter (scale) in the visible sector, as models with effective operators do.

3. STRINGS AND EXTRA DIMENSIONS

The appropriate and most convenient framework for low energy supersymmetry and grand unification is the perturbative heterotic string. Indeed, in this theory, gravity and gauge interactions have the same origin, as massless modes of the closed heterotic string, and they are unified at the string scale M_s. As a result, the Planck mass is predicted to be proportional to M_s:

$$M_P = M_s/g\,, \tag{20}$$

where g is the gauge coupling. In the simplest constructions all gauge couplings are the same at the string scale, given by the four-dimensional (4d) string coupling, and thus no grand unified group is needed for unification. In our conventions $\alpha_{\rm GUT} = g^2 \simeq 0.04$, leading to a discrepancy between the string and grand unification scale $M_{\rm GUT}$ by almost two orders of magnitude. Explaining this gap introduces in general new parameters or a new scale, and the predictive power is essentially lost. This is the main defect of this framework, which remains though an open and interesting possibility.

The other other perturbative framework that has been studied extensively in the more recent years is type I string theory with D-branes. Unlike in the heterotic string, gauge and gravitational interactions have now different origin. The latter are described again by closed strings, while the former emerge as excitations of open strings with endpoints confined on D-branes [20]. This leads to a braneworld description of our universe, which should be localized on a hypersurface, i.e. a membrane extended in p spatial dimensions, called p-brane (see Fig. 1). Closed strings propagate in all nine dimensions of string theory: in those extended along the p-brane, called parallel, as well as in the transverse ones. On the contrary, open strings are attached on the p-brane. Obviously, our p-brane world must have at least the

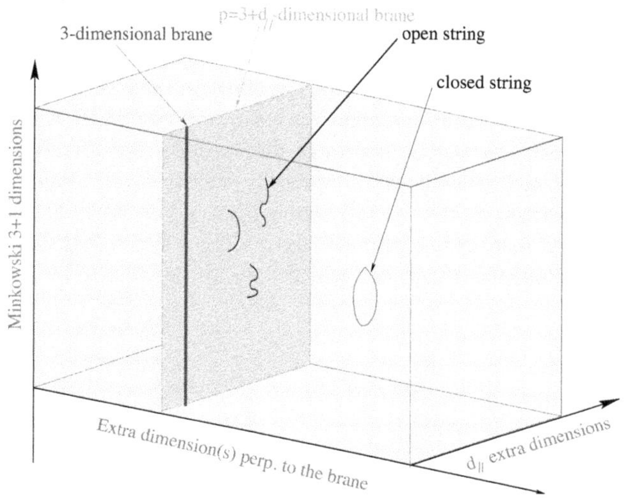

FIG. 1: D-brane world universe in type I string framework.

three known dimensions of space. But it may contain more: the extra $d_{\|} = p - 3$ parallel dimensions must have a finite size, in order to be unobservable at present energies, and can be as large as $\text{TeV}^{-1} \sim 10^{-18}$ m [21]. On the other hand, transverse dimensions interact with us only gravitationally and experimental bounds are much weaker: their size should be less than about 0.1 mm [22]. In the following, I review the main properties and experimental signatures of low string scale models [23].

3.1. Framework of low scale strings

In type I theory, the different origin of gauge and gravitational interactions implies that the relation between the Planck and string scales is not linear as (20) of the heterotic string. The requirement that string theory should be weakly coupled, constrain the size of all parallel dimensions to be of order of the string length, while transverse dimensions remain unrestricted. Assuming an isotropic transverse space of $n = 9 - p$ compact dimensions of common radius $R_{\perp}$, one finds:

$$M_P^2 = \frac{1}{g_s^2} M_s^{2+n} R_{\perp}^n \,, \qquad g_s \simeq g^2 \,. \tag{21}$$

where g_s is the string coupling. It follows that the type I string scale can be chosen hierarchically smaller than the Planck mass [23, 24] at the expense of introducing extra large transverse dimensions felt only by gravity, while keeping the string coupling small [23]. The weakness of 4d gravity compared to gauge interactions (ratio M_W/M_P) is then attributed to the largeness of the transverse space $R_\perp$ compared to the string length $l_s = M_s^{-1}$.

An important property of these models is that gravity becomes effectively $(4+n)$-dimensional with a strength comparable to those of gauge interactions at the string scale. The first relation of Eq. (21) can be understood as a consequence of the $(4+n)$-dimensional Gauss law for gravity, with $M_*^{(4+n)} = M_s^{2+n}/g^4$ the effective scale of gravity in $4+n$ dimensions. Taking $M_s \simeq 1$ TeV, one finds a size for the extra dimensions $R_\perp$ varying from 10^8 km, .1 mm, down to a Fermi for $n = 1, 2$, or 6 large dimensions, respectively. This shows that while $n = 1$ is excluded, $n \geq 2$ is allowed by present experimental bounds on gravitational forces [22, 25]. Thus, in these models, gravity appears to us very weak at macroscopic scales because its intensity is spread in the "hidden" extra dimensions. At distances shorter than $R_\perp$, it should deviate from Newton's law, which may be possible to explore in laboratory experiments.

3.2. Large number of species

Here, we open a parenthesis to describe that low scale gravity with large extra dimensions is actually a particular case of a more general framework, where the ultraviolate (UV) cutoff is lower than the Panck scale due to the existence of a large number of particle species coupled to gravity [26]. Indeed, it was shown that the effective UV cutoff M_* is given by

$$M_*^2 = M_P^2/N\,, \tag{22}$$

where the counting of independent species N takes into account all particles which are not broad resonances, having a width less than their mass. The derivation is based on black hole evaporation but here we present a shorter argument using quantum information storage [27]. Consider a pixel of size L containing N species storing information. The energy required to localize N wave functions is then given by N/L, associated to a Schwarzschild radius $R_s = N/LM_P^2$. The latter must be less than the pixel size in order to avoid the collapse of such a system to a black hole, $R_s \leq L$, implying a minimum size $L \geq L_{min}$ with $L_{min} = \sqrt{N}/M_P$ associated precisely to the effective UV cutoff $M_* = L_{min}$ given in eq. (22). Imposing $M_* \simeq 1$ TeV, one should then have $N \sim 10^{32}$ particle species below about the TeV scale!

In the string theory context, there are two ways of realizing such a large number a particle species by lowering the string scale at a TeV:

1. In large volume compactifications with the SM localized on D-brane stacks, as described in the previous section. The particle species are then the Kaluza-Klein (KK) excitations of the graviton (and other possible bulk modes) associated to the large extra dimensions, given by $N = R_\perp^n l_s^n$, up to energies of order $M_* \simeq M_s$.

2. By introducing an infinitesimal string coupling $g_s \simeq 10^{-16}$ with the SM localized on Neveu-Schwarz NS5-branes in the framework of little strings [28]. In this case, the particle species are the effective number of string modes that contribute to the black hole bound [29]: $N = 1/g_s^2$ and gravity does not become strong at $M_s \sim \mathcal{O}(\text{TeV})$.

Note the both TeV string realizations above are compatible with the general expression (21), but in the second case there is no relation between the string and gauge couplings.

3.3. Experimental implications in accelerators

We now turn to the experimental predictions of TeV scale strings. Their main implications in particle accelerators are of four types, in correspondence with the four different sectors that are generally present:

1. New compactified parallel dimensions; In this case $RM_s \gtrsim 1$, and the associated compactification scale $R_{\parallel}^{-1}$ would be the first scale of new physics that should be found increasing the beam energy [21, 30]. The main consequence is the existence of KK excitations for all SM particles that propagate along the extra parallel dimensions. These can be produced on-shell at LHC as new resonances [31].

2. New extra large transverse dimensions and low scale quantum gravity,. The main experimental signal is gravitational radiation in the bulk from any physical process on the world-brane [32].

3. Genuine string and quantum gravity effects. Direct production of string resonances in hadron colliders leads generically to a universal deviation from Standard Model in jet distribution [33]. In particular, the first Regge excitation of the gluon has spin 2 and a width an order of magnitude lower than the string scale, leading to a characteristic peak in dijet production; similarly, the first excitations of quarks have spin 3/2.

4. Extra $U(1)$'s arising generically in D-brane models as part of unitary gauge group factors. They obtain in general masses due to four- or higher-dimensional anomalies, via the so-called Green-Schwarz anomaly cancellation mechanism involving axionic fields from the closed string sector. The resulting masses are therefore suppressed by a loop factor compared to the string scale. From the low energy point of view, they gauge global symmetries of the Standard Model, such as the baryon and lepton number. An important property of the anomaly cancellation mechanism is that the anomalous $U(1)$ gauge bosons acquire masses leaving behind the corresponding global symmetries unbroken in perturbation theory. Thus, this is a way to guarantee proton stability (from unbroken baryon number) and avoid large Majorana neutrino masses (from unbroken lepton number) due to dimension-5 operators involving two higgses and two leptons that are suppressed only by the TeV string scale. Such extra $U(1)$s have interesting properties and distinct experimental signatures [34–36].

5. Concerning possible micro-black hole production, note that a string size black hole has a horizon radius $r_H \sim 1$ in string units, while the Newton's constant behaves as $G_N \sim g_s^2$. It follows that the mass of a d-dimensional black hole is [37]: $M_{\rm BH} \sim r_H^{d/2-1}/G_N \simeq 1/g_s^2$. Using the value of the SM gauge couplings $g_s \simeq g^2 \sim 0.1$, one finds that the energy threshold $M_{\rm BH}$ of micro-black hole production is about four orders of magnitude higher than the string scale, implying that one would produce 10^4 string states before reaching $M_{\rm BH}$.

3.4. Electroweak symmetry breaking

Non-supersymmetric TeV strings offer also a framework to realize gauge symmetry breaking radiatively. Indeed, from the effective field theory point of view, one expects quadratically divergent one-loop contributions to the masses of scalar fields. The divergences are cut off by M_s and if the corrections are negative, they can induce electroweak symmetry breaking and explain the mild hierarchy between the weak and a string scale at a few TeV, in terms of a loop factor [38]. More precisely, in the minimal case of one Higgs doublet H, the scalar potential is:

$$V = \lambda(H^\dagger H)^2 + \mu^2(H^\dagger H)\,, \tag{23}$$

where λ arises at tree-level. Moreover, in any model where the Higgs field comes from an open string with both ends fixed on the same brane stack, it is given by an appropriate truncation of a supersymmetric theory. On the other hand, μ^2 is generated at one loop:

$$\mu^2 = -\varepsilon^2\, g^2\, M_s^2\,, \tag{24}$$

where ε is a loop factor that can be estimated from a toy model computation and varies in the region $\epsilon \sim 10^{-1} - 10^{-3}$.

Indeed, consider for illustration a simple case where the whole one-loop effective potential of a scalar field can be computed. We assume for instance one extra dimension compactified on a circle of radius $R > 1$ (in string units). An interesting situation is provided by a class of models where a non-vanishing VEV for a scalar (Higgs) field ϕ results in shifting the mass of each KK excitation by a constant $a(\phi)$:

$$M_m^2 = \left(\frac{m + a(\phi)}{R}\right)^2, \tag{25}$$

with m the KK integer momentum number. Such mass shifts arise for instance in the presence of a Wilson line, $a = q\oint \frac{dy}{2\pi} gA$, where A is the internal component of a gauge field with gauge coupling g, and q is the charge of a given state under the corresponding generator. A straightforward computation shows that the ϕ-dependent part of the one-loop effective potential is given by [39]:

$$V_{eff} = -Tr(-)^F \frac{R}{32\,\pi^{3/2}} \sum_n e^{2\pi i n a} \int_0^\infty dl\; l^{3/2} f_s(l)\; e^{-\pi^2 n^2 R^2 l} \tag{26}$$

where $F = 0, 1$ for bosons and fermions, respectively. We have included a regulating function $f_s(l)$ which contains for example the effects of string oscillators. To understand its role we will consider the two limits $R >> 1$ and $R << 1$. In the first case only the $l \to 0$ region contributes to the integral. This means that the effective potential receives sizable contributions only from the infrared (field theory) degrees of freedom. In this limit we would have $f_s(l) \to 1$. For example, in the string model considered in [38]:

$$f_s(l) = \left[\frac{1}{4l}\frac{\theta_2}{\eta^3}(il + \frac{1}{2})\right]^4 \to 1 \qquad \text{for} \qquad l \to 0, \tag{27}$$

and the field theory result is finite and can be explicitly computed. As a result of the Taylor expansion around $a = 0$, we are able to extract the one-loop contribution to the coefficient of the term of the potential quadratic in the Higgs field. It is given by a loop factor times the compactification scale [39]. One thus obtains $\mu^2 \sim g^2/R^2$ up to a proportionality constant which is calculable in the effective field theory. On the other hand, if we consider $R \to 0$, which by T-duality corresponds to taking the extra dimension as transverse and very large, the one-loop effective potential receives contributions from the whole tower of string oscillators as appearing in $f_s(l)$, leading to squared masses given by a loop factor times M_s^2, according to eq. (24).

More precisely, from the expression (26), one finds:

$$\varepsilon^2(R) = \frac{1}{2\pi^2}\int_0^\infty \frac{dl}{(2\,l)^{5/2}}\frac{\theta_2^4}{4\eta^{12}}\left(il+\frac{1}{2}\right)R^3\sum_n n^2 e^{-2\pi n^2 R^2 l}\,, \tag{28}$$

which is plotted in Fig. 2. For the asymptotic value $R \to 0$ (corresponding upon T-duality

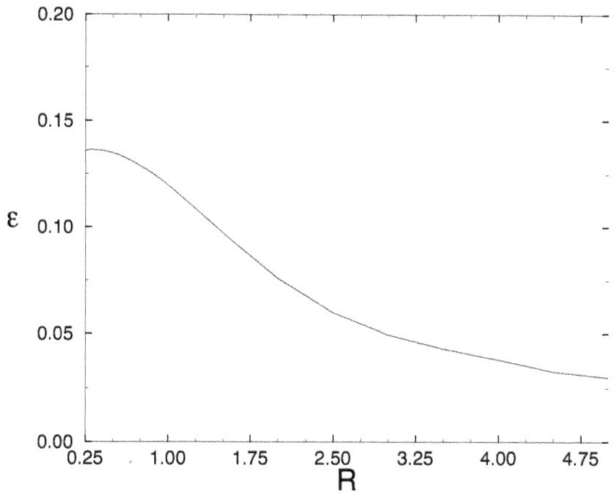

FIG. 2: The coefficient ε of the one loop Higgs mass (24).

to a large transverse dimension of radius $1/R$), $\varepsilon(0) \simeq 0.14$, and the effective cut-off for the mass term is M_s, as can be seen from Eq. (24). At large R, $\mu^2(R)$ falls off as $1/R^2$, which is the effective cut-off in the limit $R \to \infty$, as we argued above, in agreement with field theory results in the presence of a compactified extra dimension [40, 41]. In fact, in the limit $R \to \infty$, an analytic approximation to $\varepsilon(R)$ gives:

$$\varepsilon(R) \simeq \frac{\varepsilon_\infty}{M_s\,R}\,, \qquad\qquad \varepsilon_\infty^2 = \frac{3\,\zeta(5)}{4\,\pi^4} \simeq 0.008\,. \tag{29}$$

The potential (23) has the usual minimum, given by the VEV of the neutral component of the Higgs doublet $v = \sqrt{-\mu^2/\lambda}$. Furthermore, from (24), one can compute M_s in terms of the Higgs mass $m_H^2 = -2\mu^2$:

$$M_s = \frac{m_H}{\sqrt{2}\,g\varepsilon}\,, \tag{30}$$

yielding naturally values in the TeV range.

Acknowledgments

Work supported in part by the European Commission under the ERC Advanced Grant 226371.

---

[1] D. M. Ghilencea and H. M. Lee, JHEP **0509** (2005) 024.
[2] D. M. Ghilencea, Phys. Rev. D **70** (2004) 045018.
[3] S. Groot Nibbelink, M. Hillenbach, Phys. Lett. B **616** (2005) 125.
[4] S. Groot Nibbelink, M. Hillenbach, Nucl.Phys. B **748** (2006) 60.
[5] J. F. Oliver, J. Papavassiliou and A. Santamaria, Phys. Rev. D **67** (2003) 125004.
[6] I. Antoniadis, E. Dudas, D. Ghilencea, P. Tziveloglou, Nucl. Phys. B **808** (2009) 155.
[7] I. Antoniadis, E. Dudas, D. Ghilencea, P. Tziveloglou, Nucl. Phys. B **831** (2010) 133.
[8] H. E. Haber, J. D. Mason, Phys. Rev. D **77** 115011 (2008).
[9] S. P. Martin, Phys. Rev. D **61** 035004 (2000).
[10] M. S. Carena, H. E. Haber, H. E. Logan, S. Mrenna, Phys. Rev. D **65** 055005 (2002).
[11] D. M. Pierce, J. A. Bagger, K. T. Matchev, R. J. Zhang, Nucl. Phys. B **491** 3-67 (1997).
[12] L. J. Hall, R. Rattazzi and U. Sarid, Phys. Rev. D **50** 7048-7065 (1994).
[13] M. Dine, N. Seiberg, S. Thomas, Phys. Rev. D **76** (2007) 095004.
[14] S. Cassel, D. Ghilencea, G. G. Ross, Nucl. Phys. B **825** (2010) 203.
[15] K. Blum, C. Delaunay and Y. Hochberg, arXiv:0905.1701 [hep-ph].
[16] M. Carena, K. Kong, E. Ponton and J. Zurita, arXiv:0909.5434 [hep-ph]. M. Carena, E. Ponton and J. Zurita, arXiv:1005.4887 [hep-ph].
[17] D. Piriz and J. Wudka, Phys. Rev. D **56** (1997) 4170.
[18] A. Brignole, J. A. Casas, J. R. Espinosa and I. Navarro, Nucl. Phys. B **666** (2003) 105.
[19] I. Antoniadis, E. Dudas, D. M. Ghilencea and P. Tziveloglou, arXiv:1006.1662 [hep-ph].
[20] C. Angelantonj and A. Sagnotti, *Phys. Rept.* **371** (2002) 1 [Erratum-ibid. **376** (2003) 339] [arXiv:hep-th/0204089].
[21] I. Antoniadis, *Phys. Lett.* B **246** (1990) 377.
[22] D. J. Kapner, T. S. Cook, E. G. Adelberger, J. H. Gundlach, B. R. Heckel, C. D. Hoyle and H. E. Swanson, Phys. Rev. Lett. **98** (2007) 021101.
[23] N. Arkani-Hamed, S. Dimopoulos and G. R. Dvali, *Phys. Lett.* B **429** (1998) 263 [arXiv:hep-ph/9803315]; I. Antoniadis, N. Arkani-Hamed, S. Dimopoulos and G. R. Dvali, *Phys. Lett.* B **436** (1998) 257 [arXiv:hep-ph/9804398].
[24] J. D. Lykken, *Phys. Rev.* D **54** (1996) 3693 [arXiv:hep-th/9603133].
[25] J.C. Long and J.C. Price, *Comptes Rendus Physique* **4** (2003) 337; R.S. Decca, D. Lopez, H.B. Chan, E. Fischbach, D.E. Krause and C.R. Jamell, *Phys. Rev. Lett.* **94** (2005) 240401; R.S. Decca et al., arXiv:0706.3283 [hep-ph]; S.J. Smullin, A.A. Geraci, D.M. Weld, J. Chiaverini, S. Holmes and A. Kapitulnik, arXiv:hep-ph/0508204; H. Abele, S. Haeßler and A. Westphal, in 271th WE-Heraeus-Seminar, Bad Honnef (2002).
[26] G. Dvali, arXiv:0706.2050 [hep-th]; Int. J. Mod. Phys. A **25** (2010) 602 [arXiv:0806.3801 [hep-th]]; G. Dvali and M. Redi, Phys. Rev. D **77** (2008) 045027 [arXiv:0710.4344 [hep-th]];

R. Brustein, G. Dvali and G. Veneziano, JHEP **0910** (2009) 085 [arXiv:0907.5516 [hep-th]].

[27] G. Dvali and C. Gomez, Phys. Lett. B **674** (2009) 303.

[28] I. Antoniadis and B. Pioline, *Nucl. Phys.* B **550** (1999) 41; I. Antoniadis, S. Dimopoulos and A. Giveon, JHEP **0105** (2001) 055 [arXiv:hep-th/0103033]; I. Antoniadis, A. Arvanitaki, S. Dimopoulos and A. Giveon, arXiv:1102.4043.

[29] G. Dvali and D. Lust, arXiv:0912.3167 [hep-th]; G. Dvali and C. Gomez, arXiv:1004.3744 [hep-th].

[30] I. Antoniadis and K. Benakli, *Phys. Lett.* B **326** (1994) 69.

[31] I. Antoniadis, K. Benakli and M. Quirós, *Phys. Lett.* B **331** (1994) 313 and *Phys. Lett.* B **460** (1999) 176; P. Nath, Y. Yamada and M. Yamaguchi, *Phys. Lett.* B **466** (1999) 100 T. G. Rizzo and J. D. Wells, *Phys. Rev.* D **61** (2000) 016007; T. G. Rizzo, *Phys. Rev.* D **61** (2000) 055005; A. De Rujula, A. Donini, M. B. Gavela and S. Rigolin, *Phys. Lett.* B **482** (2000) 195.

[32] G.F. Giudice, R. Rattazzi and J.D. Wells, *Nucl. Phys.* B **544** (1999) 3; E.A. Mirabelli, M. Perelstein and M.E. Peskin, *Phys. Rev. Lett.* **82** (1999) 2236; T. Han, J.D. Lykken and R. Zhang, *Phys. Rev.* D **59** (1999) 105006; K. Cheung, W.-Y. Keung, *Phys. Rev.* D **60** (1999) 112003; C. Balázs *et al.*, *Phys. Rev. Lett.* **83** (1999) 2112; J.L. Hewett, *Phys. Rev. Lett.* **82** (1999) 4765.

[33] L.A. Anchordoqui, H. Goldberg, D. Lust, S. Nawata, S. Stieberger and T.R. Taylor, Phys. Rev. Lett. **101** (2008) 241803 [arXiv:0808.0497 [hep-ph]].

[34] I. Antoniadis, E. Kiritsis and T. N. Tomaras, Phys. Lett. B **486**, 186 (2000) [arXiv:hep-ph/0004214]; I. Antoniadis, E. Kiritsis and J. Rizos, Nucl. Phys. B **637**, 92 (2002) [arXiv:hep-th/0204153]; I. Antoniadis, E. Kiritsis, J. Rizos and T. N. Tomaras, Nucl. Phys. B **660**, 81 (2003) [arXiv:hep-th/0210263].

[35] G. Shiu and S.-H. H. Tye, *Phys. Rev.* D **58** (1998) 106007; Z. Kakushadze and S.-H. H. Tye, *Nucl. Phys.* B **548** (1999) 180; L. E. Ibáñez, C. Muñoz and S. Rigolin, *Nucl. Phys.* B **553** (1999) 43.

[36] L. A. Anchordoqui, I. Antoniadis, H. Goldberg, X. Huang, D. Lust and T. R. Taylor, Phys. Rev. D **85** (2012) 086003 [arXiv:1107.4309 [hep-ph]].

[37] G.T. Horowitz and J. Polchinski, Phys. Rev. D **55** (1997) 6189.

[38] I. Antoniadis, K. Benakli and M. Quirós, *Nucl. Phys.* B **583** (2000) 35.

[39] I. Antoniadis, K. Benakli and M. Quiros, *New Jour. Phys.* **3** (2001) 20.

[40] I. Antoniadis, C. Muñoz and M. Quirós, *Nucl. Phys.* B **397** (1993) 515; I. Antoniadis and M. Quirós, *Phys. Lett.* B **392** (1997) 61; A. Pomarol and M. Quirós, *Phys. Lett.* B **438** (1998) 225; I. Antoniadis, S. Dimopoulos, A. Pomarol and M. Quirós, *Nucl. Phys.* B **544** (1999) 503; A. Delgado, A. Pomarol and M. Quirós, *Phys. Rev.* D **60** (1999) 095008; R. Barbieri, L. J. Hall and Y. Nomura, *Phys. Rev.* D **63** (2001) 105007.

[41] I. Antoniadis, S. Dimopoulos and G. Dvali, *Nucl. Phys.* B **516** (1998) 70; S. Ferrara, C. Kounnas and F. Zwirner, *Nucl. Phys.* B **429** (1994) 589.

CHAIRMAN: I. ANTONIADIS

Scientific Secretaries: P. Pietrulewicz, A. Rothkopf

DISCUSSION

- *Karananas:*

Are higher derivative terms expected to be dangerous in the context of higher order operators in the MSSM?

- *Antoniadis:*

In principle yes. There are also more derivatives in the kinetic terms. Are you worried about unitarity?

- *Karananas:*

Yes.

- *Antoniadis:*

The way to treat these operators is as perturbations in the MSSM Lagrangian; they do not modify the propagators, they are just effective interactions. What I was interested in to show in my talk, are the corrections of these operators to the scalar potential.

- *Novikov:*

I would like to inquire about constraints on large extra dimensions. As you talked about constraints from torsion pendulum experiments, I would like to know whether large extra dimensions are constrained from cosmological arguments. Just yesterday Prof. Mukhanov gave us a value for short scales up to where general relativity has been validated.

- *Antoniadis:*

Yes there are also some bounds from cosmology. They are however weaker than others. Stronger bounds come from astrophysics. For instance, supernovae provide stronger constraints for the case of two extra dimensions than accelerator bounds because supernovae can emit gravitons and can cool down. This puts a bound on the size of the extra dimensions.

- *Novikov:*

And what about Planck data?

- *Antoniadis:*

I am not aware of any new bounds that did not exist before

- E. Krylov

Is supersymmetry and your model of SUSY compatible with the different mechanism of neutrino oscillations?

- *Antoniadis:*

Yes, usually neutrino oscillations are associated with some high scale, which is related to the see-saw mechanism. In other words, there is a right-handed neutrino that acquires a large mass and mixes with the left-handed neutrino, so that the resulting mass matrix has a very light eigenstate and a heavy eigenstate. In extra dimensions there are no high scales, but there exists another possibility, if the right-handed neutrino lives in the bulk of large extra dimensions together with gravity. The smallness of the neutrino mass is then related to the fact that the bulk is large, in the same way that gravity becomes weak, the neutrino masses become small.

- *Schröder:*

In particular, before the LHC started, experimentalists often gave the argument that there are good reasons for TeV supersymmetric particles due to the hierarchy problem, i.e. corrections to the Higgs mass that need to be on a natural level. Are there any other well-motivated reasons for SUSY particles to be present at this scale, so that we can observe them at the LHC?

- *Antoniadis:*

The main motivation for low scale supersymmetry is in fact the hierarchy problem. There are some secondary reasons, e.g. dark matter. If supersymmetry explains dark matter the lightest SUSY particle needs to be light in the TeV range. Also gauge coupling unification suggests that the fermionic superpartners, such as the gauginos and higgsinos should be light for unification to occur. Of course these are just indications, not necessity for low scale supersymmetry.

- *de Palma:*

Let us suppose that the LHC during its lifecycle does not see any SUSY particles. Which models will be ruled out and which will remain? Lets assume everything we see is in agreement with the standard model.

- *Zichichi:*

Then we go home!

- *Antoniadis:*

I would however say that we need to study precisely also the couplings of the Higgs particle and to see if there is any deviation in these coupling from the standard model. Personally, I believe that LHC will discover new physics. I refer to Prof. Zichichi's presentation from this morning, where he referred to the situation he was involved, just below the threshold of the J/Psi particle. I belive we are now in a similar situation.

- Wymant

You showed some models with tan[β]=1. I would have expected such models to contain a Landau pole in the top Yukawa coupling?

- *Antoniadis:*

I do no think that there is a Landau pole below the Planck mass.

- *Duff*

This is just a comment: You mentioned that the main motivation for supersymmetry is the gauge hierarchy problem. In my opinion the main motive for SUSY is gravity. If Einstein had not discovered relativity, SUSY would force us to invent it because the local SUSY transformation is the square root of a local coordinate transformation. This is the most compelling reason to work on it in my mind, whether or not it solves the gauge hierarchy problem.

- *Antoniadis:*

It is a compelling reason but not a necessity. You can have gravity without supersymmetry.

- *'t Hooft*

To me it sounds like a fairly poor motivation for SUSY, because we can think of many other ways of handling quantum gravity. Why is SUSY so necessary for QG? It seems to be convenient and happens to be a scheme with a gravitational sector but that does not mean that gravity requires SUSY, it is rather the other way around. SUSY needs gravity.

- *Duff:*

No. I said the reason why I find SUSY compelling, because it implies gravity. It is of course possible that the world does not take advantage of that but I find it compelling nonetheless.

- *'t Hooft:*

In principle it is questionable whether supersymmetry plays any role in the world. It proved to play an important mathematical role in all our models, where we have supersymmetric limits, e.g. SUSY black holes. That is well understood. But the converse, that physics needs SUSY, has never been shown.

- *Zichichi:*

The supersymmetric fellows claim that SUSY is needed, since without gravitinos, the coupling between gravity and electromagnetism diverges and hence the sun could not emit light. On the other hand with gravitinos it can emit light. Of course you can say that you have another way of avoiding such a divergence.

- *'t Hooft:*

We still do not even have proof that supergravity theories are finite. It is may be plausible but we still have a divergent perturbative expansion, which has not been replaced with something finite, and a lct is missing. I do not believe the argument that gravitinos are needed to make radiation finite. That sounds very strange to me. We understand gravity worse than some people believe and much better than others believe.

- *Zichichi:*

Very interesting! We have to discuss this further once Sergio Ferrara returns. He had another engagement.

- *Antoniadis:*

Supersymmetry improves the ultraviolet behavior but it does not make the theory finite

- *Zichichi:*

That is his claim, I am just repeating Ferrara's claim but Gerard is right that it is not proved.

- *'t Hooft:*

I think the question posed earlier was a very important one: Suppose that the LHC indeed during its lifetime does not find SUSY, which could correspond to the case of Harald's (Fritzsch) theory.

- *Zichichi:*

Harald would become famous!

- *'t Hooft:*

Harald would be very happy about that. His theory has been around for some time and other people would also be happy, but these theories also do not work very well so we would still have a problem. If nothing is observed at LHC it would still be very interesting from the viewpoint of SUSY, whether the onset of supersymmetry has to be postponed. Either we say that SUSY apparently works only from 10 TeV onwards and it has some fine tuning, but never mind. Or we go on and do something else.

- *I. Antoniadis:*

Well it really depends on what the LHC finds. If it finds nothing, it also means that the standard model is valid up to quite high energies.

- *'t Hooft:*

Of course the SUSY proponents will say it starts at 10 TeV, so we need a bigger LHC and look for it again.

- *Antoniadis:*

But then the reason is not the hierarchy problem anymore. You can also argue that SUSY is a symmetry of string theory, so that if string theory is realized at some point because of quantum gravity, then SUSY will be there also and be broken at the Plank scale or at a lower scale such as 10^11 GeV.

- *M. Duff:*

An analogy would be with gravitational waves. Einstein predicted them over 100 years ago but it is only recently that experimentalists have devised apparatuses that stand a chance to detect them. This underscores that theorists for the last 90 years have not wasted their time by working on something that could not be tested.

- *G. 't Hooft:*

That is a very different thing. General relativity as such, all its equations, have been tested in a number of ways and it predicts without doubt the existence of gravitational waves. Thus no physicist disputes the existence of these waves. It is just that they are so weak that they cannot be detected. If this were the case for SUSY no one would complain. But SUSY proponents have very strongly argued that it should be found at LHC and that is a different story. And with gravitational waves the experiments are coming, the detectors are gaining slowly the required sensitivity necessary for their detection, so within a couple of years we may have a signal.

- *Zichichi:*

Mike, I think this is a very weak argument.

- *Duff:*

No, I am addressing a common argument against what theorists are doing, which is if something cannot be tested in the near future it is not worth pursuing.

- *'t Hooft:*

I do not think anyone is saying that.

- *Duff*

Oh, I hear it all the time!

- *'t Hooft:*

This is because SUSY theorists have been claiming for a long time that it is just around the corner.

Status of Perturbative Approach to Supergravity

Z. Bern

Department of Physics and Astronomy
University of California at Los Angeles
Los Angeles, CA 90095
USA

Abstract

In this lecture we review recent developments in the perturbative approach to supergravity. A recently y conjectured duality between color and kinematics reveals a surprising connection between gravity and gauge theory allowing us to obtain gravity amplitudes directly from gauge theory ones. This duality has been used to demonstrate that supergravity theories are much better behaved in the ultraviolet than more standard symmetry arguments suggest. In particular, we summarize a recent calculation showing that half-maximal $\mathcal{N} = 4$ supergravity in four spacetime dimensions is ultraviolet finite at three loops, contrary to expectations from standard symmetry arguments.

1 Introduction

Recent years have taught us that gauge and gravity scattering amplitudes have a surprisingly rich structure beyond those based on the standard Lagrangian symmetries. Perhaps the most spectacular such symmetry is dual conformal invariance obeyed by planar $N = 4$ super-Yang-Mills theory [1]. Another surprising structure is the duality between color and kinematics [2, 3]. This links the planar sector to the nonplanar sector and then allows us to express corresponding supergravity integrands in terms of the gauge theory ones. The existence of these new symmetries and structures in gauge and gravity theories makes a compelling case to reexamine basic properties of these theories.

One basic property is the ultraviolet behavior of quantum gravity field theories. It is widely believe that such theories must necessarily be ultraviolet divergent, based on our current understanding of standard symmetries. However, the existence of these new symmetries and structures makes a compelling case for reexamining basic properties of supergravity theories.

In this talk we review new work demonstrating that certain supergravity theories are remarkably tame in the ultraviolet and perhaps may even even ultraviolet finite. A new approach for doing so uses the duality between color and kinematics. This greatly simplifies the construction of multiloop supergravity scattering amplitudes [4, 5, 6, 7, 8, 9]. This allows us to directly study their ultraviolet structure. The direct connection between gravity and gauge theory at tree level was first understood 25 years ago in string theory [12]. Today's simpler description of this connection [2], allows a straightforward extension to loop level [3].

In this lecture we review how the new ideas relating gauge and gravity theories are leading to significant progress in understanding the ultraviolet properties of supergravity theories, especially the cases of $\mathcal{N} = 8$ [13] and $\mathcal{N} = 4$ supergravity [14]. More generally, there have also been major advances in developing powerful tools for carrying out the explicit computations of scattering amplitudes. These include the unitarity method [10, 11, 15] and its refinement known as the method of maximal cuts [16], used in the constructions described here. These tools have played a central role in uncovering the duality between color and kinematics [2, 3]. There has also been enormous progress in carrying out the required loop integration needed to determine the explicit values of potential divergences, from the constructed amplitudes (see, for example, ref. [17]). The obtained results has also stimulated significant new progress in understanding the role of supersymmetry and duality symmetry in constraining potential counterterms in supergravity theories [18].

Even with the new progress, the question of whether supergravity theories diverge in perturbation theory is still open. While conventional counterterm symmetry considerations suggest that all supergravity amplitudes will diverge at some loop order, studies of multiloop scattering amplitudes suggest that additional ultraviolet cancellations exist beyond these [19]. Interestingly, even pure Einstein gravity at one loop exhibits remarkable cancellations as the number of external legs increases [20]. The double-copy structure also implies that gravity scattering amplitudes more constrained than implied by standard symmetries

The case of half-maximal $\mathcal{N} = 4$ supergravity [14] at three loops is especially interesting. A recent calculation [6], as well as string arguments [21], shows that the potential three-loop R^4 divergence [22] is absent, contrary to earlier symmetry expectations [23]. While the possible existence of a certain harmonic superspace may lead to an explanations [24], further studies are needed. Other more exotic potential explanations are based on the double-copy property of gravity [7, 25] and on a hidden superconformal invariance [26]. A recent four-loop calculation [9], does find a four loop divergence, but opens more questions than it answers. In particular, the structure of the divergence links it to an anomaly [27, 28] that is not present in $\mathcal{N} \geq 5$ supergravity theories, suggesting that such theories will be free of these types of divergences.

2 Connection between gravity and gauge theory

We start by comparing gravity to gauge theory using standard off-shell methods. Feynman rules are generated from the Yang-Mills and Einstein-Hilbert Lagrangians,

$$\mathcal{L}_{\text{YM}} = -\frac{1}{4} F^a_{\mu\nu} F^{a\,\mu\nu}\,, \qquad\qquad \mathcal{L}_{\text{EH}} = \frac{2}{\kappa^2}\sqrt{-g}R\,. \tag{1}$$

From the viewpoint of Feynman diagrams, these two Lagrangians have rather different properties. With standard gauge choices gauge theories have three- and four-point interactions, while gravity has an infinite number of contact interactions. Perhaps more striking than the infinite number of interactions is their complexity.

To be more explicit, consider the three-gluon vertex in Feynman gauge,

$$V^{abc}_{3\mu,\nu,\sigma}(k_1,k_2,k_3) = g f^{abc}\Big[(k_1-k_2)_\sigma \eta_{\mu\nu} + \text{cyclic}\Big], \tag{2}$$

where g is the coupling, f^{abc} the usual group theory structure constants, $\eta_{\mu\nu}$ the flat metric and the k_i the momenta of the vertex. This vertex is relatively simple. We may compare this to the three-graviton interaction in, for example, de Donder gauge,

$$G_{3\mu\alpha,\nu\beta,\sigma\gamma}(k_1,k_2,k_3) = i\frac{\kappa}{2}\Big[-\frac{1}{2}k_1\cdot k_2 \eta_{\mu\alpha}\eta_{\nu\beta}\eta_{\sigma\gamma} - \frac{1}{2}k_{1\nu}k_{1\beta}\eta_{\mu\alpha}\eta_{\sigma\gamma} + \cdots\Big], \tag{3}$$

where we have displayed two terms out of about 100. Here the gravitational coupling κ is related to Newton's constant by $\kappa^2 = 32\pi^2 G_N$. The precise form of the vertex depends on the gauge, but in any case the three vertex is a rather involved and unenlightening object. The complete expression can be found in refs. [29, 30].

Comparing the vertex in eq. (2) to the one in eq. (3), it certainly appears that gravity is much more complicated than gauge theory. Moreover, there does not appear to be any simplicity or obvious relation between the gauge and gravity vertices. The former leads to complicated diagrams, but the latter appears hopelessly complicated. One can do somewhat better with special gauge choices and appropriate field redefinitions [31, 32], considerably simplifying the Feynman rules. Still, multiloop Feynman diagram calculations in (super) gravity are extremely difficult, and generally out of reach using even the most powerful computers.

Now let us reconsider the same process but using on-shell vertices. If we take the three-graviton vertex in eq. (3) and dot the three legs with physical polarizations tensors satisfying the physical-state conditions, $k_i^2 = 0$, $\varepsilon_i^{\mu\nu} k_{i\mu} = \varepsilon_i^{\mu\nu} k_{i\nu} = \varepsilon^\mu{}_\mu = 0$, we obtain a greatly simplified vertex,

$$sG_3(k_1,k_2,k_3) = -i\kappa \varepsilon_1^{\mu\alpha}\varepsilon_2^{\nu\beta}\varepsilon_3^{\sigma\gamma}\Big[(k_1)_\sigma\eta_{\mu\nu} + \text{cyclic}\Big]\Big[(k_1)_\gamma\eta_{\alpha\beta} + \text{cyclic}\Big]. \tag{4}$$

Remarkably, up to overall factors, this is just a double copy of the kinematic part of the on-shell Yang-Mills vertex,

$$V_3^{abc}(k_1,k_2,k_3) = 2\varepsilon_1^\mu\varepsilon_2^\nu\varepsilon_3^\sigma g f^{abc}\Big[(k_1)_\sigma\eta_{\mu\nu} + \text{cyclic}\Big], \tag{5}$$

where the polarization vector satisfies $\varepsilon_i^\mu k_{i\mu} = 0$. To make the comparison, we identify the graviton polarization tensor as a product of gluon polarization vectors, $\varepsilon_i^{\mu\nu} = \varepsilon_i^\mu \times \varepsilon_i^\nu$. Similar considerations allow us to express all three-point vertices in supergravity as products of super-Yang-Mills vertices. Using on-shell recursion [33], these three vertices are sufficient to construct any tree-level gauge or gravity amplitude. The unitarity method then allows us to construct any loop amplitude.

Clearly, there is a rather striking relationship between gravity and gauge theory, but to make it visible we need to keep external states on shell. As we shall see below, the double-copy structure in eq. (4) is not accidental, but appears likely to extend to all loop orders. As such, it reflects a profound and important property of quantum gravity, pointing to unification of the two theories, perhaps along the lines of string theory.

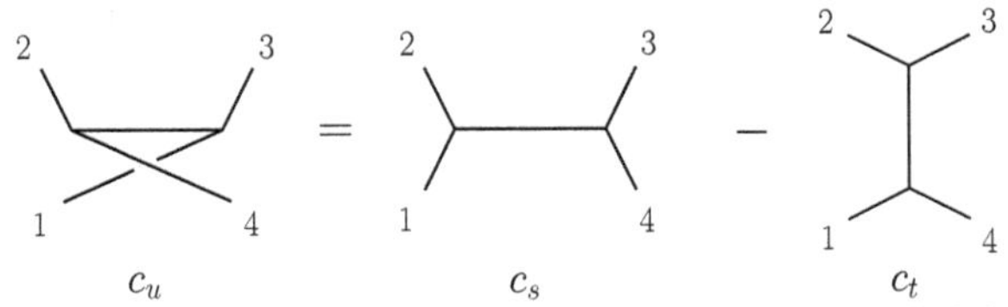

Figure 1: The Jacobi relation for at four points for the three channels labeled by s, t and u. These diagram represent either color factors or kinematic numerators.

3 A duality between color and kinematics

The above relation between gravity and gauge theory is a special case of the recently discovered duality between color and kinematics [2, 3]. In general, we can write any m-point tree-level gauge-theory amplitude with all particles in the adjoint representation as,

$$\mathcal{A}^{\text{tree}}(1,2,3,\ldots,m) = g^{m-2} \sum_i \frac{n_i\, c_i}{\prod_{\alpha_i} p^2_{\alpha_i}}\,, \tag{6}$$

where the sum runs over the set of m-point L-loop diagrams with only cubic vertices. These include distinct permutations of external legs. The product in the denominator runs over all propagators of each cubic diagram. The c_i are the color factors obtained by dressing every three vertex with an $\tilde{f}^{abc} = i\sqrt{2} f^{abc}$ structure constant, the n_i are kinematic numerator factors depending on momenta, polarizations and spinors and g is the coupling constant. The form (6) can be obtained straightforwardly, for example, from Feynman diagrams, by representing all contact terms as inverse propagators in the kinematic numerators that cancel propagators. For supersymmetric amplitudes expressed in superspace, there are also Grassmann parameters in the numerators.

The color factors satisfy Jacobi identities. In fig. 1 the basic four-point identity is shown, where the diagrams represent color factors obtained by dressing each vertex with an $\tilde{f}^{abc}$. The duality requires there to exists a way of reorganizing the diagrams so that the numerators n_i satisfy equations in one-to-one correspondence with the Jacobi identities of the color factors. That is, for each triplet of color factors satisfying a Jacobi identity, the corresponding numerators satisfy a similar identity:

$$c_i = c_j - c_k \;\Rightarrow\; n_i = n_j - n_k\,. \tag{7}$$

This duality is conjectured to hold to all multiplicity at tree level in a large variety of theories, including supersymmetric extensions of Yang-Mills theory. In fig. 1 we display the Jacobi relation at four points. The duality conjecture states that there exists representations of the amplitude, such that the color factors and numerators of the diagrams satisfy the relations. As explained in some detail in refs. [34, 5], the numerator relations are functional equations, depending nontrivially on the momenta and external polarizations and spinors. For four-point tree amplitudes such relations were noticed long ago using Feynman diagrams [35]. Beyond the four-point tree level, the relations are highly nontrivial and hold only after appropriate rearrangements of the amplitudes.

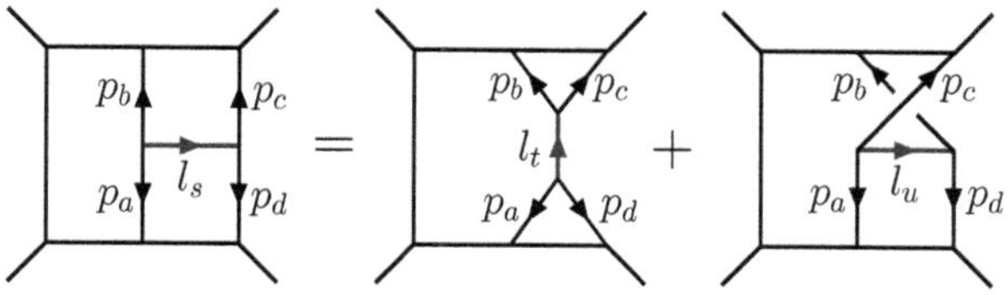

Figure 2: An example of a duality relation satisfied by numerators for diagrams of the three-loop four-point amplitude. Both color factors and numerator factors satisfy these relations.

At tree level, explicit forms of amplitudes satisfying the duality have been found for an arbitrary number of legs [36]. The duality has also been studied in string theory [37, 38]. An interesting consequence of the duality is that color-ordered partial tree amplitudes satisfy nontrivial relations [2], proven in gauge theory and in string theory [39]. Although we do not yet have a satisfactory Lagrangian understanding, some progress in this direction can be found in refs. [40, 41]. The duality (7) has also been expressed in terms of an alternative trace-based representation [42]. Progress has also been made in understanding the underlying infinite-dimensional Lie algebra [41, 43] responsible for the duality and in finding new representations that satisfy the duality [44]. The duality between color and kinematics also appears to hold in three-dimensional theories based on three algebras [45], as well as in some cases with higher-dimension operators [46].

Perhaps more surprising than the duality itself is a related conjecture that once the gauge-theory amplitudes are arranged into a form satisfying the duality (7), corresponding gravity amplitudes can be obtained simply by replacing the c_i color factor in eq. (6) with a second copy of a numerator factor $\tilde{n}_i$ [2, 3],

$$\mathcal{M}^{\text{tree}}(1,2,\ldots,m) = i\Big(\frac{\kappa}{2}\Big)\sum_i \frac{n_i\,\tilde{n}_i}{\prod_{\alpha_i} p^2_{\alpha_i}}\,, \tag{8}$$

where n_j and $\tilde{n}_j$ are kinematic numerator factors from gauge-theory amplitudes and κ is the gravitational coupling. The sum runs over the same set of diagrams with cubic vertices, as in eq. (6). This is expected to hold in a large class of gravity theories, including theories that are the low-energy limits of string theories. At tree level, this double-copy property encodes what are known as KLT relations between gravity and gauge-theory tree amplitudes [12]. The double-copy formula (8) has been proven via on-shell recursion [33] for pure gravity and for $\mathcal{N}=8$ supergravity tree amplitudes, when the duality (7) holds in the corresponding gauge theories [40]. It turns out that for the double-copy formula to hold, only one gauge-theory copy needs to satisfy the duality (7) because a generalized gauge invariance gives us the ability to rearrange the numerators of the second copy [3, 40].

More recently, the above relations have been conjectured to hold at loop level [3]. Any m-point L-loop gauge-theory amplitude with all particles in the adjoint representation can be written in the form,

$$\mathcal{A}_m^{L-\text{loop}} = i^L g^{m-2+2L} \sum_j \int \prod_{l=1}^{L} \frac{d^D p_l}{(2\pi)^D} \frac{1}{S_j} \frac{n_j c_j}{\prod_{\alpha_j} p_{\alpha_j}^2}, \tag{9}$$

where the sum labeled by j runs over the set of distinct m-point L-loop graphs with only cubic vertices, including distinct relabelings of external legs. The factor S_j is the symmetry factor of graph j that removes overcounts from internal symmetry. In this representation contact terms are absorbed into graphs with cubic vertices by multiplying and dividing by appropriate inverse propagators. The integrals are over L independent D-dimensional loop momenta. The product in the denominator runs over all Feynman propagators of graph j. As at tree level, the c_j are the color factors of the diagrams, and the n_j are kinematic numerators of graph j depending on momenta, polarizations and spinors. According to the duality conjecture of ref. [3], a representation of L-loop m-point amplitudes should exist where kinematic numerators satisfy the same algebraic properties as corresponding color factors (7).

The duality has been confirmed to hold up to four loops for the four-point amplitudes of $\mathcal{N} = 4$ super-Yang-Mills (sYM) theory [3, 5], and for the five-point one- and two-loop amplitudes of this theory [47]. Moreover, the infrared singularities do appear to be consistent with the duality to all loop orders [48]. The duality is also known to hold for the identical-helicity one- and two-loop four-point amplitudes of pure Yang-Mills theory [3]. It has also been tested in one and two-loop gravity examples in cases with fewer supersymmetries than the maximum [4] and for form factors [49]. There has also been progress in understanding more general one-loop amplitudes [50].

As at tree level, gravity amplitudes can be obtained from gauge theory amplitudes simply by replacing the color factors with a second gauge-theory numerator factor, $c_j \to \tilde{n}_j$. That is, associated with the conjectured duality between color and kinematics is a double-copy formula for m-point L-loop gravity amplitudes [3]

$$\mathcal{M}_m^{\text{loop}} = i^{L+1} \left(\frac{\kappa}{2}\right)^{m-2+2L} \sum_j \int \prod_{l=1}^{L} \frac{d^D p_l}{(2\pi)^D} \frac{1}{S_j} \frac{n_j \tilde{n}_j}{\prod_{\alpha_j} p_{\alpha_j}^2}. \tag{10}$$

The particular gravity theory obtained by the double-copy formula (10) is dictated by the choice of gauge theories. As at tree level, in eq. (10) only one of the two copies needs to satisfy the duality (7) [3, 40]. The other gauge-theory amplitude can be in another convenient representation arranged into graphs with only cubic vertices. While the duality between color and kinematics remains a conjecture, the unitarity method is used to confirm explicit constructions of multiloop amplitudes.

4 A three-loop surprise

The case of $\mathcal{N} = 4$ half-maximal supergravity is an especially interesting one, because of the expected appearance of a three-loop R^4 divergence based on symmetry consideration [23]. By constructing the three-loop four-point amplitude we can test this. To obtain pure supergravity amplitudes with no additional matter, we take one component gauge theory to be $\mathcal{N} = 4$ sYM theory and the second component to be nonsupersymmetric pure Yang-Mills theory [4, 14]. The $\mathcal{N} = 4$ sYM copy is the one discussed above [3], making the duality between color and kinematics manifest. What representation should we use for the nonsupersymmetric pure Yang-Mills amplitude? As explained in ref. [6], for the pure Yang-Mills copy, a convenient representation is ordinary Feynman-gauge Feynman diagrams. Any contact contributions are assigned to diagrams with only cubic vertices according to their color factor. Enormous simplifications follow from the fact that in this construction most Feynman graphs are irrelevant because they get multiplied by vanishing $\mathcal{N} = 4$ sYM graph numerators. In fact, at three loops only twelve graphs have nonvanishing numerators on the $\mathcal{N} = 4$ sYM side. Using these observations the complete three-loop four-point integrand of $\mathcal{N} = 4$ supergravity was obtained in ref. [6] for all external states of the theory.

Once the integrand of the amplitude is obtained, systematic methods for extracting the ultraviolet divergences from the constructed integrand can be applied [51]. We expand in large loop momenta or equivalently in small external momenta. This gives a set of vacuum graphs containing both infrared and ultraviolet divergences. To separate out the infrared divergences, a uniform mass m is introduced following the procedure in ref. [51]. Although ultimately there are no one- and two-loop ultraviolet divergences in $\mathcal{N} = 4$ supergravity, individual integrals generally will contain subdivergences, so it is best to systematically subtract these to ensure that no regulator dependence remains. Extractions of ultraviolet divergences in higher-dimensional $\mathcal{N} = 8$ supergravity were discussed recently in refs. [52, 5].

The next task is to calculate the ultraviolet divergences of the vacuum integrals. To evaluate these integrals, Lorentz invariance is used to replace tensors composed of loop momenta in the numerators by linear combinations of products of metric tensors $\eta^{\mu\nu}$ and dot products of momenta amongst themselves. (See ref. [5] for a more detailed discussion of simplifying tensor vacuum integrals.) A reduction of the scalar integrals to a basis, is carried out using integration-by-parts identities, as implemented in the program FIRE [53]. The resulting basis scalar integrals have been evaluated in refs. [54]. Combining this give the ultraviolet divergence of the amplitude [6]. Surprisingly, the sum over the contributions of all graphs vanish, so the three-loop four-graviton amplitude is ultraviolet finite. In fact, this cancellation holds for all external states of the theory [6], demonstrating that there are no three-loop four-point divergences in pure $\mathcal{N} = 4$ supergravity.

A key question is whether this result can be understood in terms of supersymmetry and duality symmetry, or whether a new mechanism is involved. While there has been a valiant attempt to explain three-loop finiteness of $\mathcal{N} = 4$ supergravity in terms of an off-shell 16 supercharge superspace [24], predictions for the behavior when matter multiplets are added disagree with explicit calculations. It is interesting that at least in a relatively simple case of half-maximal supergravity at two loops in $D = 5$ what seems to be mysterious cancellations from the standard symmetries of supergravity can be explained using the duality between color and kinematics. Ref. [7] demonstrated that the same cancellations that prevent divergences in pure nonsupersymmetric Yang-Mills from appearing in forbidden color tensors are those responsible for eliminating the divergences in the corresponding half-maximal supergravity amplitudes. It is an important problem to test this at higher loops.

5 Conclusions and outlook

In this talk, we described a new duality between color and kinematics and how it offers a new way to study the structure of supergravity scattering amplitudes, including their ultraviolet properties. In particular, we summarized the recent demonstration that the expected three-loop four-point divergence of $\mathcal{N} = 4$ supergravity [6] is not present. This example of an ultraviolet cancellation in supergravity has not yet been understood from more traditional symmetry considerations [23]. Some recent conjectures, point to possible explanations [26, 24]. New results on the three and four-loop behavior show that these proposals need to be modified [8, 9]. These results emphasize the importance of carrying out further explicit computations.

One possible explanation of the surprising ultraviolet cancellations is that the divergences of supergravity are linked to those of gauge theory [7]. When divergences in supergravity correspond to forbidden ones in gauge theory, the supergravity amplitudes are ultraviolet finite. This has been shown to be the case for half-maximal supergravity in $D = 5$ at two loops. The case of higher loops is more involved, and is currently under intense investigation.

An important outstanding problem is to unravel the underlying origin and structure of this duality. (See refs. [37, 40, 41, 43, 44] for recent progress on this issue.) At present such numerators are found by guessing their form and then using unitarity to prove the correctness. It would instead be important to devise systematic constructive rules for building duality satisfying numerators at loop level.

In summary, the new advances in our understanding of gravity amplitudes give us powerful new means for understanding fundamental properties of gravity theories, including their potential ultraviolet divergences. The rapid pace of progress suggests that in the coming years we will be able to fully unravel these properties.

The ideas presented in this talk was carried out in various papers in collaborations with John Joseph Carrasco, Scott Davies, Tristan Dennen, Lance Dixon, Yu-tin Huang, Henrik Johansson, Josh Nohle and Radu Roiban. This research was supported in part by the US Department of Energy under contract DESC0009937. We are also grateful for support from Mani Bhaumik.

References

[1] J. M. Drummond, J. Henn, V. A. Smirnov and E. Sokatchev, JHEP **0701**, 064 (2007) [hep-th/0607160]; Z. Bern, M. Czakon, L. J. Dixon, D. A. Kosower and V. A. Smirnov, Phys. Rev. D **75**, 085010 (2007) [hep-th/0610248];
J. M. Drummond, J. Henn, G. P. Korchemsky and E. Sokatchev, Nucl. Phys. B **795**,

52 (2008) [arXiv:0709.2368 [hep-th]]; Nucl. Phys. B **826**, 337 (2010) [arXiv:0712.1223 [hep-th]].

[2] Z. Bern, J. J. M. Carrasco and H. Johansson, Phys. Rev. D **78**, 085011 (2008) [arXiv:0805.3993 [hep-ph]].

[3] Z. Bern, J. J. M. Carrasco and H. Johansson, Phys. Rev. Lett. **105**, 061602 (2010) [arXiv:1004.0476 [hep-th]].

[4] Z. Bern, C. Boucher-Veronneau and H. Johansson, Phys. Rev. D **84**, 105035 (2011) [arXiv:1107.1935 [hep-th]]; C. Boucher-Veronneau and L. J. Dixon, JHEP **1112**, 046 (2011) [arXiv:1110.1132 [hep-th]]; J. J. M. Carrasco, M. Chiodaroli, M. Gunaydin and R. Roiban, arXiv:1212.1146 [hep-th].

[5] Z. Bern, J. J. M. Carrasco, L. J. Dixon, H. Johansson and R. Roiban, arXiv:1201.5366 [hep-th].

[6] Z. Bern, S. Davies, T. Dennen and Y.-t. Huang, Phys. Rev. Lett. **108**, 201301 (2012) [arXiv:1202.3423 [hep-th]].

[7] Z. Bern, S. Davies, T. Dennen and Y.-t. Huang, arXiv:1209.2472 [hep-th];

[8] Z. Bern, S. Davies and T. Dennen, Phys. Rev. D **88**, 065007 (2013) [arXiv:1305.4876 [hep-th]].

[9] Z. Bern, S. Davies, T. Dennen, A. V. Smirnov and V. A. Smirnov, Phys. Rev. Lett. **111**, 231302 (2013) [arXiv:1309.2498 [hep-th]].

[10] Z. Bern, L. J. Dixon, D. C. Dunbar and D. A. Kosower, Nucl. Phys. B **425**, 217 (1994) [arXiv:hep-ph/9403226]; Nucl. Phys. B **435**, 59 (1995) [arXiv:hep-ph/9409265].

[11] Z. Bern and A. G. Morgan, Nucl. Phys. B **467**, 479 (1996) [hep-ph/9511336]; Z. Bern, L. J. Dixon, D. C. Dunbar and D. A. Kosower, Phys. Lett. B **394**, 105 (1997) [hep-th/9611127]; Z. Bern, L. J. Dixon and D. A. Kosower, JHEP **0001**, 027 (2000) [hep-ph/0001001]; Z. Bern, L. J. Dixon and D. A. Kosower, Nucl. Phys. B **513**, 3 (1998) [hep-ph/9708239]; Z. Bern, V. Del Duca, L. J. Dixon and D. A. Kosower, Phys. Rev. D **71**, 045006 (2005) [hep-th/0410224]; R. Britto, F. Cachazo and B. Feng, Nucl. Phys. B **725**, 275 (2005) [hep-th/0412103].

[12] H. Kawai D. C. Lewellen and S.-H. H. Tye, Nucl. Phys. B **269**, 1 (1986); Z. Bern, Living Rev. Rel. **5**, 5 (2002) [gr-qc/0206071].

[13] E. Cremmer and B. Julia, Nucl. Phys. B **159**, 141 (1979).

[14] E. Cremmer, J. Scherk and S. Ferrara, Phys. Lett. B **74**, 61 (1978).

[15] Z. Bern, J. S. Rozowsky and B. Yan, Phys. Lett. B **401**, 273 (1997) [arXiv:hep-ph/9702424]; Z. Bern, L. J. Dixon, D. C. Dunbar, M. Perelstein and J. S. Rozowsky, Nucl. Phys. B **530**, 401 (1998) [arXiv:hep-th/9802162]; Z. Bern, J. J. Carrasco, L. J. Dixon, H. Johansson, D. A. Kosower and R. Roiban, Phys. Rev. Lett. **98**,

161303 (2007) [arXiv:hep-th/0702112]; Z. Bern, J. J. Carrasco, L. J. Dixon, H. Johansson and R. Roiban, Phys. Rev. Lett. **103**, 081301 (2009) [arXiv:0905.2326 [hep-th]].

[16] Z. Bern, J. J. M. Carrasco, H. Johansson and D. A. Kosower, Phys. Rev. D **76**, 125020 (2007) [arXiv:0705.1864 [hep-th]].

[17] V. A. Smirnov, *Feynman integral calculus,* Berlin, Germany: Springer (2006) 283 p

[18] M. B. Green, J. G. Russo and P. Vanhove, JHEP **1006**, 075 (2010)[arXiv:1002.3805 [hep-th]]; J. Bjornsson and M. B. Green, JHEP **1008**, 132 (2010) [arXiv:1004.2692 [hep-th]]. H. Elvang and M. Kiermaier, JHEP **1010**, 108 (2010) [arXiv:1007.4813 [hep-th]]; G. Bossard, P. S. Howe and K. S. Stelle, JHEP **1101**, 020 (2011) [arXiv:1009.0743 [hep-th]]; N. Beisert, H. Elvang, D. Z. Freedman, M. Kiermaier, A. Morales and S. Stieberger, Phys. Lett. B **694**, 265 (2010) [arXiv:1009.1643 [hep-th]].

[19] Z. Bern, L. J. Dixon and R. Roiban, Phys. Lett. B **644**, 265 (2007) [arXiv:hep-th/0611086].

[20] Z. Bern, J. J. Carrasco, D. Forde, H. Ita and H. Johansson, Phys. Rev. D **77**, 025010 (2008) [arXiv:0707.1035 [hep-th]].

[21] P. Tourkine and P. Vanhove, Class. Quant. Grav. **29**, 115006 (2012) [arXiv:1202.3692 [hep-th]].

[22] S. Deser, J. H. Kay and K. S. Stelle, Phys. Rev. Lett. **38**, 527 (1977);
S. Deser and J. H. Kay, Phys. Lett. B **76**, 400 (1978);
M. T. Grisaru and W. Siegel, Nucl. Phys. B **201**, 292 (1982) [Erratum-ibid. B **206**, 496 (1982)].

[23] G. Bossard, P. S. Howe, K. S. Stelle and P. Vanhove, Class. Quant. Grav. **28**, 215005 (2011) [arXiv:1105.6087 [hep-th]].

[24] G. Bossard, P. S. Howe and K. S. Stelle, Phys. Lett. B **719**, 424 (2013) [arXiv:1212.0841 [hep-th]]; JHEP **1307**, 117 (2013) [arXiv:1304.7753 [hep-th]].

[25] R. H. Boels and R. S. Isermann, arXiv:1212.3473 [hep-th].

[26] S. Ferrara, R. Kallosh and A. Van Proeyen, arXiv:1209.0418 [hep-th].

[27] N. Marcus, Phys. Lett. B **157**, 383 (1985).

[28] J. J. M. Carrasco, R. Kallosh, R. Roiban and A. A. Tseytlin, JHEP **1307**, 029 (2013) [arXiv:1303.6219 [hep-th]].

[29] G. 't Hooft and M. J. Veltman, *Annales Poincare Phys. Theor.* A **20**, 69 (1974);

[30] B.S. DeWitt, Phys. Rev. 162:1239 (1967); M. Veltman, in *Les Houches 1975, Proceedings, Methods In Field Theory*, eds. R. Balian and J. Zinn-Justin (North-Holland, Amsterdam, 1976); S. Sannan, Phys. Rev. D34:1749 (1986).

[31] A. E. M. van de Ven, *Nucl. Phys.* B **378**, 309 (1992).

[32] Z. Bern and A. K. Grant, Phys. Lett. B **457**, 23 (1999) [hep-th/9904026].

[33] R. Britto, F. Cachazo, B. Feng and E. Witten, Phys. Rev. Lett. **94**, 181602 (2005) [hep-th/0501052].

[34] J. J. M. Carrasco and H. Johansson, J. Phys. A **44**, 454004 (2011) [arXiv:1103.3298 [hep-th]].

[35] D. Zhu, Phys. Rev. D **22**, 2266 (1980); C. J. Goebel, F. Halzen and J. P. Leveille, Phys. Rev. D **23**, 2682 (1981).

[36] N. E. J. Bjerrum-Bohr, P. H. Damgaard, T. Sondergaard and P. Vanhove, JHEP **1101**, 001 (2011) [1010.3933 [hep-th]]; C. R. Mafra, O. Schlotterer and S. Stieberger, 1104.5224 [hep-th].

[37] N. E. J. Bjerrum-Bohr, P. H. Damgaard, T. Sondergaard and P. Vanhove, arXiv:1003.2403 [hep-th];

[38] S.-H. H. Tye and Y. Zhang, JHEP **1006**, 071 (2010) [arXiv:1003.1732 [hep-th]].

[39] N. E. J. Bjerrum-Bohr, P. H. Damgaard and P. Vanhove, Phys. Rev. Lett. **103**, 161602 (2009) [0907.1425 [hep-th]]; S. Stieberger, 0907.2211 [hep-th]. B. Feng, R. Huang and Y. Jia, Phys. Lett. B **695**, 350 (2011) [arXiv:1004.3417 [hep-th]]. Y. X. Chen, Y. J. Du and B. Feng, JHEP **1102**, 112 (2011) [1101.0009 [hep-th]]; F. Cachazo, arXiv:1206.5970 [hep-th].

[40] Z. Bern, T. Dennen, Y.t. Huang and M. Kiermaier, Phys. Rev. D **82**, 065003 (2010) [arXiv:1004.0693 [hep-th]].

[41] R. Monteiro and D. O'Connell, JHEP **1107**, 007 (2011) [arXiv:1105.2565 [hep-th]].

[42] Z. Bern and T. Dennen, 1103.0312 [hep-th].

[43] N. E. J. Bjerrum-Bohr, P. H. Damgaard, R. Monteiro and D. O'Connell, JHEP **1206**, 061 (2012) [arXiv:1203.0944 [hep-th]].

[44] J. Broedel and J. J. M. Carrasco, Phys. Rev. D **84**, 085009 (2011) [arXiv:1107.4802 [hep-th]]; A. S. Vera, E. S. Campillo and M. A. Vazquez-Mozo, arXiv:1212.5103 [hep-th]; C. -H. Fu, Y. -J. Du and B. Feng, arXiv:1212.6168 [hep-th].

[45] T. Bargheer, S. He and T. McLoughlin, arXiv:1203.0562 [hep-th]; Y.-t. Huang and H. Johansson, arXiv:1210.2255 [hep-th].

[46] J. Broedel and L. J. Dixon, [arXiv:1208.0876 [hep-th]].

[47] J. J. Carrasco and H. Johansson, Phys. Rev. D **85**, 025006 (2012) [arXiv:1106.4711 [hep-th]].

[48] S. Oxburgh and C. D. White, arXiv:1210.1110 [hep-th].

[49] R. H. Boels, B. A. Kniehl, O. V. Tarasov and G. Yang, arXiv:1211.7028 [hep-th].

[50] N. E. J. Bjerrum-Bohr, P. H. Damgaard, H. Johansson and T. Sondergaard, JHEP **1105**, 039 (2011) [arXiv:1103.6190 [hep-th]]; R. H. Boels and R. S. Isermann, JHEP **1203**, 051 (2012) [arXiv:1110.4462 [hep-th]]; E. Y. Yuan, arXiv:1210.1816 [hep-th].

[51] A. A. Vladimirov, Theor. Math. Phys. **43**, 417 (1980) [Teor. Mat. Fiz. **43**, 210 (1980)]; N. Marcus and A. Sagnotti, Nuovo Cim. A **87**, 1 (1985).

[52] Z. Bern, J. J. M. Carrasco, L. J. Dixon, H. Johansson and R. Roiban, Phys. Rev. D **82**, 125040 (2010) [arXiv:1008.3327 [hep-th]].

[53] A. V. Smirnov, JHEP **0810**, 107 (2008) [arXiv:0807.3243 [hep-ph]].

[54] K. G. Chetyrkin, M. Misiak and M. Munz, Nucl. Phys. B **518**, 473 (1998) [hep-ph/9711266]; M. Czakon, Nucl. Phys. B **710**, 485 (2005) [hep-ph/0411261].

CHAIRMAN: Z. BERN

Scientific Secretaries: S. Davies, J. Nohle

DISCUSSION

- Cerchiai:

What structure might explain the color-kinematics duality? It seems that it does not matter which gauge group is used for the color; only the Jacobi identity matters. What might be behind the kinematics?

- Bern:

That's correct. In the color-kinematics duality, on the color side, the only thing that is being used is a basic group-theory identity: the Jacobi identity. What the group is on the kinematic side is not well understood. For the specific case of self-dual Yang-Mills, which describes only a particular helicity sector of Yang-Mills, there is a nice paper from O'Connell and Monteiro. In the self-dual sector they uncover the infinite-dimensional Lie algebra that controls the group-theory structure of the kinematic numerators: it is an area-preserving diffeomorphism invariance. Unfortunately it appears nontrivial to go beyond the self-dual case, but I'm sure there will be progress on this issue.

- Cerchiai:

What is the relation of your approach with the approach of P. Mastrolia, who is in Munich, and collaborators using an algebraic geometry approach, including a Grobner basis, to compute a basis for the polynomials of the integrands?

- Bern:

Yes, there is some connection. However, they are interested in a different class of problems to the ones we're considering here. One difference is that they're interested in problems with a large number of legs. We're more interested in going higher in loop order, but with fewer numbers of legs. Furthermore, we're interested in the ultraviolet region, so the classification of the types of functions they find which is for a more general problem is not so useful to us here.

- Slepukhin:

Could you please explain the link between using on-shell methods and the absence of a higher-point vertex? Why is there only a three-point vertex?

- Bern:

The fact that higher-point vertices are unnecessary sounds too amazing to be true. If you look at the Yang-Mills Lagrangian in the context of Feynman rules, taking out the four-vertex simply gives you the incorrect result. It's a highly nontrivial fact that would take some time to explain, but there's a paper by Britto, Cachazo, Feng and Witten that explains this point. The key to the explanation involves complex deformations of the amplitude. If you introduce a complex variable z on which the amplitude depends and can prove that the amplitude vanishes as z approaches infinity, it's quite easy to demonstrate that only three-point vertices are required. Using Cauchy's theorem, you integrate over z with a contour closed off at infinity. Because the amplitude vanishes as z approaches infinity, the integral vanishes and you get an equality between the amplitude and the residues. The residues which determine the full amplitudes are controlled only by the values of the three-point vertices. A useful reference that explains this is Part N in Tony Zee's second edition of "Quantum Field Theory in a Nutshell."

- *de Palma:*

Is it in principle possible to prove that N=8 supergravity is finite or will we only be able to prove it to be finite up to a certain number of loops?

- *Bern:*

That's a great question. Every morning I wake up and ask myself, "Why can't I prove that N=8 supergravity is finite?" Unfortunately the answer is always the same: we can't make a proof we don't understand five loops. When we try to construct a proof in the context of the unitarity method, there is a problem that starts at five loops. We are confident that there is a set of cancellations that we haven't understood properly. When you think about proofs of finiteness in N=4 super-Yang-Mills, the used formalism, such as light-cone gauge or a superspace formalism, such that every diagram is manifestly finite. If you don't have that, but instead there was a cancellation between two divergent contributions, it would be extremely difficult to make a proof. This is essentially what is blocking us. I believe that if we can understand this issue at five loops, then we will be able to go forward and construct the proof.

- *Melia:*

Is there a particular topology at five loops that is problematic?

- *Bern:*

Yes, there is a whole set of these trouble makers. We can prove that in any covariant formalism will always have contributions with bad power counting. So, the only way that you can find a manifest cancellation is to have a non-Lorentz covariant formalism. But, those are generally much harder to work with and to make matters worse we don't yet have the appropriate formalism. So, it's an inherent problem in the structure of the amplitudes that makes the proof very hard. This is why we really need to understand five loops in N=8 supergravity before we can proceed in either direction: to prove it divergences or to prove it it ultraviolet finitene.

- *Wever:*

Is there anything that you can say about infrared divergences by using these color-kinematics dualities?

- *Bern:*

In Yang-Mills theory, there is an important problem at high loop orders. This has to do with issues in phenomenology--Sudakov factors--where you need to understand the infrared singularities of the theory. At three, four loops, in the non-planar contributions, there are some issues that have to be understood. In principle, we could apply these techniques to extract that. We have the integrands. And, by the way, N=4 super-Yang-Mills theory is all you need to solve this issue in QCD because the infrared structures are universal between the different theories. There are simple connections between N=4 super-Yang-Mills theory and QCD. If we were able to do the integrals, which are nontrivial, then we would be able to resolve that problem. In fact, there has been great progress in doing the integrals. I think within a year we'll have a lot of information about infrared singularities, and new information about infrared singularities in QCD in the non-planar contributions. For the case of gravity, it turns out that the infrared singularities are very simple; they are very well understood. There are no mysteries. The only use of infrared singularities would be to make sure that a calculation was done correctly because they are known a priori before doing the calculation.

- *Wymant:*

What is the jump in complexity at five loops that makes you think that, once you've understood that, you will be in a position to make a statement about *all* loop orders? Why don't you think that there will be another unexpected jump in complexity at any higher order beyond five loops?

- *Bern:*

One can never know for sure until we have a full understanding but we have extensive experience with this problem, suggesting we understand the sources of difficulty. We know for a fact that there is a jump in complexity at five loops and we understand why it happens. The basic point of the complexity has to do with the fact that we can prove that there is no manifest Lorentz covariant formalism that will make the scattering amplitudes finite manifestly, without any hidden cancellations. At higher loops it is possible new things can happen but we are pretty confident that the five loop issue is the key one.

- *Nikitin:*

You have used unitarity to calculate diagrams, but this relation only gives the imaginary part of diagrams. Is this a restriction on the applicability of the calculation?

- *Bern:*

If you look at the way unitarity was used in the 1960's, there are limitations that come from the fact that you construct the imaginary part, you need dispertion relations, there might be subtractions which are not under complete control, and to even get all the channels to work simultaneously is a nontrivial problem. But, we use extra information compared to the 1960's to do the reconstruction of the complete amplitude. This comes from the fact that we know that it had to come from a Feynman diagram. So, we understand a lot about the analytic structure of the scattering amplitude. All of the problems mentioned are basically bypassed because what we do is construct the *integrand,* not the integrated expression, by using the unitarity relations. The integrand is equivalent to the one you would get from Feynman diagrams, but its structure is much simpler. This point is nontrivial. In a sense, it's the secret of the unitarity method.

The Pedagogic Higgs - or Somebody's - Boson

Frank Close*
*Rudolf Peierls Centre for Theoretical Physics, University of Oxford,
1 Keble Road, Oxford, OX1 3NP, UK*

(Dated: Version 1.0 - 1 October 2013)

This talk gives a pedagogic example of how a photon can appear massive in the presence of a scalar field, and illustrates how this led to the discovery of the "mass mechanism" in relativistic quantum field theory. I also illustrate why the "Higgs mechanism" is misnamed but the "Higgs boson" is not.

I. INTRODUCTION

Particles would have a mass unless there's a reason for them not to. By "reason" we refer to a symmetry; such as chiral symmetry implying massless fermions and gauge symmetry leading to massless gauge bosons. The fundamental question underlying much of the work that has culminated in discovery - "beyond reasonable doubt" - of Higgs' boson deals with the question: can one keep symmetry but hide its consequences? In the case of interest: can one maintain gauge symmetry but have massive gauge bosons?

Today we know the answer is yes.

A pedagogic example of hiding symmetry can be found with the shapes of galaxies. The force of gravity is 3D (three-dimensional) symmetric and thus one might anticipate that it would lead to structures with spherical symmetry. The sun is approximately so, and there are spherically symmetric galaxies of stars; but there are lots of spiral galaxies which are more nearly 2-dimensional structures. A reason is that the 2D structure is more stable (in the presence of perturbations from other galaxies) and one has given up the underlying symmetry for stability. This example also has a further feature: there are billions of spiral galaxies and although each individually is roughly a 2D plate, the orientations of these discs cover all directions: the fundamental 3D symmetry is present globally - or in this case "universally"!

The pedagogic dynamical model of the hidden symmetry of the Goldstone-Higgs et al potential, familiar to all particle physics students, is the Mexican hat. A ball placed at the central peak sets up a situation that is 2D rotationally symmetric. It is however metastable. Due to quantum uncertainty it is impossible in principle for this situation to persevere and the ball rolls into the valley. The final state has a ball at one spot, and hence has broken the original symmetry.

There is an infinity of such states and it costs no energy to rotate from one to another. In relativistic quantum field theory including electric charge this corresponds to a massless scalar boson, electrically charged, and such "Goldstone bosons" do not exist in nature. It was the problem of this unwanted angular or rotational mode that caused so much trouble in the early 1960s. However, if one includes a long range field, such as implied by Maxwell's equations, this mode becomes the "missing" longitudinal mode of a massive vector boson. The gauge field has become massive but the fundamental gauge symmetry remains, albeit hidden.

This was discovered independently by several theorists in and around 1964 [1–5]. My purpose is not to review that solution to the Goldstone problem but to discuss the massive scalar boson that is named after Higgs[3, 6, 7].

In the dynamical model just discussed, there is a radial degree of freedom, where the ball oscillates up and down the walls at the valley's base. The frequency of oscillation depends on the steepness of the valley, and that is not determined in general. It can be shallow or steep, corresponding to high or low frequency, and in the field theory analogue, to low or high mass scalar boson. Higgs alone drew attention to this massive mode in 1964, and if that were the totality it would be relatively trivial. We shall see later, however, that in 1966 Higgs took an important further step[6].Indeed, Goldstone had already identified this mode in his seminal 1961 paper - see the equation below fig 7 which explicitly describes a massive boson.

This was, however, not the question of primary interest at that time. His 1961 paper had given examples of the massless angular mode, and this was elevated to a proof by Goldstone, Salam and Weinberg in 1962[9]. None of this

*E-mail: f.close1@physics.ox.ac.uk

included a gauge field in the mix. That turns out to be the crucial feature.

II. THE MASSIVE - "HIGGS" - BOSON

The renormalisability of QED had been demonstrated in 1947, and one of the critical features was that the photon is massless. Whether the fundamental feature was gauge invariance or masslessness was a technical issue, but as the two were thought to be intimately linked, this was regarded by most as a fine detail at that time - at least in the standard lectures on QED in the subsequent decades. In 1956 Schwinger was toying with the possibility that the weak force could be married with the electromagnetic, the dimensionful parameter of weak interactions - the Fermi constant - being potentially the dimensionless quantity $\alpha \sim 1/137$ moderated by a large mass W boson. This we now know to be the essence of the modern $SU2XU1$ structure of electroweak theory. However at the time Schwinger had a worry: a massive W boson would, apparently, spoil the renormalisation arguments that had worked so well in QED.

There were two significant developments that arose from this. One is the role of Schwinger's student - Sheldon Glashow - in formulating the $SU2XU1$ model, with its Z boson and neutral weak current. The other leads into our main theme in this talk: what is the link between gauge invariance and masslessness? Schwinger had demonstrated a link in perturbation theory, but had been unable to do so in general. Today we know why: there is no general link as the gauge symmetry can be hidden - the W boson can be massive within a gauge invariant theory, as was proved in the 1960s by several authors [1–5].

In 1962 Schwinger had shown that in a strong coupling theory, a massive boson can arise. This is described in Professor Cao's lecture here so I shall not say more. I am interested in what Anderson did following this: he made a pedagogic model that is gauge invariant with an effective massive photon. It is a perfect example that illustrates the basic ideas behind the so called Higgs-mechanism, and is legitimate physics, unlike some popular examples such s balls falling through treacle, which 're dissipative and do not exhibit the true features at all. Anderson's pedagogic model is the standard example of an electromagnetic wave interacting with a plasma.

The basic phenomenon is that an electromagnetic waves only penetrate the plasma if its frequency is higher than the plasma frequency. The physics behind this is familiar to anyone who has studied basic electromagnetic theory, so if you use it for a general audience some further motivation may be required. One possible route, if they are old enough to remember hearing "old-fashioned" radio receivers, is to remind them of the phenomenon where long-wave radio (i.e. low frequency electromagnetic waves) bounce off the Appleton layer, where by you may hear stations from far away bounced from the plasma above us into your receiver from the other side of the world. This illustrates the rejection of low-frequency waves by the plasma. Meanwhile, it is possible to see the stars shining through it: optical radiation, high frequency relative to radio waves, are transmitted through the plasma. That is a summary of what happens when an electromagnetic wave hits a plasma: low frequency rejected; high frequency transmitted.

So now let's return to Anderons's use of this phenomenon.

He imagined a creature that lives inside a plasma, and has never experienced anything other than this environment. That creature will only experience electromagnetic waves above the critical plasma frequency. Thus it will perceive photons to occur with a minimum frequency, hence, by $E = h\nu$, a minimum energy. A minimum energy is the property of a particle with mass: $E = mc^2$. Thus the creature inside the plasma will interpret the photons as being massive.

In field theory a massive vector has three degrees of freedom - the standard transverse polarisations, familiar for the massless photon, and a longitudinal one also. Anderson's plasma example has this feature too. The high frequency wave that penetrates the plasma gives rise to a compression mode of the plasma itself. This is the "missing" third longitudinal mode of the massive vector field. Thus, the electromagnetic wave inside a plasma has all the features of a massive vector field. And the theory is gauge invariant.

There is one final point to make in this model. The plasma can be coherently excited if energy is applied, forming in quantum theory a plasmon. Now take this example over to the entire universe and imagine it is filled with an "electroweak plasma". The W and Z modes act as massive quanta vector fields; the analogue of the plasmon is the higgson - the massive Higgs boson.

As this pedagogic example was present with Anderson in 1963, why do we credit the assorted authors in 1964? The reason is that if one extends Anderson's non relativistic example to the universe at large, one has reinvented the non relativistic ether, that was dismissed over a century ago. Making a relativistic analogue is non trivial, and that in essence is what was achieved in 1964. Indeed, I quote from Higgs [3]: "This phenomenon is just the relativistic analogue of the plasmon phenomenon to which Anderson as drawn attention".

Brout and Englert were first to publish the relativistic mass mechanism in relativistic quantum field theory[1]. Higgs alone drew attention to the presence of a massive boson, of which more in a moment. Kibble made two significant contributions; first as one of the trio that also discovered the mechanism in 1964[4] and then in 1967 with his paper that set the scene for the real world. As there has been much confusion about the historical priorities, let me now look at this.

The massive boson was already present in Goldstone's 1961 paper[8]. That was not, however, the thrust of that paper, nor was a gauge field present. The 1964 papers show that a gauge field can gain mass, and Higgs' paper shows that the massive boson survives unscathed. The equation of motion is eq(2b) in [3]. An analogous equation is in [4] but in their approximation the boson is massless. They mention this: "While one sees by inspection there is a massless particle in the theory it is easily seen that it completely decoupled from the other (massive) excitation and has nothing to do with the Goldstone theorem". This highlights what was the main concern at that time: demonstrating that the Goldstone theorem - the presence of a massless (unwanted) mode - is bypassed when gauge fields are included.

III. THE CRUCIAL STEPS: 1966 AND 1967

In 1966 Higgs took a more profound step. This paper includes a section "Decay of a Scalar Boson into two Vector bosons". Equation 17 gives the matrix element for this decay and it is explicitly proportional to the mass that has been induced in the vector meson by the mass mechanism. The observation of the Higgs decay to 4 leptons via the intermediate role of a pair of virtual Z or W bosons is a direct example of this amplitude; the decays to pairs of photons, which take place by the action of virtual loops, also test this.

In his 1966 paper, Higgs also extended the mechanism to fermions. Around eqs 25 and 26 he considered massless fermions ("baryons") which couple to the scalar field with a Yukawa interaction. This leads to "spontaneous breakdown of the symmetry" for the fermions to an extent proportional to the Yukawa coupling. Thus Higgs has shown that spontaneous symmetry breaking of massless gauge bosons and massless fermions in the presence of a scalar field leads to massive vectors and massive fermions; the decay of the massive "Higgs" boson then provides a means to test the theory. It is this feature, in my opinion, that sets this work apart in that it enables a theoretical construct to be tested experimentally and established, beyond reasonable doubt, as reality.

The role of Kibble in 1967 is also significant, both for making contact with the real world and in its historical context for the work of Salam and Weinberg.

Kibble took the basic ideas that he and others had developed in 1964, and showed how it is possible to incorporate a group theory structure such that some gauge fields become massive where others need not. He considers a specific model example, where the gauge bosons can have masses $m_1, m_2, m_1 - m_2$ and, in effect, $(m_1 + m_2)/2$. A specific example would be where $m_1 = m_2$, in which case one has three massive degenerate gauge bosons and one massless. This is realised in nature with the W and Z massive, and the massless photon.

Kibble has created a general framework, but does not make the specific example of $SU2XU1$ explicit. He has shown the way, however and stimulated both Salam and Weinberg at this time. As I explain in detail in ref.[7], Kibble's incorporation of group theory resonated with Salam, for whom group theory was a favourite area in his opus. In 1964, Salam had missed the work of Kibble et al, which was taking place in his own department at Imperial College, and failed to complete the $SU2XU1$ picture that he and John Ward had worked on[13]. However, Kibble's work in 1967 made Salam take note. He advertised this in a talk in Sweden in 1968[14], and by claim of having lectured on this in 1967, he was regarded by the Nobel Committee as independent discoverer of the importance of electroweak theory, where the $SU2XU1$ structure is extended by what Salam idiosyncratically always called the "Higgs-Kibble mechanism".

Weinberg also was inspired by Kibble, as is clear from reference 4 in Weinberg's seminal paper "A model of leptons", in 1967 [15]. Here Weinberg explicitly used Kibble's general construct to formulate the successful description of electromagnetic and weak interactions, with massive W and Z bosons and massless photon.

Thus Kibble has played a special role in the development of these ideas. He was also involved in 1964, as co-author of the trio that were scooped[4]. Thus I have suggested that if the Nobel committees were to choose three out of the several theorists, a plausible case could be made for Englert, Higgs and Kibble. This is because Englert is the sole survivor of the Brout-Englert duo who first published the mass mechanism in relativistic quantum field theory. Higgs uniquely drew attention to the massive boson, whose pattern of decays can test the mechanism. Kibble is a candidate as I just discussed.

However, the role of Goldstone is profound in this history, and the fact that the experimental proof has been found also gives a major role to the experimental teams and CERN itself.

After the lectures were given, the award for 2013 was announced in October. The Nobel Prize for physics went to

Francois Englert and Peter Higgs. The citation read:

" For the theoretical discovery of a mechanism that contributes to our understanding of the origin of mass of subatomic particles..and which was recently confirmed through the discovery of the predicted fundamental particle by ATLAS and CMS at CERN LHC."

In particular this recognises the important role played by the experimental teams in finding the boson. It is notable that the option of including a third winner was not taken up. There will, as often, continue to be controversy as to whether other(s) should have been included. Nonetheless, in a profound way, I feel that the decision was appropriate as Robert Brout, had he been alive, would undoubtedly have been included alongside Francois Englert. So, for me, the award was implicitly made to a trio.

Now that the existence of the boson is confirmed, and the mechanism verified, we look to the future. We may claim to understand (almost) everything about some 5% of the material universe. What controls the 95% that remains hidden in the dark is hopefully soon going to be answered at the LHC. However, why the particles and forces have the specific masses and properties that enable life to have emerged, is wide open. At the moment I dont see any clues as to how this will be resolved. Hopefully one of you will come back decades hence to tell future generations.

Acknowledgements

I am grateful to Nino Zichichi and the Ettore Majorana Centre for enabling me to return and lecture some four decades after I was here as a student. And I am grateful to the Nobel Committees for having enabled me to complete the story in October 2013.

[1] F Englert and R Brout, Phys Rev Letters 13, 321 (1964)
[2] P W Higgs Physics Letters 12, 132 (1964)
[3] P W Higgs Phys Rev Letters 13, 508 (1964)
[4] G Guralnik, C Hagen and T W B Kibble, Phys Rev Letters 13, 585 (1964)
[5] MIgdal and Polyakov Soviet Physics JETP, 24, 91 (1967)
[6] P W Higgs Phys Rev145, 1156 (1966)
[7] Frank Close, The Infinity Puzzle, (Oxford,UK, 2012; Basic Books USA, 2012)
[8] J Goldstone, Nuovo Cimento 19, 154 (1961)
[9] J Goldstone, A Salam and S Weinberg, Phys Rev 127, 965 (1962)
[10] Close F.E. Introduction to Quarks and Partons (Academic Press; New York and London)
[11] J Schwinger Phys Rev 125, 397 (1962)
[12] P Anderson, Phys Rev 112, 1900 (1958)
[13] A Salam and J C Ward, Physics Letters 13,168, (1964)
[14] A Salam, Proc of Nobel Symposium, (1968)
[15] S Weinberg, Phys Rev Letters 19, 1264 (1967)
[16] T W B Kibble, Phys Rev 155, 1554 (1967)

Magic Supergravity from Squaring Yang Mills

M. J. Duff[1,*]

[1] *Theoretical Physics, Blackett Laboratory, Imperial College London, London SW7 2AZ, United Kingdom*

We give a division algebra $\mathbb{R}, \mathbb{C}, \mathbb{H}, \mathbb{O}$ description of $D = 3$ Yang-Mills with $\mathcal{N} = 1, 2, 4, 8$ and hence, by tensoring left and right multiplets, a magic square $\mathbb{RR}$, $\mathbb{CR}$, $\mathbb{CC}$, $\mathbb{HR}$, $\mathbb{HC}$, $\mathbb{HH}$, $\mathbb{OR}$, $\mathbb{OC}$, $\mathbb{OH}$, $\mathbb{OO}$ description of $D = 3$ supergravity with $\mathcal{N} = 2, 3, 4, 5, 6, 8, 9, 10, 12, 16$.

I. INTRODUCTION

The octonions occupy a privileged position as the largest of the division algebras $\mathbb{A}$: reals $\mathbb{R}$, complexes $\mathbb{C}$, quaternions $\mathbb{H}$ and octonions $\mathbb{O}$. They provide an intuitive basis for the exceptional Lie groups. For example, the smallest exceptional group G_2 can be understood as the set of automorphims preserving the octonionic product. Efforts to understand the remaining exceptional groups geometrically in terms of octonions resulted in the Freudenthal-Rozenfeld-Tits magic square [1–6] presented in Table I. Despite much effort, however, it is fair to say that the ultimate physical significance of octonions and the magic square remains an enigma.

$\mathbb{A}_L/\mathbb{A}_R$	$\mathbb{R}$	$\mathbb{C}$	$\mathbb{H}$	$\mathbb{O}$
$\mathbb{R}$	SL(2, $\mathbb{R}$)	SU(2, 1)	USp(4, 2)	$F_{4(-20)}$
$\mathbb{C}$	SU(2, 1)	SU(2, 1) × SU(2, 1)	SU(4, 2)	$E_{6(-14)}$
$\mathbb{H}$	USp(4, 2)	SU(4, 2)	SO(8, 4)	$E_{7(-5)}$
$\mathbb{O}$	$F_{4(-20)}$	$E_{6(-14)}$	$E_{7(-5)}$	$E_{8(8)}$

TABLE I: Magic square

In apparently different developments, a recurring theme in attempts to understand the quantum theory of gravity is the idea of "Gravity as the square of Yang-Mills". This idea of tensoring left (L) and right (R) multiplets appears in many different (but sometimes overlapping) guises: KLT relations in string theory [10], $D = 10$ dimensional Type IIA and IIB supergravity (SG) multiplets from $D = 10$ super Yang-Mills (SYM) multiplets [15], asymmetric orbifold contructions [16], gravity anomalies from gauge anomalies [17], (super)gravity scattering amplitudes from those of (super) Yang-Mills [11–14] in various dimensions etc.

In the supersymmetric context it is not difficult to see that the amount of supersymmetry is given by

$$[\mathcal{N}_L \text{ SYM}] \otimes [\mathcal{N}_R \text{ SYM}] \rightarrow [\mathcal{N} = \mathcal{N}_L + \mathcal{N}_R \text{ SG}], \tag{1}$$

but it is harder to see how the other gravitational symmetries arise from those of Yang-Mills. In particular, supergravities are characterized by non-compact global symmetries G (the so-called U-dualities) with local compact subgroups H, for example $G = E_{7(7)}$ and $H = \text{SU}(8)$ for $\mathcal{N} = 8$ supergravity in $D = 4$; whereas the Yang-Mills we start with has global R-symmetries, for example R $= \text{SU}(4)$ for $\mathcal{N} = 4$ in $D = 4$. See [18] for an approach linking SU(4) to SU(8) based on scattering amplitudes.

In the present paper we focus initially on $D = 3$. This is not only intrinsically interesting [14, 19–21], but also throws light on higher-dimensional theories to which it is related by dimensional reduction. First we give a division algebra $\mathbb{R}, \mathbb{C}, \mathbb{H}, \mathbb{O}$ description of $D = 3$ Yang-Mills with $\mathcal{N} = 1, 2, 4, 8$, which is of interest in its own right. More remarkable, however, is that tensoring left and right multiplets yields a magic square $\mathbb{RR}$, $\mathbb{CR}$, $\mathbb{CC}$, $\mathbb{HR}$, $\mathbb{HC}$, $\mathbb{HH}$, $\mathbb{OR}$, $\mathbb{OC}$, $\mathbb{OH}$, $\mathbb{OO}$ description of $D = 3$ supergravity with $\mathcal{N} = 2, 3, 4, 5, 6, 8, 9, 10, 12, 16$, as presented in Table II. For $\mathcal{N} > 8$ the multiplets are those of pure supergravity; for $\mathcal{N} \leq 8$ supergravity is coupled to matter. In both cases the field content is such that the U-dualities exactly match the groups of Table I.

Thus not only do $D = 3$ supergravities fill out a magic square but their field content and, hence, symmetries are *derived* from squaring Yang-Mills.

* m.duff@imperial.ac.uk

	$\mathbb{R}$	$\mathbb{C}$	$\mathbb{H}$	$\mathbb{O}$
$\mathbb{R}$	$\mathcal{N}=2, f=4$ $G=\mathrm{SL}(2,\mathbb{R})$, dim 3 $H=\mathrm{SO}(2)$, dim 1	$\mathcal{N}=3, f=8$ $G=\mathrm{SU}(2,1)$, dim 8 $H=\mathrm{SU}(2)\times\mathrm{SO}(2)$, dim 4	$\mathcal{N}=5, f=16$ $G=\mathrm{USp}(4,2)$, dim 21 $H=\mathrm{USp}(4)\times\mathrm{USp}(2)$, dim 13	$\mathcal{N}=9, f=32$ $G=F_{4(-20)}$, dim 52 $H=\mathrm{SO}(9)$, dim 36
$\mathbb{C}$	$\mathcal{N}=3, f=8$ $G=\mathrm{SU}(2,1)$, dim 8 $H=\mathrm{SU}(2)\times\mathrm{SO}(2)$, dim 4	$\mathcal{N}=4, f=16$ $G=\mathrm{SU}(2,1)^2$, dim 16 $H=\mathrm{SU}(2)^2\times\mathrm{SO}(2)^2$, dim 8	$\mathcal{N}=6, f=32$ $G=\mathrm{SU}(4,2)$, dim 35 $H=\mathrm{SU}(4)\times\mathrm{SU}(2)\times\mathrm{SO}(2)$, dim 19	$\mathcal{N}=10, f=64$ $G=E_{6(-14)}$, dim 78 $H=\mathrm{SO}(10)\times\mathrm{SO}(2)$, dim 46
$\mathbb{H}$	$\mathcal{N}=5, f=16$ $G=\mathrm{USp}(4,2)$, dim 21 $H=\mathrm{USp}(4)\times\mathrm{USp}(2)$, dim 13	$\mathcal{N}=6, f=32$ $G=\mathrm{SU}(4,2)$, dim 35 $H=\mathrm{SU}(4)\times\mathrm{SU}(2)\times\mathrm{SO}(2)$, dim 19	$\mathcal{N}=8, f=64$ $G=\mathrm{SO}(8,4)$, dim 66 $H=\mathrm{SO}(8)\times\mathrm{SO}(4)$, dim 34	$\mathcal{N}=12, f=128$ $G=E_{7(-5)}$, dim 133 $H=\mathrm{SO}(12)\times\mathrm{SO}(3)$, dim 69
$\mathbb{O}$	$\mathcal{N}=9, f=32$ $G=F_{4(-20)}$, dim 52 $H=\mathrm{SO}(9)$, dim 36	$\mathcal{N}=10, f=64$ $G=E_{6(-14)}$, dim 78 $H=\mathrm{SO}(10)\times\mathrm{SO}(2)$, dim 46	$\mathcal{N}=12, f=128$ $G=E_{7(-5)}$, dim 133 $H=\mathrm{SO}(12)\times\mathrm{SO}(3)$, dim 69	$\mathcal{N}=16, f=256$ $G=E_{8(8)}$, dim 248 $H=\mathrm{SO}(16)$, dim 120

TABLE II: Magic square of $D=3$ supergravity theories. The first row of each entry indicates the amount of supersymmetry $\mathcal{N}$ and the total number of degrees of freedom f. The second (third) row indicates the U-duality group G (the maximal compact subgroup $H\subset G$) and its dimension. The scalar fields in each case parametrise the coset G/H, where $\dim_{\mathbb{R}}(G/H)=f/2$.

	$\mathbb{R}$	$\mathbb{C}$	$\mathbb{H}$	$\mathbb{O}$
$\mathbb{R}$	SO(2)	SO(3) × SO(2)	SO(5) × SO(3)	SO(9)
$\mathbb{C}$	SO(3) × SO(2)	[SO(3) × SO(2)]^2	SO(6) × SO(3)	SO(10) × SO(2)
$\mathbb{H}$	SO(5) × SO(3)	SO(6) × SO(3)	SO(8) × SO(4)	SO(12) × SO(3)
$\mathbb{O}$	SO(9)	SO(10) × SO(2)	SO(12) × SO(3)	SO(16)

TABLE III: Magic square of maximal compact subgroups.

II. MAGIC SQUARE

Magic squares are based on the four division algebras, $\mathbb{R}, \mathbb{C}, \mathbb{H}$ and $\mathbb{O}$, which are of dimension $1, 2, 4$ and 8, respectively. One can also use their split (non-division) cousins to obtain different real forms. See [7, 22] and the reference therein for details. They can be built, one-by-one, using the Cayley-Dickson doubling procedure starting with $\mathbb{R}$. The reals are ordered, commutative and associative. With each doubling one such property is lost: $\mathbb{C}$ is commutative and associative, $\mathbb{H}$ is associative, $\mathbb{O}$ is *non-associative*.

An element $x\in\mathbb{O}$ may be written $x=x^a e_a$, where $a=0,\dots,7$, $x^a\in\mathbb{R}$ and $\{e_a\}$ is a basis with one real $e_0=1$ and seven $e_i, i=1,\dots,7$, imaginary elements. The octonionic conjugation is denoted by e_a^*, where $e_0^*=e_0$ and $e_i^*=-e_i$. The octonionic multiplication rule is,

$$e_a e_b = (\delta_{a0}\delta_{bc}+\delta_{0b}\delta_{ac}-\delta_{ab}\delta_{0c}+C_{abc})\,e_c, \tag{2}$$

where C_{abc} is totally antisymmetric such that $C_{0bc}=0$. The non-zero C_{ijk} are given by the Fano plane, see [23].

A natural inner product on $\mathbb{A}$ is defined by

$$\langle x|y\rangle := \frac{1}{2}(x\overline{y}+y\overline{x}) = x^a y^b \delta_{ab}. \tag{3}$$

To understand the symmetries of the magic square and its relation to SYM we shall need in particular two algebras defined on $\mathbb{A}$. First, the norm-preserving algebra,

$$so(\mathbb{A}) := \{D\in \mathrm{Hom}_{\mathbb{R}}(\mathbb{A}) | \langle Dx|y\rangle + \langle x|Dy\rangle = 0\}, \tag{4}$$

isomorphic to $so(\dim_{\mathbb{R}}\mathbb{A})$. Second, the *triality* algebra

$$tri(\mathbb{A}) := \{(A,B,C)|A(xy)=(Bx)y+x(Cy)\} \tag{5}$$

where $A,B,C\in so(\mathbb{A})$. For $\mathbb{A}=\mathbb{R},\mathbb{C},\mathbb{H},\mathbb{O}$ we have $tri(\mathbb{A})\cong\emptyset, so(2)\oplus so(2), so(3)\oplus so(3)\oplus so(3), so(8)$ [22].

There are a variety of magic squares in which different real forms appear. See [7] for a comprehensive account in the context of supergravity. Famously, the $\mathbb{C},\mathbb{H}$, and $\mathbb{O}$ rows of one example describe the U-dualities of the aptly named "magic" supergravities in $D=5,4,3$ respectively [8, 9]. In this paper we instead demonstrate the novel appearance of the magic square of Table I in conventional $D=3$ supergravities. The specific magic square presented in Table I was first obtained in [7] using a version of the Tits construction based on a *Lorentzian* Jordan algebra. The two algebras $\mathbb{A}_L, \mathbb{A}_R$ enter this definition on distinct footings; the "magic" of the square is its symmetry under the exchange $\mathbb{A}_L \leftrightarrow \mathbb{A}_R$, which is obscured by their undemocratic treatment.

For the purposes of squaring SYM a manifestly $\mathbb{A}_L \leftrightarrow \mathbb{A}_R$ symmetric formulation of the square is required. This is achieved by adapting the *triality algebra* construction introduced by Barton and Sudbery [22]. Our definition of Table I is given by,

$$L_3(\mathbb{A}_L, \mathbb{A}_R) \cong tri(\mathbb{A}_L) \oplus tri(\mathbb{A}_R) + 3(\mathbb{A}_L \otimes \mathbb{A}_R). \tag{6}$$

Note, this is not quite the triality construction as defined in [22]. We will not present the details here, but it can be easily obtained by making a slight modification to the commutators in $3(\mathbb{A}_L \otimes \mathbb{A}_R)$ w.r.t. those appearing in [22]. Here, we have used $\oplus$ and $+$ to distinguish the direct sum between Lie algebras and vector spaces, i.e. only if $[g, h] = 0$ do we use $g \oplus h$.

We shall also need a magic square of the maximal compact subalgebras of Table I, given in Table III. This is given by the *reduced* triality construction,

$$L_1(\mathbb{A}_L, \mathbb{A}_R) := tri(\mathbb{A}_L) \oplus tri(\mathbb{A}_R) + (\mathbb{A}_L \otimes \mathbb{A}_R), \tag{7}$$

which is easily obtained from (6).

$\mathbb{A}_L/\mathbb{A}_R$	$A_\mu(R) \in \text{Re}\mathbb{A}_R$	$\phi(R) \in \text{Im}\mathbb{A}_R$	$\lambda(R) \in \mathbb{A}_R$
$A_\mu(L) \in \text{Re}\mathbb{A}_L$	$g_{\mu\nu} + \varphi \in \text{Re}\mathbb{A}_L \otimes \text{Re}\mathbb{A}_R$	$\varphi \in \text{Re}\mathbb{A}_L \otimes \text{Im}\mathbb{A}_R$	$\Psi_\mu + \chi \in \text{Re}\mathbb{A}_L \otimes \mathbb{A}_R$
$\phi(L) \in \text{Im}\mathbb{A}_L$	$\varphi \in \text{Im}\mathbb{A}_L \otimes \text{Re}\mathbb{A}_R$	$\varphi \in \text{Im}\mathbb{A}_L \otimes \text{Im}\mathbb{A}_R$	$\chi \in \text{Im}\mathbb{A}_L \otimes \mathbb{A}_R$
$\lambda(L) \in \mathbb{A}_L$	$\Psi_\mu + \chi \in \mathbb{A}_L \otimes \text{Re}\mathbb{A}_R$	$\chi \in \mathbb{A}_L \otimes \text{Im}\mathbb{A}_R$	$\varphi \in \mathbb{A}_L \otimes \mathbb{A}_R$

TABLE IV: Tensor product of left/right ($\mathbb{A}_L/\mathbb{A}_R$) SYM multiplets, using SO(1,2) spacetime reps and dualising all p-forms.

III. $\mathbb{R}, \mathbb{C}, \mathbb{H}, \mathbb{O}$ DESCRIPTION OF $D = 3, \mathcal{N} = 1, 2, 4, 8$ YANG-MILLS

The $D = 3$, $\mathcal{N} = 8$ SYM Lagrangian is given by

$$\begin{aligned}\mathcal{L} = &-\frac{1}{4}F^A_{\mu\nu}F^{A\mu\nu} - \frac{1}{2}D_\mu\phi^A_i D^\mu\phi^A_i + i\bar{\lambda}^A_a\gamma^\mu D_\mu\lambda^A_a \\ &-\frac{1}{4}g^2 f_{BC}{}^A f_{DE}{}^A\phi^B_i\phi^D_i\phi^C_j\phi^E_j - g f_{BC}{}^A\phi^B_i\bar{\lambda}^{Aa}\Gamma^i_{ab}\lambda^{Cb},\end{aligned}$$

where Γ^i_{ab}, $i = 1, \ldots, 7$, $a, b = 0, \ldots, 7$, belongs to the SO(7) Clifford algebra. The key observation is that this gamma matrix can be represented by the octonionic structure constants,

$$\Gamma^i_{ab} = i(\delta_{bi}\delta_{a0} - \delta_{b0}\delta_{ai} + C_{iab}), \tag{8}$$

which allows us to rewrite the action over octonionic fields. If we replace $\mathbb{O}$ with a general division algebra $\mathbb{A}$, the result is $\mathcal{N} = 1, 2, 4$ 8 over $\mathbb{R}, \mathbb{C}, \mathbb{H}, \mathbb{O}$:

$$\begin{aligned}\mathcal{L} = &-\frac{1}{4}F^A_{\mu\nu}F^{A\mu\nu} - \frac{1}{2}D_\mu\phi^{*A}D^\mu\phi^A + i\bar{\lambda}^A\gamma^\mu D_\mu\lambda^A \\ &-\frac{1}{4}g^2 f_{BC}{}^A f_{DE}{}^A\langle\phi^B|\phi^D\rangle\langle\phi^C|\phi^E\rangle \\ &+\frac{i}{2}g f_{BC}{}^A\left((\bar{\lambda}^A\phi^B)\lambda^C - \bar{\lambda}^A(\phi^{*B}\lambda^C)\right),\end{aligned}$$

where $\phi = \phi^i e_i$ is an Im$\mathbb{A}$-valued scalar field, $\lambda = \lambda^a e_a$ is an $\mathbb{A}$-valued two-component spinor and $\bar{\lambda} = \bar{\lambda}^a e^*_a$. Note, since λ^a is anti-commuting we are dealing with the algebra of octonions defined over the Grassmanns.

The supersymmetry transformations in this language are given by

$$\begin{aligned}\delta\lambda^A &= \frac{1}{2}(F^{A\mu\nu} + \varepsilon^{\mu\nu\rho}D_\rho\phi^A)\sigma_{\mu\nu}\epsilon + \frac{1}{2}g f_{BC}{}^A\phi^B_i\phi^C_j\sigma_{ij}\epsilon, \\ \delta A^A_\mu &= \frac{i}{2}(\bar{\epsilon}\gamma_\mu\lambda^A - \bar{\lambda}^A\gamma_\mu\epsilon), \\ \delta\phi^A &= \frac{i}{2}e_i[(\bar{\epsilon}e_i)\lambda^A - \bar{\lambda}^A(e_i\epsilon)],\end{aligned} \tag{9}$$

where ϵ is an $\mathbb{A}$-valued two-component spinor and $\sigma_{\mu\nu}$ are the generators of $SL(2,\mathbb{R}) \cong SO(1,2)$. The σ_{ij} generate $SO(\text{Im}\mathbb{A})$ and are proportional to the identity as 2×2 matrices, but act as operators on $\mathbb{A}$ itself. In the octonionic case these operators are best expressed by decomposing $so(\text{Im}\mathbb{O})$ into its g_2 subalgebra to give $\sigma_{ij} = \Gamma_{ij} + \Sigma_{ij}$, with

$$\begin{aligned}\Gamma_{ij} &= \left(\frac{1}{2}[e_i, e_j, \ \cdot\ ] - \frac{1}{6}[[e_i, e_j], \ \cdot\ ]\right)\mathbb{1}, \\ \Sigma_{ij} &= -\frac{1}{12}[[e_i, e_j], \ \cdot\ ]\mathbb{1},\end{aligned} \tag{10}$$

where $[\cdot,\cdot,\cdot]$ is the associator: $[a,b,c] := (ab)c - a(bc)$. Under $SO(7) \supset G_2$, $\mathbf{21} \to \mathbf{14}+\mathbf{7}$, the Γ_{ij} and Σ_{ij} and correspond to the $\mathbf{14}$ and $\mathbf{7}$, respectively. For $\mathbb{A} = \mathbb{H}$ the associator vanishes and the σ_{ij} generate SO(3), the automorphism group of the quaternions; for $\mathbb{A} = \mathbb{R}, \mathbb{C}$ the σ_{ij} trivially vanish.

IV. SQUARING YANG-MILLS

Having cast the magic square in terms of a manifestly $\mathbb{A}_L \leftrightarrow \mathbb{A}_R$ symmetric triality algebra construction, and having written $\mathcal{N} = 1, 2, 4, 8$ SYM in terms of fields valued in $\mathbb{R}, \mathbb{C}, \mathbb{H}, \mathbb{O}$ we shall now obtain the magic square of supergravities in Table II, with symmetries G (Table I) and H (Table III), by "squaring" $\mathcal{N} = 1, 2, 4, 8$ SYM.

Taking a left SYM multiplet $\{A_\mu(L) \in \text{Re}\mathbb{A}_L, \phi(L) \in \text{Im}\mathbb{A}_L, \lambda(L) \in \mathbb{A}_L\}$ and tensoring it with a right multiplet $\{A_\mu(R) \in \text{Re}\mathbb{A}_R, \phi(R) \in \text{Im}\mathbb{A}_R, \lambda(R) \in \mathbb{A}_R\}$ we obtain the field content of a supergravity theory valued in both $\mathbb{A}_L$ and $\mathbb{A}_R$. See Table IV. Note, the left/right SYM R-symmetries act on each slot of the $\mathbb{A}_L, \mathbb{A}_R$ tensor products.

Grouping spacetime fields of the same type we find,

$$g_{\mu\nu} \in \mathbb{R}, \quad \Psi_\mu \in \begin{pmatrix}\mathbb{A}_L \\ \mathbb{A}_R\end{pmatrix}, \quad \varphi, \chi \in \begin{pmatrix}\mathbb{A}_L \otimes \mathbb{A}_R \\ \mathbb{A}_L \otimes \mathbb{A}_R\end{pmatrix}. \tag{11}$$

The $\mathbb{R}$-valued graviton and $\mathbb{A}_L \oplus \mathbb{A}_R$-valued gravitino carry no degrees of freedom. The $(\mathbb{A}_L \otimes \mathbb{A}_R)^2$-valued scalar and Majorana spinor each have $2(\dim \mathbb{A}_L \times \dim \mathbb{A}_R)$ degrees of freedom.

As we have already mentioned, the $\mathcal{N} > 8$ supergravities in $D = 3$ are unique, all fields belonging to the gravity multiplet, while those with $\mathcal{N} \leq 8$ may be coupled to k additional matter multiplets [20, 21]. The real miracle is that tensoring left and right SYM multiplets yields the field content of $\mathcal{N} = 2, 3, 4, 5, 6, 8$ supergravity with $k = 1, 1, 2, 1, 2, 4$: just the right matter content to produce the U-duality groups appearing in Table I.

The largest linearly realised global symmetry of these theories is H, which has Lie algebra given by the reduced triality construction (7). Consequently, we expect the fields in (11) to carry linear representations of H. The metric is a singlet, while Ψ_μ, φ and χ transform as a vector, spinor and conjugate spinor, respectively. Fortunately, $\mathbb{A}_L \oplus \mathbb{A}_R$ and $(\mathbb{A}_L \otimes \mathbb{A}_R)^2$ are precisely the representation spaces of the vector and (conjugate) spinor. For example, in the maximal case of $\mathbb{A}_L, \mathbb{A}_R = \mathbb{O}$, we have the $\mathbf{16}, \mathbf{128}$ and $\mathbf{128}'$ of SO(16). The distinction between spinor and conjugate spinor in terms of $(\mathbb{O}_L \otimes \mathbb{O}_R)^2$ is encoded in the division-algebraic realisation of the Lie algebra action, which is inherited from the left/right SYM. For example, consider $x, y \in \mathbb{O}$ transforming respectively as the $\mathbf{8}_s$ and $\mathbf{8}_c$ of SO(8). A subset of SO(8) generators are given by left/right multiplications with elements $a \in \text{Im}\mathbb{O}$ under which $x \mapsto ax$ implies $y \mapsto ya$.

The U-dualities G are realised non-linearly on the scalars, which parametrise the symmetric spaces G/H. This can be understood using the remarkable identity relating the projective planes over $(\mathbb{A}_L \otimes \mathbb{A}_R)^2$ to G/H,

$$(\mathbb{A}_L \otimes \mathbb{A}_R)\mathbb{P}^2 \cong G/H. \tag{12}$$

The scalar fields may be regarded as points in division-algebraic projective planes. The tangent space at any point of $(\mathbb{A}_L \otimes \mathbb{A}_R)\mathbb{P}^2$ is just $(\mathbb{A}_L \otimes \mathbb{A}_R)^2$, the required representation space of H. The *Cayley plane* $\mathbb{O}\mathbb{P}^2$, with isometry group $F_{4(-52)}$, is a classic example: $F_{4(-52)}/\text{Spin}(9) \cong (\mathbb{R} \otimes \mathbb{O})\mathbb{P}^2 = \mathbb{O}\mathbb{P}^2$. The tangent space at any point of $\mathbb{O}\mathbb{P}^2$ is $\mathbb{O}^2$, the spinor of Spin(9) as required. Note, the cases $(\mathbb{C} \otimes \mathbb{O})\mathbb{P}^2, (\mathbb{H} \otimes \mathbb{O})\mathbb{P}^2, (\mathbb{O} \otimes \mathbb{O})\mathbb{P}^2$ are not strictly speaking projective spaces, but may nevertheless be identified with G/H [3, 23, 24].

V. MAGIC PYRAMID

We have given an $\mathbb{R}, \mathbb{C}, \mathbb{H}, \mathbb{O}$ description of $\mathcal{N} = 1, 2, 4, 8$ SYM. Tensoring left($\mathbb{A}_L$)/right($\mathbb{A}_R$) multiplets we built an $(\mathbb{A}_L, \mathbb{A}_R)$ magic square of $D = 3$, $\mathcal{N} = 2, 3, 4, 5, 6, 8, 9, 10, 12, 16$ supergravity.

However, it has also been known for some time now that $\mathcal{N} = 1$ SYM in $D = 3, 4, 6, 10$ has a concise division-algebraic formulation [25–28]. Indeed, a simple alternative way to obtain the $\mathcal{N} = 1, 2, 4, 8$ theories over $\mathbb{A} = \mathbb{R}, \mathbb{C}, \mathbb{H}, \mathbb{O}$ in $D = 3$ is to start from this description of $\mathcal{N} = 1$ SYM in $D = 3, 4, 6, 10$ and dimensionally reduce on a torus. The theories in $D = \dim(\mathbb{A}) + 2$ exploit the $so(1, \dim(\mathbb{A}) + 1) \cong sl(2, \mathbb{A})$ Lie algebra isomorphisms (in the sense of [29]). Reducing to $D = 3$ decomposes $\mathrm{SL}(2, \mathbb{A}) \supset \mathrm{SL}(2, \mathbb{R}) \times \mathrm{SO}(\mathrm{Im}\mathbb{A})$; the algebra that started out life determining D now fixes $\mathcal{N}$.

Hence, we may specify a $(D, \mathcal{N})$ SYM theory using a pair of division algebras $\mathbb{A}^2 = (\mathbb{A}_1, \mathbb{A}_2)$, which are constrained to satisfy $\dim \mathbb{A}_1 + \dim \mathbb{A}_2 \leq 9$. Therefore, tensoring $(\mathbb{A}_L^2)/(\mathbb{A}_R^2)$ SYM multiplets, we can build supergravity theories in $D = 3, 4, 6, 10$ with $\mathcal{N} = 16, \ldots, 2$.

Since the first slots of $(\mathbb{A}_L^2)$ and $(\mathbb{A}_R^2)$ must be correlated and $\dim \mathbb{A}_{1L/R} + \dim \mathbb{A}_{2L/R} \leq 9$, we obtain a *supergravity magic pyramid*, of which the square described here is only the base. The tip is type II supergravity in $D = 10$. The details of this pyramid will be presented elsewhere.

Acknowledgments

The work was carried out in collaboration with L. Borsten, L. J. Hughes and S. Nagy. Supported by the STFC under rolling grant ST/G000743/1. I am grateful to A. Zichichi for hospitality in Erice and to Zvi Bern for his encouragement.

---

[1] H. Freudenthal, Nederl. Akad. Wetensch. Proc. Ser. **57**, 218 (1954).
[2] H. Freudenthal, Nederl. Akad. Wetensch. Proc. Ser. **A62**, 466 (1959).
[3] H. Freudenthal, Adv. Math. **1**, 145 (1964).
[4] B. A. Rosenfeld, Dokl. Akad. Nauk. SSSR **106**, 600 (1956).
[5] J. Tits, Indag. Math. **28**, 223 (1966).
[6] E. B. Vinberg, Tr. Semin. Vektorn. Tr. Semin. Vektorn. Tensorn. Anal. **13** (1966).
[7] S. L. Cacciatori, B. L. Cerchiai, and A. Marrani, (2012), arXiv:1208.6153 .
[8] M. Günaydin, G. Sierra, and P. K. Townsend, Nucl. Phys. **B242**, 244 (1984).
[9] M. Günaydin, G. Sierra, and P. K. Townsend, Phys. Lett. **B133**, 72 (1983).
[10] H. Kawai, D. Lewellen, and S. Tye, Nucl.Phys. **B269**, 1 (1986).
[11] Z. Bern, J. Carrasco, and H. Johansson, Phys.Rev. **D78**, 085011 (2008), arXiv:0805.3993 .
[12] Z. Bern, J. J. M. Carrasco, and H. Johansson, Phys.Rev.Lett. **105**, 061602 (2010), arXiv:1004.0476 .
[13] Z. Bern, T. Dennen, Y.-t. Huang, and M. Kiermaier, Phys.Rev. **D82**, 065003 (2010), arXiv:1004.0693 .
[14] Y.-t. Huang and H. Johansson, (2012), arXiv:1210.2255 .
[15] M. B. Green, J. H. Schwarz, and E. Witten, *Superstring Theory vol. 1: Introduction*, Cambridge Monographs on Mathematical Physics (Cambridge University Press, Cambridge, UK, 1987) 469 p.
[16] A. Sen and C. Vafa, Nucl. Phys. **B455**, 165 (1995), arXiv:hep-th/9508064 .
[17] I. Antoniadis, E. Gava, and K. Narain, Phys.Lett. **B283**, 209 (1992), arXiv:hep-th/9203071 .
[18] M. Bianchi, H. Elvang, and D. Z. Freedman, JHEP **0809**, 063 (2008), arXiv:0805.0757 .
[19] S. Deser, R. Jackiw, and S. Templeton, Annals Phys. **140**, 372 (1982).
[20] N. Marcus and J. H. Schwarz, Nucl.Phys. **B228**, 145 (1983).
[21] B. de Wit, A. Tollsten, and H. Nicolai, Nucl.Phys. **B392**, 3, hep-th/9208074 .
[22] C. H. Barton and A. Sudbery, Adv. in Math. **180**, 596 (2003), arXiv:math/0203010 .
[23] J. C. Baez, Bull. Amer. Math. Soc. **39**, 145 (2001), arXiv:math/0105155 .
[24] J. Landsberg and L. Manivel, Journal of Algebra **239**, 477 (2001).
[25] T. Kugo and P. K. Townsend, Nucl. Phys. **B221**, 357 (1983).
[26] M. Duff, Class.Quant.Grav. **5**, 189 (1988).
[27] J. M. Evans, Nucl. Phys. **B298**, 92 (1988).
[28] J. C. Baez and J. Huerta, in *Superstrings, Geometry, Topology, and C*-Algebras, eds. R. Doran, G. Friedman and J. Rosenberg, Proc. Symp. Pure Math*, Vol. 81 (2009) pp. 65–80, arXiv:0909.0551 .
[29] A. Sudbery, J. Phys. **A17**, 939 (1984).

CHAIRMAN: M.J. DUFF

Scientific Secretaries: G. De Palma, S. Downes

DISCUSSION I

- Nohle:

Are the symmetries that you discuss R-symmetries? Can similar methods be used to investigate matter-coupled theories (Kahler manifold symmetries) or non-SUSY double copies?

- Duff:

Yes, today I have talked about the R-symmetries of Yang-Mills theories, and tomorrow the subject of my talk will be the relationship between R-symmetries on the Yang-Mills side and the so-called U-dualities on the gravitational side. For example how Yang-Mills knows about the E7(7) U-duality on the gravity side when we square it. Octonions, quaternions, complexes and reals are respectively tied to N = 8, 4, 2, 1, so we do not have much to say about non-SUSY theories in this context. Tomorrow I will look both at pure gravity supermultiplets and gravity supermultiplets coupled to matter; on the Yang-Mills side I will look only at pure Y-M without matter multiplets.

- Davies:

We know that if we square Chern-Simons theory we get three-dimensional supergravity. What is the role of Octonions in Chern-Simons theory?

- Duff:

I have asked my students to look at this question. Chern-Simons is conformal in three dimensions, Yang-Mills in four and there are tensor theories conformal in six. One may ask whether squaring conformal theories has some extra feature. We are discovering that there are indeed some interesting features; hopefully there will be a paper on that soon.

- Csanàd:

There is a huge number of non-isomorphic division algebras not isomorphic to R, C, H or O. What is the role of these particular algebras that you are using and what would happen with other ones?

- Duff:

We have not found any use for these other algebras. As far as dimension is concerned, d = 3, 4, 6 and 10 have a special role since the Lorentz groups SO(2,1), SO(3,1), SO(5,1) and SO(9,1) are described by the SL(2) over the division algebras, *i.e.* SL(2,R), SL(2,C), SL(2,H) and SL(2,O), respectively. Besides, tomorrow I will show that the magic square is just the base of a magic pyramid, where there are squares also for d = 4, 6 and 10.

- Moroz:

Is there any division algebra with dimension three? If it exists, what are its applications to field theory and physics in general?

- Duff:

\- There is a theorem which says that there are only four normed division algebras R,C,H and O with dimensions 1,2,4 and 8. Your construction would require dropping the requirement of a multiplicative identity. We have not looked at such algebras.

- *Ivanytskyi*:

Complex numbers can be interpreted as two-dimensional vectors. Do the octonions also have a similar interpretation as a vector space? If this is the case, what are their transformation laws?

- *Duff:*

Yes, in some sense they can. There is a rotation group associated with each of the four division algebras. SO(n) for n=1,2,4,8. In that sense you can think of it in the way you describe. But there are more interesting algebras associated to them, for example the triality algebras, which are not so straightforward. It happens to be SO(8) for the case of the Octonions, but it is not so obvious in the other cases. This symmetry will have a role to play when we square the Yang-Mills theories.

- *Mojaza*:

Some of the division algebras can be interpreted as Clifford Algebras. Therefore, we can derive a spin structure associated to them. How does this generalize to the octonions?

- *Duff:*

The correspondence between the special orthogonal and special linear groups is not, strictly speaking, true for the octonions. The true correspondence is subtle. Barton and Sudbery [Barton:2003] have a compensating procedure for the "missing generators" that arise in the octonionic case.

- *Sastre*:

Are there nontrivial subalgebras of the octonions that are associative?

- *Duff:*

Yes R, C and H are associative.

- *Minkowski*:

The Fano quadrangles defined in terms of the four-index object Q_{ijkl} involve indices which run from one to seven. Q also involves a seven index alternating symbol. Can these be interpreted in terms of indices which run through the full eight-dimensional space?

- *Duff:*

You can define a Q that runs over eight indices, if you put the Q_{0ijk} element to be the C_{ijk}, so this four index Q now has each index running over eight indices, rather than seven, and it is anti-self dual. So I hope that answers his question.

- *Korthals-Altes*:

How does the mathematical construction of the Magic Square work?

- *Duff:*

We shall discuss this tomorrow. The reason the square was called magic was because the original Freudenthal construction was not manifestly symmetric, although the square was actually symmetric. Barton's construction – the one shown in the slides – is manifestly symmetric and hence suitable for comparison with squaring Yang-Mills. To derive the magic square, you take the triality algebra from division algebra on the left, and the triality algebra from the division algebra on the right, and add three times their product. Let me give you an example. E_8 has dimension 248. How does that correspond to O times O? The triality algebra of O is SO(8). The adjoint of SO(8) has dimension twenty-eight. Two of those have fifty-six. Three times the product of SO(8) gives the square of the vector, spinor and conjugate spinor representations. This gives three times sixty-four, which added to fifty-six yields 248.

CHAIRMAN: M.J. DUFF

Scientific Secretaries: A. Leithes, M.Mariti

DISCUSSION II

- *Hartmann:*

I have a question regarding the physical aspects of your model; trying to understand physically what it means. So, you said if you square your super Yang-Mills theory you get your supergravity. So you have two super Yang-Mills field spaces into eight dimensional representations and via the Kronecker products you get all the various representations of your supergravity. But, does that mean that your fields in your supergravity are composite states of your super Yang-Mills theory?

- *Duff:*

Excellent question because that is the issue which we are currently writing up. Since we have not yet finished the paper, I will not attempt to give you a half-baked explanation, but the short answer is yes, they can be regarded as composite in some sense.

- *Carlucci:*

Does it make sense to continue the magic squares beyond the 4x4 ones?

- *Duff:*

No, it doesn't. A mathematician will be limited to 4x4 because there are just the 4 normed division algebras, namely R, C, H and O. They don't extend beyond O. A physicist will also be limited to 4 x 4 because the corresponding Yang-Mills theories in 3 dimensions are N=1,2,4,8. They don't extend beyond N= 8. So from but mathematical and physical reasons we are limited us to 4x4.

- *Cerchiai:*

There are other magic squares with other real forms of the groups. Would you be able to get them in this way, or might they be relevant?

- *Duff:*

Yes, the symbol E8 for example covers a multitude of different E8's - there are complex and real and the real ones can have various numbers of non-compact generators. The one I showed-- with E8(8)-- is the one we get by squaring Yang-Mills. So although mathematically there are other magic squares that one could consider, this is the one you get by squaring Yang-Mills. Now, we were conventional in our signature, we looked at a spacetime with one time and the rest space. If you wanted to consider exotic signatures you might generate other non-compact forms, but with just the conventional one time dimension this is what you get.

- *Downes*:

Are there any implications of this construction for heterotic supergravity?

- *Duff:*

Good question, but we haven't really thought about that in the octonionic context. Heterotic supergravity theory has one just one supersymmetry in D=10. That will require an N=1 Yang-Mills on the left side and a non supersymmetric one on the right.

- *Downes:*

Have you looked at the paper from Radu Roiban and his collaborators?

- M.J. Duff:

Yes, but we just focused on the cases where both sides are supersymmetric.

CERN-PH-TH/2013-302

Electric-Magnetic Duality and Supersymmetry

Paolo Aschieri[1], Sergio Ferrara[2,3] and Alessio Marrani[4]

[1] *Dipartimento di Scienze e Innovazione Tecnologica, Università del Piemonte Orientale, and INFN, Sezione di Torino, Viale T. Michel 11, 15121 Alessandria, Italy*
aschieri@to.infn.it

[2] *Physics Department, Theory Unit, CERN, CH 1211, Geneva 23, Switzerland*

[3] *INFN - Laboratori Nazionali di Frascati, Via Enrico Fermi 40,00044 Frascati, Italy*
sergio.ferrara@cern.ch

[4] *Instituut voor Theoretische Fysica, KU Leuven, Celestijnenlaan 200D, B-3001 Leuven, Belgium*
alessio.marrani@fys.kuleuven.be

Contribution to the Proceedings of the International School of Subnuclear Physics, 51th Course: "Reflections on the Next Step for LHC", Erice, Italy, June 24 - July 3, 2013
Based on lectures given by S. Ferrara

Abstract

In these lectures we discuss non linear theories of U(1) gauge fields and their duality properties. In particular we study the "constitutive relations", defining the equations of motion of a general nonlinear theory, in the formulation devised by Schrödinger for the Born-Infeld theory. This investigation is motivated by the search for actions describing Goldstino interactions in the situation of spontaneously broken supersymmetry as well as in the study of duality constraints on higher loop counter-terms in extended supergravities.

1 Gauge Theory and Electric-Magnetic Duality

Electric-magnetic (e.m.) duality is one of the most fascinating symmetries of non-linear (and thus interacting) theories of gauge fields. Free Maxwell theory is the prototype of e.m. duality invariant theory. In the vacuum, the Eqs. of motion read

$$\partial_\mu F^{\mu\nu} = 0, \quad \partial_\mu {}^*F^{\mu\nu} = 0 \tag{1.1}$$

where

$${}^*F_{\mu\nu} := \frac{1}{2}\epsilon_{\mu\nu\rho\sigma}F^{\rho\sigma} , \tag{1.2}$$

and they are invariant under $SO(2)$ rotations:

$$\begin{pmatrix} F \\ {}^*F \end{pmatrix}' = \begin{pmatrix} \cos\alpha & -\sin\alpha \\ \sin\alpha & \cos\alpha \end{pmatrix} \begin{pmatrix} F \\ {}^*F \end{pmatrix} . \tag{1.3}$$

This latter transformation, in terms of electric and magnetic fields $F_{\mu\nu} \to \left(\vec{E}, \vec{B}\right)$ reads

$$\begin{pmatrix} \vec{E} \\ \vec{B} \end{pmatrix}' = \begin{pmatrix} \cos\alpha & -\sin\alpha \\ \sin\alpha & \cos\alpha \end{pmatrix} \begin{pmatrix} \vec{E} \\ \vec{B} \end{pmatrix} . \tag{1.4}$$

Also in this notation it is immediate to see that Maxwell Equations

$$\vec{\nabla} \cdot \vec{E} = 0, \quad \partial_t \vec{E} = \vec{\nabla} \times \vec{B}, \tag{1.5}$$
$$\vec{\nabla} \cdot \vec{B} = 0, \quad \partial_t \vec{B} = -\vec{\nabla} \times \vec{E} \tag{1.6}$$

are invariant.

Note that the Hamiltonian

$$\mathcal{H} = \frac{1}{2}\left(\left|\vec{E}\right|^2 + \left|\vec{B}\right|^2\right) \tag{1.7}$$

is e.m. duality invariant, while the Lagrangian

$$\mathcal{L} = \frac{1}{2}\left(\left|\vec{E}\right|^2 - \left|\vec{B}\right|^2\right) \tag{1.8}$$

is not. It should also be pointed out that the "duality symmetry" is *not* an internal symmetry, since it rotates tensors with pseudo-tensors.

Electric-magnetic dualities are transformations among two-form field strengths and their duals. These transformations extend to $(p+2)$-form field strengths and their duals, which, in D space-time dimensions, are $(D-p-2)$-forms (for $D = 4$ and $p = 0$, one obtains that $p+2 = D-p-2$). In order to generalize the e.m. duality to p-forms, one demands that $D/2 = p+2$, so that one has $(p+1)$-form gauge fields coupled to sources, which are p-extended objects: $p = 0$ (pointlike), $p = 1$ (string), $p = 2$ (membrane), $p = 3$ (three-brane), *etc.*:

$$F_{\mu_1\ldots\mu_{p+2}} = \partial_{[\mu_1}A_{\mu_2\ldots\mu_{p+2}]} . \tag{1.9}$$

When $D/2 = p+2$ we have that $D-p-2 = p+2$, and p-dimensional extended objecs can source both electric and magnetic fields. We thus have *dyons*, extended objects that carry both an electric and a magnetic charge. In all other cases, electric and magnetic objects extends in different space dimensions, $(p,\ D-p-4)$ respectively, so dyons *cannot* exist.

The Dirac-Schwinger-Zwanziger quantization condition reads as follows:

- for $D/2 \neq p+2$:

$$em' = 2\pi k, \quad k \in \mathbb{Z}, \tag{1.10}$$

- for $D/2 = p+2$

$$em' - e'm = 2\pi k, \quad k \in \mathbb{Z}, \tag{1.11}$$

Note that the latter condition is not only invariant under $SO(2)$, but also under $Sp(2,\mathbb{R})$, isomorphic to $SL(2,\mathbb{R})$.

In this talk, we will mainly consider dualities in $D = 4$ dimensions, especially in view of the fact that maximally supersymmetric $\mathcal{N} = 8$ supergravity has an e.m. duality based on the exceptional group $E_{7(7)}$. $E_{7(7)}$ duality seems to severely restrict the ultraviolet divergences of the theory, *even if higher-loop divergences may evade such restrictions.* Also, $\mathcal{N} = 8$ supergravity has a fascinating *"double-copy" structure* in terms of $\mathcal{N} = 4$ super-Yang-Mills theory, which is the maximally susy theory without gravity (*cfr. e.g.* M. Duff's and Z. Bern's contributions to the Proceedings of the School).

2 Non-Linear Electric-Magnetic Duality

The very idea of duality is to generalize the duality of free Maxwell theory to the case of interactions. Interactions introduce non-linearities, which can be due either to the Maxwell field itself, or to other matter fields (as in electrodynamics) which interact with the e.m. field.

Non-linear (pure) electromagnetism can be seen as an effective theory of Maxwell theory in a medium that has a nonlinear response to the electric and magnetic fields. These nonlinearities are captured by the relations expressing the electric displacement $\vec{D}$ and the magnetic field $\vec{H}$ in terms of the magnetic induction $\vec{B}$ and the electric field $\vec{E}$. The field equations are thus divided in a set of linear ones that are independent from the medium

$$\vec{\nabla} \cdot \vec{D} = 0, \quad \partial_t \vec{D} = \vec{\nabla} \times \vec{H}, \tag{2.1}$$

$$\vec{\nabla} \cdot \vec{B} = 0, \quad \partial_t \vec{B} = -\vec{\nabla} \times \vec{E}, \tag{2.2}$$

and a set describing the nonlinearities, i.e. the properties of the medium, hence these equations are called constitutive relations,

$$\vec{D} = \vec{D}\left(\vec{E}, \vec{B}\right) = \vec{E} + ..., \tag{2.3}$$

$$\vec{H} = \vec{H}\left(\vec{E}, \vec{B}\right) = \vec{B} + ..., \tag{2.4}$$

Maxwell theory is recovered when

$$\vec{D} = \vec{E}, \quad \vec{H} = \vec{B} \ . \tag{2.5}$$

The dots ... in eq. (2.4) stand for some higher powers of the electric field $\vec{E}$ and of the magnetic induction $\vec{B}$, so that the nonlinear theories we consider are deformations of linear electromagnetism because (2.3) and (2.4) reduce to (2.5) for weak fields.

In a theory based on a Lagrangian density $\mathcal{L} = \mathcal{L}\left(\vec{E}, \vec{B}\right)$, one has the Eqs.

$$\vec{D} = \frac{\delta \mathcal{L}}{\delta \vec{E}}, \quad \vec{H} = -\frac{\delta \mathcal{L}}{\delta \vec{B}}. \tag{2.6}$$

Note that Eqs. (2.1) are invariant under the linear transformations

$$\begin{pmatrix} \vec{B} \\ -\vec{D} \end{pmatrix}' = \begin{pmatrix} A & B \\ C & D \end{pmatrix} \begin{pmatrix} \vec{B} \\ -\vec{D} \end{pmatrix}; \tag{2.7}$$

$$\begin{pmatrix} \vec{E} \\ \vec{H} \end{pmatrix}' = \begin{pmatrix} A & B \\ C & D \end{pmatrix} \begin{pmatrix} \vec{E} \\ \vec{H} \end{pmatrix}. \tag{2.8}$$

Hence the full set of equations of motion are invariant under the $\begin{pmatrix} A & B \\ C & D \end{pmatrix}$ duality rotations if the constitutive relations are compatible with these rotations, i.e. if the transformations $\vec{D}(\vec{E}, \vec{B}) \to \vec{D}(\vec{E}', \vec{B}')$ and $\vec{H}(\vec{E}, \vec{B}) \to \vec{H}(\vec{E}', \vec{B}')$ are the same as the rotations (2.7), (2.8).

It is conveninent to rewrite the nonlinear equations and the duality rotations using a relativistic formalism. We set as usual $F_{\mu\nu} = \left(\vec{E}, \vec{B}\right)$ and $G_{\mu\nu} = \left(\vec{H}, -\vec{D}\right)$, so that in the vacuum:

$$\vec{D} = \vec{E}, \quad \vec{H} = \vec{B} \quad \Leftrightarrow G_{\mu\nu} = \ ^*F_{\mu\nu}. \tag{2.9}$$

The dynamical equations then read $dF = 0, dG = 0$ i.e.,

$$\partial_\mu {}^*F^{\mu\nu} = 0, \quad \partial_\mu {}^*G^{\mu\nu} = 0 \ , \tag{2.10}$$

while the constitutive relations become

$$G_{\mu\nu} = G_{\mu\nu}\left(F, {}^*F\right), \tag{2.11}$$

$$G_{\mu\nu} = {}^*F_{\mu\nu} + \ldots \ , \tag{2.12}$$

and, if we have a Lagrangian,

$${}^*G^{\mu\nu} = 2\frac{\delta \mathcal{L}\left(F\right)}{\delta F_{\mu\nu}}. \tag{2.13}$$

The duality rotations correspond to transform

$$\begin{pmatrix} F \\ G \end{pmatrix} \to \begin{pmatrix} F \\ G \end{pmatrix}' = \begin{pmatrix} A & B \\ C & D \end{pmatrix} \begin{pmatrix} F \\ G \end{pmatrix}, \tag{2.14}$$

and are a symmetry of the theory if the transformation $G(F) \to G(F') = G(AF + BG)$ is the same as the rotation $G(F) \to CF + DG(F)$. This highly non-linear constraint restricts the possible forms of G (the constitutive relations), and also the linear transformations (2.14).

Gaillard and Zumino [1] have proven that the most general e.m. duality rotation for a non-linear theory, also depending on extra matter fields (such as fermions, scalars, and possibly in a curved gravitational background) is $Sp(2n, \mathbb{R})$, where n is the number of vector field strengths F^Λ ($\Lambda = 1, ..., n$). This means that the matrix

$$\mathcal{S} := \begin{pmatrix} A & B \\ C & D \end{pmatrix} \tag{2.15}$$

instead of being a $GL(2n, \mathbb{R})$ matrix is a simplectic matrix, indeed it satisfies

$$\mathcal{S}^T \Omega \mathcal{S} = \Omega, \quad \Omega := \begin{pmatrix} 0 & -\mathbb{I} \\ \mathbb{I} & 0 \end{pmatrix}, \tag{2.16}$$

which equivalently reads

$$A^T D - C^T B = \mathbb{I}, \quad A^T C = C^T A, \quad B^T D = D^T B.$$

In absence of scalars, or if scalars are inert under duality rotations, the e.m. duality group is at most $U(n)$ (in particular, for $n = 1$, one retrieves Maxwell e.m. duality group: $U(1) \sim SO(2)$). The constraints on the n field strength F^Λ and G^Λ in this case read

$$^*F^\Lambda F^\Sigma + {}^*G^\Lambda G^\Sigma = 0; \tag{2.17}$$

$$F^\Lambda\, {}^*G^\Sigma - {}^*F^\Sigma G^\Lambda = 0. \tag{2.18}$$

For $n = 1$ (namely, $U(1)$), one gets $^*FF + {}^*GG = 0$.

For general e.m. theories where also matter fields are present the constitutive relations will also depend on these fields, that we generically denote χ^i, (and on their derivatives that we omit writing) so that we have $G_{\mu\nu} = G_{\mu\nu}(F, \chi^i)$. Duality rotations now will include also a (nonlinear) transformation on the matter fields χ^i.

As shown by Gaillard and Zumino [1] (and reviewed in [2]) in this general case, under an infinitesimal duality rotation parametrized by

$$s = \begin{pmatrix} a & b \\ c & d \end{pmatrix} \in \mathfrak{sp}(2n, \mathbb{R}), \text{ i.e. } \; a = -d^T,\; b = b^T, c = c^T\;, \tag{2.19}$$

the Lagrangian varies by

$$\delta\mathcal{L} \;=\; \frac{1}{4}\left(Fc\, {}^*F + Gb\, {}^*G\right), \tag{2.20}$$

where $\delta L \equiv L(F + \delta F, \chi^i + \delta\chi^i) - L(F, \chi^i)$ is the total infinitesimal variation of the Lagrangian under the duality rotations of the field strengths and of the matter fields. If we expand δF in terms of F and G using the infinitesimal version of the duality rotation (2.14) then we obtain

$$\delta_\chi L \equiv L(F, \chi^i + \delta\chi^i) - L(F, \chi^i) = \frac{1}{4}\left(Fc\, {}^*F - Gb\, {}^*G\right) + \frac{1}{2} Fd\, {}^*G. \tag{2.21}$$

If there are no matter fields the left hand side is zero, the duality rotation group reduces then to $U(n)$, and in this case the infinitesimal parameters satisfy $c = -b$ and $d = -d^T$. Then Eq. (2.21) reproduces just the previous Eqs. (2.17) and (2.18) for $U(n)$.

Note that for e.m. duality rotations the total variation $\delta\mathcal{L}$ never vanishes. If $c \neq 0, b = 0$ the infinitesimal variation $\delta\mathcal{L}$ is a total derivative because of the equations of motion $dF = 0$. If $c = 0, b \neq 0$ it is a total derivative because of the equations of motion $dG = 0$. If both $b \neq 0$ and $c \neq 0$ we have to use both equations of motion $dF = 0$ and $dG = 0$. We therefore see that the Lagrangians of theories with duality rotation symmetry change by a term that is a total derivative only on shell of the equations of motion $dF = 0$ and $dG = 0$.

We also have that under a finite duality rotation the action changes by a term that is a total derivative only on shell of both $dF = 0$ and $dG = 0$ equations of motions. Indeed under a finite $U(1)$ rotation of angle α we have [3]

$$\mathcal{L}(F') - \mathcal{L}(F) = \frac{1}{8}\Big(sin(2\alpha)\,(F\,{}^*F - G\,{}^*G) - 4\,sin^2\alpha\,F\,{}^*G\Big)\,. \tag{2.22}$$

We notice that in the special case of Maxwell theory $G = {}^*F$ and therefore the infinitesimal variation $\mathcal{L}(F') - \mathcal{L}(F)$ is a total derivative just on shell of the relation $G = {}^*F$ and of the equation $dF = 0$ (i..e without using the equation of motion $dG = 0$). In other words, if we solve the equation $dF = 0$ by introducing a gauge potential so that $F = dA$ and hence see the Lagrangian as dependent on A, then for infinitesimal angle α the variation (2.22) is a total derivative *off shell* of the equations of motion for the gauge potential A. This off shell infinitesimal symmetry of Maxwell theory (that acts nonlocally on the gauge potential A) was studied in [4]; however it does not correspond to a finite *off shell* symmetry. As we see from the second term in (2.22) in order for $F\,{}^*G$ to be a total derivative we have to use also the equations of motion $dG = 0$, that in Maxwell theory simply read $d\,{}^*F = 0$.

Two important results are obtained from the previous formulæ for general nonlinear theories.

1. *Any* Lagrangian density can be written as $\frac{1}{4}F\,{}^*G$ plus a duality invariant Lagrangian $\mathcal{L}_{inv}$,

$$\mathcal{L} = \frac{1}{4}F\,{}^*G + \mathcal{L}_{inv}\;, \quad \delta\mathcal{L}_{inv} = 0. \tag{2.23}$$

   The term $\mathcal{L}_{inv}$ provides information about the constitutive relations (*i.e.*, the medium), in particular $\mathcal{L}_{inv} = 0$ vanishes in linear Maxwell theory without coupling to matter. On the other hand, for (quadratic) theories coupled to matter $\mathcal{L}_{inv}$ provides information about some matter couplings (such as Pauli-like terms). In supergravity it is in fact possible to construct pairs of antisymmetric Lorentz tensors $(H_{\mu\nu}(\chi), I_{\mu\nu}(\chi))$ (typically bilinear in fermions) which transforms as the couple (F, G). In this case, one has

$$\mathcal{L}_{inv} = \frac{1}{4}(FI - GH) + \mathcal{L}_{inv}(\chi)\,. \tag{2.24}$$

2. The energy-momentum tensor

$$\Theta^{\mu}_{\lambda} = -\partial_{\lambda}\chi^{i}\frac{\partial\mathcal{L}}{\partial(\partial_{\mu}\chi^{i})} + \delta^{\mu}_{\lambda}\mathcal{L} + \widetilde{G}^{\Lambda|\mu\nu}F_{\Lambda|\nu\lambda} \tag{2.25}$$

is duality invariant if duality is a symmetry of the theory. In the absence of matter fields the trace of the energy momentun tensor is proportional to $\mathcal{L}_{inv}$, $\frac{1}{4}\Theta^{\mu}_{\mu} = \mathcal{L}_{inv}$.

3 Supersymmetry and the Higgs Effect

After the discovery of the Higgs particle at LHC at CERN, and the 2013 Nobel Prize for Physics bestowed to F. Englert and P. Higgs, a fundamental issue should draw the attention of the scientific community committed to the search for supersymmetry: if supersymmetry is a fundamental law of Nature, then the supersymmetric version of the Higgs mechanism should be at work.

The $\mathcal{N}$-extended supersymmetry algebra in $D = 4$ reads ($A, B = 1, ..., \mathcal{N}$, $\alpha, \dot{\alpha} = 1, 2$)

$$\begin{aligned}\left\{Q^{A}_{\alpha}, Q_{\dot{\alpha}B}\right\} &= (\sigma_{\mu})_{\alpha\dot{\alpha}}P^{\mu}\delta^{A}_{B}; &(3.1)\\ \left\{Q^{A}_{\alpha}, Q^{B}_{\beta}\right\} &= \epsilon_{\alpha\beta}Z^{AB}, &(3.2)\end{aligned}$$

where Z^{AB} is a generally scalar-dependent skew-symmetric complex matrix, named *central charge* matrix.

Supersymmetry multiplets fall in three species [5]:

- massless multiplets ($M = 0$): the maximal helicity reads $\lambda_{\max} = \mathcal{N}/4$ (for $\mathcal{N}$ even) and $\lambda_{\max} = (\mathcal{N} - 1)/4$ (for $\mathcal{N}$ odd);
- *long* massive ($M \neq 0$) multiplets, with maximal spin $j_{\max} = \mathcal{N}/2$;
- *short* ($\frac{q}{\mathcal{N}}$-BPS-saturated) massive multiplets, with maximal spin $j_{\max} = (\mathcal{N} - q)/2$, where q denotes the number of *skew-eigenvalues* [6] $\mathcal{Z}_i$ of Z^{AB} which saturate the BPS bound ($i = 1, ..., q$; $q = 1, ..., \mathcal{N}/2$ for $\mathcal{N}$ even, and $q = 1, ..., (\mathcal{N} - 1)/2$ for $\mathcal{N}$ odd):
$$|\mathcal{Z}_i| = M. \tag{3.3}$$

For instance, one can consider the massive spin-1 multiplets:

- $\mathcal{N} = 1$: one long multiplet;
- $\mathcal{N} = 2$: one long, and one short ($\frac{1}{2}$-BPS) multiplet;
- $\mathcal{N} = 3$: one short ($\frac{1}{3}$-BPS) multiplet;
- $\mathcal{N} = 4$: one short ($\frac{1}{2}$-BPS) multiplet.

On the other hand, considering massive spin-3/2 multiplets, one obtains:

- $\mathcal{N} = 1$: one long multiplet;

- $\mathcal{N}=2$: one long, and one short ($\frac{1}{2}$-BPS) multiplet;
- $\mathcal{N}=3$: one long, and one short ($\frac{1}{3}$-BPS) multiplet;
- $\mathcal{N}=4$: two short ($\frac{1}{2}$-BPS and $\frac{1}{4}$-BPS) multiplets;
- $\mathcal{N}=5$: one short ($\frac{2}{5}$-BPS) multiplet;
- $\mathcal{N}=6$: one short ($\frac{1}{2}$-BPS) multiplet;
- $\mathcal{N}=8$: none.

From the classification mentioned above, one can give model-independent restrictions on the spontaneous breaking of gauge symmetries and the Higgs mechanism. In presence of supersymmetry there are two kinds of Higgs effects:

1. *The supersymmetric Higgs effect*: spontaneous breaking of an ordinary Yang-Mills symmetry without symmetry breaking. Vectors become massive by absorption of a scalar mode. In $\mathcal{N}=1$ an example is the following one:

$$\mathcal{N}=1: \underbrace{\left(1,\frac{1}{2}\right)+\left(\frac{1}{2},2\,(0)\right)}_{\text{massless}} \longrightarrow \underbrace{\left(1,2\left(\frac{1}{2}\right),0\right)}_{\text{massive}}, \tag{3.4}$$

in which a massive vector multiplet is obtained by absorption of a chiral multiplet. In $\mathcal{N}=2$ there are two possibilities: **i)** in the Abelian Higgs model, it holds that

$$\mathcal{N}=2\text{ Abelian}: \underbrace{\left(1,2\left(\frac{1}{2}\right),2(0)\right)+\left(2\left(\frac{1}{2}\right),4\,(0)\right)}_{\text{massless}} \longrightarrow \underbrace{\left(1,4\left(\frac{1}{2}\right),5\,(0)\right)}_{\text{massive}}, \tag{3.5}$$

namely, a massive vector multiplet is obtained by absorption of an hypermultiplet; **ii)** in non-Abelian theories, there is instead a transmutation of charged vector multiplets:

$$\mathcal{N}=2\text{ non-Abelian}: \underbrace{2\left(1,2\left(\frac{1}{2}\right),2(0)\right)}_{\text{massless}} \longrightarrow \underbrace{2\left(1,2\left(\frac{1}{2}\right),0\right)}_{\text{massive}}. \tag{3.6}$$

The same occurs *e.g.* in $\mathcal{N}=4$:

$$\mathcal{N}=4:\ \underbrace{2\left(1,4\left(\frac{1}{2}\right),6(0)\right)}_{\text{massless}} \longrightarrow 2\left(1,4\left(\frac{1}{2}\right),5\,(0)\right), \tag{3.7}$$

in which there is only one case, because, from the previous discussion, $\mathcal{N}=4$ massive spin-1 multiplets must be necessarily (1/2-)BPS saturated.

2. *The super-Higgs effect*: spontaneous breaking of local supersymmetry. Spin-3/2 *gravitinos* become massive by absorption of a spin-1/2 Goldstino mode. A prototypical case is the $\mathcal{N}=2\to\mathcal{N}=1$ super-Higgs mechanism, in which a massive

$\mathcal{N}=1$ spin-3/2 gravitino multiplet is generated, with 6 bosonic and 6 fermionic degrees of freedom:

$$\mathcal{N}=1 \text{ massive}: \left(\frac{3}{2}, 2(1), \frac{1}{2}\right). \tag{3.8}$$

When the partial breaking is implemented by $\mathcal{N}=2$ *linear* multiplets (as in the cases mentioned above), the minimal $\mathcal{N}=2$ Higgs branch is given by the following massless multiplets [7]:

$$\begin{array}{llcl} & \mathcal{N}=2 \text{ massless} & \longrightarrow & \mathcal{N}=1 \text{ massless} \\ \text{gravity} & \left(2, 2\left(\frac{3}{2}\right), 1\right) & \longrightarrow & \left(2, \frac{3}{2}\right) + \left(\frac{3}{2}, 1\right) \\ \text{hyper} & \left(2\left(\frac{1}{2}\right), 4(0)\right) & \longrightarrow & \left(\frac{1}{2}, 2(0)\right) + \left(\frac{1}{2}, 2(0)\right) \\ \text{vector} & \left(1, 2\left(\frac{1}{2}\right), 2(0)\right) & \longrightarrow & \left(1, \frac{1}{2}\right) + \left(\frac{1}{2}, 2(0)\right). \end{array} \tag{3.9}$$

The massless spin 0 sector, namely the $\mathcal{N}=2$ non-linear sigma model of the scalars respectively of the vector multiplet, that parametrize a special Kähler manifold, and of the hypermultiplet, that parametrize a quaternionic manifold, reads[1]

$$\mathcal{N}=2 \ \ (n_V=1, n_H=1): \frac{SU(1,1)}{U(1)} \times \frac{SO(1,4)}{SO(4)} \tag{3.10}$$

The $\mathcal{N}=2 \to \mathcal{N}=1$ super-Higgs mechanism yields the massive gravitino multiplet (3.8) as follows:

$$\mathcal{N}=1: \underbrace{\left(\frac{3}{2}, 1\right) + \left(1, \frac{1}{2}\right) + \left(\frac{1}{2}, 2(0)\right)}_{\text{massless}} \longrightarrow \underbrace{\left(\frac{3}{2}, 2(1), \frac{1}{2}\right)}_{\text{massive}}. \tag{3.11}$$

From the massless branching (3.9) it thus follows that some $\mathcal{N}=1$ multiplets remain massless after such super-Higgsing, namely:

$$\mathcal{N}=1 \text{ massless mults.}: \left(2, \frac{3}{2}\right) + 2\left(\frac{1}{2}, 2(0)\right), \tag{3.12}$$

where the $\mathcal{N}=1$ non-linear sigma model spanned by the 2 complex (Kähler) scalars of the chiral multiplets reads

$$\mathcal{N}=1 \ \ (n_c=2): \frac{SU(1,1)}{U(1)} \times \frac{SU(1,1)}{U(1)}. \tag{3.13}$$

An important subtlety concerns the so-called *anti-Higgs effect*: namely, the fact that a 2-form field can describe a spin-0 *or* a spin-1 field, depending on the very Lagrangian density under consideration:

[1] Note that $\frac{SO(1,4)}{SO(4)} \sim \frac{USp(2,2)}{USp(2)\times USp(2)} =: \mathbb{HP}^1$ (the *projective quaternionic plane*); this is actually the first element of an infinite sequence $\mathbb{HP}^n := \frac{USp(2,2n)}{USp(2)\times USp(2n)}$ of symmetric quaternionic spaces, which is the *unique* class of symmetric quaternionic spaces which are *not* in the c-map [8] of (symmetric) special Kähler spaces.

- in the case it describes a spin-0, the general form of the Lagrangian reads

$$\mathcal{L}_{s=0} \sim \partial_\mu \tilde{F}^{\mu\nu} B_\nu + (B_\mu)^2 , \tag{3.14}$$

such that integrating B_μ, resp. $\tilde{F}_{\mu\nu}$, out one respectively obtains:

$$\mathcal{L}_{s=0}|_{B_\mu \text{ integrated out}} \sim \left(\partial_\mu \tilde{F}^{\mu\nu}\right)^2 ; \tag{3.15}$$

$$\mathcal{L}_{s=0}|_{\tilde{F}_{\mu\nu} \text{ integrated out}} \sim (\partial_\mu \phi)^2 , \quad B_\mu = \partial_\mu \phi. \tag{3.16}$$

- in the case it describes a spin-1, the general form of the Lagrangian reads

$$\mathcal{L}_{s=1} \sim \partial_\mu \tilde{F}^{\mu\nu} B_\nu + (F_{\mu\nu})^2 , \tag{3.17}$$

such that integrating B_μ, resp. $F_{\mu\nu}$, out one respectively obtains:

$$\mathcal{L}_{s=1}|_{B_\mu \text{ integrated out}} \sim (F_{\mu\nu})^2 , \quad \partial_\mu \tilde{F}^{\mu\nu} = 0; \tag{3.18}$$

$$\mathcal{L}_{s=1}|_{F_{\mu\nu} \text{ integrated out}} \sim (\partial_\mu B_\nu)^2 . \tag{3.19}$$

By joining $\mathcal{L}_{s=0}$ and $\mathcal{L}_{s=1}$ one thus obtains:

$$\mathcal{L}_{\text{massive}} \sim \partial_\mu \tilde{F}^{\mu\nu} B_\nu + (B_\mu)^2 + (F_{\mu\nu})^2 . \tag{3.20}$$

In the *Higgs* perspective $(B_\mu)^2$ is interpreted as a mass term for spin-1 field; whereas, in the *anti-Higgs* perspective, $(F_{\mu\nu})^2$ is interpreted as a mass term for spin-0 field.

4 Goldstino Non-Linear Theories and Born-Infeld (BI) Actions

The BI model [9] is the simplest non-linear e.m. duality invariant theory. It was introduced to remove the divergence of the electrons self-energy in classical electrodynamics. This is obtained by having an upper bound on the electric field.

The BI action is the bosonic part of the $\mathcal{N} = 1$ non-linear Goldstino Lagrangian for the aforementioned partial supersymmetry breaking $\mathcal{N} = 2 \to \mathcal{N} = 1$ [10, 11, 12]. Denoting the Goldstino by λ_α, and its $\mathcal{N} = 1$ superpartner as $F_{\mu\nu}$ ($U(1)$ gauge field strength), it indeed holds that

$$\mathcal{L}_{Goldstino}\left(\lambda_\alpha, F_{\mu\nu}\right)|_{\lambda_\alpha=0} = \mathcal{L}_{BI}\left(F_{\mu\nu}\right) . \tag{4.1}$$

The equations of motion of BI theory are a particular realization of the constitutive relations

$$^*G_{\mu\nu} = {}^*G_{\mu\nu}\left(F, g\right) , \tag{4.2}$$

(here g is a dimensionful coupling constant tipically present in nonlinear theories, which was previously omitted for sake of brevity). They satisfy the e.m. duality conditions

$$F_{\mu\nu} \, {}^*F^{\mu\nu} + G_{\mu\nu} \, {}^*G^{\mu\nu} = 0 . \tag{4.3}$$

The explicit dependence of G on F is

$$G_{\mu\nu} = \frac{{}^*F_{\mu\nu} + \frac{1}{4}g^2 F\, {}^*F F_{\mu\nu}}{\sqrt{1 + \frac{1}{2}g^2 F^2 - \frac{1}{16}g^4 \left(F\, {}^*F\right)^2}}, \tag{4.4}$$

and, recalling that ${}^*G^{\mu\nu} = 2\frac{\partial \mathcal{L}}{\partial F_{\mu\nu}}$, it can be traced back to the Lagrangian density

$$\begin{aligned} \mathcal{L}_{BI} &= \frac{1}{g^2}\left(1 - \sqrt{-\det\left(\eta + gF\right)}\right) && (4.5) \\ &= \frac{1}{g^2}\left(1 - \sqrt{1 + \frac{g^2}{2}F^2 - \frac{g^4}{16}\left(F\, {}^*F\right)^2}\right). && (4.6) \end{aligned}$$

Furthermore, by expanding in power series of the coupling constant g we have

$${}^*G_{\mu\nu}\left(F, g\right) = -F_{\mu\nu} + ... \tag{4.7}$$

and hence for $g = 0$, i.e., for weakly varying field, we recover Maxwell theory.

If one rewrites the Lagrangian in terms of the variables $\left(\vec{E}, \vec{B}\right)$, one obtains

$$\mathcal{L}_{BI} = \frac{1}{g^2}\left(1 - \sqrt{1 + g^2\left(\left|\vec{B}\right|^2 - \left|\vec{E}\right|^2\right) - g^4\left(\vec{E}\cdot\vec{B}\right)^2}\right). \tag{4.8}$$

For vanishing $\vec{B}$ this Lagrangian yields the bound

$$\left|\vec{E}\right|^2 \leqslant \frac{1}{g^2}\,. \tag{4.9}$$

The Hamiltonian depends on the variables $\left(\vec{D}, \vec{B}\right)$, and is obtained by a Legendre transform with respect to $\vec{E}$, namely by defining

$$\vec{D} = \frac{\delta\mathcal{L}\left(E, B\right)}{\delta\vec{E}}\,, \quad \mathcal{H} = \vec{D}\cdot\vec{E} - \mathcal{L}\,. \tag{4.10}$$

It is worth stressing that the Hamiltonian is perfectly regular for all values of these fields variables:

$$\mathcal{H}_{BI} = \frac{1}{g^2}\left(\sqrt{1 + g^2\left(\left|\vec{D}\right|^2 + \left|\vec{B}\right|^2\right) + g^4\left|\vec{D}\wedge\vec{B}\right|^2} - 1\right). \tag{4.11}$$

In particular in the limits of weakly and of strongly valued fields:

$$\begin{aligned} \lim_{g\to 0}\mathcal{H}_{BI} &= \frac{1}{2}\left(\left|\vec{D}\right|^2 + \left|\vec{B}\right|^2\right)\,; && (4.12) \\ \lim_{g\to\infty}\mathcal{H}_{BI} &= \left|\vec{D}\wedge\vec{B}\right|. && (4.13) \end{aligned}$$

The general study and classification of constitutive relations admitting e.m. duality symmetry can be attacked by adopting complex and chiral variables, indeed these variables simply transform with a phase under duality symmetry rotations. We therefore define:

$$T := F - iG \Leftrightarrow \overline{T} := F + iG, \tag{4.14}$$

so that the e.m. duality condition $F\,{}^*F + G\,{}^*G = 0$ equivalently reads

$${}^*\overline{T}T = 0. \tag{4.15}$$

The corresponding chiral combinations read

$$T^{\pm} := \frac{1}{2}\left(F^{\pm} - iG^{\pm}\right) \Leftrightarrow \overline{T^{\mp}} := \frac{1}{2}\left(F^{\mp} + iG^{\mp}\right), \tag{4.16}$$

where $F^{\pm} = \frac{1}{2}(F \pm i\,{}^*F)$, $G^{\pm} = \frac{1}{2}(G \pm i\,{}^*G)$. The equations of motion of linear electromagnetism in these variables read

$$T^{+} = 0 \Leftrightarrow (F - iG) + i\left({}^*F - i\,{}^*G\right) = 0 \Leftrightarrow G = {}^*F. \tag{4.17}$$

The constitutive relations in these new variables, studied in [13], and indipendently and more generally in [14, 15], read as follows:

$$T^{+} = \frac{\partial I\left(T^{-},\overline{T^{-}}\right)}{\partial \overline{T^{-}}}, \quad \overline{T^{+}} = \frac{\partial I\left(T^{-},\overline{T^{-}}\right)}{\partial T^{-}}. \tag{4.18}$$

These are six relations if the arbitrary function I (the "Lagrangian" of these new variables) is real. Moreover, duality invariance is obtained if I is invariant under duality rotations. This is the case if

$$I\left(T^{-},\overline{T^{-}},g\right) = I(u), \tag{4.19}$$

where

$$u = 2g^2|(T^-)^2| = g(|T^2| + |T\,{}^*T|)\,, \quad \text{i.e.,} \quad u^2 = 4g^4(T^-)^2(\overline{T}^-)^2\,. \tag{4.20}$$

We note that the function I is $U(1)$-duality-invariant because the Lorentz scalar $(T^-)^2(\overline{T^-})^2$ is duality invariant, indeed under a $U(1)$ duality rotation of angle α, the fields T and $\overline{T}$ transform with opposite phases $e^{-i\alpha}$ and $e^{i\alpha}$.

By use of a Legendre transform [13, 16, 3] between the functions $\mathcal{L}(F)$ and $I(T^-,\overline{T^-})$, the constitutive relations (4.18) in the T-variables can be shown to be equivalent to the original constitutive relations (2.13).

To summarize, the freedom in writing a self-dual and non-linear theory of electromagnetism can be traced back to a real function I of a real (Lorentz-invariant and $U(1)$-duality invariant) variable $(T^-)^2\left(\overline{T^-}\right)^2$, or its square root u. Arbitrary functions $I(u)$ are in one-to-one correspondence with Lagrangians $\mathcal{L}(F)$ such that the equations (4.18) are equivalent to the original constitutive relations and self duality condition

$${}^*G^{\mu\nu} = 2\frac{\delta\mathcal{L}(F)}{\delta F_{\mu\nu}}\,, \quad G\,{}^*G + F\,{}^*F = 0\,. \tag{4.21}$$

Furthermore, it was shown by Schrödinger [17], see also [18], that the constitutive relations for the BI theory are

$$ {}^*T_{\mu\nu} = -\frac{T^2}{{}^*TT}T_{\mu\nu} - \frac{g^2}{8}\,{}^*TT\,\overline{T}_{\mu\nu}. \tag{4.22} $$

In the general case the following constitutive relations à la Schrödinger define a nonlinear and self-dual theory [16]

$$ {}^*T_{\mu\nu} = -\frac{T^2}{{}^*TT}T_{\mu\nu} - \frac{g^2}{8}\frac{f(u)}{u}\,{}^*TT\,\overline{T}_{\mu\nu}, \tag{4.23} $$

where we recall that $u = 2|(T^-)^2| = |T^2| + |T\,{}^*T|$, and where we require $lim_{u\to 0}f(u) = 0$, so to recover Maxwell theory for weak fields. For example the duality conditions (4.15) follow contracting (4.23) with ${}^*T^{\mu\nu}$. Contracting (4.23) with $T^{\mu\nu}$ and with $\overline{T}^{\mu\nu}$ it can also be shown that

$$ \left(1 - t^2\right)^2 f(u) = 32t; \tag{4.24} $$

where

$$ t = \frac{T\overline{T}}{|T^2| + |T\,{}^*T|}\,. \tag{4.25} $$

This quartic equation determines $f(u)$ in terms of t, explicitly the solution that for weak fields is Maxwell theory is given by [3]

$$ t(u) = \frac{f(u)}{16}\sum_{k=0}^{\infty}(-1)^k\frac{(4k+1)!}{(3k+2)!k!}\left(\frac{f(u)^2}{4^5}\right)^k. \tag{4.26} $$

The relation between the constitutive relations à la Schrödinger and those in terms of the $I(u)$ function follows from the relations between the T variables and the chiral $T^{\pm}$ variables. This allows to determine $f(u)$ from $I(u)$ and vice versa. Explicitly, given $f(u)$, one obtains $t = t(u)$ from (4.26); the function $I(u)$ is then given by

$$ I\left(u\right) = \frac{1}{2}\int t(u)du. \tag{4.27} $$

5 Conclusions and Perspectives

In this lectures we have presented instances in which electric-magnetic duality plays a fundamental role. We focused on non-linear theories of electromagnetism, where e.m. duality restricts the form of the non-linear theory. The same happens in supergravity, where e.m. duality yields important information on the form of the matter couplings to gauge fields.

One of the main motivations to study duality symmetry is that it also restricts the form of counterterms in supergravity [19], and in fact its imposition gives rise to an infinite sequence of terms [20], similar in spirit to the BI action.

It would be interesting to see if the quantum corrected supergravity Lagrangian density gets a functional form of the BI type, with the dimensionful BI coupling constant replaced by $\hbar$ (so that the expansion is order by order in the number of loops).

Acknowledgments

The work of PA is partially supported by the MIUR- PRIN contract 2009-KHZKRX.

The work of SF is supported by the ERC Advanced Grant no. 226455 *SUPERFIELDS*. He is on leave of absence from Department of Physics and Astronomy, University of California, Los Angeles, USA.

The work of AM is supported in part by the FWO - Vlaanderen, Project No. G.0651.11, and in part by the Interuniversity Attraction Poles Programme initiated by the Belgian Science Policy (P7/37).

References

[1] M.K. Gaillard and B. Zumino, *Duality Rotations for Interacting Fields*, Nucl. Phys. **B193**, 221 (1981).

[2] P. Aschieri, S. Ferrara, B. Zumino, *Duality Rotations in Nonlinear Electrodynamics and in Extended Supergravity*, Nuovo Cimento, vol. **31**, n.11, 625 (2008), `arXiv:0807.4039 [hep-th]`.

[3] P. Aschieri, S. Ferrara and S. Theisen, *Constitutive relations, off shell duality rotations and the hypergeometric form of Born-Infeld theory*, `arXiv:1310.2803 [hep-th]`.

[4] S. Deser and C. Teitelboim, *Duality Transformations of Abelian and Nonabelian Gauge Fields*, Phys. Rev. D **13** (1976) 1592.

[5] S. Ferrara, C. A. Savoy, B. Zumino, *General Massive Multiplets in Extended Supersymmetry*, Phys. Lett. **B100** (1981) 393.

[6] L. K. Hua, *On the Theory of Automorphic Functions of a Matrix Variable. I: Geometrical Basis*, Amer. J. Math. **66**, 470 (1944). C. Bloch and A. Messiah, *The Canonical Form of an Antisymmetric Tensor and its Application to the Theory of Superconductivity*, Nucl. Phys. **B39**, 95 (1962). B. Zumino, *Normal Forms of Complex Matrices*, J. Math. Phys. **3**, 1055 (1962).

[7] S. Ferrara, L. Girardello, M. Porrati, *Spontaneous breaking of $\mathcal{N}=2$ to $\mathcal{N}=1$ in rigid and local supersymmetric theories*, Phys. Lett. **B376** (1996) 275, `hep-th/9512180`.

[8] S. Cecotti, S. Ferrara, L. Girardello, *Geometry of Type II Superstrings and the Moduli of Superconformal Field Theories*, Int. J. Mod. Phys. **A4** (1989) 2475.

[9] M. Born and L. Infeld, *Foundations of the New Field Theory*, Proc. R. Soc. London sez. **A 144** (1934), 425.

[10] S. Deser, R. Puzalowski, "Supersymmetric Nonpolynomial Vector Multiplets and Causal Propagation," J. Phys. A13 (1980) 2501

[11] S. Cecotti and S. Ferrara, "Supersymmetric Born-Infeld Lagrangians," Phys. Lett. B **187** (1987) 335.

[12] J. Bagger and A. Galperin, *A New Goldstone multiplet for partially broken supersymmetry*, Phys. Rev. D **55**, (1997) 1091 `hep-th/9608177`.

[13] E. A. Ivanov and B. M. Zupnik, *New approach to nonlinear electrodynamics: Dualities as symmetries of interaction*, Phys. Atom. Nucl. **67** (2004) 2188 [Yad. Fiz. **67** (2004) 2212] `hep-th/0303192`.

[14] G. Bossard and H. Nicolai, "Counterterms vs. Dualities," JHEP **1108**, 074 (2011) `arXiv1105.1273 [hep-th]`.

[15] J. J. M. Carrasco, R. Kallosh, R. Roiban, *Covariant procedures for perturba- tive non-linear deformation of duality-invariant theories*, Phys. Rev. D**85** (2012) 025007 `arxiv1108.4390 [hep-th]`. W. Chemissany, R. Kallosh and T. Ortin, *Born-Infeld with Higher Derivatives*, Phys. Rev. D **85** (2012) 046002 `arXiv1112.0332 [hep-th]`.

[16] P. Aschieri and S. Ferrara, *Constitutive relations and Schroedinger's formulation of nonlinear electromagnetic theories*, JHEP **1305** (2013) 087 `arXiv:1302.4737 [hep-th]`

[17] E. Schrödinger, *Contributions to Born's New Theory of the Electromagnetic Field*, Proc. R. Soc. London, sez. **A 150** (1935), 465.

[18] M. K. Gaillard and B. Zumino, "Nonlinear electromagnetic self-duality and Legendre transformations," `hep-th/9712103`.

[19] G. Bossard, P.S. Howe, K.S. Stelle, *The Ultra-violet question in maximally supersymmetric field theories*, Gen. Rel. Grav. **41** (2009) 919, `arXiv:0901.4661 [hep-th]`. G. Bossard, P.S. Howe, K.S. Stelle, *On duality symmetries of supergravity invariants*, JHEP **1101** (2011) 020, `arXiv:1009.0743 [hep-th]`.

[20] R. Kallosh, *The Ultraviolet Finiteness of $\mathcal{N}= 8$ Supergravity*, JHEP **1012** (2010) 009, `arXiv:1009.1135 [hep-th]`. R. Kallosh, *$E_{7(7)}$ Symmetry and Finiteness of $\mathcal{N}= 8$ Supergravity*, JHEP **1203** (2012) 083, `arXiv:1103.4115 [hep-th]`. R. Kallosh, *$\mathcal{N}= 8$ Counterterms and $E_{7(7)}$ Current Conservation*, JHEP **1106** (2011) 073, `arXiv:1104.5480 [hep-th]`.

CHAIRMAN: S. FERRARA

Scientific Secretaries: I. Danilenko, V. Slepukhin

DISCUSSION

- *Lazslo:*

Hello. My name is Andras Lazslo from Wigner Research for Physics, Budapest. You mentioned the SO(2) hidden symmetry of vacuum field theory and that had just triggered the following question in my mind: if one has such a hidden symmetry in the Lagrangian theory at a classical field level, is there any way to deduce symmetries from the Lagrangian instead of looking at the energy-momentum tensor or at the field equation?

- *Ferrara:*

Are you asking whether this symmetry reflects some property of the Lagrangian?

- *Lazslo:*

Yes.

- *Ferrara:*

Let me start by saying that the electromagnetic duality is not internal symmetry. It is a space-time symmetry, because it rotates electric to magnetic field, so it changes the parity. So, it cannot be an internal symmetry. Secondly, any theory, which changes electric field into magnetic field, has a Lagrangian, which is not invariant, but the energy-momentum tensor is. In other words, the Hamiltonian is invariant. In the lectures I presented the simplest example of this fact. In the case of Maxwell theory the Lagrangian is $L = E^2 - H^2$, whereas the Hamiltonian is $H^2 = E + H^2$, and that is 00 component of the energy-momentum tensor. An example is provided by black holes. The square of the mass of the extremal black hole is $M^2 = (P^2 + Q^2)\ M_{Planck}^2$, where P and Q are electric and magnetic charges. The reason why this combination appears due to the fact that the theory has a rotational symmetry between Q and P, the electric and magnetic charges. This is the condition for which the black hole to have zero temperature, an extremality condition.

- *'t Hooft:*

So, one may say that actually a lack of symmetry is in our notations, not in the real physics. Think of harmonic oscillator, there is something similar: the energy is $p^2 + q^2$, but the lagrangian is $p^2 - q^2$.

- *Ferrara:*

Exactly the same.

- *'t Hooft:*

Whereas, of course, if you wait for one quarter of a period, you would change p and q. We certainly call that a symmetry. It is just our notation, which is not symmetric.

- *Ferrara:*

But your notation in some sense is the natural notation for the quantum mechanics. You can't make both invariant, that's what I'm saying. The natural notation, where the invariance is manifest, is the Hamiltonian formulation.

-'t Hooft:

There is an interesting question, whether you can use this symmetry to have a conservation law. I think there is…

- Ferrara:

There is…

- 't Hooft:

Ok, that was just a remark.

- Mojaza:

My name is Matin Mojaza and my question is on more general non-Abelian electro-magnetic dualities, and the attempts to construct such dualities. As far as I know, it is not an easy task and I only know the construction of N=4 SYM electro-magnetic duality of Montonen-Olive. I was wondering what is the framework for supergravity and, in particular, how Born-Infeld approach looks like.

- Ferrara:

It looks like, in some sense, Montonen-Olive. It has in common with the duality YM theory that it applies to the non-Abelian gauge theories. So, it is a duality between gauge group and dual gauge group. Thus, this duality is also connected to the monopole spectrum of the YM theory. Let me consider, as an example, N=8 supergravity. In N=8 supergravity, you find that there are 8 spin-3/2, 28 spin-1, 56 spin-1/2 and 70 scalar fields. The non-linear Lagrangian for the scalar fields is a non-linear sigma model E _7(7)/SU(8). As explained in the seminar we had today about symmetries of supergravities, the actual symmetry of the non-linear sigma model extends to the entire Lagrangian. The way it extends to the vector sector is as follows. The fundamental representation of E_7(7) is precisely 56-dimensional, so when you make a E_7(7) rotation in N=8 supergravity you have to exchange the 28 electric fields with the 28 magnetic fields. This operation can't be done at the Lagrangian level, because in the Lagrangian you can't put both electric and magnetic in the same time. But this is a symmetry of the equations of motion, which have this E_7(7) symmetry. This is the generalization in supergravity case of the Montonen-Olive duality, in the sense that this is a symmetry, which rotates the electric and magnetic fields of the supergravity multiplet, instead of the YM multiplet. This is interesting, because this duality controls, for instance, the black hole solutions. If you have a black hole in N=8, its entropy is proportional to the area, which in term is proportional to the square root of the quartic invariant you construct with a 56-dimensional representation of E_7(7). So, the E_7(7) symmetry reflects, for instance, in the horizon area of the black hole, which is given in terms of the electric and magnetic charges. In some sense, the electro-magnetic duality of supergravity is related to the vector fields which appear in the supergravity multiplet.

- Zichichi:

But the fact is that 70 scalars have never been seen and the 28 vectors, as well…

- Ferrara:

The 70 scalars are precisely the coordinates of the manifold E_7(7)/SU(8). E_7(7) has dimension 133, if you subtract 63 (dimension of SU(8)), you get 70.

- Zichichi:

Ok, but where are they?

- *Ferrara:*

These are the scalar fields, which exist in the N=8 supergravity multiplet. This is the actual theory we are considering.

- *Zichichi:*

In order to exist, they must be experimentally observable. So, where are they?

- *Ferrara:*

Well, unfortunately none of these fields, with the exception of the graviton, is a field, with a meaning from view of particle physics. So the only observable here is the gravity field.You could ask the same question for N=4 YM. But as you know, N=4 YM has been a crucial ingredient for the understanding of holography and of AdS/CFT correspondence. So, I mean, sometimes you start with a theory which doesn't look to have a direct physical meaning, but, because of the symmetries this theory possesses, it just led to some progress in understanding theoretical physics. N=8 supergravity, I would say, stands on the some footing as N=4 YM. We still don't understand exactly where the bridge to real world is. However, it is certainly a theory of gravity, which has maximal space-time symmetry, because it has the maximal supersymmetry that you can have in a field theory (with maximum spin 2).

- *Zichichi:*

These speculations are very exciting, and we are fascinated by all of them, but you should not forget that it is very important to pay some attention to project all these speculations into the real world.

- *Ferrara:*

Julian Schwinger discovered dyons looking at the possible relativistic symmetries of the Maxwell equations. You may say that nobody had ever seen a dyon, but in some sense, if you believe in monopoles, you should also believe in dyons. And these are constructs which come from the fundamental invariance of relativistic field theories. That's something you have to live with.

- *Cerchiai:*

Hi, Bianca Cerchiai from Milan. I wanted to ask, whether you could speculate a bit more about the last thing you said, at the end of the conclusion, when you spoke about the quantum corrected supergravity lagrangian, maybe being of the functional form of the Born-Infeld type?

- *Ferrara:*

Let me just say a few words about duality symmetries. If there are no anomalies, they are useful, because only very few possible corrections to the classical action impose conservation of electromagnetic duality. That yields Ward identities for the amplitude, which in principle restrict the possible forms of counterterms.

- *Borsten*:

Hi, Leron Borsten. You can make these dualities manifest at the Lagrangian level if you append the lagrangian a constraint. Now, is there any prospect for quantizing a lagrangian, which is constrained with these self-duality conditions? And a tightly related question is. I wondered. If you follow the work by Perry and Berman controlling manifestly U-duality invariant metrics and building lagrangians out of them, where the metrics actually live in a extended space-time, so in the case of E_7 space-time is 56 - dimensional and they imposed some sectional conditions to reduce it back down to 4 dimensions of N=8 supergravity. Where does the electro-magnetic duality story live in that formulation of the theory?

- *Ferrara:*

Well, I don't know enough to answer your question, which I see to involve what is “double geometry.

- *Korthals-Altes:*

Your classical Lagrangians had continuous symmetries, rotations. Now, if you would quantize these theories, then, I suppose, this continuous group becomes discrete, right?

- *Ferrara:*

Correct. The condition of quantization of black hole charges would break this continuous symmetry into a discrete symmetry. Correct.

- *Zichichi:*

This is very important point…

- *Ferrara:*

Yes, I considered the classical limit, in which the continuous symmetry is retrieved for a large value of the charges.

- *Zichichi:*

What exactly do you mean by “large”, Sergio?

- *Ferrara:*

Well, when the charges are large, then the discrete becomes a continuum.

Composite Weak Bosons at the LHC

Harald Fritzsch
Department für Physik
Ludwig-Maximilians-Universität
München, Germany

Abstract

In a composite model of the weak bosons the excited bosons, in particular the p-wave bosons, are studied. The state with the lowest mass is identified with the boson, which has been discovered recently at the "Large Hadron Collider" at CERN. Specific properties of the excited weak bosons are studied, in particular their decays into weak bosons and into photons.

In the Standard Model of the electroweak interactions the masses of the weak bosons and of the leptons and quarks are generated by a spontaneous breaking of the electroweak symmetry. Besides the weak bosons a scalar boson must exist ("Higgs boson"). Recently one has observed effects at the *LHC*, which might be due to the production of the Higgs boson and the subsequent decay into two weak bosons or into two photons (ref.(1,2)). The mass of this new boson is about 126 GeV.

In this lecture I discuss the possibility that the weak bosons are composite particles, i.e. particles with a finite size. The new scalar boson, observed at the *LHC*, would be an excited Z-boson, in which the constituents are in a p-wave.

In the Standard Theory of particle physics the leptons, quarks and the weak bosons are pointlike particles without any substructure. In the experiments at SLAC, DESY, Fermilab and CERN one has searched for a substructure of the electron, of the myon and of the light quarks, but nothing has been found. The present limit on the substructure of the electron, the muon and the quarks is about 0.001 fm.

Baryons and mesons are composed of quarks. Their extension is of the order of 1 fm. Their masses are directly linked to the substructure and can be calculated in the theory of quantum chromodynamics. The proton mass is due to the field energy of the gluons and quarks inside the proton, if the quark masses are set to zero. The field energy of the gluons provides about 70% of the proton mass, 30% are due to the field energy of the quarks. The mass of the proton is given by the scale parameter of QCD, multiplied by a dimensionless constant, which can be calculated: $M(p) = const. \cdot \Lambda_c$.

In the Standard Theory the masses of the weak bosons, leptons and quarks are generated by the spontaneous symmetry breaking. A doublet of scalar fields is introduced, which breaks the weak isospin symmetry spontaneously and develops a non-zero vacuum expectation valus. The weak bosons absorb three of the four scalar fields and obtain a mass, which is proportional to the vacuum expectation value. The remaining neutral scalar boson is the "Higgs" boson, which might be the new boson, observed recently at the *LHC*.

Thus in the Standard Theory there are two different ways to generate masses. The masses of the hadrons are generated dynamically - the masses of the weak bosons and of the leptons and quarks originate from a spontaneous symmetry breaking. I prefer a theory, in which also the masses of the weak bosons and of the leptons and quarks are generated dynamically. This is possible, if the weak bosons as well as the leptons and quarks are composite particles.

About 60 years ago the ρ-mesons were considered to be pointlike gauge bosons, providing the strong interaction between the hadrons. The masses of the ρ-mesons can be introduced by a spontaneous symmetry breaking, analogous to the introduction of the masses of the weak bosons in the electroweak theory. A doublet of scalar fields is required, which describes four scalar particles. The three ρ-mesons absorb three scalars and become massive.

Such a mechanism for the mass generation of the ρ-mesons is now excluded. Their masses are dynamical masses as the nucleon mass. The ρ-mesons are not elementary gauge bosons, but quark-antiquark bound states. Besides the ρ-mesons there is also the iso-singlet ω- meson, which could not obtain its mass through the spontaneous symmetry breaking, involving a scalar doublet.

In QCD the three ρ-mesons are degenerate in mass, if the electromagnetic interaction is switched off and the two light quark masses are zero. Once the electromagnetic interaction is introduced, the charged mesons receive an additional small contribution to the mass, which is due to the Coulomb self energy. In addition the neutral ρ-meson mixes with the photon and its mass increases. This mass shift can be calculated. It depends on a mixing parameter μ, which is determined by the electric charge, the decay constant F_ρ and the mass of the ρ-meson:

$$\mu = e\frac{F_\rho}{M_\rho} . \tag{1}$$

The mass shift due to the mixing is given by:

$$M_{\rho^0}^2 = M_{\rho^+}^2 \left(\frac{1}{1-\mu^2} \right) . \tag{2}$$

The decay constant is measured to about 220 MeV - it is about equal to the QCD scale parameter Λ_c. One obtains $\mu \approx 0.09$ - it leads to a mass shift of about 3 MeV. Due to this mass shift the mass difference between the charged and neutral ρ-meson is rather small.

We assume that the weak bosons are composite particles. They consist of a lefthanded fermion and its antiparticle, which are denoted as "haplons". A theory of this type was proposed in 1981 (see ref.(3) and ref.(4,5,6,7,8)). The new confining chiral gauge theory is denoted as QHD. The QHD mass scale is given by a mass parameter Λ_h, which determines the size of the weak bosons. The haplons interact with each other through the exchange of massless gauge bosons, which we denote as "glutons" (the Latin expression for "glue" is "gluten").

Two types of haplons are needed as constituents of the weak bosons, denoted by α and β. Their electric charges in units of e are:

$$h = \begin{pmatrix} +\frac{1}{2} \\ -\frac{1}{2} \end{pmatrix} . \tag{3}$$

The three weak bosons have the following internal structure:

$$\begin{aligned} W^+ &= \overline{\beta}\alpha , \\ W^- &= \overline{\alpha}\beta , \\ W^3 &= \frac{1}{\sqrt{2}} \left(\overline{\alpha}\alpha - \overline{\beta}\beta \right) . \end{aligned} \tag{4}$$

In the absence of electromagnetism the weak bosons are degenerate in mass. If the electromagnetic interaction is introduced, the mass of the neutral boson increases due to the mixing with the photon (ref. (9,10)).

In the Standard Theory the mixing is generated by the scalar fields. Both the photon and the Z-boson are mixtures of the $SU(2)$ and $U(1)$ gauge bosons. Here the mixing is a dynamical mixing, analogous to the mixing of ρ - mesons. It is described by the mixing parameter m, determined by the decay constant of the weak boson:

$$m = e\frac{F_W}{M_W} \,. \tag{5}$$

One finds for the mass difference beween the charged and the neutral weak boson:

$$M_Z^2 \; = \; M_{W^+}^2 \left(\frac{1}{1-m^2}\right) \,. \tag{6}$$

In the standard electroweak theory there is a similar equation - the mixing parameter m must be replaced by $\sin\theta_w$. According to the experiments the mixing parameter m is about 0.485, i. e. about five times larger than the mixing paramter for the ρ-mesons. Using the experimental value, one can determine the decay constant for the weak bosons:

$$F_W \approx 125 \, GeV. \tag{7}$$

As in QCD it is expected that the decay constant of the weak boson and the QHD mass scale are related. The decay constant of the ρ-meson and the QCD mass scale are about the same - in QHD the weak decay constant and Λ_h should be of the same order of magnitude. Details will depend in particular on the gauge group of QHD. We expect that Λ_h is in the range between 0.12 TeV and 1 TeV.

At low energies our composite model and the standard elektroweak gauge theory are essentially identical. But at high energies large deviations are expected. In particular a new isoscalar neutral weak boson X should exist, which is analogous to the ω-meson:

$$X = \frac{1}{\sqrt{2}}\left(\overline{\alpha}\alpha + \overline{\beta}\beta\right) . \tag{8}$$

The present limit on the mass of this isoscalar weak boson is about 0.4 TeV. It will decay mainly into weak bosons, e.g. into two charged weak bosons or into two Z-bosons. Decays of the X-boson into two fermions are expected to be strongly suppressed. We expect that the mass of the X-boson is less than 1 TeV, thus it should be observed soon at the LHC.

Above the energy of 1 GeV exist many excited states of the baryons and mesons. We expect similar effects in the electroweak sector. The QHD mass scale is about thousand times larger than the QCD mass scale. Thus at energies above 0.1 TeV there should in particular exist excited weak bosons, which will decay mainly into several weak bosons.

The weak bosons consist of pairs of haplons, which are in an s-wave. The spins of the two haplons are aligned, as the spins of the quarks in a ρ-meson. The first excited states are those, in which the two haplons are in a p-wave. We describe the quantum numbers of these states by $I(J)$. The $SU(2)$-representation is denoted by I - the symbol J describes the total angular momentum. There are three $SU(2)$ singlets, which we denote by S(0), S(1) and S(2):

$$\begin{aligned} S(0) &= [0(0)], \\ S(1) &= [0(1)], \\ S(2) &= [0(2)], \end{aligned} \tag{9}$$

and three triplet states:

$$\begin{aligned} T(0) &= [1(0)], \\ T(1) &= [1(1)], \\ T(2) &= [1(2)]. \end{aligned} \tag{10}$$

These three states have the internal structure:

$$
\begin{aligned}
T^{+} &= \overline{\beta}\alpha\ , \\
T^{-} &= \overline{\alpha}\beta\ , \\
T^{3} &= \frac{1}{\sqrt{2}}\left(\overline{\alpha}\alpha - \overline{\beta}\beta\right)\ .
\end{aligned} \tag{11}
$$

We compare the spectrum of these excited bosons with the spectrum of the mesons in strong interaction physics, in which the quarks are in a p-wave. The lowest meson is the scalar σ-meson (mass $\sim 600-800$ MeV). This meson can easily mix with glue mesons, i.e. the σ-meson would be a superposition of a quark-antiquark-meson and a glue meson. This might explain the relatively low mass of this meson.

The σ-meson is the QCD analogue of the boson $S(0)$. The spin one meson with isospin zero is $h_1(1170)$, analogous to the excited boson S(1). The corresponding spin two meson is $f_2(1270)$. It is analogous to the excited boson S(2):

$$
\begin{aligned}
S(0) &\sim \sigma, \\
S(1) &\sim h_1, \\
S(2) &\sim f_2.
\end{aligned} \tag{12}
$$

We expect that the scalar state $S(0)$ with the quantum numbers $[0(0^+)]$ mixes with gluton bosons, analogous to the mixing of the σ-meson with glue mesons. Due to this mixing the mass of $S(0)$ is much lower than the masses of the other p-wave states.

Now we consider the isospin triplet mesons, the scalar meson $a_0(980)$, the vector meson $b_1(1235)$ and the spin two meson $a_2(1320)$. These mesons are analogous to the bosons T(0), T(1) and T(2) respectively:

$$
\begin{aligned}
T(0) &\sim a_0, \\
T(1) &\sim b_1, \\
T(2) &\sim a_2.
\end{aligned} \tag{13}
$$

What is the mass spectrum of the excited weak bosons, in particular of the S and T bosons? The boson $S(0)$ should have a relatively small mass, due to the mixing with the gluton bosons. We identify this state with the particle, which has been observed at CERN (ref. (1,2)):

$$
M(S(0)) = 126 \; GeV. \tag{14}
$$

In analogy to QCD we expect that the masses of the other p-wave states are in the range 0.3 - 0.5 TeV. The mixing of the bosons $S(1)$ and $S(2)$ with the gluton bosons is not as strong as the mixing for the $S(0)$. Thus we expect that the mass of the $S(1)$ - boson is just above 0.3 TeV, and the mass of the $S(2)$ - boson would be between 0.4 and 0.5 TeV.

The $SU(2)$ - triplet bosons T cannot mix with gluton bosons. For this reason their masses should be larger than the masses of the S - bosons. We compare the spectrum of these bosons with the spectrum of the corresonding mesons in QCD. Thus the mass of the $T(0)$ - boson should be about 0.3 TeV, the mass of the $T(1)$ - boson just above 0.4 TeV, and the mass of the $T(0)$ - boson should be in the range 0.5 - 0.6 TeV.

The $S(0)$ - boson will decay mainly into two charged weak bosons or into two Z-bosons (one of them virtual respectively), into a photon and a Z-boson and into two photons. The Z-boson is the boson W^3, mixed with the photon. The mixing angle is the weak angle, measured to about 28.7 degrees. Using this angle, we can calculate the branching ratios BR for the various decays - the branching ratio for the decay into charged weak bosons is denoted by B.

First we assume that the $S(0)$ - boson has a mass, which is more than twice the mass of the W - boson, in order to avoid phase space calculations. We find for the branching ratios:

$$S(0) \Longrightarrow (W^+ + W^-) \qquad \text{BR} = B,$$

$$S(0) \Longrightarrow (W^- + W^+) \qquad \text{BR} = B,$$

$$S(0) \Longrightarrow (Z + Z) \qquad \text{BR} \simeq 0.59\ B,$$

$$S(0) \Longrightarrow (Z + \gamma) \qquad \text{BR} \simeq 0.18\ B,$$

$$S(0) \Longrightarrow (\gamma + \gamma) \qquad \text{Br} \simeq 0.05\ B.$$

Now we consider the real value of the mass of the $S(0)$ - boson (126 GeV), which implies that one of the weak bosons can only be produced as a virtual particle, decaying into a fermion pair (a virtual weak boson is indicated by "W" or "Z"). Taking into account phase space corrections, one finds:

$$S(0) \Longrightarrow ("W^{+}" + W^-) \qquad \text{BR} = B,$$

$$S(0) \Longrightarrow ("W^{-}" + W^+) \qquad \text{BR} = B,$$

$$S(0) \Longrightarrow ("Z" + Z) \qquad \text{BR} \approx 0.35\ B,$$

$$S(0) \Longrightarrow (Z + \gamma) \qquad \text{BR} \approx 0.10\ B,$$

$$S(0) \Longrightarrow (\gamma + \gamma) \qquad \text{Br} \approx 0.06\ B.$$

We emphasize that the branching ratios for the decays into two Z -bosons, into two photons and into a photon and a Z - boson have rather large errors, due to the uncertainties in the phase space calculations.

We would not expect that the decay rates for the decays of $S(0)$ into letons and quarks are given by the mass of the fermion, as they are for the Higgs boson. The branching ratio for the decay into an electron pair, into a muon pair, into a tau pair or into a neutrino pair would be similar:

$$\text{BR } (S(0) \Longrightarrow (e^+ + e^-)\ ) \approx \text{BR } (S(0) \Longrightarrow (\mu^+ + \mu^-)\ )$$

$$\approx \text{BR } (S(0) \Longrightarrow (\tau^+ + \tau^-)\ ) \approx \text{BR } (S(0) \Longrightarrow (\nu + \bar{\nu})\ ),$$

(ν stands for the electron, muon and tau neutrino).

Due to the color of the quarks the branching ratio for the decays into quarks are three times larger:

$$\text{BR } (S(0) \Longrightarrow (u + \bar{u})\ ) \approx 3 \text{ BR } (S(0) \Longrightarrow (e^+ + e^-)\ )$$

$$\text{BR } (S(0) \Longrightarrow (d + \bar{d})\ ) \approx 3 \text{ BR } (S(0) \Longrightarrow (e^+ + e^-)\ )$$

$$\text{BR } (S(0) \Longrightarrow (u + \bar{u})\ ) \approx \text{BR } (S(0) \Longrightarrow (c + \bar{c})\ )$$

$$\text{BR } (S(0) \Longrightarrow (d + \bar{d})\ ) \approx \text{BR } (S(0) \Longrightarrow (s + \bar{s})\ )$$

$$\approx \text{BR } (S(0) \Longrightarrow (b + \bar{b})\ ).$$

Presumably the $S(0)$ decays mainly into two weak bosons. But decays of the $S(0)$ into muon pairs could be observed in the near future at the LHC.

The bosons $S(1)$ and $S(2)$ have a much higher mass as the $S(0)$ - boson. They will decay mainly into three or four weak bosons, e.g.:

$$S(1) \Longrightarrow (W^+ + W^- + Z),$$
$$S(1) \Longrightarrow (W^+ + W^- + W^+ + W^-),$$
$$S(1) \Longrightarrow (W^+ + W^- + \gamma).$$

Decays of $S(1)$ or $S(2)$ into two weak bosons or two photons would be suppressed.

The $SU(2)$ - triplet bosons $T(0)$, $T(1)$ and $T(2)$ will decay mainly into four or five weak bosons or photons. Decays into two weak bosons, a weak boson and a photon or two photons are strongly suppressed.

The boson T^+ or T^- can in principle decay into a charged weak boson and a Z-boson or a photon, however the branching ratio for this decay will be very small. The main decay modes of the T^+ would be:

$$T^+ \Longrightarrow (W^+ + Z + Z),$$

$$T^+ \Longrightarrow (W^+ + Z + \gamma),$$

$$T^+ \Longrightarrow (W^+ + \gamma + \gamma).$$

The properties of the new boson with a mass of 126 GeV, which has been discovered at the LHC , should be investigated in detail. If the model, discussed here, is correct and the new boson is the state $S(0)$, the other excited bosons $S(1)$, $S(2)$ and $T(0)$ should be discovered in the near future at the LHC.

References

[1] Atlas collaboration:
arXiv:1202.1414,arXiv:1202.1408, arXiv:1202.1415.

[2] CMS collaboration:
arXiv:1202.1488, arXiv:1202.1489, arXiv:1202.1487.

[3] H. Fritzsch and G. Mandelbaum, Phys. Lett. B102 (1981) 319; Phys. Lett. B 109 (1982) 224.

[4] R. Barbieri, R. Mohapatra and A. Masiero, Phys. Lett. B 105 (1981) 369.

[5] H. Fritzsch. D. Schildknecht and R. Kogerler, Phys. Lett. B 114 (1982) 157.

[6] L. F. Abbott and E. Farhi, Phys. Lett. B 101, 69 (1981).

[7] T. Kugo, S. Uehara and T. Yanagida, Phys. Lett. B 147, 321 (1984).

[8] S. Uehara and T. Yanagida, Phys. Lett. B 165, 94 (1985).

[9] H. Fritzsch, arXiv:1201.2512

[10] H. Fritzsch, Mod. Phys. Lett. A26, 2305 (2011)

CHAIRMAN: H. FRITZSCH

Scientific Secretaries: K. Doroud, J. Zennamo

DISCUSSION

- *Leontsinis:*

Have you used the latest LHC measurements to constrain the masses of your particles?

- *Fritzsch:*

I predict besides the S(0) boson eleven new bosons with masses below 1 TeV. Thus far there are no constraints from the LHC experiments. The S(1) will decay into complicated final states (4 jets, 4 Ws, etc.) and it will be difficult to discover it. Probably the best way would be to discover rare decay modes, e.g. the decay into ZZ. At present there are no limits on the S(1) mass, since the mass should be above 200 GeV.

- *Nikitin:*

If the weak bosons have internal structure, then there are additional energy levels corresponding to the internal degrees of freedom. What can you say about this excitation spectrum?

- *Fritzsch:*

As I have stated in my lecture, there should be eleven new bosons with masses below 1 TeV: S(1), S(2), T(0), T(1) and T(2). Each T-state describes three bosons, with electric charges 0, (+1) and (-1).

- *Nikitin:*

What was the mass for the S(0) in your theory, before the LHC found that the boson with a mass of 125 GeV? How do your predictions compare to the Weinberg-Salam theory?

- *Fritzsch:*

I considered this theory only after the discovery of the new boson. The mass of the S(0) should be above the mass of the Z-boson, but I cannot predict the mass. It is strange the mass of the S(0) is equal to the value of the weak decay constant – probably this is an accident.

- *Mojaza:*

I am interested in discussing the generation of mass in the quark and lepton sectors of this theory. You consider them composite by introducing scalar degrees-of-freedom which are also charged under these QHD interactions. Scalar degrees of freedom are unstable unless there is some symmetry protecting it. Can you comment on this?

- *Fritzsch:*

The electrons and quarks are composed of a mass-less spin ½ haplon and a mass-less spin 0 haplon. The masses of the leptons and quarks are generated by the QHD interaction, as the proton mass is generated by the QCD interaction.

- *Mojaza:*

Can you explain the number of generations?

- *Fritzsch:*

We do not understand why there are three dimensions of space, why there are three colors in QCD, and why there are three generations. It seems that nature prefers the number "three". In my theory the number of generations might be given by the QHD interaction.

- *Lukács:*

Does your theory have a mechanism to avoid pseudoscalar haplon bound states? Can you make them very heavy?

- *Fritzsch:*

QHD is different from QCD, since it is a chiral gauge theory. Such theories are poorly understood. I come to the conclusion that in a chiral gauge theory the vector meson is like the pion in QCD, it should be very light. There are no "pions" in the s-wave configuration, since there are only left-handed haplons.

- *Wymant:*

Is it the case that the electroweak "rho" parameter is an input here, rather than a prediction as in the SM?

- *Fritzsch:*

No, my theory is at low energy like the SM, thus the "rho" parameter is predicted as in the SM.

- *Wymant:*

Is your model strongly constrained by electroweak precision data?

- *Fritzsch:*

Since my model is at low energies identical to the SM, there are no problems.

- *Torres:*

Can we consider the heavy mass of the W^0 boson as the reason for the axial anomaly in QHD?

- *Fritzsch:*

The Axial anomaly is related to QCD. The reason is that the eta prime is heavier than it is supposed to be. This is clear since the divergence of the Axial single count in QCD has a term proportion to qq. However for W Bosons or other particles like rho mesons is not affected by axial anomaly. In addition they are not heavy; the w bosons in the QHD theory if that is right, they are actually quite light. The basic scale of QHD is of the order of 1 TeV, may be slightly less but the W boson is 80 keV, like the pion in QCD. They are light particle and the reason might be that the Gauge theory. We still do not understand fully the chiral Gauge theory. At CERN, Still the investigation in detail is on going to understand what happens in such a theory. In any case the W boson is fairly light.

- *Bondamenko:*

Is the photon composite in your theory, because of the electroweak theory?

- *Fritzsch:*

No, the photons are really fundamental particles like the gluons, are in my theory. The dynamic are mixing in my models. In the electroweak theory the photons are not fundamental and it is just the bound state.

- *Bern:*

How do you get the Quarks (S(0)) to couple to Fermions proportional to their masses?

- *Fritzsch:*

No, it is not proportional to the mass in my theory. If the LHC finds this then my theory is ruled out. The LHC should be able to find the decay of this particle in to mu+ mu- and perhaps e+e, because it is smaller than the mass. And this is the same case for the quarks, however the quarks decay is more difficult to see.

- *Minkowski:*

I would like to comment on the analogy with the glueballs and also the F0 states. The Sigma object is hard to assign a mass because it is a wide object. Therefore the theory of uds spectroscopy especially the scalar and the glueballs had to be separated from the experiments, since at the present moment they are not understood well enough both theoretically and experimentally, in a straight forward way. They are there but there is no self-explanatory experimental data.
So this is a negative quality of the current state.

- *Close:*

So this is the Hadron analogy of string theory.

- *Fritzsch:*

There is no string theory to begin with!

Gauge Forces: From QCD to Quantum Gravity

L. N. Lipatov*
St. Petersburg State University,
Petersburg Nuclear Physics Institute
Gatchina, Orlova Roshcha, 188300,St.Petersburg, Russia

Abstract

The BFKL and BKP equations for gluon composite states in leading and next-to-leading approximations are considered in QCD and in N=4 SUSY. In the leading order they are integrable at the large N_c limit. The effective action for high energy interactions in QCD gives a possibility to calculate next-to-leading corrections to the gluon trajectory and its couplings. In the $N = 4$ SUSY the intercept of the BFKL pomeron is constructed in the strong coupling expansion with the use of the AdS/CFT correspondence from the known results for the anomalous dimensions of the twist-2 operators. We discuss the effective action approach for the gravity and supergravity. In these models the high energy graviton-graviton scattering amplitude is calculated in the double-logarithmic approximation. The eikonal phase for the amplitude is constructed with the same accuracy. The analytic properties of the production amplitudes in the multi-color $N = 4$ SYM are also discussed.

1 High energy scattering and the t-channel partial waves

The essential contribution to the elastic cross-section at large energies $\sqrt{s}$ takes place at the fixed momentum transfer $|q|$

$$s = 4E^2 >> -t = \vec{q}^2\,. \tag{1}$$

In this region it is convenient to use the t-channel partial wave representation for the scattering amplitude

$$A^p(s,t) = s\int_{a-i\infty}^{a+i\infty}\frac{d\omega}{2\pi i}\left((-s)^\omega - ps^\omega\right) f^p_\omega(t)\,, \tag{2}$$

where $p = \pm 1$ is its signature. If the leading singularity of the partial wave $f^p_\omega(t)$ is the pole

$$f^p_\omega(t) = \frac{\gamma^2(t)}{\omega - \omega_p(t)}\,,\; \omega_p(t) \approx \Delta - \alpha'\vec{q}^2\,, \tag{3}$$

the amplitude has the power-like behavior [1]

$$A^p_{Regge}(s,t) = \xi_p(t)\, s^{1+\omega_p(t)}\,\gamma^2(t)\,,\; \xi_p(t) = e^{-i\pi\omega_p(t)} - p\,. \tag{4}$$

The poles generate the cut singularities of $f^p_\omega(t)$ in the ω-plane [2]. For example, in the simple case of the two reggeon exchange the Mandelstam cut contribution can be written as follows

$$A^p_{Mand}(s,t) = \xi_p\, s\int\frac{d^2k}{(2\pi)^2}\Phi^2(k,q-k)s^{\omega_{p_1}(-k^2)}s^{\omega_{p_2}(-(q-k)^2)}\,,\; p = p_1p_2\,. \tag{5}$$

*e-mail: lipatov@thd.pnpi.spb.ru

The cuts in the ω-plane appear as a result of an analytic continuation of the multi-particle t-channel unitarity relation [3]

$$\Im_t f_\omega(t) \sim \sum_n \int d\Omega_n |f_\omega^{(n)}|^2 \,. \tag{6}$$

To describe an approximately constant behavior of the total cross-sections σ a special j-plane singularity - Pomeron with vacuum quantum numbers is introduced

$$\sigma = \frac{1}{s}\,\Im A(s.0)\,, \;\; A(s,t) \approx is\, s^{\Delta - \alpha'_P q^2}\,, \tag{7}$$

where the intercept Δ and slope α'_P of its trajectory are small according to the experimental data. All Pomeron interactions are taken into account in the Gribov effective field theory [4].

In the leading logarithmic approximation (LLA)

$$\alpha_s \ln s \sim 1\,, \;\; \alpha_s = \frac{g^2}{4\pi} \ll 1 \tag{8}$$

the Born scattering amplitude in QCD is multiplied by the Regge factor

$$M_{AB}^{A'B'}(s,t) = M_{AB}^{A'B'}(s,t)|_{Born}\, s^{\omega(t)}\,. \tag{9}$$

The gluon trajectory in LLA is given below

$$\omega(-|q|^2) = -\frac{\alpha_s N_c}{4\pi^2} \int d^2k \frac{|q|^2}{|k|^2 |q-k|^2} \approx -\frac{\alpha_s N_c}{2\pi} \ln \frac{|q^2|}{\lambda^2}\,. \tag{10}$$

In the inelastic processes the most essential contribution to the total cross-section appears from the kinematics corresponding to the large pair energies $\sqrt{s_k}$ of produced gluons with fixed momentum transfers q_i and transverse components $k_{r\perp}$ of the particle momenta. The corresponding amplitude has the factorized form [5]

$$M_{2\to 2+n} \sim \frac{s_1^{\omega_1}}{|q_1|^2}\, gT_{c_2c_1}^{d_1} C(q_2,q_1)\, \frac{s_2^{\omega_2}}{|q_2|^2} ... gT_{c_{n+1}c_n}^{d_n}\, C(q_{n+1},q_n)\, \frac{s_{n+1}^{\omega_{n+1}}}{|q_{n+1}|^2}\,, \tag{11}$$

where ω_r are the gluon trajectories

$$\omega_r = -\frac{\alpha_s N_c}{2\pi}\left(\ln \frac{|q_r^2|}{\mu^2} - \frac{1}{\epsilon}\right) \tag{12}$$

and $C(q_i, q_{i-1})$ are the vertices for the produced gluon with a definite helicity

$$C(q_2,q_1) = \frac{q_2\, q_1^*}{q_2^* - q_1^*}\,. \tag{13}$$

The knowledge of quantities $M_{2\to 2+n}$ gives a possibility to construct the total cross-section and the scattering amplitude with color singlet quantum numbers in the crossing channel. This amplitude can be expressed in terms of the Pomeron wave function satisfying the BFKL equation [5]

$$E\,\Psi(\vec{\rho}_1,\vec{\rho}_2) = H_{12}\,\Psi(\vec{\rho}_1,\vec{\rho}_2)\,,\;\; \sigma_t \sim s^\Delta\,,\;\; \Delta = -\frac{\alpha_s N_c}{2\pi} E_0\,. \tag{14}$$

The operator H_{12} is the BFKL Hamiltonian and Δ is the Pomeron intercept. In the impact parameter representation the hamiltonian can be written as follows [6]

$$H_{12} = \frac{1}{p_1 p_2^*}(\ln|\rho_{12}|^2) p_1 p_2^* + \frac{1}{p_1^* p_2}(\ln|\rho_{12}|^2) p_1^* p_2 + \ln|p_1p_2|^2 - 4\psi(1)\,, \tag{15}$$

where we introduced the complex notations

$$\rho_{12} = \rho_1 - \rho_2 \,, \;\; \rho_r = x_r + iy_r \,. \tag{16}$$

The Schrödinger equation is invariant under the Möbius transformations [7]

$$\rho_k \to \frac{a\rho_k + b}{c\rho_k + d} \tag{17}$$

and the eigenfunctions and eigenvalues of H_{12} are classified by the conformal weights

$$m = \gamma + n/2 \,, \;\; \widetilde{m} = \gamma - n/2 \,, \tag{18}$$

depending on the anomalous dimension $\gamma = 1/2 + i\nu$ of twist-2 operators and on the integer conformal spin n. The energies have the property of the holomorphic separability and are expressed in terms of the logarithmic derivative $\psi(x)$ of the gamma function

$$E = 2\chi(n, \gamma) = \psi(m) + \psi(1-m) + \psi(\widetilde{m}) + \psi(1-\widetilde{m}) - 4\psi(1) \,. \tag{19}$$

and, as a result. the Pomeron intercept can be calculated in the form

$$\Delta = 4\alpha N_c \ln 2 \,/\pi \,. \tag{20}$$

These properties lead to the integrability of the BFKL dynamics in the multi-color limit [8, 9].

2 Integrability of the BKP equation

The hamiltonian in the Bartels-Kwiecinski-Praszalowicz equation for the wave function of several gluons

$$E\,\Psi = H\,\Psi \,, \;\; H = \sum_{k<l} \frac{T_k^a T_l^a}{-N_c} H_{kl} \tag{21}$$

is expressed in terms of the BFKL hamiltonian H_{kl} and the color group generators T_l^a acting on the color indices of the gluon l. Due to the holomorphic separability of H at large N_c its eigenfunction has the property of the holomorphic factorization

$$\Psi(\vec{\rho}_1, ...\vec{\rho}_n) = \sum_{r,s} a_{r,s} \Psi_r(\rho_1 ... \rho_n) \Psi_s(\rho_1^* ... \rho_n^*). \tag{22}$$

Here the coefficients $a_{r,s}$ are chosen from the condition that $\Psi(\vec{\rho}_1, ...\vec{\rho}_n)$ is a single valued function in the $\vec{\rho}$-space. The hamiltonian in the holomorphic space is invariant under the Möbius transformations. Moreover, we can introduce the monodromy matrix satisfying the Yang-Baxter equation [8]

$$t(u) = L_1(u) ... L_n(u) = \begin{pmatrix} A(u) & B(u) \\ C(u) & D(u) \end{pmatrix}, \tag{23}$$

where the L-operators are given below

$$L_k(u) = \begin{pmatrix} u + \rho_k \, p_k & p_k \\ -\rho_k^2 \, p_k & u - \rho_k \, p_k \end{pmatrix}. \tag{24}$$

The corresponding transfer matrix

$$T(u) = A(u) + D(u) \,, \;\; [T(u), h] = 0 \tag{25}$$

is the generating function for integrals of motion [8]. The holomorphic hamiltonian is invariant also under the duality transformations [9]

$$p_r \to \rho_{r,r+1} \to p_{r+1} \,. \tag{26}$$

3 Effective action for high energy interactions in QCD

Generally the BFKL approach is based on an effective field theory for the reggeized gluons similar to the Gribov Pomeron calculus. The gluon trajectory and various reggeon couplings in upper loops can be calculated from the effective lagrangian written for a cluster of quarks and gluons interacting with the reggeized gluons and having their rapidities y in some interval η

$$y = \frac{1}{2} \ln \frac{\epsilon_k + |k|}{\epsilon_k - |k|}\,, \; |y - y_0| < \eta\,, \; \eta << \ln s\,. \tag{27}$$

We introduce the anti-hermitian fields v_μ describing the usual gluons and the gauge invariant fields $A_\pm$ describing the production and annihilation of the reggeized gluons

$$v_\mu(x) = -iT^a v_\mu^a(x)\,, \; A_\pm(x) = -iT^a A_\pm^a(x)\,, \; \delta A_\pm(x) = 0\,. \tag{28}$$

The fields $A_\pm$ satisfy the kinematical constraints

$$\partial_- A_+ = \partial_+ A_- = 0\,, \tag{29}$$

corresponding to the fact, that the Sudakov variables of momenta of various clusters of particles are strongly ordered in their rapidities y_i in an accordance with the quasi-multi-Regge kinematics.

The effective action for particles inside a cluster with a given rapidity value has the form [10]

$$S = \int d^4x \left(L_{QCD} + Tr(V_+ \partial_\mu^2 A_- + V_- \partial_\mu^2 A_+) \right)\,, \tag{30}$$

where L_{QCD} is the usual QCD action and the effective currents $V_\pm$ are given below

$$V_+ = -\frac{1}{g} \partial_+ \, P \exp \left(-\frac{g}{2} \int_{-\infty}^{x^+} v_+(x') d(x')^+ \right) = v_+ - g v_+ \frac{1}{\partial_+} v_+ + \ldots . \tag{31}$$

The Feynman rules for this action are derived in ref. [11]. The effective action approach gives a possibility to construct various reggeon vertices needed to calculate next-to-leading order (NLO) corrections to the BFKL kernel in the color singlet [12, 13] and adjoint [14, 15] representations.

The NLO correction to the BFKL kernel in QCD was calculated in ref. [12]. In the case of the $N = 4$ extended super-symmetric gauge theory the two loop result has the property of the hermitian separability [13]. Moreover, in this model the eigenvalue contains only the special functions having the maximal transcendentality possible in the given order [17]. The maximal transcendentality property in $N = 4$ SUSY is valid in each order of perturbation theory also for the anomalous dimensions of the twist 2 operators [17], which allowed to calculate them up to the fifth order [18, 19, 20]. Its singular behavior at $j \to 1$, obtained from the BFKL equation in Ref. [17], is in an agreement with the direct calculation with the use of the integrability of the theory.

It is well known, that the $N = 4$ four-dimensional conformal field theory according to Maldacena is equivalent to the super-strings living on the anti-de-Sitter 10-dimensional space [21, 22, 23]. As a result, the Pomeron in this model is dual to the reggeized graviton in the anti-de-Sitter space. On can write the kernel of the BFKL equation for N=4 SUSY in a diffusion approximation in the form

$$j = 2 - \Delta - D\nu^2\,, \; \gamma = 1 + \frac{j-2}{2} + i\nu\,, \tag{32}$$

where the parameters Δ and D depend on the gauge coupling constant. According to the energy-momentum conservation we have $D = \Delta$.

It turns out, that due to the AdS/CFT correspondence [21, 22, 23] the diffusion approximation for the BFKL equation corresponds to the linear Regge trajectory

$$j = 2 + \frac{\alpha'}{2} t \, , \; t = E^2/R^2 \, , \; \alpha' = \frac{R^2}{2} \Delta \, . \tag{33}$$

of the graviton living in the 10-dimensional anti-de-Sitter space.

Because the behavior of the anomalous dimension at large coupling constants is known, one can calculate the Pomeron intercept in the strong coupling limit [18, 24, 25]

$$j = 2 - \frac{2}{\sqrt{\lambda}} - \frac{1}{\lambda} + \frac{1}{4\lambda^{-3/2}} + \frac{2+6\zeta_3}{\lambda^2} + \ldots \, , \; \lambda = g^2 N_c \, . \tag{34}$$

The slope of the anomalous dimension at $j = 2$ related to the Pomeron intercept was obtained in the perturbation theory [18] and calculated recently explicitly by Basso

$$\gamma'(2) = -\frac{\lambda}{24} + \frac{1}{2}\frac{\lambda^2}{24^2} - \frac{2}{5}\frac{\lambda^3}{24^2} + \frac{7}{20}\frac{\lambda^4}{24^4} - \frac{11}{35}\frac{\lambda^5}{24^5} + \ldots = -\frac{\sqrt{\lambda}}{4}\frac{I_3(\sqrt{\lambda})}{I_2(\sqrt{\lambda})} . \tag{35}$$

The duality between the BFKL Pomeron and reggeized graviton means, that the Pomeron calculus could be formulated in the framework of the approach based on an effective action for the reggeized graviton interactions (see Ref. [26]).

4 High energy action in gravity

One can introduce the fields $h_{\mu\nu}$ and $A_{\pm\pm}$ describing the usual graviton and reggeized graviton, respectively, with the properties

$$g_{\mu\nu} = \eta_{\mu\nu} + h_{\mu\nu} \, , \; \partial_+ A^{++} = \partial_- A^{--} = 0 \, . \tag{36}$$

In this case the effective action for the high energy gravity takes the form [26]

$$S = -\frac{1}{2\kappa} \int d^4x \left(\sqrt{-g}\, R + \Delta L \right) \, , \tag{37}$$

where

$$\Delta L = \frac{1}{2} \left(\partial_+ j^- \, \partial_\mu^2 A^{++} + \partial_- j^+ \, \partial_\mu^2 A^{--} \right) \, . \tag{38}$$

The dependence of the currents $j^\pm$ from the metric tensor $g_{\mu\nu}$ is obtained from the Hamilton-Jacobi equation

$$g^{\mu\nu} \partial_\mu \omega^\pm \partial_\nu \omega^\pm = 0 \, , \; j^\pm = 2x^\pm - \omega^\pm \, . \tag{39}$$

It is possible to find the perturbative solution for these currents

$$\partial_\pm j^\mp = h_{\pm\pm} - \left(h_{\rho\pm} - \frac{1}{2} \frac{\partial_\rho}{\partial_\pm} h_{\pm\pm} \right)^2 + \ldots \tag{40}$$

They can be obtained also from the extremum condition for the Hamilton-Jacobi functional

$$\omega = \int_{-\infty}^{\tau} d\tau \, (p_\mu \frac{\partial}{\partial \tau} x^\mu - \frac{e(\tau)}{2} \, g^{\mu\nu} p_\mu p_\nu). \tag{41}$$

The effective action gives a possibility to calculate the graviton production amplitudes in the multi-Regge kinematics

$$A_{2\to n} = -s^2 \Gamma^{\mu'\nu'}_{\mu\nu} \frac{s_1^{\omega(q_1^2)}}{q_1^2} \Gamma_{\rho_1\sigma_1} \frac{s_2^{\omega(q_2^2)}}{q_2^2} \ldots \Gamma^{\rho'\sigma'}_{\rho\sigma} \, . \tag{42}$$

The scattering vertex for gravitons

$$\Gamma_{\mu\nu}^{\mu'\nu'} = \frac{\kappa}{4} \left(\Gamma_{\mu\mu'}\Gamma_{\nu\nu'} + \Gamma_{\mu\nu'}\Gamma_{\nu\mu'} \right) \tag{43}$$

is related to the similar vertex for gluons in QCD

$$\Gamma_{\mu\mu'} = -\delta_{\mu\mu'} + \frac{p_{\mu'}^A p_\mu^B + p_\mu^{A'} p_{\mu'}^B}{p^A p^B} + \frac{q^2 p_\mu^B p_{\mu'}^B}{2(p^A p^B)^2} \,. \tag{44}$$

The effective vertex for the graviton production

$$\Gamma_{\rho\sigma} = \frac{\kappa}{4} \left(C_\rho C_\sigma - N_\rho N_\sigma \right) \tag{45}$$

is expressed in terms of the similar vertex C for the gluon production and the photon emission factor in QED

$$N = \sqrt{q_1^2 q_2^2} \left(\frac{p^A}{kp^A} - \frac{p^B}{kp^B} \right) \tag{46}$$

5 Graviton trajectory and double-logarithms

From the effective action one can calculate also the graviton Regge trajectory in one loop

$$\omega(q^2) = \frac{\alpha}{\pi} \int \frac{q^2 \, d^2k}{k^2 (q-k)^2} f(k,q) \,, \tag{47}$$

where

$$f(k,q) = (k, q-k)^2 \left(\frac{1}{k^2} + \frac{1}{(q-k)^2} \right) - q^2 + \frac{N}{2}(k, q-k) \,, \; \alpha = \frac{\kappa^2}{8\pi^2} \,. \tag{48}$$

Here we took into account the contribution of gravitinos in the N-extended supergravity with the action

$$S_{3/2} = -\frac{1}{2} \epsilon^{\mu\nu\rho\sigma} \int d^4x \sum_{r=1}^{N} \bar{\psi}_\mu^r \gamma_5 \gamma_\nu \partial_\rho \psi_\sigma^r . \tag{49}$$

The trajectory contains the infrared and ultraviolet divergencies

$$\omega(q^2) = -\alpha |q|^2 \left(\ln \frac{|q|^2}{\lambda^2} + \frac{N-4}{2} \ln \frac{|\Lambda|^2}{|q|^2} \right) \,. \tag{50}$$

One should cut the integral at large values of k^2 by the energy invariant s because the gravity is renormalizable in one loop. It leads to the appearance of the double-logarithmic terms $\sim (\alpha \ln^2 s)^n$ for the scattering amplitude [27]. To sum these terms it is convenient to extract from the amplitude the factor corresponding to the infrared divergencies

$$A(s,t) = A_{Born} \, s^{-\alpha |q|^2 \ln \frac{|q|^2}{\lambda^2}} \, \Phi(\xi) \,, \tag{51}$$

where the variable ξ is given below

$$\xi = \alpha \, |q|^2 \ln^2 \frac{s}{|q|^2} \tag{52}$$

and we perform the Mellin transformation for $\Phi(\xi)$

$$\Phi(\xi) = \int_{a-i\infty}^{a+i\infty} \frac{d\,\omega}{2\pi i \, \omega} \left(\frac{s}{|q|^2} \right)^\omega f_\omega \,, \tag{53}$$

The partial wave satisfies the infrared evolution equation [27]

$$f_\omega = 1 + \alpha|q|^2 \left(\frac{d}{d\omega}\frac{f_\omega}{\omega} - \frac{N-6}{2}\frac{f_\omega^2}{\omega^2} \right) . \tag{54}$$

Its solution can be obtained in terms of the parabolic cylinder function [27]

$$\frac{f_\omega}{\omega} = \frac{2}{6-N}\frac{1}{\sqrt{b}}\frac{d}{dx}\ln\left(e^{\frac{x^2}{4}} D_{\frac{6-N}{2}}(x)\right) , \tag{55}$$

where $x = \frac{\omega}{\sqrt{b}}$. We have the following perturbative expansion for Φ

$$\Phi(\xi) = 1 - \frac{N-4}{2}\frac{\xi}{2} + \frac{(N-4)(N-3)}{2}\frac{\xi^2}{4!} - \frac{N-4}{8}(5N^2 - 26N + 36)\frac{\xi^3}{6!} + ... \tag{56}$$

It turns out, that the supersymmetry improves the high energy behavior of scattering amplitudes in the double-logarithmic approximation. Namely, for $N > 4$ the function $\Phi(\xi)$ falls at large ξ. It decreases especially rapidly at $N = 8$, which could be related to a possible renormalizability of this super-symmetric model.

One can attempt to combine the double logarithmic summation with the eikonal approximation

$$A_{DL}(s,t) = -2is\, s^{-\alpha|q|^2 \ln\frac{|q|^2}{\mu^2}} \int d^2\rho\, e^{i\vec{q}\vec{\rho}} \left(e^{i\delta_{DL}(\vec{\rho},\ln s)} - 1\right) , \tag{57}$$

where the eikonal phase is given below

$$\delta_{DL}(\vec{\rho}, \ln s) == \frac{s}{2}\frac{\kappa^2}{(2\pi)^2}\int \frac{d^2q}{|q|^2} e^{-i\vec{q}\vec{\rho}}\,\Phi(\xi) . \tag{58}$$

Note, that the amplitude A_{DL} in the two loop approximation

$$A_4^{N=8} = \frac{\kappa^2 s^2}{|q|^2}(-i\pi s)\alpha^2|q|^2 \frac{\ln^3 \frac{s}{|q|^2}}{3} \tag{59}$$

coincides with that obtained by other authors independently.

6 Production amplitudes in $N = 4$ SYM

Several years ago Z.Bern, L.Dixon and V.Smirnov suggested an explicit expression at $N_c \to \infty$ for multi-gluon production amplitude in $N = 4$ SUSY with the maximal helicity violation [28]. It turned out, that this amplitude is not correct already in two loops [29]. The reason is that the BDS amplitude does not satisfy the Steinmann requirement of absence of simultaneous singularities in the overlapping channels [30]

$$\Delta_{s_r}\Delta_{s_{r+1}} M_{2\to 2+n} = 0 . \tag{60}$$

Moreover, it does not contain the Mandelstam cut contributions [31]. These contributions appear in the planar amplitudes starting from 6 external legs [32]. The correct amplitude differs from the BDS expression by a factor - so-called remainder function R. This function should be conformal invariant in the momentum space [33]. In LLA R can be calculated in the multi-Regge kinematics from the BFKL and BKP equations because its hamiltonian coincides with the local hamiltonian of the open integrable spin chain [34]. We consider below the analytic properties of the planar gluon production amplitudes in $N = 4$ SUSY at the multi-Regge kinematics. For 6-point amplitude such study was done in Ref. [35]). In an accordance with the Steinmann constraint for one gluon production the amplitude has the form (cf. [36])

$$\frac{M_{2\to 3}}{\Gamma(t_1)\Gamma(t_2)|\Gamma_a|} = c_R^a(-\tilde{s})^{j_2}(-s_1)^{j_1-j_2} + c_L^a(-\tilde{s})^{j_1}(-s_2)^{j_2-j_1} , \ \tilde{s} = s|k_\perp^a|^2, \tag{61}$$

where $\Gamma(t_r)$ are reggeon residues, c_R^a and c_L^a are two real vertices for the gluon a with the transverse momentum $k_\perp^a$ produced from the reggeized gluon [29]

$$c_R^a = \frac{\sin\pi\omega_{1a}}{\sin\pi\omega_{12}}\,,\ c_L^a = \frac{\sin\pi\omega_{2a}}{\sin\pi\omega_{21}} \tag{62}$$

and ω_a is the phase of the complex production vertex [29]

$$\Gamma_a(\ln|k_\perp^a|^2 - i\pi) = |\Gamma_a|\,e^{i\pi\omega_a}\,. \tag{63}$$

In the case of production of two particles a and b the expression for the multi-Regge amplitude consists of five terms

$$\frac{M_{2\to4}}{\Gamma(t_1)\Gamma(t_2)|\Gamma_a||\Gamma_b|} = c_R^a c_L^b(-\widetilde{s})^{j_2}(-s_1)^{j_1-j_2}(-s_3)^{j_3-j_2}$$
$$+c_R^a c_R^b(-\widetilde{s})^{j_3}(-\widetilde{s}_{012})^{j_2-j_3}(-s_1)^{j_1-j_2} + c_L^a c_L^b(-\widetilde{s})^{j_1}(-\widetilde{s}_{123})^{j_2-j_1}(-s_3)^{j_3-j_2}$$
$$+c_L^a c_R^b\left(k(-\widetilde{s})^{j_3}(-\widetilde{s}_{012})^{j_1-j_3}(-s_2)^{j_2-j_1} + l(-\widetilde{s})^{j_1}(-\widetilde{s}_{123})^{j_3-j_1}(-s_2)^{j_2-j_3}\right), \tag{64}$$

where the invariants $\widetilde{s}$ contain the corresponding factors depending on the transverse momenta. The quantities k and l

$$k = \frac{\sin\pi\omega_1\,\sin\pi\omega_{23}}{\sin\pi\omega_2\,\sin\pi\omega_{13}}\,,\ l = \frac{\sin\pi\omega_3\,\sin\pi\omega_{21}}{\sin\pi\omega_2\,\sin\pi\omega_{31}}\,,\ k+l=1 \tag{65}$$

are fixed from the factorization relations for the Regge poles [36]. To formulate these factorization relations it is needed to introduce the signatures $\tau_r = \pm1$ for reggeons in t_r-channels. Then for the planar case the generating function for amplitudes $2\to n+1$ in all physical regions related by twists in the corresponding t-channel lines can be written as follows

$$A_{2\to n+1}^{\tau_1\ldots\tau_n} = A + \sum_{r=1}^{n}\tau_r\,A_r + \sum_{r_1<r_2}\tau_{r_1}\tau_{r_2}\,A_{r_1r_2} + \ldots + \tau_1\tau_2\ldots\tau_n\,A_{1\ldots n}\,. \tag{66}$$

The factorized Weis expression for this expression has the following complex form [36]

$$\frac{A_{2\to n+1}^{\tau_1\ldots\tau_n}}{s\Gamma(t_1)\Gamma(t_2)|\Gamma_a||\Gamma_b|\ldots} = |s_1|^{\omega_1}\xi_1 V^{1,2}|s_2|^{\omega_2}\xi_2 V^{2,3}\ldots V^{n-1,n}|s_n|^{\omega_n}\xi_n \tag{67}$$

where

$$\xi_r = e^{-i\pi\omega_r} - \tau_r\,,\ \xi_{12} = e^{-i\pi\omega_{12}} + \tau_1\tau_2\,,\ V^{1,2} = \frac{\xi_{12}}{\xi_1}c_R^a + \frac{\xi_{21}}{\xi_2}c_L^a\,. \tag{68}$$

For the case of the transition $2\to4$ we obtain the amplitude in two physical regions, where the BDS expression is not correct,

$$\frac{A_{2\to4}^{\tau_1\ldots\tau_3}}{s|s_1|^{\omega_1}|s_2|^{\omega_2}|s_3|^{\omega_3}\Gamma(t_1)\Gamma(t_2)|\Gamma_a||\Gamma_b|} = \ldots + \left(\tau_1\tau_3\,e^{-i\pi\omega_2} + \tau_1\tau_2\tau_3\right)A + \ldots \tag{69}$$

Here

$$A = \frac{2\cos\pi\omega_2\,\sin\pi\omega_a\,\sin\pi\omega_b}{i\sin\pi\omega_2} + i\sin\pi(\omega_a+\omega_b) + \cos\pi\omega_{ab} \tag{70}$$

contains the non-physical pole $1/\sin\pi\omega_2$. But just in these physical regions there is a contribution of the Mandelstam cuts [29]. For example at $s, s_2 > 0, s_1, s_3 < 0$ we obtain the following complex structure of the cut

$$A_{cut} = i\int\frac{d\omega_2'}{2\pi i}(-s_2)^{\omega_2'}f_{\omega_2'} \tag{71}$$

with a real function $f_{\omega_2'}$. Therefore we can redefine the expression for the Regge pole contribution by subtracting from its expression two first term and including them in the redefined cut contribution [35]. In such manner we have

$$A_{pole}^{2\to 4} = \cos\pi\omega_{ab}\,. \tag{72}$$

As a result, in the region $s, s_2 > 0$, $s_1, s_3 < 0$ we obtain the following dispersion representation for the remainder function [35, 37]

$$R\,e^{i\pi\delta} = cos\,\pi\omega_{ab} + i\,\frac{a}{2}\sum_{n=-\infty}^{\infty}(-1)^n e^{i\phi n}\int_{-\infty}^{\infty}\frac{|w|^{2i\nu}d\nu}{\nu^2+\frac{n^2}{4}}\,\Phi(\nu,n)\left(\frac{-1}{\sqrt{u_2u_3}}\right)^{\omega(\nu,n)}, \tag{73}$$

where the anharmonic ratios in the momentum space are

$$u_1 = \frac{ss_2}{s_{012}s_{123}}\,,\; u_2 = \frac{s_1t_3}{s_{012}t_2}\,,\; u_3 = \frac{s_3t_1}{s_{123}t_2}\,,\; |w|^2 = \frac{u_2}{u_3}\,,\; \cos\phi = \frac{1-u_1-u_2-u_3}{2\sqrt{u_2u_3}}\,, \tag{74}$$

the BDS phases

$$\delta = \frac{\gamma_K}{8}\,\ln\frac{|w|^2}{|1+w|^4}\,,\; \omega_{ab} = \frac{\gamma_K}{8}\,\ln|w|^2\,, \tag{75}$$

the product of two impact factors [38]

$$\Phi = 1 - a\left(\frac{E_{\nu n}^2}{2} + \frac{3}{8}n^2/(\nu^2+\frac{n^2}{4})^2 + \zeta(2)\right). \tag{76}$$

The eigenvalue of the BFKL kernel is

$$\omega(\nu,n) = -aE_{\nu,n} - a^2(\epsilon_{\nu n}^{FL} + 3\zeta(3)) \tag{77}$$

in leading [32]

$$E_{\nu n} = -\frac{|n|/2}{\nu^2+\frac{n^2}{4}} + 2\Re\psi(1+i\nu+\frac{|n|}{2}) - 2\psi(1)\,, \tag{78}$$

and next-to-leading [37] approximations

$$\epsilon_{\nu n}^{FL} = -\frac{\Re}{2}\left(\psi''(1+i\nu+\frac{|n|}{2}) - \frac{2i\nu\psi'(1+i\nu+\frac{|n|}{2})}{\nu^2+\frac{n^2}{4}}\right) - \zeta(2)\,E_{\nu n} - \frac{1}{4}\,\frac{|n|\left(\nu^2-\frac{n^2}{4}\right)}{\left(\nu^2+\frac{n^2}{4}\right)^3} \tag{79}$$

The analogous dispersion relations are obtained for the 7-point amplitude in all physical regions having the Mandelstam cut contributions [39].

Acknowledgments

I thank the Hamburg University for hospitality and my collaborators J. Bartels, V.S. Fadin, A. Kormilitzin, A. Prygarin and A. Sabio Vera for helpful discussions. This work was supported by the RFFI grant.

References

[1] V. N. Gribov, *Sov. Phys. JETP* **14** 478 (1962).

[2] S. Mandelstam, *Nuovo Cim.* **30**, 1148 (1963).

[3] V. N. Gribov, I. Ya. Pomeranchuk and K. A. Ter-Martirosyan, *Phys. Rev.* **B 139**, 184 (1965).

[4] V. N. Gribov, *Sov. Phys. JETP* **26**, 414 (1968).

[5] L. N. Lipatov, *Sov. J. Nucl. Phys.* **23** 338 (1976);
V. S. Fadin, E. A. Kuraev, L. N. Lipatov, *Phys. Lett.* **B 60** 50 (1975);
E. A. Kuraev, L. N. Lipatov, V. S. Fadin, Sov. Phys. JETP **44** 443 (1976).

[6] L. N. Lipatov, *Phys. Lett.* **B 309** 394 (1993).

[7] L. N. Lipatov, *Sov. Phys. JETP* **63** 904 (1986).

[8] L. N. Lipatov *High energy asymptotics of multi-colour QCD and exactly solvable lattice models*, hep-th/9311037, unpublished.

[9] L. N. Lipatov, *Nucl. Phys.* **B 548** 328 (1999).

[10] L. N. Lipatov, *Nucl. Phys.* B **B 452**, 369 (1995); *Phys. Rept.* **286**, 131 (1997).

[11] E. N. Antonov, L. N. Lipatov, E. A. Kuraev, I. O. Cherednikov, *Nucl. Phys.* **B 721**, 111 (2005).

[12] V. S. Fadin, L. N. Lipatov, *Phys. Lett.* **B 429** 127 (1998);
M. Ciafaloni and G. Camici, *Phys. Lett.* **B 430** 349 (1998).

[13] A. V. Kotikov, L. N. Lipatov, *Nucl. Phys.* **B 582** 19 (2000).

[14] V. S. Fadin, L. N. Lipatov, *Phys. Lett.* **B 706** 470 (2012).

[15] J. Bartels, V. S. Fadin, L. N. Lipatov, G. P. Vacca, arXiv: 1210.0797 [hep-ph].

[16] L. N. Lipatov, *J. Phys.* **A 42**:304020 (2009).

[17] A. V. Kotikov, L. N. Lipatov, *Nucl. Phys.* **B 661** 19 (2003).

[18] A. V. Kotikov, L. N. Lipatov, A. I. Onishchenko, V. N. Velizhanin, *Phys. Lett.* **B 595** 521 (2004); [Erratum-ibid. **B 632** 754 (2006)].

[19] A. V. Kotikov, L. N. Lipatov, A. Rej, M. Staudacher, V. N. Velizhanin, *J. Stat. Mech.* **0710** P10003 (2007).

[20] T. Lukowski, A. Rei, V. N. Velizhanin, *Nucl. Phys.* **. B831** 105 (2010).

[21] J. M. Maldacena, *Adv. Theor. Math. Phys.* **2** 231 (1998).

[22] S. S. Gubser, I. R. Klebanov, A. M. Polyakov, *Phys. Lett.* **B 428** 105 (1998).

[23] E. Witten, *Adv. Theor. Math. Phys.* **2** 253 (1998).

[24] R. C. Brower, J. Polchinsky, M. J. Strassler, C. I. Tan, JHEP **0712** 005 (2007).

[25] A. V. Kotikov, L.N. Lipatov, preprint, arXiv: 1301.0882 [hep-th].

[26] L. N. Lipatov, *Physics of Particles and Fields*, **44** 391 (2013), arXiv: 1105.31277 [hep-ph].

[27] J. Bartels, L. N. Lipatov, A. Sabio Vera, preprint arXiv: 1208.3423 [hep-th].

[28] Z. Bern, L. J. Dixon, V. A. Smirnov, Phys. Rev. D **72**, 085001 (2005).

[29] J. Bartels, L. N. Lipatov, A. Sabio Vera, Phys.Rev.D80:045002,2009,

[30] O. Steinmann, Helv. Physica Acta **33** 257, 349 (1960).

[31] S. Mandelstam, Nuovo Cim.**30**, 1148 (1963).

[32] J. Bartels, L. N. Lipatov, A. Sabio Vera, Eur.Phys.J.C65:587-605,2010,

[33] J. M. Drummond, J. Henn, V. A. Smirnov, E. Sokatchev, JHEP **0701** 064 (2007) ; J. M. Drummond, G. P. Korchemsky, E. Sokatchev, Nucl.Phys. **B795**, 385 (2008).

[34] L. N. Lipatov, J.Phys.A42:304020,2009,

[35] L. N. Lipatov, Theor.Math.Phys. **170**, 166 (2012), hep-th 1008.1015

[36] I. T. Drummond, P. V. Landshoff, W. J. Zakrzewski, Nucl. Phys. B **11**, 383 (1969); J. H. Weis, Phys. Rev. D **4**, 1777 (1971).

[37] V. S. Fadin, L. N. Lipatov, Physics Letters, **B706**, 470-476 (2012), hep-th 1111.0782.

[38] L. N. Lipatov and A. Prygarin, Phys. Rev. **D83**:125001 (2011), hep-th 1011.2673.

[39] J. Bartels, A. Kormilitzin, L. N. Lipatov, preprint DESY 13-209. hep-th 1311.2061

CHAIRMAN: L.N. LIPATOV

Scientific Secretaries: F. Popov , N. Sopenko

DISCUSSION

- *Moroz:*

Good Evening! You considered pomeron in non-relativistic approximation. Is this approximation really needed?

- *Lipatov:*

Do you mean a phenomenological theory of pomeron?

- *Moroz:*

Yes.

- *Lipatov:*

Indeed, it' is surprising that to describe scattering amplitudes at very high energies one introduces a non-relativistic Green function of pomeron. But this result is a simple consequence of the assumption that the pomeron trajectory is linear. If it will be non-linear then the dispersion law for the relation between the t-channel angular momentum and the pomeron transverse momentum will be different. If I remember correctly, in one of his papers V. Gribov considered a relativistic reggeon propagator for fermions. But for the case of the reggeon with a linear trajectory the propagator should be non-relativistic. It doesn't mean that the Lorentz invariance is violated. With the use of such propagator we describe the relativistic hadron-hadron scattering. I remind, that in the Gribov approach the t-channel angular momentum does not coincide with energy and therefore it is not necessary to have the Lorentz symmetry for the pomeron Green function.

- *Nohle:*

On slide thirteen there was calculated the divergence of graviton Regge trajecory and you stated that if there are 4 gravitinos there will be no divergence. Could you expound upon it, please?

- *Lipatov:*

Do you ask about the ultraviolet divergence of the Regge trajectory of graviton?

- *Nohle:*

Yes.

- *Lipatov:*

I remind that in the Yang-Mills theory the gluon effective vertices and its trajectory do not contain ultraviolet divergences. The Einstein gravity is a non-renormalizable theory and generally we expect such divergences. But it is known that the pure gravity in one loop is renormalizable. The ultraviolet divergence in the graviton trajectory appears because we factorize the one loop Feynman integral at high energies in a product of longitudinal and transverse momenta integrals. Such factorization is approximate and it is not valid even in QCD at the next-to-leading approximation. In gravity the factorization is destroyed already in one loop approximation, where we should cut the divergent integral over transverse momentum by the longitudinal momentum proportional to the initial energy. As a result, the scattering amplitude in one loop contains two logarithms of energy. The leading double-logarithmic terms were calculated and summed over all orders of perturbation theory. On the other hand, the ultraviolet divergences with the use of the dimensional regularization lead to the appearance of single logarithmic terms in perturbation theory. One can neglect these contributions in comparison with double-logarithmic terms. Nevertheless we don't know the region of applicability of obtained results for scattering amplitudes in the double logarithmic approximation. Note that these DL amplitudes are in agreement with multi-loop calculations of Bern, Dixon and others although the double-logarithmic terms are not leading at high energies in comparison with the contributions appearing in the eikonal expression. But I think that the graviton reggezation could help us to work with the non-renormalizable gravity. It is known that in the super-string approach the gravity is renormalizable. What is the reason for it? A possible explanation is that here we take into account not only gravitons and gravitinos but also massive string states. Analogous cancellations of divergencies could be realized in the Einstein gravity.

- *Zichichi:*

Excuse me. We have a trouble here. Did you say renormalizable or finite? You said renormalizable.

- *Lipatov:*

Finite. String theory is finite.

- *Zichichi:*

It is a big difference.

- *Lipatov:*

I agree with you. In my opinion these nice properties appear because in the Feynman diagrams of the super-string model we have contributions of all excited states apart from graviton and gravitino. And as a result, there are cancelations of ultraviolet divergences. In a similar way at the Einstein gravity we should take into account in perturbation theory also the massive states lying on the Regge trajectories. Further, one can calculate scattering amplitudes using the dispersion relations. Their imaginary parts can be calculated from the s-channel unitarity and their real parts are restored using the dispersion relations. The non-renormalizability of gravity is related to the fact that the number of subtraction constants grows with the order of perturbation theory. One could avoid this problem if for large s and negative t the imaginary part of the scattering amplitude falls at high energies as a consequence of the graviton reggezation. In this case at negative t we can write the dispersion relations without subtractions and the subtraction constants in the region of positive t can be calculated with the use of an analytic continuation. Therefore in my opinion for the good ultraviolet behavior of gravity it is good that the graviton is reggeized.

Embedding Oscillatory Modes of Quarks in Baryons in QCD
Looking to Construct a Bridge

Peter Minkowski
Albert Einstein Center for Fundamental Physics - ITP, University of Bern

Abstract

The regularities implied at large distances by complete gauge invariance in QCD were the topic of my lectures in the two initial parts of the tri-annual '50th Birthday Celebrations' of this school . This lecture shall be devoted to oscillatory modes of (valence) u,d,s quarks in baryons (also evident in $q \, \overline{q}^{\, \prime}$ mesons) , based on the representation derived in ref. [1-1980] and extended to the discussion of paired oscillator modes and associated Bogoliubov transformations in ref. [2-2010] . The embedding into the base field dynamics of QCD with three light flavors is layed out as a bridge in its present raw construction hopefully to be completed as a task of the future .

List of contents

List of figures

1 - Introduction

In order to put the main topic of this lecture into perspective let me begin citing ref. [1-1980] , my construction of Poincaré invariant oscillatory modes of three valence quarks (u , d) restricted to the two nonstrange flavors in nonstrange baryons . A small collection of references to previous work in this direction is given there ([1-1980]) . The disappearing of perfectly gauge invariant explicit dependence on color of quark- and gauge boson-fields is by the confined nature of the oscillator wave functions – restricted to the center of mass system relative to space-time coordinates – relegated to an outer factor

$$
\begin{array}{rcl}
\varepsilon_{\, c_\alpha c_\beta c_\gamma} \; ; \; c_{1,2,3} & = & \text{red , green , blue} \\
\alpha, \beta, \gamma \; = \; 1,2,3 & : & \text{numbering individual quark } \textit{positions}
\end{array}
\tag{1}
$$

The color factor in eq. 1 : $\varepsilon_{\, c_\alpha c_\beta c_\gamma}$, fully antisymmetric in its three color indices, must be fully gauge invariant with respect to the local $SU3_c$ gauge group and thus reduced from the 3 positions $\vec{x}_{1,2,3}$ to a common space point through parallel transport $q \; q \; q$ (triple) QCD string factors. The detailed form of the QCD string factors is discussed in section 2 – premises, and in two previous Erice lectures in refs. [3-2011], [4-2012] maintaining consistency of notation . A sketch of the $q \, \overline{q}$ (bilocal-) and $q \; q \; q$ (triple-) QCD string factors or 'bond structures', is given in Fig. 1 - I below

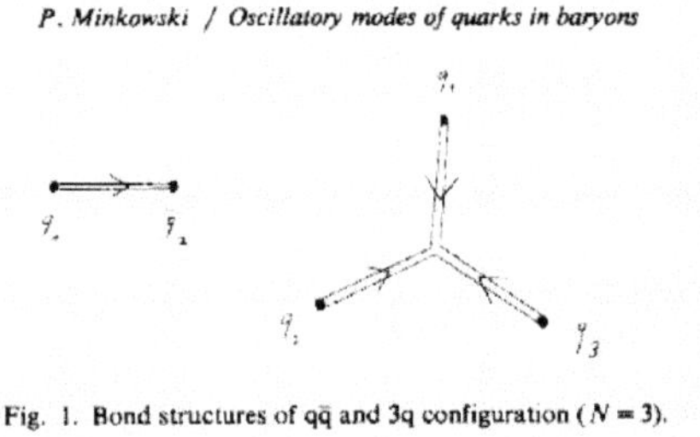

Fig. 1 - I : Bond structures of $q\overline{q}$ and $3q$ configurations ($N = 3$) from ref. [1-1980]

The next step revived questions related to oscillatory modes of a pair of independent oscillators and their eventual connection to Bogoliubov transformations in November 2010 in ref. [2-2010] .
Finally on the road leading to the present outline a discussion arose with the group of Willibald Plessas from the University of Graz during the Oberwölz Symposium 2012: Quantum Chromodynamics: History and Prospects 516. WE-Heraeus-Seminar , Oberwölz, Styria, Austria. 3. - 8. September 2012 as to whether the baryon modes of light flavors u, d, s of quarks with low total spin were presently more or less completely accounted for – according to the new PDG-review [5-2011] – contrasting with the situation in 1980 with respect to only u and d flavors .
Looking at the exhaustive tables of ref. [5-2011] *included today* the answer is obviously to the negative , yet these tables were not available in the web-version of the PDG tables [6-2012] upon my last search before the Oberwölz Symposium 2012 .
Out of this situation the challenge took shape to count the oscillatory modes in baryons in their own right in analytic ways , the content of this lecture . The notefile to it assembles the results obtained in ref. [7-2013] to date .

2-1 – Assembling elements of the QCD Lagrangean density – premises

We face the theoretical abstraction of QCD with $N_{fl} = 6$, representing strong interactions – adaptable to two or three light flavors (u , d , s) of quarks and antiquarks.

quarks : color is counted in $\pi^0 \rightarrow \gamma\gamma$ $\left(\begin{array}{l} \text{assuming global color- and} \\ \text{flavor-projections to commute} \end{array} \right)$

yet see ref. [8-2001]

spin and flavor are clearly seen in $q\overline{q}$ and $3q$, $3\overline{q}$ spectroscopy

$\left(\begin{array}{l} \text{a pre-condition} \\ \text{to count color} \end{array} \right)$

$$
\mathcal{L} = \left[\begin{array}{c} \overline{q}^{\,\dot{c}'}_{\,\dot{\mathcal{S}}'\,\dot{f}} \left\{ \begin{array}{c} \frac{i}{2}\, \overleftrightarrow{\partial}_{\mu}\, \delta_{\,c'\dot{c}} \\ + W^{\,r}_{\mu} \left(\frac{1}{2}\lambda_{\,r} \right)_{c'\dot{c}} \end{array} \right\} \gamma^{\,\mu}_{\,\dot{\mathcal{S}}'\mathcal{S}}\, q^{\,c}_{\,\mathcal{S}\,f} \\ - m_{\,f}\, \overline{q}^{\,\dot{c}}_{\,\dot{\mathcal{S}}\,\dot{f}}\, q^{\,c}_{\,\mathcal{S}\,f} \end{array} \right] - \frac{1}{4\,g^{\,2}}\, B^{\,\mu\nu\,r}\, B^{\,r}_{\,\mu\nu} + \Delta\,\mathcal{L} \tag{2}
$$

$W^{\,r}_{\mu} \equiv - v^{\,r}_{\mu}$: for identification of convention for potentials

quarks : c' , $c = 1, 2, 3$ color , $f = 1, \cdots, 6$ flavor
$\mathcal{S}', \mathcal{S} = 1, \cdots, 4$ spin , $m_{\,f}$ mass

In eq. 2 . the $\mathcal{D}$ associated gauge connection fields – where $\mathcal{D} = \mathcal{D}\,(\,\mathcal{G}\,)$ denotes a general , irreducible representation of the local gauge group $\mathcal{G} =$

$SU3_c$ – appear in the form appropriate for quarks : $\mathcal{D} = \{3\}$, and antiquarks : $\mathcal{D} = \{\overline{3}\}$ respectively

$$\boxed{\begin{array}{l} (\mathcal{W}_\mu(\mathcal{D}))_{\alpha\beta}(x) = W_\mu^r(x)\,(d_r)_{\alpha\beta} \leftrightarrow \mathcal{W}_\mu(\mathcal{D}) = -\mathcal{W}_\mu(\mathcal{D})^\dagger \\ d_r = -d_r^\dagger = \frac{1}{i} J_r \in Lie(\mathcal{D})\;;\;[d_p, d_q] = f_{pqr}\, d_r \\ r, p, q = 1, \cdots, dim\,\mathcal{G}\;;\;\alpha, \beta = 1, \cdots, dim\,\mathcal{D} \end{array}} \tag{3}$$

For $\mathcal{D}(SU3_c) = \{3\,(\overline{3})\}$ the representation matrices become (the Gell-Mann matrices [9-1964])

$$\begin{array}{l} \left(d_r(3) = \frac{1}{i}\frac{1}{2}\lambda_r\right)_{\alpha\beta}\;;\; r = 1, \cdots, 8\;;\;(\alpha, \beta) \leftrightarrow \left(\dot{c}', \dot{c}\right) = 1, \cdots, 3 \\ d_r(\overline{3}) = \overline{d}_r(3) \\ \text{with the } conventional \text{ normalization conditions : } -\,tr\, d^r d^s = \frac{1}{2}\delta^{rs} \end{array} \tag{4}$$

The quantity proportional to the gauge potentials W_μ^r for the $\overline{q}\,q$ in eq. 2 is thus identified as

$$\left[W_\mu^r\left(\tfrac{1}{2}\lambda_r\right)_{c'\dot{c}} = i\left(\mathcal{W}_\mu(\mathcal{D} = \{3\})\right)_{c'\dot{c}}\right](x) \tag{5}$$

Here we postpone the discussion of complete connections and extend the QCD Lagrangean density to include the term quadratic in the field strengths $B_{\mu\nu}^r$ and $\Delta\mathcal{L}$ in eq. 2, in Fermi gauges.

$$\begin{array}{c} \text{gauge bosons :} \qquad \mathcal{L}_B = -\frac{1}{4g^2} B^{\mu\nu\, r} B_{\mu\nu}^r \\ B_{\mu\nu}^r = \partial_\mu W_\nu^r - \partial_\nu W_\mu^r + f_{rst} W_\mu^s W_\nu^t \longleftarrow (W_\mu^r \equiv -v_\mu^r) \\ r, s, t = 1, \cdots, dim\,(\mathcal{G} = SU3_c) = 8 \\ \text{Lie algebra labels, } \left[\frac{1}{2}\lambda^r, \frac{1}{2}\lambda^s\right] = i\, f_{rst}\,\frac{1}{2}\lambda^t \\ \text{perturbative rescaling :} \\ W_\mu^r = g\, W_{\mu\, pert}^r,\; B_{\mu\nu}^r = g\, B_{\mu\nu\, pert}^r \end{array} \tag{6}$$

Degrees of freedom are seen in jets , in (e.g.) the energy momentum sum rule in deep inelastic scattering but not clearly in spectroscopy.
Completing $\Delta\mathcal{L}$ in Fermi gauges

$$\begin{array}{l} \Delta\mathcal{L} = \left\{\begin{array}{c} -\frac{1}{2\eta g^2}\left(\partial_\mu W^{\mu\, r}\right)^2 \\ +\,\partial^\mu \overline{c}^r\,(D_\mu c)^r \end{array}\right\}\;;\;\eta\;:\;\text{gauge parameter} \\ \text{ghost fermion fields : } c, \overline{c}\;;\;(D_\mu c)^r = \partial_\mu c^r + f_{rst} W_\mu^s c^t \\ \text{gauge fixing constraint : } C^r = \partial_\mu W^{\mu\, r} \end{array} \tag{7}$$

2-2 – Gauge boson binary bilocal and adjoint (here octet-) string operators

One goal is, to identify – not just some candidate resonance – gluonic mesons, binary and higher modes, and to relate them to the base quantities within QCD . Here we follow ref. [3-2011] .

$$\begin{aligned} & B_{[\mu_1 \nu_1], [\mu_2 \nu_2]}(x_1, x_2) = \\ & = B^{r}_{[\mu_1 \nu_1]}(x_1)\, U(x_1, r; x_2, s)\, B^{s}_{[\mu_2 \nu_2]}(x_2) \\ & r, s, \cdots = 1, \cdots, 8 \ ; \ \text{no flavor but spin} \end{aligned} \tag{8}$$

$B^{r}_{[\mu \nu]}(x)$ denote the local color octet of field strengths.
The quantity $U(x, r; y, s)$ in eq. (8) denotes the octet string operator, i. e. the path ordered exponential over a straight line path $\mathcal{C}$ from y to x

$$\begin{aligned} U(x,r; y,s) &= P \exp \left(\left. \int_{y}^{x} \right|_{\mathcal{C}} d z^{\mu} \tfrac{1}{i} v^{t}_{\mu}(z)\, \mathcal{F}_{t} \right) \Bigg|_{r s} \\ &= P \exp \left(- \left. \int_{y}^{x} \right|_{\mathcal{C}} d z^{\mu} W^{t}_{\mu}(z) \left(\tfrac{1}{i} \mathcal{F}_{t} \right) \right) \Bigg|_{r s} \\ (\mathcal{F}_{t})_{rs} &= i f_{r t s} \ ; \ (ad_{t})_{rs} = \tfrac{1}{i} (\mathcal{F}_{t})_{rs} = f_{r t s} \end{aligned} \tag{9}$$

The path ordered exponential as a matrix function of the argument is to be performed before the matrix elements, denoted $|_{..}$ in eq. 9, are taken. The local limit becomes

$$\begin{aligned} & B_{[\mu_1 \nu_1], [\mu_2 \nu_2]}(x_1 = x_2 = x) = (:)\, B^{r}_{[\mu_1 \nu_1]}(x)\, B^{r}_{[\mu_2 \nu_2]}(x)\, (:) \\ & \text{no flavor but spin} \end{aligned} \tag{10}$$

2-3 – $\overline{q}\, q$ bilinears and triplet-string operators

$$\begin{aligned} & B^{q}_{[\mathcal{A} f_1, \mathcal{B} f_2]}(x_1, x_2) = \\ & = \overline{q}^{\dot{c}_1}_{\mathcal{B} f_2}(x_1)\, U(x_1, c_1; x_2, \dot{c}_2)\, q^{c}_{\mathcal{A} f_1}(x_2) \end{aligned}$$

flavor *and* spin

$$\begin{aligned} U(x, c_1; y, \dot{c}_2) &= P \exp \left(\left. \int_{y}^{x} \right|_{\mathcal{C}} d z^{\mu} \tfrac{1}{i} v^{t}_{\mu}(z)\, \tfrac{1}{2} \lambda_{t} \right) \Bigg|_{c_1 \dot{c}_2} \\ &= P \exp \left(- \left. \int_{y}^{x} \right|_{\mathcal{C}} d z^{\mu} W^{t}_{\mu}(z) \left(\tfrac{1}{i} \tfrac{1}{2} \lambda_{t} \right) \right) \Bigg|_{c_1 \dot{c}_2} \end{aligned} \tag{11}$$

with the local limit

$$B^{\,q}_{\,[\,\dot{\mathcal{B}}\,\dot{f}\,2\,,\,\mathcal{A}\,f\,1\,]}\,(\,x_{\,1} \;=\; x_{\,2} \;=\; x\,) \;=\; (:)\,\overline{q}^{\,\dot{c}}_{\,\mathcal{B}\,\dot{f}\,2}\,(\,x\,)\,q^{\,c}_{\,\mathcal{A}\,f\,1}\,(\,x\,)\,(:) \tag{12}$$

The symbols (:) in eqs. 10 and 12 should indicate that normal ordering of regulating the local limits is required and further that such normal ordering is *not* unique , and dependent on quark masses in the case of the $\overline{q}\;q$ bilinears .

2-4 – Connection and curvature - forms preparing the ensuing analysis of regularity conditions

Lets begin this (sub-)section rewriting the bilocal (formally) unitary operators forming the gauge connection dependent octet- (eq. 9) and triplet (eq. 11) QCD strings , substituting an equivalent , matrix oriented notation

$$\begin{array}{ll} \text{octet . string} & : U\;(x,r;y,s) \;=\; P\;\exp\left(-\int_y^x\Big|_{\mathcal{C}}\; d\,z^{\,\mu}\,W^{\,t}_{\,\mu}\,(\,z\,)\;(\,\tfrac{1}{i}\,\mathcal{F}_{\,t}\,)\right)\Bigg|_{rs} \\ \text{triplet string} & :U\;(x,c_{\,1};y,\dot{c}_{\,2}) = P\;\exp\left(-\int_y^x\Big|_{\mathcal{C}}\; d\,z^{\,\mu}\,W^{\,t}_{\,\mu}\,(z)\;(\tfrac{1}{i}\tfrac{1}{2}\lambda_{\,t})\right)\Bigg|_{c_1\,\dot{c}_2} \end{array}$$

with the substitutions $\longrightarrow$

$$\left.\begin{array}{ll} \text{octet string} & : \;U\;(\,x,r\;;\;y,s\,) \;\rightarrow\; \left(U\left(x \overset{\mathcal{C}}{\leftarrow} y\right)\right)_{rs} \\ \text{triplet string} & : \;U\;(\,x,c_1\;;\;y,\dot{c}_2\,) \;\rightarrow\; \left(U\left(x \overset{\mathcal{C}}{\leftarrow} y\right)\right)_{c\,1\dot{c}\,2} \end{array}\right\}\rightarrow$$

$$\left(U\;(\,x,C,y\;;\;\mathcal{D}\,)\right)_{\alpha\beta} \;\in \mathcal{D}\;(\mathcal{G})$$

with $\mathcal{G}$ = simple compact gauge group

$\mathcal{D}$: general irreducible representation of $\mathcal{G}$

(13)

Here $\mathcal{G} \;=\; SU3_{\,c}$ and $\mathcal{D}$ is the octet- , triplet representation for the respective QCD $\mathcal{D}$ - strings.

Further let us consider matrix valued connection 1-forms , which define the bilocal matrix valued operators $(\,U\;(\,x\;,C\;,y\;;\;\mathcal{D}\,)\,)_{\alpha\beta} \in \mathcal{D}\,(\,\mathcal{G}\,)$ as given in eq. 13 . To this end the form of octet and triplet strings in eq. 13 is repeated below

$$\begin{array}{ll} \text{octet string} & : \;U\;(x,r;y,s) \;=P\;\exp\left(-\int_y^x\Big|_{\mathcal{C}}\; d\,z^{\,\mu}\,W^{\,t}_{\,\mu}\,(z)\;(\tfrac{1}{i}\,\mathcal{F}_{\,t})\right)\Bigg|_{rs} \\ \text{triplet string} & :U\;(x,c_1;y,\dot{c}_2) \;=P\;\exp\left(-\int_y^x\Big|_{\mathcal{C}}\; d\,z^{\,\mu}\,W^{\,t}_{\,\mu}\,(z)\;(\tfrac{1}{i}\tfrac{1}{2}\lambda_{\,t})\right)\Bigg|_{c_1\dot{c}_2} \end{array} \tag{14}$$

The two matrices in brackets to the right of the integrand expressions in eq. 14 form an antihermitian basis of the Lie algebra representation *Lie* ($\mathcal{D}$)

for $\mathcal{D}$ = adjoint and $\mathcal{D}$ = triplet representations of $\mathcal{G} = SU3_c$ respectively

$$d_t \equiv d_t(\mathcal{D}) \leftrightarrow (d_t)_{\alpha\beta} = \begin{cases} \left(\frac{1}{i}\mathcal{F}_t\right)_{rs} & \text{for} Lie(\mathcal{D}) = \text{adjoint} \\ \left(\frac{1}{i}\frac{1}{2}\lambda_t\right)_{c_1\dot{c}_2} & \text{for} Lie(\mathcal{D}) = \text{triplet} \end{cases}$$
$$d_t = -d_t^{\dagger}\ ;\ t = 1,\cdots,dim\,\mathcal{G}\ ;\ \alpha\ ,\ \beta = 1,\cdots,dim\,\mathcal{D}\ \text{for general}\ \mathcal{D} \tag{15}$$

From eqs. 14 and 15 we construct a – hopefully – consistent notation as appropriate for matrix valued $\mathcal{D}$ connections, 1-forms and strings, as well as derived 2- and higher forms. First eq. 15 is subject to the (matrix-) commutation relations

$$\begin{array}{l} [d_r\ ,\ d_s] = f_{rst}\, d_t\ ;\ \forall\ \mathcal{D}(\mathcal{G})\ ;\ r,s,t = 1,\cdots,dim\,\mathcal{G} \longrightarrow \\ (d_t(\mathcal{D} = \text{adjoint representation}))_{sr} = (ad_t)_{sr} = f_{str}\ : \\ \quad \text{independent of}\ \mathcal{D} \\ f_{str}\ :\ \text{totally antisymmetric, real structure constants of}\ Lie(\mathcal{G}) \end{array} \tag{16}$$

In physics the antihermitian matrix code with respect to the representations $Lie(\mathcal{D})$ is (mostly) replaced by the hermitian one [1]

$$\begin{array}{l} \left(d_t \equiv \frac{1}{i}h_t\right)\ (Lie(\mathcal{D}))\,|_{\alpha\beta}\ ; \\ h_t = h_t^{\dagger}\ ;\ [h_r\ ,\ h_s] = i\,f_{rst}\,h_t\ ;\ \alpha,\beta = 1,\cdots,dim\,\mathcal{D} \end{array} \tag{17}$$

Eq. 16 serves to define matrix valued connections built from a basis of $Lie(\mathcal{D})$ representation matrices as defined in eq. 16 for general irredcible representations $\mathcal{D}(\mathcal{G})$ denoted $\mathcal{W}_\mu(z\ ,\ \mathcal{D})$

$$\begin{array}{l} \mathcal{W}_\mu(z,\mathcal{D})\,|_{\alpha\beta} = W_\mu^r(z)\,(d_r)_{\alpha\beta\ Lie(\mathcal{D})}\ ;\ \left[\begin{array}{l} r = 1,\cdots,dim(\mathcal{G}) \\ \alpha,\beta = 1,\cdots,dim(\mathcal{D}) \end{array}\right] \\ \mathcal{W}_\mu(z,\mathcal{D})|_{\alpha\beta} \longrightarrow \mathcal{W}_\mu\ \text{for compact matrix notation} \end{array} \tag{18}$$

In the following it is to be understood, that $\mathcal{W}_\mu$ is extended to a general collection of representations $\bigcup \mathcal{D}$ – thought to be carried by real and spurious spin $\frac{1}{2}$ fields – care beeing taken that asymptotic freedom in the ultraviolet is not upset.

From eq. 18 we define the associated matrix valued connection 1-form displayed alongside the base definition repeated from eq. 18 in eq. 19 below

$$\begin{array}{l} \mathcal{W}^{(1)}(z\ ,\ \mathcal{D})\,|_{\alpha\beta} = d\,z^\mu\ \mathcal{W}_\mu(z\ ,\ \mathcal{D})\,|_{\alpha\beta} \longrightarrow \mathcal{W}^{(1)} \\ \mathcal{W}_u(z\ ,\ \mathcal{D})\,|_{\alpha\beta} = W_\mu^r(z)\,(d_r)_{\alpha\beta\ Lie(\mathcal{D})} \longrightarrow \mathcal{W}_\mu \end{array} \tag{19}$$

[1] A (partial) collection of historical and textbook references to the topics pertaining to 'Continous transformation groups and differential geometry' is presented under R-H references and labelled by the symbols 1H , 2H ⋯ .

and the matrix valued field-strength tensor

$$\mathcal{W}_{\mu\nu}(z,\mathcal{D})\,|_{\alpha\beta} = \left\{ \begin{matrix} \partial_\mu \mathcal{W}_\nu(z,\mathcal{D}) - \partial_\nu \mathcal{W}_\mu(z,\mathcal{D}) + \\ + [\mathcal{W}_\mu(z,\mathcal{D}), \mathcal{W}_\nu(z,\mathcal{D})] \end{matrix} \right\}_{\alpha\beta} \longrightarrow \mathcal{W}_{\mu\nu} \tag{20}$$

together with their associated curvature 2-form

$$\mathcal{W}^{(2)}(z,\mathcal{D})\big|_{\alpha\beta} = \tfrac{1}{2}\, d z^{\mu} \wedge d z^{\mu}\, \mathcal{W}_{\mu\nu}(z,\mathcal{D})\big|_{\alpha\beta} \qquad \longrightarrow \mathcal{W}^{(2)}$$

$$\mathcal{W}_{\mu\nu}(z,\mathcal{D})\big|_{\alpha\beta} = \left\{ \begin{matrix} \partial_\mu \mathcal{W}_\nu(z,\mathcal{D}) - \partial_\nu \mathcal{W}_\mu(z,\mathcal{D}) + \\ + [\mathcal{W}_\mu(z,\mathcal{D}), \mathcal{W}_\nu(z,\mathcal{D})] \end{matrix} \right\}_{\alpha\beta} \longrightarrow \mathcal{W}_{\mu\nu}$$

$$= W^{r}_{\mu\nu}(z)\,(d_r)_{\alpha\beta\ Lie\ (\mathcal{D})}$$

$$W^{r}_{\mu\nu} = \partial_\mu W^{r}_{\nu} - \partial_\nu W^{r}_{\mu} + f_{rst}\, W^{s}_{\mu} W^{t}_{\nu} \;; \text{ independent of } \mathcal{D} \tag{21}$$

Two remarks are in place here

1) In order to distinguish field strength from potentials (connections) the following equivalent but different notations for the field strength shall be used

$$\mathcal{W}_{\mu\nu} \equiv \mathcal{B}_{\mu\nu} \;;\; \mathcal{W}^{(2)} \equiv \mathcal{B}^{(2)} \;;\; W^{r}_{\mu\nu} \equiv B^{r}_{\mu\nu} \tag{22}$$

2) From the last relation in eq. 21 it may appear redundant to extend connections and curvatures to matrix valued form with respect to a wide collection of irreducible representations $\mathcal{D}(\mathcal{G})$. This however is tantamount to neglecting nontrivial global regularity conditions in the infrared .

We end this subsection (2-4) displaying the bilocal (parallel transport-) operators defined in eq. 13 using the shorthand notation in eq. 21

$$(U(x,C,y;\mathcal{D}))_{\alpha\beta} = P \exp\left(- \int_{y}^{x}\Big|_{C} \mathcal{W}^{(1)}(z,\mathcal{D}) \right)\Bigg|_{\alpha\beta}$$

$$\downarrow \qquad\qquad\qquad\qquad \downarrow$$

$$U(x,C,y) = P \exp\left(- \int_{y}^{x}\Big|_{C} \mathcal{W}^{(1)} \right) \quad [\text{ for } (\bigcup \mathcal{D})] \tag{23}$$

2-5 – The U1- or singlet axial current anomaly

The U1-axial central anomaly involves the local chiral current projections from $B^{q}_{[\dot{\mathcal{B}}\dot{f}_2,\,\mathcal{A} f_1]}(x)$ in eq. 12

$$(j^{\pm}_{\mu})_{\dot{f}_2 f_1}(x) = B^{q}_{[\dot{\mathcal{B}}\dot{f}_2,\,\mathcal{A} f_1]}(x)\left(\gamma_\mu \tfrac{1}{2}(\P \pm \gamma_5)\right)_{\mathcal{B}\dot{\mathcal{A}}}$$
$$= (:)\, \overline{q}^{\,\dot{c}}_{\dot{f}_2}\, \gamma^{\pm}_{\mu}\, q^{\,c}_{f_1}(x)\,(:)$$
$$\gamma_5 = \gamma_{5R} = \tfrac{1}{i}\gamma_0\gamma_1\gamma_2\gamma_3 \;;\; \gamma^{\pm}_{\mu} = \gamma_\mu \tfrac{1}{2}(\P \pm \gamma_5) \tag{24}$$

The equations of motion for the fermion fields are *and superficially imply* (upon $f_1 \leftrightarrow f_2$)

$$
\begin{array}{l}
\not{\partial}\, q^{\,c}_{f_2} = \frac{1}{i}\left(\not{v}^{\,c\,\dot{c}'} + \delta^{\,c\dot{c}'}\, m_{f_2}\right) q^{\,c'}_{f_2} \\
\overline{q}^{\,\dot{c}}_{\dot{f}_1}\, \overleftarrow{\not{\partial}} = \overline{q}^{\,\dot{c}'}_{\dot{f}_1}\, \frac{1}{i}\left(-\not{v}^{\,c'\dot{c}} - \delta^{\,c'\dot{c}}\, m_{f_1}\right)
\end{array}
\; ; \text{ no sums over } \dot{f}_1, f_2 \rightarrow
$$

$$
\begin{array}{l}
\partial^{\,\mu} (j^{\,\pm}_{\mu})_{\dot{f}_1 f_2} = \frac{1}{2i}\left((m_{f_2} - m_{f_1})\, S_{\dot{f}_1 f_2} \mp (m_{f_2} + m_{f_1})\, P_{\dot{f}_1 f_2}\right) \\
S_{\dot{f}_1 f_2} = (:)\, \overline{q}^{\,\dot{c}}_{\dot{f}_1}\, q^{\,c}_{f_2}\, (:)\ ,\ P_{\dot{f}_1 f_2} = (:)\, i\, \overline{q}^{\,\dot{c}}_{\dot{f}_1}\, \gamma_5\, q^{\,c}_{f_2}\, (:)
\end{array}
\tag{25}
$$

In eq. 25 m_f denotes the *real* , *nonnegative* quark mass for flavor f. From eq. 25 the relations for vector and axial vector currents *superficially* follow

$$
\begin{array}{lll}
(j_{\mu})_{\dot{f}_1 f_2} & = & (j^{\,+}_{\mu})_{\dot{f}_1 f_2} + (j^{\,-}_{\mu})_{\dot{f}_1 f_2} \\
(j^{\,5}_{\mu})_{\dot{f}_1 f_2} & = & (j^{\,+}_{\mu})_{\dot{f}_1 f_2} - (j^{\,-}_{\mu})_{\dot{f}_1 f_2} \\
\partial^{\,\mu}\, (j_{\mu})_{\dot{f}_1 f_2} & = & \frac{1}{i}\, (m_{f_2} - m_{f_1})\, S_{\dot{f}_1 f_2} \\
\partial^{\,\mu}\, (j^{\,5}_{\mu})_{\dot{f}_1 f_2} & = & (m_{f_2} + m_{f_1})\, i\, P_{\dot{f}_1 f_2}
\end{array}
\tag{26}
$$

As it follows from the original derivation by Adler and Bell and Jackiw [10-1969] in QED, the vecor current Ward identities in eq. 26 can be implemented also in QCD , leaving the axial current ones reduced to the flavor non-singlet case , leaving the U1 axial current divergent anomalous

$$
\partial^{\,\mu}\, (j_{\mu})_{\dot{f}_1 f_2} = \frac{1}{i}\, (m_{f_2} - m_{f_1})\, S_{\dot{f}_1 f_2} \ \surd
$$

$$
\left\{ \begin{array}{c} j^{\,5}_{\mu} \\ P \end{array} \right\}^{NS}_{\dot{f}_1 f_2} = \left\{ \begin{array}{c} j^{\,5}_{\mu} \\ P \end{array} \right\}_{\dot{f}_1 f_2} - \frac{1}{N_{fl}}\, \delta_{\dot{f}_1 f_2} \sum_f \left\{ \begin{array}{c} j^{\,5}_{\mu} \\ P \end{array} \right\}_{\dot{f} f}
\tag{27}
$$

and similarly

$$
\left\{ \begin{array}{c} j^{\,5}_{\mu} \\ P \end{array} \right\}^{S}_{\dot{f}_1 f_2} = \sum_f \left\{ \begin{array}{c} j^{\,5}_{\mu} \\ P \end{array} \right\}_{\dot{f} f}
\tag{28}
$$

2-6 – Quark masses and splittings : m_f **and** $\Delta\, m_f = m_f - \langle\, m\, \rangle$

In the subtitle above $\langle\, m\, \rangle$ stands for the mean quark mass

$$
\langle\, m\, \rangle = \frac{1}{N_{fl}} \sum_f m_f
\tag{29}
$$

The identities for vector currents in eqs. 26 and 27 can be extended separating the conributions proportional to $\Delta\, m_{\,f}$ and $\langle\, m\,\rangle$

$$\begin{aligned}
\partial^{\,\mu}\,(j_{\,\mu})_{\,\dot{f}_{\,1}\,f_{\,2}} &= \tfrac{1}{i}\,(\Delta\, m_{f\,2} - \Delta\, m_{f\,1})\, S_{\,\dot{f}_{\,1}\,f_{\,2}} \ \surd \\
\partial^{\,\mu}\,(j^{\,5}_{\,\mu})^{\,NS}_{\,\dot{f}_{\,1}\,f_{\,2}} &= (\Delta\, m_{\,f\,2} + \Delta\, m_{\,f\,1})\, i\, P^{\,NS}_{\,\dot{f}_{\,1}\,f_{\,2}} \ \surd \\
\partial^{\,\mu}\,(j^{\,5}_{\,\mu})^{\,S}_{\,\dot{f}_{\,1}\,f_{\,2}} &= 2\,\langle\, m\,\rangle\, i\, P^{\,S} \ \not{\surd}\ [\longrightarrow + \delta_{\,5}\,] \\
\delta_{\,5} = (\,2\, N_{\,fl}\,)\, \tfrac{1}{32\pi^{\,2}}\, B^{\,r}_{\mu\nu}\, \widetilde{B}^{\mu\,\nu\,r}\Big|_{\to ren.gr.inv} &;\ \widetilde{B}^{\,r}_{\,\mu\,\nu} = \tfrac{1}{2}\,\varepsilon_{\mu\nu\sigma\tau}\, B^{\sigma\,\tau\,r}
\end{aligned} \tag{30}$$

[2]

We shall return to the question of how the local operator $ch_2\,(\,B\,) \equiv \frac{1}{32\pi^{\,2}}\,(\,:\,)\, B^{\,r}_{\,\mu\,\nu}\, \widetilde{B}^{\,\mu\,\nu\,r}\,(:)$ is to be normalized and rendered renormalization group invariant [11-1991] . Here we just assume this to have been achieved and denote the U1-axial anomaly, the first of the central two, in its general form (eq. 30)

$$\begin{aligned}
\left\{ \partial^{\,\mu}\,(\,j^{\,5}_{\,\mu}\,)^{\,S} = 2\,\langle\, m\,\rangle\, i\, P^{\,S} + \delta_{\,5} \right\}(\,x\,) \\
\delta_{\,5} = (\,2\, N_{\,fl}\,)\, \tfrac{1}{32\pi^{\,2}}\,(:)\, B^{\,r}_{\,\mu\,\nu}\, \widetilde{B}^{\,\mu\,\nu\,r}\,(:)\,\Big|_{\to ren.gr.inv}
\end{aligned} \tag{31}$$

From here it is conceptually clear how the scale- (or trace-) anomaly arises but strictly within QCD . The renormalizability of a field theory in the limit of uncurved space-time gives rise to a local , symmetric and *conserved* energy momentum tensor, implying exact Poincaré invariance

$$\begin{aligned}
\{\,\vartheta_{\,\mu\,\nu} = \vartheta_{\,\nu\,\mu}\,\}\,(\,x\,) \\
\partial^{\,\nu}\,\vartheta_{\,\mu\,\nu} = 0
\end{aligned} \tag{32}$$

In connection with the normal ordering questions it is important to admit in the precise form of the energy momentum tensor a nontrivial vacuum expected value , which in view of exact Poincaré invariance must be of the form

$$\begin{aligned}
&\langle\,\Omega\,|\,\vartheta_{\,\mu\,\nu}\,(\,x\,)\,|\,\Omega\,\rangle = \tfrac{1}{4}\,\eta_{\,\mu\,\nu}\,\tau \\
&\left\{ \begin{array}{c} \eta_{\,\mu\,\nu} = diag\,(\,1\,,\,-1\,,\,-1\,,\,-1\,) \\ \tau \end{array} \right\} \text{independent of } x \longrightarrow \\
&\Delta\,\vartheta_{\mu\,\nu}\,(x) = \vartheta_{\mu\,\nu}\,(x) - \langle\,\Omega\,|\,\vartheta_{\mu\,\nu}\,(x)\,|\,\Omega\,\rangle \times \left\{ \begin{array}{l} \widehat{\P} \\ \text{or } |\Omega\,\rangle\,\langle\,\Omega| \end{array} \right. \\
&\text{with } \partial^{\,\nu}\,\Delta\,\vartheta_{\,\mu\,\nu}\,(\,x\,) = 0\ ;\ \langle\,\Omega\,|\,\Delta\,\vartheta_{\,\mu\,\nu}\,(\,x\,)\,|\,\Omega\,\rangle = 0
\end{aligned} \tag{33}$$

In eq. 33 $\widehat{\P}$ denotes the unit operator in the entire Hilbert space of states, while $P_{\,\Omega} = |\,\Omega\,\rangle\,\langle\,\Omega\,|$ stands for the projector on the ground state .

[2] $\delta_{\,5}$ was – as far as I know – introduced by Murray Gell-Mann in lectures ~ 1970 in Hawaii .

Furthermore from the two local, *conserved* tensors in eq. 33 only $\Delta \vartheta_{\mu \nu}(x)$ with vanishing vacuum expected value is acceptable as representing the conserved 4 momentum *operators* in the integral form

$$\widehat{P}_{\mu} = \int_{t} d^{3} x \, \Delta \vartheta_{\mu 0}(t, \vec{x}) \tag{34}$$

All these arguments *notwithstanding* to subtract any eventual vacuum expected values of local operators , often put forward as mathematical prerequisites , it is wise *not to do so* in the presence of spontaneous parameters , the dynamical origin of spontaneous symmetry breaking, e.g. chiral symmetries in the limit or neighbourhood of some $m_{f} \rightarrow 0$.
Using the (classical) equations of motion associated to the Lagrangean in eqs. 2 - 7

$$\begin{array}{l}
(D_{\nu} B^{\mu\nu})^{r} = j^{\mu\, r}(\overline{q}, q) \; ; \; B \rightarrow B_{pert} \\
(D_{\varrho} B^{\mu\nu})^{r} = \partial_{\varrho} B^{\mu\nu\, r} + f_{rst} W_{\varrho}^{s} B^{\mu\nu\, t} \\
j_{\mu}^{r}(\overline{q}, q) = g\, \overline{q}_{\dot{A}\, f}^{\dot{c}} (\gamma_{\mu})_{\dot{A} B} \tfrac{1}{2} (\lambda^{r})_{c c'} q_{A\, f}^{c'} \\
i\, (\gamma^{\mu} D_{\mu} q)_{A\, f}^{c} = m_{f}\, q_{A\, f}^{c} \text{ and } q \rightarrow \overline{q} \\
(D_{\mu} q)_{A\, f}^{c} = \left[\partial_{\mu} \delta_{c c'} + \tfrac{1}{i} W_{\mu}^{t} \tfrac{1}{2} (\lambda^{t})_{c c'} \right] q_{A\, f}^{c'} \\
\hline
W_{\mu}^{r} \equiv - v_{\mu}^{r} = g\, (W_{\mu}^{r})_{pert} \equiv - g\, (v_{\mu}^{r})_{pert}
\end{array} \tag{35}$$

the form of the energy momentum becomes

$$\vartheta_{\mu\nu}^{(cl)} = \left[\begin{array}{l} \frac{1}{4 g^{2}} \left[B_{\mu\varrho}^{t} B_{\nu}^{\varrho\, t} - \frac{1}{4} \eta_{\mu\nu} B_{\sigma\varrho}^{t} B^{\varrho\sigma\, t} \right] + \\ \quad + \frac{1}{2} \left[\overline{q}_{\dot{f}} \gamma_{\mu} \frac{i}{2} \overleftrightarrow{D}_{\nu} q_{f} + \mu \leftrightarrow \nu \right] \end{array} \right] \tag{36}$$

and using once more the fermion part of the equations of motion the trace of the classical energy momentum tensor becomes

$$\begin{array}{l}
\vartheta^{\mu}{}_{\mu}^{(cl)} = \sum_{f} m_{f} S_{\dot{f} f} \\
S_{\dot{f}_{1} f_{2}} = (:)\, \overline{q}_{\dot{f}_{1}}^{\dot{c}} q_{f_{2}}^{c} \,(:)
\end{array} \tag{37}$$

2-7 – The scale- or trace- anomaly

From the classical soft fermionic contribution to the trace of the energy momentum tensor there is a clear conjecture, also by Murray Gell-Mann , of the anomalous contribution, which subsequently became the scale- or trace-anomaly within QCD

$$\begin{array}{l}
\vartheta^{\mu}{}_{\mu} = \sum_{f} m_{f} S_{\dot{f} f} + \delta_{0} \\
\delta_{0} = - \left(-2 \beta(g) / g^{3} \right) \left[\tfrac{1}{4} (:)\, B_{\mu\nu}^{t} B^{\mu\nu\, t} (:) \right]_{\rightarrow ren.gr.inv}
\end{array} \tag{38}$$

2-8 – The two central anomalies alongside : scale- or trace- and U1-axial anomaly

We collect the two anomalous identities in eqs. 38 and 20

$$\begin{array}{l} \left\{ \vartheta^{\mu}{}_{\mu} \quad = \sum_f m_f \, S_{\dot{f} f} + \delta_0 \right\} (x) \\ \left\{ \partial^{\mu} \left(j^5_{\mu} \right)^S = 2 \, \langle m \rangle \, i \, P^S + \delta_5 \right\} (x) \\ \delta_0 = - \left(-2 \, \beta(g) / g^3 \right) \left[\tfrac{1}{4} (:) \, B^t_{\mu\nu} \, B^{\mu\nu t} (:) \right]_{\rightarrow \, ren.gr.inv} \\ \delta_5 = \quad \left(2 \, N_{fl} \right) \tfrac{1}{8\pi^2} \left[\tfrac{1}{4} (:) \, B^t_{\mu\nu} \, \widetilde{B}^{\mu\nu t} (:) \right]_{\rightarrow ren.gr.inv} \\ \hline - \beta / g^3 = \tfrac{1}{16\pi^2} \, b_0 + O(X) \; ; \; X = g^2 / (16 \pi^2 \\ \beta(g) \; : \; \text{Callan-Symanzik rescaling function in QCD} \end{array} \tag{39}$$

The qualification 'central' for the anomalies in eq. 39 stands for the property that in rendering the square coupling constant and the associated ϑ – parameter in the gauge boson *renormalized* Lagrangean density x dependent

$$\begin{array}{l} \mathcal{L}_{g.b.} = - \tfrac{1}{g^2} \tfrac{1}{4} (:) \, B^t_{\mu\nu} \, B^{\mu\nu t} (:) + \vartheta \, \tfrac{1}{8\pi^2} \tfrac{1}{4} (:) \, B^t_{\mu\nu} \, \widetilde{B}^{\mu\nu t} \longrightarrow \\ g^2 \rightarrow g^2(x) \; ; \; \vartheta \rightarrow \vartheta(x) \end{array} \tag{40}$$

maintains perturbative renormalizability and acts together with suitable boundary- – more generally – regularity conditions as external sources for the scalar and pseudoscalar local field strength bilinears

$$\tfrac{1}{4} (:) \, B^t_{\mu\nu} \, B^{\mu\nu t} (:) \; , \; \tfrac{1}{4} (:) \, B^t_{\mu\nu} \, \widetilde{B}^{\mu\nu t} \tag{41}$$

We will use the following definitions relative to the rescaling function β

$$\begin{array}{l} - \beta / g = X \, B(X) \; ; \; B(X) = b_0 \, A(X) \\ B(X) \sim \sum_{n=0}^{\infty} b_n \, X^n \; , \; A(X) \sim \sum_{n=0}^{\infty} a_n \, X^n \\ \kappa = g^2 / (16 \pi^2) \text{ generic } \longrightarrow X , Y \\ b_0 = \tfrac{1}{3} \left(33 - 2 N_{fl} \right) \; , \; a_0 = 1 \; , \; a_n = b_n / b_0 \\ b_1 = \tfrac{2}{3} \left(153 - 19 N_{fl} \right) \\ b_2 = \tfrac{1}{54} \left(77139 - 15099 N_{fl} + 325 N^2_{fl} \right) \\ b_3 \sim 29243 - 6946.3 N_{fl} + 405.089 N^2_{fl} + 1.49931 N^3_{fl} \end{array} \tag{42}$$

References in conjunction with this section (2 - premises) are presented in five (partial) collections :

1 : (R) directly related to the two central anomalies

2 : (rBsquare) establishing the one renormalization group invariant quantity of dimension $\left[M^4 \right]$

3 : (r-sp-1) a recent paper by Guido Altarelli and references cited therein

4 : (r-A2x) a selection of papers and textbooks for the entire realm of QCD

5 : (r-concx) : Condensation phenomena and field theory realizations

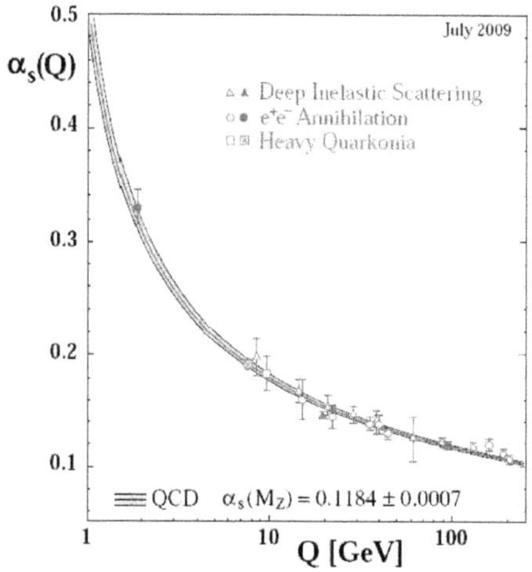

Fig. A21 : $\alpha_s(Q) = 4\pi\,\kappa_{\overline{\mu}=Q}$ **from ref. [12-2009] .**

3-1 – Perturbative renormalization rescaling

We begin with the renormalization of external operaters and gauge boson fields, denoting unrenormalized fields and operators by the suffix $^{(0)}$

$$\begin{array}{lll} J_\alpha = (Z_J)^{-1} J_\alpha^{(0)} & , & \mathcal{O} = (Z_{\mathcal{O}})^{-1} \mathcal{O}^{(0)} \\ g = (Z_3)^{3/2} (Z_1)^{-1} g^{(0)} & , & \eta = (Z_3)^{-1} \eta^{(0)} \\ W^r_{\mu\,pert} = (Z_3)^{-\frac{1}{2}} \left(W^r_{\mu\,pert}\right)^{(0)} & & \end{array} \tag{43}$$

and continue with the ghost fields c^r , $\overline{c}^s$ as defined in $\Delta\,\mathcal{L}$ in eq. 7

$$\begin{array}{lll} c^r = \left(\widetilde{Z}_2\right)^{-\frac{1}{2}} (c^r)^{(0)} & , & \overline{c}^r = \left(\widetilde{Z}_2\right)^{-\frac{1}{2}} (\overline{c}^r)^{(0)} \\ g = (Z_3)^{1/2}\, \widetilde{Z}_2 \left(\widetilde{Z}_1\right)^{-1} g^{(0)} & & \end{array} \tag{44}$$

The renormalization constant $Z_1 = Z_{1\,(3W)}$ refers to the 3 gauge boson vertex of the derivative type $W\,W\,\partial\,W$ as shown in figure 1, while $\widetilde{Z}_1 = \widetilde{Z}_1\,(\,\overline{c}\,c\,W\,)$ refers to the ghost-W vertex, shown in figure 2. The 4 gauge boson vertex involves the analogous renormalization constant $Z_{1(4W)}$ and the $\overline{q}\,q\,W$ vertex $Z_{1(\overline{q}qW)}$, shown in figures 3 and 4 respectively below .

4 irreducible vertex diagrams

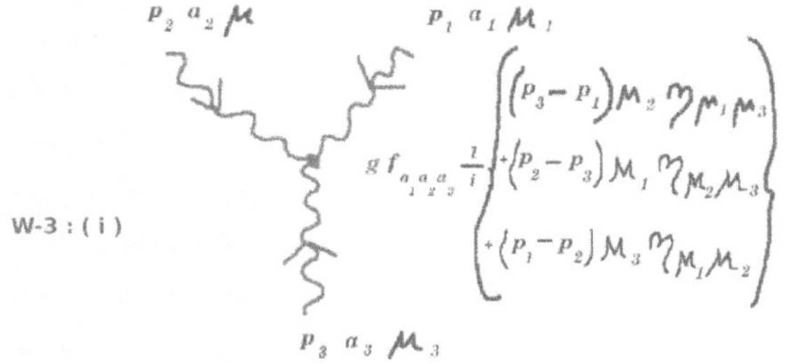

Fig. 1 : $Z_1 \leftrightarrow$ **W-3 vertex of the type** $W\ W\ \partial\ W$

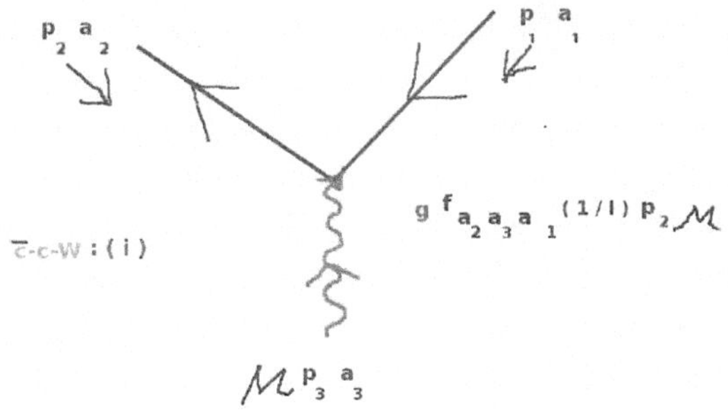

Fig. 2 : $\widetilde{Z}_1 \leftrightarrow$ **cbar-c W-3 vertex of the type** $\partial\ \bar{c}\ c\ W$

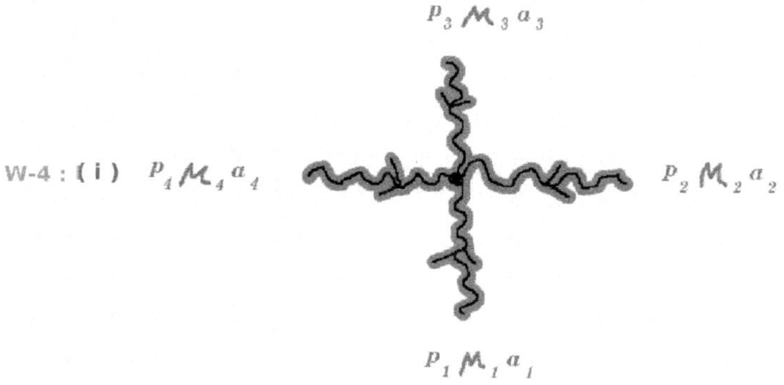

Fig. 3 : $Z_{1(4W)} \leftrightarrow$ **W-4 vertex for gauge fields**

$$
g^{2}\left(\begin{array}{ll}
r_{[a_1 a_2][a_3 a_4]} & \left[\begin{array}{l} \eta_{\mu_1 \mu_4}\eta_{\mu_2 \mu_3} \\ -\eta_{\mu_1 \mu_3}\eta_{\mu_2 \mu_4} \end{array}\right] \\
+r_{[a_1 a_3][a_2 a_4]} & \left[\begin{array}{l} \eta_{\mu_1 \mu_4}\eta_{\mu_2 \mu_3} \\ -\eta_{\mu_1 \mu_2}\eta_{\mu_3 \mu_4} \end{array}\right] \\
+r_{[a_1 a_4][a_2 a_3]} & \left[\begin{array}{l} \eta_{\mu_1 \mu_3}\eta_{\mu_2 \mu_4} \\ -\eta_{\mu_1 \mu_2}\eta_{\mu_3 \mu_4} \end{array}\right]
\end{array}\right)^{\to r}
$$

$$
r_{[a_1 b_1][a_2 b_2]} = f_{d a_1 b_1} f_{d a_2 2_2}
$$

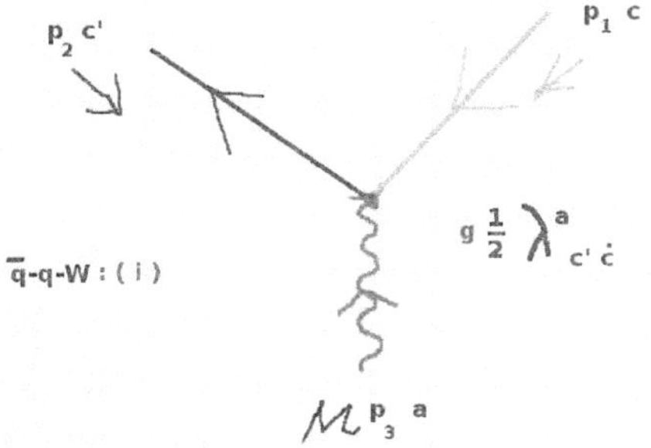

Fig. 4 : $Z_{1(\overline{q}qW)} \leftrightarrow$ **W-3 vertex for** $\overline{q}\,q$ **to gauge fields**

3-2 – The complete coupling constant renormalization equations

The minimal renormalization constants defined in eqs. 43 and 44, referring beyond the vertices shown in figures 1-4 to the two point functions for gauge connections , Z_3 , and ghost fields , $\widetilde{Z}_2$, thus need extension (also) to $q, \overline{q}$ fields and their two point function, Z_2 , as shown below

$$
\begin{array}{ll}
g = (Z_3)^{1/2} \, (Z_3 / Z_1) \, g^{(0)} \; ; W_{\mu\,pert}^{\,r} = (Z_3)^{-\frac{1}{2}} \, \left(W_{\mu\,pert}^{\,r} \right)^{(0)} \\
g = (Z_3)^{1/2} \left(\widetilde{Z}_2 / \widetilde{Z}_1 \right) g^{(0)} \; ; (c \, , \, \overline{c})^{\,r} = \left(\widetilde{Z}_2 \right)^{-\frac{1}{2}} \left((c \, , \, \overline{c})^{\,r} \right)^{(0)} \\
g^2 = Z_3 \, \left(Z_3 / Z_{1(4W)} \right) \left(g^{(0)} \right)^2 \\
\hline
g = (Z_3)^{1/2} \left(Z_2 / Z_{1(\overline{q}qW)} \right) g^{(0)} \; ; q^{\,c}_{\,\mathcal{S}\,f} = (Z_2)^{-\frac{1}{2}} \left(q^{\,c}_{\,\mathcal{S}\,f} \right)^{(0)} \\
\qquad\qquad\qquad\qquad\qquad\qquad \overline{q}^{\,\dot{c}'}_{\,\dot{\mathcal{S}}'\,\dot{f}} = (Z_2)^{-\frac{1}{2}} \left(\overline{q}^{\,\dot{c}'}_{\,\dot{\mathcal{S}}'\,\dot{f}} \right)^{(0)} \\
\hline
\end{array}
\tag{45}
$$

Maintaining gauge invariance through renormalization thus implies the relations

$$
Z_3 / Z_1 \; = \; \widetilde{Z}_2 / \widetilde{Z}_1 \; = \; Z_2 / Z_{1(\overline{q}qW)} \; ; \; Z_3 / Z_{1(4W)} \; = \; (Z_3 / Z_1)^2 \tag{46}
$$

It remains to include quark mass renormalization and associated rescaling.

3-3 – Renormalization equations for quark masses m_f

We perform quark mass renormalization 'at zero mass' [A218-1973] , with the following relations , using the $q, \overline{q}$ fields and their field renormalization constant, Z_2 , as defined in eqs. 45 and 46, through the extension

$$
(\mathcal{L}_m)^{(0)} \; = \; - \sum_f Z_2 \, Z_m \, m_f \; ; \; m_f \; = \; (Z_m)^{-1} \, m_f^{(0)} \tag{47}
$$

The coefficients of the quark mass rescaling functions are known to four loops as calculated by Chetyrkin and independently by others [13-1997] . They are given in eq. 106 in Appendix 1.

Here we subsum the complete set of renormalized rescaling functions, which follows from the renormalization constants defined in eqs. 44 - 46

$$
\begin{pmatrix} \mu^2 \, \partial_{\mu^2} + (\beta(g) / g) \, g^2 \, \partial_{g^2} + \\ + \gamma_m \, m_f \, \partial_{m_f} - 2 \, \gamma_3 (\eta \, \partial_\eta) - \gamma_{J\,\mathcal{O}} \end{pmatrix} C_{J\,\mathcal{O}}^{\,T\,(\Pi)} \; (z; \mu, g, m_\beta, \eta) \; = \; 0
$$

$$
\left\{ \begin{array}{l} \beta(g) \; = \; - \, g \, b \, (g^2) \\ \gamma_m (g^2) \; \equiv \; - \, \chi_m \\ \gamma_{J\,\mathcal{O}} (g^2, (\eta)) \\ \gamma_3 \, (g^2, \eta) \end{array} \right\} \; = \; \mu^2 \, d / d \, \mu^2 \left\{ \begin{array}{c} 2 \, \log \left((Z_3)^{3/2} \, (Z_1)^{-1} \right) \\ \log \left((Z_m)^{-1} \right) \\ \log \left(Z_{\mathcal{O}} / Z_J^2 \right) \\ \log \left((Z_3)^{1/2} \right) \end{array} \right\} \tag{48}
$$

We can always return to the standard form of the renormalization group equation

$$\begin{pmatrix} \mu\,\partial_{\mu} + \beta\,(g)\partial_{g} + (2\,\gamma_{m})\,m_{f}\,\partial_{m_{f}} \\ -2\,(2\,\gamma_{3})\,(\eta\,\partial_{\eta}) - (2\,\gamma_{J\,\mathcal{O}}) \end{pmatrix} C_{J\,\mathcal{O}}^{T\,(\Pi)}\;(z;\mu,g,m_{\beta},\eta) = 0 \tag{49}$$

without changing the definition of β, transforming the lower relation in eq. 48, which amounts to double the coefficients of the quantities γ_{m}, γ_{3}, $\gamma_{J\,\mathcal{O}}$, as displayed in eq. 49

$$\left\{\begin{array}{l} \beta\,(g) = -\,g\,b\,(g^{2}) \\ 2\,\gamma_{m}\,(g^{2}) \equiv -\,2\,\chi_{m} \\ 2\gamma_{J\,\mathcal{O}}\,(g^{2},\,(\eta)) \\ 2\,\gamma_{3}\,(g^{2},\,\eta) \end{array}\right\} = \mu\,d\,/\,d\,\mu \left\{\begin{array}{l} \log\left((Z_{3})^{3/2}\,(Z_{1})^{-1}\right) \\ \log\left((Z_{m})^{-1}\right) \\ \log\,(Z_{\mathcal{O}}\,/\,Z^{2}_{J}) \\ \log\left((Z_{3})^{1/2}\right) \end{array}\right\} \tag{50}$$

rescaling equations in the $\overline{\textbf{MS}}$ scheme for coupling constant g and quark masses m_{f}
$\beta\,(g)$ and $\gamma_{m}\,(\kappa)$, $\kappa = g^{2}\,/\,(16\,\pi^{2})$

We turn to the scale (-square) evolution of the rationalized coupling constant and quark masses, introducing the variables

$$\begin{array}{l} \tau = \log\,(\mu^{2}\,/\,\mu_{0}^{2})\;;\;\kappa = g^{2}\,/\,(16\,\pi^{2}) \rightarrow \\ \left\{\,\widehat{\beta} = (\beta\,/\,g)\,\kappa = \widehat{\beta}\,(\kappa),\,\gamma_{m}\,(\kappa)\,\right\} \end{array} \tag{51}$$

The renormalization group equation in the form given in eq. 48 then becomes

$$\begin{pmatrix} \partial_{\tau} + \widehat{\beta}\,\partial_{\kappa} + \gamma_{m}\,m_{f}\,\partial_{m_{f}} \\ -2\,\gamma_{3}\,(\eta\,\partial_{\eta}) - \gamma_{J\,\mathcal{O}} \end{pmatrix} C_{J\,\mathcal{O}}^{T\,(\Pi)}\;(z;\mu,g,m_{\beta},\eta) = 0 \tag{52}$$

The partial derivative terms in brackets in eq. 52 determine the sliding scale equations with respect to the quantities τ, $\overline{\kappa}$, $\overline{m}_{f}$, $\overline{\eta}$ in the sense of its initial value structure. We restrict the discussion to τ, $\overline{\kappa}\,(\tau)$, $\overline{m}_{f}\,(\tau)$ here for simplicity.

$$\begin{array}{l} \overset{\bullet}{\overline{\kappa}} = \widehat{\beta}\,(\overline{\kappa}),\;\overset{\bullet}{\overline{m}}_{f} = \gamma_{m}\,(\overline{\kappa})\,\overline{m}_{f}\;;\;\bullet = d\,/\,d\,\tau \\ \overline{\kappa}\,(\tau = 0) = \kappa,\;\overline{m}_{f}\,(\tau = 0) = m_{f} \end{array} \tag{53}$$

$$\text{sliding coupling constant}:\quad \overset{\bullet}{\overline{\kappa}} = \widehat{\beta}\,(\overline{\kappa})$$

We associate integration variables, initial values and endpoint variables in the following way

$$\theta \leftrightarrow [\,\tau\,,\,0\,]\,,\;\lambda \leftrightarrow [\,\overline{\kappa}\,,\,\kappa\,] \tag{54}$$

The integration of the rescaling differential equation becomes

$$\int_0^{\tau} d\,\theta \;=\; \tau \;=\; \int_{\kappa}^{\overline{\kappa}} d\,\lambda \,\left(\widehat{\beta}(\,\lambda\,) \right)^{-1} \;\longrightarrow\; \tau \;=\; \int_{\overline{\kappa}}^{\kappa} d\,\lambda \,\left(-\,\widehat{\beta}\,(\,\lambda\,) \right)^{-1} \tag{55}$$

The first two coefficients in the expansion of $\widehat{\beta}$ in powers of λ necessitate a twofold subtraction for $\lambda \downarrow 0$ for the integral on the right hand side of eq. 55 to converge at the lower limit of integration

$$\widehat{\beta}\,(\,\lambda\,) \;=\; \sum_{n=0}^{\infty} \widehat{\beta}_{\,n}\,\lambda^{\,n+2} \;;\; \widehat{\beta}_{\,n} \;\equiv\; -\,b_{\,n} \tag{56}$$

The first four coefficients $b_{\,n}\;;\;n \;=\; 0,\cdots,3$ are known , (ref. [14-1997]) and given in eq. 105 in Appendix 1 .
We perform this subtraction splitting $\widehat{\beta}$

$$\widehat{\beta} \;=\; \widehat{\beta}_{\,(2)} \;+\; \Delta_{\,(2)}\,\widehat{\beta}\;;\;\left[\begin{array}{lcl} \widehat{\beta}_{\,(2)} & = & -\,\left(b_{\,0}\lambda^{\,2} \;+\; b_{\,1}\,\lambda^{\,3} \right) \\ \Delta_{\,(2)}\,\widehat{\beta} & = & -\,\sum_{n=2}^{\infty} b_{\,n}\,\lambda^{\,n+2} \end{array}\right. \tag{57}$$

The subtraction, displayed in eq. 57
gives rise to the substitutions

$$\begin{array}{l} \zeta_{\,(2)} \;=\; \dfrac{\Delta_{\,(2)}\,\widehat{\beta}}{\widehat{\beta}_{\,(2)}} \;=\; b_{\,2}\,b_{\,0}^{-1}\,\lambda^{\,2}\,\left(\,1\;+\;O\,(\,\lambda\,)\,\right) \\ \left(-\,\widehat{\beta} \right)^{-1} \;=\; \left(-\,\widehat{\beta}_{\,(2)} \right)^{-1} \Big/ \left(\,1\;+\;\zeta_{\,(2)}\,\right) \;=\; \left(-\,\widehat{\beta}_{\,(2)} \right)^{-1} \;-\; \psi_{\,(2)} \\ \psi_{\,(2)} \;=\; \left(-\,\widehat{\beta}_{\,(2)} \right)^{-1} \dfrac{\zeta_{\,(2)}}{1\;+\;\zeta_{\,(2)}} \;=\; b_{\,2}\,\left(\,b_{\,0}\,\right)^{-2}\,\left(\,1\;+\;O\,(\,\lambda\,)\,\right) \end{array} \tag{58}$$

$\psi_{\,2}$ as defined in eq. 58 has a regular power series expansion in λ , fully determined by $\widehat{\beta}$ and hence the sought subtractions involve only $\widehat{\beta}_{\,(2)}$, i.e. the beta function truncated to the first two terms.
We proceed in the reduction of $\left(-\,\widehat{\beta} \right)^{-1}$ using the substitutions in eq. 58

$$\begin{array}{lcl} \left(-\,\widehat{\beta} \right)^{-1} & = & \left(-\,\widehat{\beta}_{\,(2)} \right)^{-1} \;-\; \psi_{\,(2)} \\ \left(-\,\widehat{\beta}_{\,(2)} \right)^{-1} & = & b_{\,0}^{-1}\,\lambda^{\,-2}\,\left(1\;+\;b_{\,1}^{\,(0)}\,\lambda\right)^{-1}\;;\; b_{\,1}^{\,(0)} \;=\; b_{\,1}\,/\,b_{\,0} \\ & = & b_{\,0}^{-1}\,\lambda^{\,-2}\,\left(1\;-\;b_{\,1}^{\,(0)}\,\lambda\right)\,\left(1\;-\;\left(b_{\,1}^{\,(0)}\,\lambda\right)^{2}\right)^{-1} \end{array} \tag{59}$$

The last expression in eq. 59 decomposes into

$$\left(-\widehat{\beta}_{(2)}\right)^{-1} = b_0^{-1}\lambda^{-2} - b_0^{-2} b_1 \lambda^{-1} + \phi_{(2)}$$

$$\phi_{(2)} = b_0^{-1}\left(b_1^{(0)}\right)^2\left(1 - \left(b_1^{(0)}\lambda\right)^2\right)^{-1} \; ; \; b_1^{(0)} = b_1 / b_0 \tag{60}$$

$\phi_{(2)}$ defined in eq. 60 , as ψ_2 defined in eq. 58 , has a regular power series expansion in λ .
Thus we rearrange the expressions in eq. 59

$$\left(-\widehat{\beta}\right)^{-1} = \left(-\widehat{\beta}_{sing.}\right)^{-1} + \left(-\widehat{\beta}_{reg.}\right)^{-1} \; ; \; \widehat{\beta} = \widehat{\beta}_{(2)} + \Delta_{(2)}\widehat{\beta}$$

$$\left(-\widehat{\beta}_{sing.}\right)^{-1} = b_0^{-1}\lambda^{-2} - b_0^{-2} b_1 \lambda^{-1}, \; \left(-\widehat{\beta}_{reg.}\right)^{-1} = \phi_{(2)} - \psi_{(2)}$$

$$\phi_{(2)} = b_0^{-1}\left(b_1^{(0)}\right)^2\left(1 - \left(b_1^{(0)}\lambda\right)^2\right)^{-1} \; ; \; b_1^{(0)} = b_1 / b_0$$

$$\psi_{(2)} = \left(-\widehat{\beta}_{(2)}\right)^{-1} \frac{\zeta_{(2)}}{1 + \zeta_{(2)}} = b_2 (b_0)^{-2} (1 + O(\lambda))$$

$$\zeta_{(2)} = \frac{\Delta_{(2)}\widehat{\beta}}{\widehat{\beta}_{(2)}} = b_2 b_0^{-1} \lambda^2 (1 + O(\lambda)) \tag{61}$$

We recall that in the decomposition $\widehat{\beta} = \widehat{\beta}_{(2)} + \Delta_{(2)}\widehat{\beta}$, retained in eq. 61 , $\widehat{\beta}_{(2)}$ and $\Delta_{(2)}\widehat{\beta}$ denote the beta function truncated to the first wo terms and the remainder term respectively . It further follows from eq. 61

$$\left(-\widehat{\beta}_{reg.}\right)^{-1} = b_0^{-1}\left((b_1/b_0)^2 - b_2/b_0\right)(1 + O(\lambda)) \tag{62}$$

We return to eq. 55 and integrate the singular part $\left(-\widehat{\beta}_{sing.}\right)^{-1}$

$$\begin{aligned} \tau &= \int_{\overline{\kappa}}^{\kappa} d\lambda \left(-\widehat{\beta}(\lambda)\right)^{-1} \\ &= \int_{\overline{\kappa}}^{\kappa} d\lambda \left\{\left(-\widehat{\beta}_{reg.}(\lambda)\right)^{-1} + \left(-\widehat{\beta}_{sing.}(\lambda)\right)^{-1}\right\} \end{aligned}$$

$$\int_{\overline{\kappa}}^{\kappa} d\lambda \left(-\widehat{\beta}_{sing.}(\lambda)\right)^{-1} = b_0^{-1}\begin{bmatrix} \overline{\kappa}^{-1} - \kappa^{-1} \\ - b_0^{-1} b_1 \left(\log\left(\overline{\kappa}^{-1}\right) - \log\left(\kappa^{-1}\right)\right)\end{bmatrix} \tag{63}$$

Eq. 55 thus allows the separation of variables $\overline{\kappa}$ and κ

$$\begin{aligned} \tau &= F(\overline{\kappa}) - G_{reg.}(\overline{\kappa}) - (F(\kappa) - G_{reg.}(\kappa)) \\ F(\overline{\kappa}) &= b_0^{-1}\overline{\kappa}^{-1} - \left(b_1/b_0^2\right)\log\left(\overline{\kappa}^{-1}\right) \\ G_{reg}(\overline{\kappa}) &= \int_0^{\overline{\kappa}} d\lambda \left(-\widehat{\beta}_{reg.}(\lambda)\right)^{-1} \end{aligned} \tag{64}$$

the substitution $\tau \rightarrow t = \tau - \log \left(\mu_0^2 / \Lambda^2 \right)$ **and inverting the functional relation** $t = t(\overline{\kappa}) \longleftrightarrow \overline{\kappa} = \overline{\kappa}(t)$

We rewrite eq. 64 separating sliding scale parts and associated scale μ and initial value parts associated with scale μ_0, as indicated in eqs. 51 - 53

$$\begin{bmatrix} \tau + b_0^{-1} \kappa^{-1} - \left(b_1 / b_0^2 \right) \log \left(b_0^{-1} \kappa^{-1} \right) - G_{reg.}(\kappa) + \\ + \left(b_1 / b_0^2 \right) \log \left(b_0^{-1} \overline{\kappa}^{-1} \right) + G_{reg.}(\overline{\kappa}) \end{bmatrix} = b_0^{-1} \overline{\kappa}^{-1} \tag{65}$$

The substitutions $\kappa \rightarrow b_0 \kappa$ and $\overline{\kappa} \rightarrow b_0 \overline{\kappa}$ in the inverse of the arguments of the logarithm terms in eq. 65 cancels out in the difference. Next we adjust the scale parameter $\mu_0 \rightarrow \Lambda$ setting

$$\begin{array}{l} -b_0^{-1} \kappa^{-1} + \left(b_1 / b_0^2 \right) \log \left(b_0^{-1} \kappa^{-1} \right) + G_{reg.}(\kappa) = \log \left(\mu_0^2 / \Lambda^2 \right) \\ t = \tau - \log \left(\mu_0^2 / \Lambda^2 \right) = \log \left(\Lambda^2 / \mu^2 \right) \\ x = b_0 \overline{\kappa} \; ; \; G_{reg.}(x / b_0) = G(x) \; ; \; b_1 / b_0^2 = a \end{array} \tag{66}$$

Further we substitute the sliding scale variable $\overline{\kappa} \longrightarrow b_0 \overline{\kappa} = x$. With these substitutions eq. 65 becomes

$$\begin{array}{c} t + a \log \left(x^{-1} \right) + G(x) = x^{-1} \\ \hline G(x) = x \sum_{n=0}^{\infty} G_n x^n \end{array} \tag{67}$$

In order to establish the asymptotic ultraviolet expansion for $t \rightarrow \infty$ and $Y = x^{-1} \rightarrow \infty$, we introduce the notation inivolving a change of variables $t = \exp(L) \; ; \; L = \log(t)$

$$\begin{array}{l} \log(t) = L \; ; \; x = x(L) \longrightarrow Y = Y(L) \equiv (x(L))^{-1} \\ \log(Y) = Z(L) \equiv \log \left((x(L))^{-1} \right) \end{array} \tag{68}$$

Eq. 67 takes the form

$$\begin{array}{c} L + \log \left[1 + \dfrac{a Z(L) + G(x = Y^{-1})}{\exp(L)} \right] = Z(L) \\ \hline t = \exp(L) \; ; \; Y = \exp(Z) \longleftarrow Z = Z(L) \end{array} \tag{69}$$

Eq. 69 can now be solved by iteration in the asymptotic limit $L \rightarrow \infty$

$$\begin{array}{l} L = \log \left[\log \left(\Lambda^{2} / \mu^{2} \right) \right] \rightarrow \infty \\ Z (L) = lim_{n \rightarrow \infty} Z_{n} (L) \\ L + \log \left[1 + \frac{a Z_{n} (L) + G \left(x_{n} = Y_{n}^{-1} \right)}{\exp (L)} \right] = Z_{n+1} (L) \\ \hline \text{anchor} : Z_{0} (L) = L \end{array} \tag{70}$$

The first anchor takes care of the largest contribution to to Z , i.e. L . Thus we obtain $Z_1 (L)$ and $Y_1 = \exp (Z_1)$ from rq. 70

$$\begin{array}{l} Z_{1} (L) = L + \log \left[1 + a L \exp (- L) + \frac{G (\exp (- L))}{\exp (L)} \right] \\ Y_{1} (L) = e^{L} + a L + G (\exp (- L)) \; ; \; t = e^{L} \end{array} \tag{71}$$

Comparing with the recursive equations (eq. 70) , repeated below

$$\begin{array}{l} L = \log \left[\log \left(\mu^{2} / \Lambda^{2} \right) \right] \rightarrow \infty \\ Z (L) = lim_{n \rightarrow \infty} Z_{n} (L) \\ L + \log \left[1 + \frac{a Z_{n} (L) + G \left(x_{n} = Y_{n}^{-1} \right)}{\exp (L)} \right] = Z_{n+1} (L) \\ \hline \text{anchor} : Z_{0} (L) = L \end{array} \tag{72}$$

it follows that all contributions from the function G , i.e. resulting from the subleading terms proportinal in the beta function truncated by the first to terms $b_0 \kappa^2 + b_1 \kappa^3$ give vanishing contributions to $Y (L) \equiv (b_0 \overline{\kappa})^{-1} (L)$ for $L \rightarrow \infty$. This can be verified recursively.
Thus modulo terms vanishing for $L \rightarrow \infty$ we can solve the simpler functional equation, omitting G from eqs. 67, 71 and 72

$$t + a \log \left(\widetilde{Y} \right) = \widetilde{Y} \; ; \; \widetilde{Y} = x^{-1} \; ; \; Y - \widetilde{Y} \rightarrow 0 \text{ for } t \rightarrow \infty \tag{73}$$

or equivalently

$$L + \log \left(1 + a \widetilde{Z} e^{-L} \right) = \widetilde{Z} \; ; \; \widetilde{Y} = e^{\widetilde{Z}} \; ; \; t = e^{L} \tag{74}$$

The same argument which led to the elimination of the term proportional to G in eq. 72 as far as nonvanishing contributions for Y are concerned in the limit $L \rightarrow \infty$ imply that expanding the logarithm on the left hand side of eq. 74 only the first term need be retained

$$\log \left(1 + a \widetilde{Z} e^{-L} \right) \sim a \widetilde{Z} e^{-L} \tag{75}$$

leading to the equivalent approximate functional equation simplifying eq. 74

$$\begin{aligned} & Z \sim \widetilde{Z} \sim L / \left(1 - a e^{-L} \right) \sim L + a e^{-L} L \longrightarrow \\ & Y \sim \widetilde{Y} \sim e^{L} \left(1 + a e^{-L} L \right) = e^{L} + a L + [\,0\,] \text{ for } L \rightarrow \infty \end{aligned} \tag{76}$$

In eq. 76 the $\sim$ symbol is meant to imply modulo additive terms to Y, vanishing for $L \rightarrow \infty$, proving the remarkable fact that the potential constant in the asymptotic expansion for Y vanishes .

We conclude this subsection comparing side by side the functional dependence of $t \equiv \log \left(\mu^{2} / \Lambda^{2} \right) = t \left(\overline{\kappa} \left(\mu \right) \right)$ and $Y \equiv \left(b_{0} \overline{\kappa} \right)^{-1} = Y \left(t \right)$.

The first relation is given in eq. 67

$$\begin{aligned} & t = x^{-1} - a \log \left(x^{-1} \right) - G \left(x \right) \\ \hline & G \left(x \right) = x \sum_{n=0}^{\infty} G_{n} x^{n} \; ; \; x = b_{0} \overline{\kappa} \left(\mu^{2} \right) \\ & t = \log \left(\mu^{2} / \Lambda^{2} \right) \longrightarrow \\ \hline & \Lambda^{2} = \mu^{2} \left[\exp \left(- \frac{1}{b_{0} \overline{\kappa}} \right) \right] \left[\frac{1}{b_{0} \overline{\kappa}} \right]^{a} \exp \left[G \left(b_{0} \overline{\kappa} \right) \right] \\ \hline & a = b_{1} / b_{0}^{2} \end{aligned} \tag{77}$$

The second relation is displayed in eq. 76 , as an asymptotic expansion for $t = e^{L} \rightarrow \infty$

$$\begin{aligned} \hline & Y{=}Y \left(L \right) \equiv \left(b_{0} \overline{\kappa} \right)^{-1} \left(L \right) \text{ for } L = \log \left[\log \left(\mu^{2} / \Lambda^{2} \right) \right] \rightarrow \infty \\ & \sim e^{L} + a L + [\,0\,] \\ \hline \end{aligned} \tag{78}$$

Eqs. 77 and 78 constitute the essence of this subsection, giving rise to some remarks :

1) The central scale Λ

   with all subtle properties as defined, in the $\overline{\text{MS}}$ scheme, constitutes a renormalization group invariant finite mass scale, given a region of sliding scale coupling constant $\kappa = g^{2} / \left(16 \pi^{2} \right)$ generically or $\overline{\kappa}$ in the perturbative region here, nonvanishing but appropriately small.

This scale does *not* depend on quark masses , only on the number of quark flavors , while derivations from measurable quantities (in the asymptiotic region) suffer from systematic errors , which do depend on quark massses .

The combined analysis of the sliding scale coupling constant , as of 2009 by Bethke , is shown in Fig. A21 .

I quote here a discussion of Λ for $N_{fl} = 3$ by Bodenstein et al., ref. [15-2011] , from data in the mass scale region of the τ lepton

$$\begin{array}{l} \Lambda_{N_{fl}=3} = 382 \pm 24 \text{ MeV} \longleftrightarrow \\ \alpha_s = 4\pi\kappa \left(\mu = m_\tau \right) = 0.344 \pm 0.014 \end{array} \tag{79}$$

The reconstruction of Λ from the 'perturbatively accessible region' implies a hard breaking of scale invariance , and hence equivalently a nonvanishing trace of the co-renormalizable and renormalized energy momentum density operator. This shall be discussed in the next subsection.

2) sliding scale quark masses

The rescaling equations (eqs. 51 - 53) extend to the quark mass parameters , again universally and quark mass independently rescaling m_f , keeping ratios invariant

$$r_{f_1 f_2} = m_{f_1} / m_{f_2} \tag{80}$$

Here no detailed discussion of extracting sliding scale quark masses including heavy flavors c , b is given. For heavy flavor comparisons with initially the reaction $e^+ e^- \to f \overline{f}-$ flavored hadrons , the QCD sum rule approach is the main tool , as pioneered by Shifman, Vainshtein and Zakharov [16-1979] , together with lattice QCD .

The sliding quark masses as for 3 flavors are shown in figure 6 below , restricted to two loop approximation . References [15-2011] - [17-2011] are representative for present refinement to four loops in the rescaling functions and derived results , e.g.

$$\begin{array}{l} \overline{m}_b \left(\overline{m}_b \right) = \left\{ \begin{array}{lll} 4163 \pm & 16 \text{ MeV} & \text{ref. [18-2010]} \\ 4171 \pm & 7 \text{ MeV} & \text{ref. [19-2011]} \\ 4177 \pm & 11 \text{ MeV} & \text{ref. [17-2011]} \end{array} \right. \\ \overline{m}_c \left(3 \text{ GeV} \right) = 986 \pm 13 \text{ MeV} , \text{ ref. [18-2010]} \\ \overline{m}_c \left(\overline{m}_c \right) = 1262 \pm 17 \text{ MeV} , \text{ ref. [17-2011]} \end{array} \tag{81}$$

3-4 – Reinterpreting the central anomalies

$$\frac{1}{4}\left[B^{\,r}_{\mu\nu}\, B^{\,\mu\nu\, r} \right]_{\infty} = \mathcal{B}^{\,+}_{\infty} \;\&\; \frac{1}{4}\left[B^{\,r}_{\mu\nu}\, \widetilde{B}^{\,\mu\nu\, r} \right]_{\infty} = \mathcal{B}^{\,-}_{\infty}$$

The central anomalies (eq. 39) become, using the field strength bilinears (re-)normalized at $\mu = \infty$, and separating gauge- and quark antiquatrk parts in the lowest order constant $b_{\,0} = b_{\,0\; g.b.} - \frac{2}{3} N_{\,fl}$ of the coupling constant rescaling function $\beta\,(\,g\,)$

$$\left\{ \begin{array}{ll} \vartheta^{\,\mu}{}_{\mu} & = \sum_{q} \left[m_{\,q}\,(\mu)\,\overline{q}\,q\,(\mu) + \frac{1}{12\pi^{\,2}}\,\mathcal{B}^{+}_{\infty} \right] - b_{\,0\;g.b.}\,\frac{1}{8\pi^{\,2}}\,\mathcal{B}^{\,+}_{\infty} \\ \partial^{\,\nu}\,(\overline{q}\gamma_{\nu}\gamma_{5}q) = 2m_{\,q}\,(\mu)\overline{q}\,i\,\gamma_{5}\,q\,(\mu) & \qquad\qquad + \;\; 2\,\frac{1}{8\,\pi^{\,2}}\,\mathcal{B}^{-}_{\infty} \end{array} \right\} (x)$$

$$-\,\beta/g^{\,3} = b_{\,0}\,/\,(\,16\,\pi^{\,2}\,) + O\,(\,\kappa\,)\,;\; \kappa = g^{\,2}\,/\,(\,16\,\pi^{\,2}\,)$$

$$b_{\,0} = b_{\,0\; g.b.} - \frac{2}{3} N_{\,fl}\,;\; b_{\,0\; g.b.} = \frac{11}{3}\, C_{\,2}\,(\, adj\;(Lie - \;SU3_{c})\,) = \frac{11}{3}\, 3 \qquad (82)$$

In eq. 82 normal ordering symbols are omitted for brevity of notation. Furthermore the renormalization scale for the field strengths bilinears

$$\frac{1}{4}\left[B^{\,r}_{\mu\nu}\, B^{\,\mu\nu\, r} \right]_{\infty} = \mathcal{B}^{\,+}_{\infty}\;;\; \frac{1}{4}\left[B^{\,r}_{\mu\nu}\, \widetilde{B}^{\,\mu\nu\, r} \right]_{\infty} = \mathcal{B}^{\,-}_{\infty}$$

is set to $\mu = \infty$ which does lead to a conceptual clarification of these quantities most importantly for a selfdual or anti-selfdual classical background field configuration.

In figure 5 , below , the behaviour of $m_{\,u}$, $\frac{1}{2}\,(\,m_{\,u} + m_{\,d}\,)$, $m_{\,d}$ as functions of the sliding scale Q is shown in a range of euclidean values of Q, believed to be within the perturbatively accessible region. The curves are all proportional to *one* universal function .

We return to the main theme of this discussion in the next section

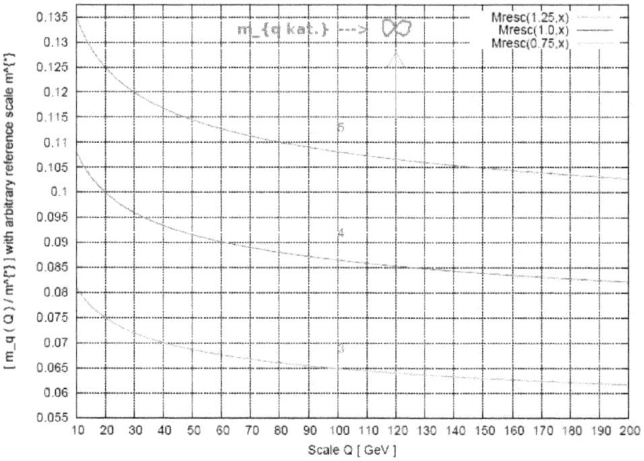

Fig. 5 : $m_{\,q}\,(\,Q\,)\,/\,m^{\,*}$ **with fixed ratio of rescaled quark masses** $m_{\,u} : \frac{1}{2}\,(\,m_{\,d} + m_{\,u}\,) : m_{\,d} = 3 : 4 : 5$

4 – Ideas forging and foregoing - the dynamics of genuinely oscillatory modes [1-1980]

It is worth noting, that Erwin Schrödinger turned to the discussion of oscillatory modes of single- and by reduction of c.m. coordinates – of a pair mode of oscillatory motion – in the last (4th) paper in ref. [20-1926]. However in the above paper he (E.S.) makes the assumption, that associated forces arise from the exchange of a photon, i.e. involve a local electromagnetic exchange interaction as giving rise to an equally local second order wave equation, responsible for oscillatory (pair-) modes . This is incorrect, contrary to the structure embedded in QCD, up to the present level of completion, which remains a future task .
Thus we concentrate on the present topic and lay out the ideas in ref. [1-1980] and relate the bond structure of quark-antiquark (meson) and $N-$ quark (baryon) systems, subject to an SU(N) unbroken color gauge group , to the long-range dynamics involving the oscillatory modes in the phase space of the center of mass position and momentum variables
$[N \equiv N_c \rightarrow 3]$. These canonical barycentric 3-vectors are shown in eq. 83

$$\begin{array}{lll}
\vec{\pi}_1 = \frac{1}{\sqrt{2}} \left(\vec{p}_1 - \vec{p}_2 \right) & , & \vec{z}_1 = \frac{1}{\sqrt{2}} \left(\vec{x}_1 - \vec{x}_2 \right) \\
\vec{\pi}_2 = \frac{1}{\sqrt{6}} \left(\vec{p}_1 + \vec{p}_2 - 2\,\vec{p}_3 \right) & , & \vec{z}_2 = \frac{1}{\sqrt{6}} \left(\vec{x}_1 + \vec{x}_2 - 2\,\vec{x}_3 \right) \\
\cdot & & \cdot \\
\vec{\pi}_\nu = \left(\nu(\nu+1) \right)^{-1/2} \begin{pmatrix} \sum_{\alpha=1}^{\nu} \vec{p}_\alpha \\ -\nu\,\vec{p}_{\nu+1} \end{pmatrix} & , & \vec{z}_\nu = \left(\nu(\nu+1) \right)^{-1/2} \begin{pmatrix} \sum_{\alpha=1}^{\nu} \vec{x}_\alpha \\ -\nu\,\vec{x}_{\nu+1} \end{pmatrix} \\
\cdot & & \cdot \\
\vec{\pi}_{N-1} = \cdots & , & \vec{z}_{N-1} = \cdots \\
\hline
\vec{\pi}_N = N^{-1/2} \sum_{\alpha=1}^{N} \vec{p}_\alpha \rightarrow 0 & , & \vec{z}_N = N^{-1/2} \sum_{\alpha=1}^{N} \vec{x}_\alpha \rightarrow 0
\end{array} \tag{83}$$

The last line in eq. 83 refers to c.m. momentum and position .
The bond structures of $q\overline{q}$ and $3q$ are repeated in Fig. 1-I below

P. Minkowski / Oscillatory modes of quarks in baryons

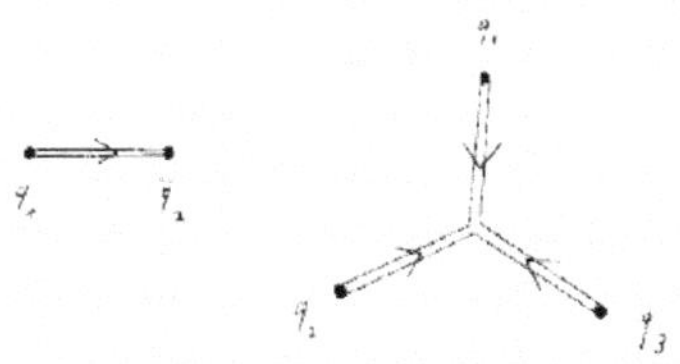

Fig. 1. Bond structures of $q\bar{q}$ and 3q configuration ($N = 3$).

Fig. 1 - I : Bond structures of $q\overline{q}$ and $3q$ configurations ($N = 3$)

Of course we are interested in $N = 3$ but one goal of this investigation is to shed light on the N dependence of the ratio of baryonic to mesonic inverse Regge slopes

$$\Lambda_N / \Lambda \; (= 1) \text{ for vanishing quark masses } M_\alpha \to 0 \; ; \; \alpha = 1, \cdots, N \tag{84}$$

$$\begin{aligned} \Delta \mathcal{M}^{\,2}_{baryon} &= 2 \, \Lambda_N \, \textstyle\sum_{\alpha=1}^{3N-3} \Delta \, \nu_\alpha \\ & \qquad \qquad \Delta \, \nu_\alpha = 0 \; , \; \pm 1 \; , \; \pm 2 \; , \cdots \\ \Delta \mathcal{M}^{\,2}_{meson} &= 2 \, \Lambda \quad \textstyle\sum_{\alpha=1}^{3} \Delta \, \nu_\alpha \end{aligned} \tag{85}$$

In eq. 84 quark masses are denoted M_α ; α : quark flavor , as throughout section 4, in order to distinguish them from – the oscillator state configuration space, variable, dependent masses – denoted m , $\overline{m}$ $\cdots$, and the connection of their ratio to the intermediary range quark-antiquark potential – mainly studied for the $c\overline{c}$ charmonium (binding) spectrum – determined by the conditions

$$Min_{\,\alpha=1,\cdots,N} \, M_\alpha \ll \left| \, V_{NR} \, (\, z_\beta \,) \, \right| \ll Min_{\,\gamma=1,\cdots,N-1} \, \left| \, z_\gamma \, \right|^{\,-1} \tag{86}$$

We reformulate as starting point the snowball effect for a $q\overline{q}$ equal mass pair , described (in the c.m. system) by the Lagrangean

$$\begin{aligned} \mathcal{L}_{\,(2)} &= - \, m_1 \, \left(1 - v_1^{\,2} \right)^{-1/2} - \, m_2 \, \left(1 - v_2^{\,2} \right)^{-1/2} \\ &= - \, 2 \, m \, \left(1 - v^{\,2} \right)^{-1/2} \\ \vec{v}_j &= \dot{\vec{x}}_j \; ; j = 1, 2 \; ; \; m = m \, (\, z \,) \text{ for (just here) } M_{q_1} = M_{\overline{q}_2} = 0 \end{aligned} \tag{87}$$

Eq. 87 is only valid in the c.m. frame , where the analog of energy conservation takes the form

$$\begin{aligned} \mathcal{H}_{\,(2)} &= \vec{v} \, \mathcal{L}_{\,(2)\,,\,\vec{v}} - \mathcal{L}_{\,(2)} = \frac{2 \, m}{\sqrt{1 - v^{\,2}}} \\ (\vec{p}\,)_1 - (\vec{p}\,)_2 &= 2 \, \vec{p}_{\,c.m.} = \mathcal{L}_{\,(2)\,,\,\vec{v}} = \mathcal{H}_{\,(2)} \, \vec{v} \\ \left(\mathcal{H}_{\,(2)} \right)^2 v^{\,2} &= \left(\mathcal{H}_{\,(2)} \right)^2 - 4 \, m^{\,2} = 4 \, p^{\,2}_{\,c.m.} \end{aligned} \tag{88}$$

Eqs. 87 and 88 are understood as approximations for large distances . They can be interpreted classically or quantum mechanically .

We note that in the discussion ongoing of $q\overline{q}$ oscillator modes, we do not use the orthogonal normalization as displayed in eq. 83 . This is so because other conventions had been used before, as for myself to the year 1976, while working at Caltech. The definitions used are shown in the next equation

$$\overline{m} = 2\,m \ , \ -\Delta_{\,z} + \overline{m}^{\,2}\,(\,z\,) = H^{\,2}_{\,(2)}$$

$$(\,\mathcal{H}_{\,(2)}\,\vec{v}\,)^{\,\cdot} = \mathcal{H}_{\,(2)}\,\tfrac{1}{2}\,\ddot{\vec{y}} = -\,\frac{1}{\mathcal{H}_{\,(2)}}\,grad_{\,y}\,\overline{m}^{\,2} \ ; \ \vec{y} = \vec{x}_{\,1} - \vec{x}_{\,2}$$

$$\longrightarrow \mathcal{H}^{\,2}_{\,(2)}\,\tfrac{1}{4}\,(\,\dot{\vec{y}}\,)^{\,2} = 4\,p^{\,2}_{\,c.m.} + \overline{m}^{\,2} = \mathcal{M}^{\,2} = \text{constant} \tag{89}$$

adopting the long range approximate nature of the harmonic oscillator relations – for the $q\overline{q}$ – bond

$$\overline{m}^{\,2} \sim_{|y|\to\infty} (\tfrac{1}{2}\,\Lambda\,)^{\,2}\,y^{\,2}\,[\,1 + O\,(\,M_{\,q}\,/\,|\,y\,|\,) + \cdots\,] \tag{90}$$

In eq. 89 we have substituted the variable $\vec{y}$ for $z \rightarrow \vec{z}$.
A few remarks shall follow

1) The lessons from the $q\overline{q}$ – bond are limited

The exclusive relative *distance*-dependence contained in the asymptotic term in eq. 90

$$\overline{m}^{\,2} \sim \tfrac{1}{2}\,\Lambda^{\,2}\,y^{\,2} \tag{91}$$

generates genuine oscillatory modes for the $q\overline{q}$ – bond , yet no multi-position dependent generalization can accomplish the same for the $3\,q$ – ($N\,q$ –) bonds .

2) but not empty

For the $q\overline{q}$ – bond it follows

$$\ddot{\vec{y}} = -\left(\frac{\Lambda}{\mathcal{H}_{\,(2)}}\right)\,y \ , \Lambda\,\left[\,-\,\Delta_{\,\xi} + \xi^{\,2}\,\right] = H^{\,2}_{\,(2)}$$

$$\xi = \tfrac{1}{2}\,(\,\Lambda\,)^{\,1/2}\,y$$

$$\longrightarrow \omega_{\,cl} = \frac{\Lambda}{\mathcal{H}_{\,(2)}} \ , H^{\,2}_{\,(2)}\,(\,\{\,\nu\,\}\,) = 2\,\Lambda\,\textstyle\sum^{\,3}_{\,\alpha=1}\,\nu_{\,\alpha} + 3\,\Lambda$$

$$\nu_{\,\beta} = 0\,,\,1\,,\,\cdots \ ; \ \beta = 1,2,3 \ ; \ \text{oscillator occupation numbers} \tag{92}$$

3) The color quantum number has vanished from the description

Vacuum - vacuum amplitedes of two colored local operators are not gauge invariant, provided local gauge invariance is not conserved 'completely' , to be defined including appropriate generalized boundary conditions, in QCD .

The wave functions on the other hand are not local.

4) Which are the dependences on quantum numbers like (light-) flavors and spin ?

The quantity Λ in eq. 91 , of dimension mass ^{2} is considered universal, i.e. does not depend on any other quantum numbers neither on quark masses, except on the occupation numbers of the oscillatory modes at hand.

Extension to include the $N\ q$ – bond

The key idea arose upon a discussion initiated by H. R. Dicke , concerning the feasibility and appropriateness to envisage a revision of the errors , as established by Loránd (Roland v.) Eötvös in 1918 , in his famous experiments in ref. [21-1918-1961] measuring the equality of inertial and gravitational mass with the help of rotating springs , to which test bodies are attached , while the springs are fixed to one point , say atop a rotating rod . It should be added that the springs must be elongating under the centrifugal force only in one longitudinal direction .
Dicke reports in ref. [21-1918-1961] that Eötvös , in his description of the experiments mentions an important obstacle to overcome , consisting in a precise separation of the mass of the test bodies from a combination of a part of the spring mass with them.
Hence the conclusion from the above situation to the question envisaged was, that in presence of position dependent mass this mass and inertial mass were *not the same* .
This led to the Ansatz, as I followd in ref. [1-1980]

$$\begin{array}{lll} \mathcal{L}_{N} = - \sum_{\alpha=1}^{N} \left[m_{\alpha}^{2} - Q_{\beta\gamma}^{\alpha}\, \vec{v}_{\beta} \cdot \vec{v}_{\gamma} \right]^{1/2} & & \\ \vec{v}_{\alpha} = \dot{\vec{x}}_{\alpha} & & \\ m_{\alpha} = m_{\alpha} \left[\vec{z}_{1}, \cdots, \vec{z}_{N-1} \right] & , & \text{gravitational effective masses} \\ Q_{\beta\gamma}^{\alpha} = Q_{\beta\gamma}^{\alpha} \left[\vec{z}_{1}, \cdots, \vec{z}_{N-1} \right] & , & \text{inertial effective masses} \end{array} \tag{93}$$

valid in the c.m. system of the N quarks .
The external quark masses M_{q} , appropriate multipliers of the scalar densities $\overline{q}q$ composing the mass term in the local (QCD) Lagrangian

$$- \mathcal{L}_{m} = \sum_{flavors} \frac{z_{q}}{z_{M}} M_{q}\, \overline{q}^{\,c} q^{\,c} \tag{94}$$

appear as constants in the gravitational mass

$$\begin{array}{l} m_{\alpha} \left[M_{q} , \underline{z} \right] = M_{\alpha} + m_{\alpha} \left[M_{q} = 0 , \underline{z} \right] \\ \underline{z} = \left(\vec{z}_{1} , \cdots , \vec{z}_{N=1} \right) \end{array} \tag{95}$$

whereas a consistent non-relativistic limit demands

$$
\begin{aligned}
\left[Q^{\alpha}_{\alpha\alpha} \left(M_q , \underset{\sim}{z} \right) \right]^{1/2} & \underset{M_q \to \infty}{\longrightarrow} M_\alpha + O \left[m_\beta \left[M_q = 0 , \underset{\sim}{z} \right] \right] \\
Q^{\alpha}_{\beta\gamma} , \ \beta,\gamma \neq \alpha & \underset{M_q \to \infty}{\longrightarrow} O \left[m_\beta \left[M_q = 0 , \underset{\sim}{z} \right] \right]
\end{aligned}
\tag{96}
$$

From eqs. 95 and 96 we recognize the problem of separation of mass and binding energy, as relevant, e.g., to the gravitational interaction of the whole N-quark system, appearing.
In the following m- , Q- are approximated by the corresponding quantities for $M_q = 0$, i.e., in the chiral limit with respect to all the quark flavors composing the $N \, q -$ bond .
Then in the harmonic long-range limit Q- is determined from m- through the relation

$$
Q^{\alpha}_{\beta\gamma} = \frac{1}{K_N} \left(m_\alpha \right)^2 \delta_{\beta\gamma} \ ; \ K_N \ : \ \text{constant}
\tag{97}
$$

The kinetic term for the quark – a – depends on all the velocities $\vec{v}_\beta$ and the Lagrangean $\mathcal{L}_N$ in eq. 93 takes the simplified form

$$
\begin{aligned}
& \mathcal{L}_N = - \overline{m} \left[1 - \frac{1}{K_N} \Sigma_\beta \left(\vec{v}_\beta \right)^2 \right]^{1/2} \ ; \ \overline{m} = \Sigma_{\alpha=1}^{N} m_\alpha = \overline{m} \left(\vec{x}_\beta - \vec{X} \right) \\
& \vec{X} = \frac{1}{N} \Sigma_{\alpha=1}^{N} \vec{x}_\alpha \to 0
\end{aligned}
\tag{98}
$$

The meaning of the constant K_N is the following : under the constraint $\Sigma_\alpha \vec{v}_\alpha = 0$ the maximum any individual $(\vec{v}_\beta)^2$ can assume for given $\Sigma_\gamma (\vec{v}_\gamma)^2$ is for the so-called $\lambda -$ mode , shown in Fig. 6 below .
An inequality for any individual square velocity follows

$$
\begin{aligned}
v_\alpha & = \left((\vec{v}_\alpha)^2 \right)^{1/2} \longrightarrow \\
v_\alpha^2 & \leq \frac{N - 1}{N} \Sigma_\gamma v_\gamma^2 \leq \frac{N-1}{N} K_N c^2 \quad (c = 1)
\end{aligned}
\tag{99}
$$

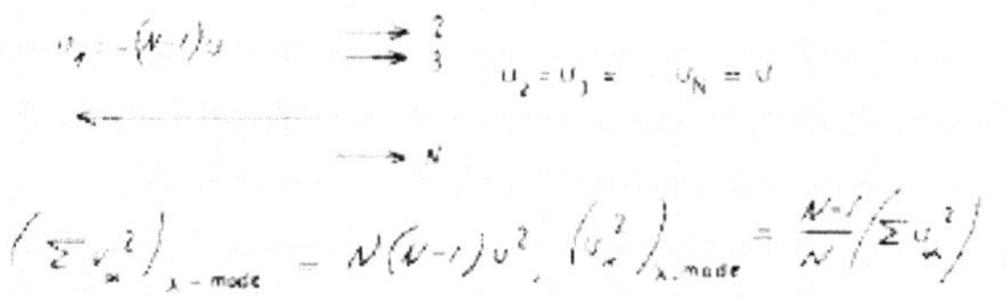

Fig. 6 : $\lambda -$ **mode for N quark bond** $\longleftrightarrow$

$\mathcal{L}_N$ in eqs. 98 and 99 delimits its validity *consistently* to physical values of v_α^2 for $K_N = N/(N-1)$

$$v_\alpha^2 \leq c^2 \longleftrightarrow K_N = \frac{N}{N-1} \tag{100}$$

The equation for the conserved energy (eq. 88 for $\mathcal{L}_{(2)}$) for $\mathcal{L}_N$ becomes

$$\begin{aligned} \mathcal{H}_N &= \vec{v}_\alpha \vec{p}_\alpha - \mathcal{L}_N \\ \vec{p}_\beta &= \mathcal{L}_{N,\vec{v}_\beta} = \frac{\overline{m}}{K_N} \left(\left[1 - \omega^2 \right]^{-1/2} \right)_{,\vec{v}_\beta} = \frac{\mathcal{H}_N}{K_N} \vec{v}_\beta \\ \omega^2 &= \frac{1}{K_N} \sum_{\gamma=1}^N v_\gamma^2 \; ; \; K_N = \frac{N}{N-1} \end{aligned} \tag{101}$$

From eq. 101 we obtain in canonically conjugate oscillator variables

$$\begin{aligned} (\mathcal{H}_N)^2 &= \left[K_N \sum_{\alpha=1}^N (\vec{p}_\alpha)^2 \Big|_{\sum_{\beta=1}^N \vec{p}_\beta = 0} + \overline{m}^2 (x_\gamma - X) \right] \\ &= \left[\begin{array}{l} K_N \sum_{\alpha=1}^N (\vec{p}_\alpha)^2 \Big|_{\sum_{\beta=1}^N \vec{p}_\beta = 0} + \\ + \dfrac{\Lambda^2}{K_N} \sum_{\alpha=1}^N (\vec{x}_\alpha)^2 \Big|_{\sum_{\beta=1}^N \vec{x}_\beta = 0} \end{array} \right] \end{aligned} \tag{102}$$

$\mathcal{H}_N$ is a constant of the motion by the relations displayed in eq. 101 , but it is $(\mathcal{H}_N)^2 \equiv \mathcal{M}_N^2$ becomes the genuinely canonical dynamic operator or in the classical framework 'Hamiltonian function' , in the genuinely relativistic situation .
The structure of $\mathcal{M}_N^2$ is derived straightforwardly from eqs. 101 and 102

$$\begin{aligned} \mathcal{M}_N^2 &= \left[\begin{array}{l} K_N \sum_{\alpha=1}^N (\vec{p}_\alpha)^2 \Big|_{\sum_{\beta=1}^N \vec{p}_\beta = 0} + \\ + \dfrac{\Lambda^2}{K_N} \sum_{\alpha=1}^N (\vec{x}_\alpha)^2 \Big|_{\sum_{\beta=1}^N \vec{x}_\beta = 0} \end{array} \right] \\ &= \left[K_N \sum_{\alpha=1}^{N-1} (\vec{\pi}_\alpha)^2 + \frac{\Lambda^2}{K_N} \sum_{\alpha=1}^{N-1} (\vec{z}_\alpha)^2 \right] \\ & (\vec{\pi}_\beta, \vec{z}_\beta) \; : \; \text{barycentric coordinates defined in eq. 83} \; ; \; K_N = \frac{N}{N-1} \end{aligned} \tag{103}$$

The mass-square spectrum according to $\mathcal{M}_N^2$ in eq. 103 is given by

$$\begin{aligned} \mathcal{M}_N^2 \Big|_{spectrum} &= 2\Lambda \sum_{\alpha=1}^{3N-3} \nu_\alpha + 3\Lambda (3N-3) \\ \nu_\beta &= 0, 1, \cdots \; ; \; \beta = 1, 2, \cdots, 3N-3 \end{aligned} \tag{104}$$

Except for the zero point contribution , i. e. for the oscillation level splittings, it is universal, independent of N, proving the validity of the universal relation in eq. 84 ($\Lambda_N / \Lambda = 1$) .
Taking the $\nu = \sum_{\eta=1}^{6} \nu_\eta = 2$, $P = +$ nonstrange baryon states, i. e. for $N_{fl} = 2$, $N = 3$ and counting in a reduced way, all corresponding oscillatory modes with positive parity , compatible with overall Bose symmetry, neglecting color, we obtain the following table of 21 states, not counting isospin and spin degrees of freedom separately

TABLE
πN-partial waves and associated number of baryon resonances with $\nu = 2$, $P = +$

πN partial wave	Number of states with $\nu = 2$, $P = +$	Candidates
P11 $N(\frac{1}{2}^+)$	4	2
P13 $N(\frac{3}{2}^+)$	5	1-2
F15 $N(\frac{5}{2}^+)$	3	2
F17 $N(\frac{7}{2}^+)$	1	1
P31 $\Delta(\frac{1}{2}^+)$	2	1-2
P33 $\Delta(\frac{3}{2}^+)$	3	1
F35 $\Delta(\frac{5}{2}^+)$	2	1
F37 $\Delta(\frac{7}{2}^+)$	1	1
	21	**10-12**

Fig. 7 : Nonstrange baryons with $\nu = 2$, $P = +$ in 1980 $\longleftrightarrow$

The candidates were collected in ref. [1-1980] from the PDG tables valid in 1980 . At least then almost 50 % of the resonances so characterized were missing, using this way of counting.

5 – Epilogue and outlook

The present outline has its root in my discussion of the oscillatory modes of quarks in baryons in ref. [1-1980] , as well as the presentation of two hadron resonance collections used in ref. [22-2010] , worked out subsequently in a notefile similar to this one in ref. [23-2011] .
In comparing the two collections used in ref. [22-2010], denoted Ntype=65 $\supset$ Ntype=26 , turn out both to be too small to account for the hadron abundances , as measured at RHIC and LHC , comparing with the Hadron Resonance Gas (HRG) approach (see e.g. ref. [24-2010]) , with noninteracting hadrons.
The detailed description of the collection of hadron resonances used in thermal fits to all , including RHIC and LHC hadron abundances is explicitely presented in ref. [25-2009] . The selection consists of including all hadron resonances in the PDG tables of 2012 [6-2012] and the review within the PDG tables [26-2012] .
Here we propose to consider the three light flavors of quark u , d , s , extending the modes discussed in ref. [1-1980] as a first step .
We define as final goal of the present study , based also on the 'Notes 22.12.2012 $\longrightarrow$ 17.06.2013' in ref. [27-2012] to illustrate eventual new data

relative to 1980 and the situation of missing states as it persists today according to the PDG particle tables . The recent inclusion of 1- and 2-star hadron resonances also used in evaluating the Hadron Resonance Gas states in heavy ion collisions [25-2009] , shall be scrutinized as well and compared with spectroscopic theoretical valence quark counting of oscillatory modes in hadrons.

Appendix 1 - Expansion coefficients of the rescaling functions $\widehat{\beta}$, γ to four loops

$$
\begin{array}{l}
- \beta / g = X B(X) \; ; \; B(X) = b_0 A(X) \\
B(X) \sim \sum_{n=0}^{\infty} b_n X^n \; , \; A(X) \sim \sum_{n=0}^{\infty} a_n X^n \\
\kappa = g^2 / (16 \pi^2) \text{ generic } X \\
b_0 = \frac{1}{3} (33 - 2 N_{fl}) \; , \; a_0 = 1 \; , \; a_n = b_n / b_0 \\
b_1 = \frac{2}{3} (153 - 19 N_{fl}) \\
b_2 = \frac{1}{54} \left(77139 - 15099 N_{fl} + 325 N_{fl}^2 \right) \\
b_3 \sim 29243 - 6946.3 N_{fl} + 405.089 N_{fl}^2 + 1.49931 N_{fl}^3
\end{array}
\tag{105}
$$

$$
\begin{array}{l}
-\gamma_m^0 = 4 \, , \, -\gamma_m^1 = \dfrac{202}{3} - n_{fl} \dfrac{20}{9} \qquad \left| \, -\gamma_m^l \equiv \chi_m^l \right. \\
-\gamma_m^2 = 1249 - \left[\dfrac{2216}{27} + \dfrac{160}{3} \zeta(3) \right] N_{fl} - \dfrac{140}{81} N_{fl}^2 \\
-\gamma_m^3 = \left\{ \begin{array}{l}
\dfrac{4603055}{162} + \dfrac{135680}{27} \zeta(3) - 8800 \zeta(5) + \\
+ \left[- \dfrac{91723}{27} - \dfrac{34192}{9} \zeta(3) + 880 \zeta(4) + \dfrac{18400}{9} \zeta(5) \right] N_{fl} + \\
+ \left[\dfrac{5242}{243} + \dfrac{800}{9} \zeta(3) - \dfrac{160}{3} \zeta(4) \right] N_{fl}^2 + \\
+ \left[- \dfrac{332}{243} + \dfrac{64}{27} \zeta(3) \right] N_{fl}^3
\end{array} \right\}
\end{array}
\tag{106}
$$

References

[1-1980] P. Minkowski, 'On oscillatory modes of quarks in baryons', Nucl.Phys. B174 (1980) 258-268 .

[2-2010] P. Minkowski, 'Oscillations of three Majorana neutrinos and a possible connection to QCD', http://www.mink.itp.unibe.ch in 'Lectures and talks' , file : singafl2010.pdf .

[3-2011] P. Minkowski, 'Infrared Instability in QCD' , file : QCDinfrared-instab2011.pdf, Lecture prepared for the
'INTERNATIONAL SCHOOL OF SUBNUCLEAR PHYSICS, 49th Course: Searching for the Unexpected at LHC and Status of our Knowledge Erice, 24 June 3 July 2011, 50th Anniversary Celebrations 2011-2012-2013 , ETTORE MAJORANA FOUNDATION AND CENTRE FOR SCIENTIFIC CULTURE .

URL : http:/www.ccsem.infn.it/

[4-2012] P. Minkowski, 'Glueballs a key consequence of QCD' $\longrightarrow$ 'Gluemesons : their conception needs all of QCD in the infrared' , files : QCDggb2012n.pdf , QCDggb2012n-A4.pdf , Lecture prepared for the INTERNATIONAL SCHOOL OF SUBNUCLEAR PHYSICS, 50th Course: What we would like LHC to give us, Erice, 23. June - 2. July 2012, 50th Anniversary Celebrations 2011-2012-2013 , "ETTORE MAJORANA" FOUNDATION AND CENTRE FOR SCIENTIFIC CULTURE , to appear in the Proceedings .

URL : http:/www.ccsem.infn.it/

[5-2011] C Amsler , T. De Grand and B. Krusche , 'Quark Model',
URL : http://pdg.lbl.gov/2012/reviews/rpp2012-rev-quark-model.pdf .

[6-2012] J. Beringer et al. (Particle Data Group), Phys. Rev. D86 (2012) 010001 .

[7-2013] P. Minkowski , 'Oscillatory modes of quarks in baryons for 3 quark flavors u , d , s' , file : oscimodes-fl6.pdf .

[8-2001] O. Bär and U.J. Wiese, 'Can one see the number of colors?', MIT-CTP-3144, May 2001, 24 pp., Nucl.Phys. B609 (2001) 225-246, hep-ph/0105258 .

[9-1964] M. Gell-Mann and Y. Neeman, 'The eightfold way', W. A. Benjamin, Inc., New York 1964 .

[10-1969] S. L. Adler, 'Axial vector vertex in spinor electrodynamics', Phys.Rev.177 (1969) 2426-2438 .

J. Bell and R. Jackiw, 'A PCAC puzzle: $\pi^0 \to \gamma\gamma$ in the sigma model', Nuovo Cim.A60 (1969) 47-61 .

[11-1991] P. Minkowski, 'Proton spin and quark spin', in Proceedings of the "Workshop on Effective Field Theories of the Standard Model", Dobcgókõ, Hungary August 22 - 26, 1991 ; U. Meissner edt., World Scientific Singapore 1992 , BUTP-91-2, Aug 1991. 31pp. .

[12-2009] S. Bethke, 'The 2009 Wolrd Average of $\alpha_s(M_Z)$ ', arXiv:0908.1135 [hep-ph] .

[13-1997] K.G. Chetyrkin, 'Quark mass anomalous dimension to $O\ (\alpha\ _s^4\ )$ ', March 1997, 7 pp., MPI-PHT-97-019, Phys.Lett. B404 (1997) 161-165, hep-ph/9703278 , and references cited therein .

[14-1997] T. van Ritbergen, J.A.M. Vermaseren and S.A. Larin, 'The Four loop beta function in quantum chromodynamics', UM-TH-97-01, NIKHEF-97-001, Jan 1997. 9pp.,
Phys.Lett.B400 (1997) 379 , hep-ph/9701390 , and references cited therein .

[15-2011] S. Bodenstein, C.A. Dominguez, S.I. Eidelman, H. Spiesberger and K. Schilcher, 'Confronting electron-positron annihilation into hadrons with QCD: An Operator product expansion analysis', Oct 2011, 17 pp., UCT-TP-282-10,MZ-TH-10-40, JHEP 1201 (2012) 039, arXiv:1110.2026 [hep-ph] , and references cited therein .

[16-1979] M.A. Shifman, A.I. Vainshtein and V.I. Zakharov, 'QCD and resonance physics. Theoretical Foundations.' , Nucl. Phys. B147 (1979) 385 , 448 .

[17-2011] S. Narison, 'Gluon Condensates and $m_{c,b}$ from QCD-Moments and their ratios to Order α_s^3 and $< G^4 >$ ', Phys.Lett. B706 (2012) 412-422, arXiv:1105.2922 [hep-ph] .

[18-2010] K. Chetyrkin, J.H. Kühn, A. Maier, P. Maierhöfer, P. Marquard, M. Steinhauser and C. Sturm, 'Precise Charm- and Bottom-Quark Masses: Theoretical and Experimental Uncertainties', Theor.Math.Phys. 170 (2012) 217-228, arXiv:1010.6157v2 [hep-ph] , and references cited therein .

[19-2011] S. Bodenstein, J. Bordes, C.A. Dominguez, J. Penarrocha and K. Schilcher, 'Bottom-quark mass from finite energy QCD sum rules', Nov 2011, UCT-TP-288-11, MZ-TH-11-37, Phys.Rev. D85 (2012) 034003, arXiv:1111.5742 [hep-ph] .

[20-1926] E. Schrdinger:

Quantisierung als Eigenwertproblem I, Annalen der Physik 79 (1926), 361-376 ,

Quantisierung als Eigenwertproblem II, Annalen der Physik 79 (1926), 489-527 ,

Quantisierung als Eigenwertproblem III, Annalen der Physik 80 (1926), 734-756 ,

Quantisierung als Eigenwertproblem IV, Annalen der Physik 81 (1926), 109-139 .

[21-1918-1961] R. v. Eötvös, Die Naturwissenschaften 6 (1918) 445; 7 (1919) 149; 327 ,
R.H. Dicke, 'The Eötvös Experiment', Scientific American (Dec., 1961) p. 84 .

[22-2010] S. Kabana and P. Minkowski, 'On the thermal phase structure of QCD at vanishing chemical potentials', CERN-PH-TH-2010-002, Jan 2010. 7pp., Int.J.Mod.Phys. A26 (2011) 3035-3050 , arXiv:1001.0707 [hep-ph] .

[23-2011] P. Minkowski, in the notefile phaseQCD2011.pdf , unpublished .

[24-2010] J. Cleymans, Recent developments around chemical equilibrium, J. Phys. G: Nucl. Part. Phys. 37 (2010) 094015 .

[25-2009] A. Andronic, P. Braun-Munzinger and J. Stachel, 'Thermal hadron production in relativistic nuclear collisions: the hadron mass spectrum, the horn, and the QCD phase transition', Phys. Lett. B673 (2009) 142 , arxiv:0812.1186v3 .

[26-2012] 'The Review of Particle Physics', PDG , J. Beringer et al , Review : 14. QUARKMODEL , Revised August 2011 by C. Amsler (University of Zurich), T. DeGrand (University of Colorado, Boulder) and B. Krusche (University of Basel) .

[27-2012] P. Minkowski, 'On the oscillatory modes of quarks in baryons for 3 quark flavors u, d, s', 'Notes 22.12.2012 $\longrightarrow$ 17.06.2013' , work to be continued .

Historical and textbook references to 'Continous transformation groups and differential geometry' – labelled 'xH

[1H-1893] S. Lie and G. Scheffers, 'Vorlesungen über continuierliche Gruppen, mit geometrischen und anderen Anwendungen', Leipzig 1893 .

[2H-1890] W. Killing, 'Die Zusammensetzung der stetigen endlichen Transformationsgruppen', Mathematische Annalen 31 (1888) 252-290; 33 (1888) 1-48; 34 (1889) 57-122; 36 (1890) 161-189 .

[3H-1894] É. Cartan, 'Sur la structure des groupes de transformations finis et continues', Thèse, Paris 1884 , IIème édition 1933 .

[4H-1897] F. Klein and A. Sommerfeld, 'Über die Theorie des Kreisels', 4 volumes, Teubner Verlag 1897 .

[5H-1903] L. Bianchi, 'Lezioni sulla teoria dei gruppi continui finiti di trasformazioni', Pisa 1903 , IIa edizione 1918 .

[6H-1927] F. Peter und H. Weyl, 'Die Vollständigkeit der primitiven Darstellungen einer geschlossenen kontinuierlichen Gruppe' , Ann. Math. 97 (1927) 737-755 .

[7H-1933] A. Haar, 'Der Massbegriff in der Theorie der kontinuierlichen Gruppen', Ann. Math. (2) 34 (1933) 147-169 .

[8H-1933] H. Weyl, 'The structure and representation of continous groups', I and II, Princeton 1933 and 1934-35.

[9H-1933] L.P. Eisenhart, 'Continuous groups of tranformations', Princeton 1933.

[10H-1941-1986] L. S. Pontryagin, 'Topological Groups', Gordon and Breach Sci. Publ., New York/London/Paris 1986 , first edition in russian : 1941 .

[11H-1951-1961] G. Racah, 'Group theory and spectroscopy', CERN Yellow report 61-8, 1961, URL : http://doc.cern.ch/cernrep/1961/1961-008/1961-008.html , reprinted from lectures dlivered (by the author) at the Institute for Advanced Study in Princeton, 1951 ; and references therein quoted below.

[12H-1953] G. W. Mackey, 'Induced Representations of Locally Compact Groups II. The Frobenius Reciprocity Theorem', The Annals of Mathematics, Second Series, Vol. 58, No. 2 (Sep., 1953) 193-221 .

[13H-1962] S. Helgason, 'Differential geometry and Symmetric Spaces', Academic Press New York and London 1962 .

[14H-1963] S. Kobayashi and K. Nomizu, 'Foundations of Differential Geometry', 2 volumes, John Wiley and Sons, Inc., Hoboken, NJ 1963, Wiley Classics Library 2009 .

End of historical and textbook references to 'Continous transformation groups and differential geometry – labelled 'xH'

References directly related to the two central anomalies – labelled 'R'

[R-1-1969] S. L. Adler, 'Axial vector vertex in spinor electrodynamics', Phys.Rev.177 (1969) 2426-2438 .

J. Bell and R. Jackiw, 'A PCAC puzzle: $\pi^0 \to \gamma\gamma$ in the sigma model', Nuovo Cim.A60 (1969) 47-61 .

[R-2-1991] P. Minkowski, 'Proton spin and quark spin', in Proceedings of the "Workshop on Effective Field Theories of the Standard Model", Dobogókõ, Hungary August 22 - 26, 1991 ; U. Meissner edt., World Scientific Singapore 1992 , BUTP-91-2, Aug 1991. 31pp. .

[R-3-1976] P. Minkowski, 'On the anomalous divergence of the dilatation current in gauge theories', Bern preprint 1976 ,
URL : http://www.mink.itp.unibe.ch .

[R-4-1931] G. de Rham, 'Sur l'analysis situs des variétés à n dimensions', J. Math. Pures Appl. (9) 10 (1931) 115-200 .

[R-5-1985] J. Manes, R. Stora and B. Zumino, 'Algebraic Study of Chiral Anomalies' LBL-19138, UCB-PTH-85/7, LAPP-TH-137, Feb 1985. 32pp., Commun.Math.Phys.102 (1985) 157 , scanned version from KEK library .

[R-6-1991] A. S. Kronfeld and U.-J. Wiese, '$SU(N)$ gauge theories with C-periodic boundary conditions – (I). Topological structure' , Nucl. Physics B357 (1991) 521-533 .

End of references directly related to the two central anomalies – labelled 'R'

References to 'Establishing the one renormalization group invariant quantity of dimension $\left[M^{4} \right]$ ' – labelled 'rBsquare'

[Bsquare-1975] H. Kluberg-Stern and J. B. Zuber, 'Ward Identities and Some Clues to the Renormalization of Gauge Invariant Operators', Phys.Rev. D12 (1975) 467 ,
'Renormalization of Nonabelian Gauge Theories in a Background Field Gauge. 1. Green Functions', Phys.Rev. D12 (1975) 482 and
'Renormalization of Nonabelian Gauge Theories in a Background Field Gauge. 2. Gauge Invariant Operators', Phys.Rev.D12 (1975) 3159-3180.

End of references to 'Establishing the one renormalization group invariant quantity of dimension $\left[M^{4} \right]$ ' – labelled 'rBsquare'

References to 'A recent paper by Guido Altarelli and references cited therein' – labelled 'r-sp-1'

[r-sp-1-2011] G. Altarelli, 'The Early Days of QCD (as seen from Rome)', Comments: 8 pages, 2 figures. Talk in honour of Mario Greco presented at the Rencontres de la Vallee d'Aoste 2011, La Thuile, Italy, arXiv:1106.3189 [hep-ph] .

End of references to 'A recent paper by Guido Altarelli and references cited therein' – labelled 'r-sp-1'

References to 'A selection of papers and textbooks for the entire realm of QCD' – labelled 'r-A2x'

[A21-197(3)4] D. Politzer, ' Reliable Perturbative Results for Strong Interactions?', Phys. Rev. Lett. 30 (1973) 1346 ,
H. Georgi and D. Politzer, 'Electroproduction scaling in an asymptotically free theory of strong interactions', Phys. Rev. D9 (1974) 416 .

[A22-197(3)4] D.J. Gross and F. Wilczek,
'Asymptotically free gauge theories. 1',Phys.Rev.D8 (1973) 3633,
'Asymptotically free gauge theories. 2',Phys.Rev.D9 (1974) 980-993.

[A23-1974] H. Georgi and D. Politzer, 'Electroproduction scaling in an asymptotically free theory of strong interactions', Phys. Rev. D9 (1974) 416 and ref. [A21-197(3)4] .

[A24-1975] H. Kluberg-Stern and J. B. Zuber, 'Ward Identities and Some Clues to the Renormalization of Gauge Invariant Operators', Phys.Rev. D12 (1975) 467 ,
'Renormalization of Nonabelian Gauge Theories in a Background Field Gauge. 1. Green Functions', Phys.Rev. D12 (1975) 482 and
'Renormalization of Nonabelian Gauge Theories in a Background Field Gauge. 2. Gauge Invariant Operators', Phys.Rev.D12 (1975) 3159-3180.

[A25] C. Itzykson and J. B. Zuber, 'Quantum field theory', McGraw-Hill, New York 1980.
M. E. Peskin, D. V. Schroeder, 'An Introduction to quantum field theory', Addison-Wesley, Reading, MA 1995 .
J. Zinn-Justin, 'Quantum field theory and critical phenomena', The International Series of Monographs on Physics, Vol. 113 , Oxford University Press, Oxford 2002 .

[A26-1982] G. Altarelli, 'Partons in quantum chromodynamics', NFN-ROME-783, Jul 1981. 211pp., Phys.Rept.81 (1982) .

[A27-1975] H. Fritzsch and P. Minkowski, 'Psi Resonances, Gluons and the Zweig Rule', Caltech preprint CALT-68-492, Mar 1975. 61pp., Nuovo Cim.A30 (1975) 393 .

[A28-1970] C. G. Callan, 'Broken scale invariance in scalar field theory', Phys.Rev.D2 (1970) 1541-1547 .

[A29-1970] K. Symanzik, 'Small distance behavior in field theory and power counting', Commun.Math.Phys.18 (1970) 227-246 .

[A210-1971(2)] G. 't Hooft, 'Renormalizable Lagrangians for Massive Yang-Mills Fields', Nucl.Phys.B35 (1971) 167-188 ,
G. 't Hooft and M.J.G. Veltman, 'Regularization and Renormalization of Gauge Fields', Nucl.Phys.B44 (1972) 189-213 .

[A211-1969] I. B. Khriplovich, 'Greens functions in theories with a non-abelian gauge group' , Yad. Fiz. 10 (1969) 409 , engl. transl. Sov. J. Nucl. Phys. 10 (1970) 235 .

[A212-1972] G. 't Hooft, Remarks on the ultraviolet stable sign of the beta function at the Colloquium on renormalization of Yang-Mills fields and applications to particle physics, Marseille 1972 .

[A213-1988] S. G. Gorishnii , A. L. Kataev and S. A. Larin, 'Next-To-Leading $O(\alpha_s^3)$ QCD Correction to Σ_t ($e^+ e^- \to$ Hadrons) : Analytical Calculation and Estimation of the Parameter Lambda (MS)', JINR-E2-88-254, Apr 1988. 18pp., Phys.Lett.B212 (1988) 238-244 , also in the same proceedings as ref. [cond-11-1988] .

[A214-1997] T. van Ritbergen, J.A.M. Vermaseren and S.A. Larin, 'The Four loop beta function in quantum chromodynamics', UM-TH-97-01, NIKHEF-97-001, Jan 1997. 9pp.,
Phys.Lett.B400 (1997) 379 , hep-ph/9701390 .

[A215-2009] S. Bethke, 'The 2009 Wolrd Average of $\alpha_s(M_Z)$',
arXiv:0908.1135 [hep-ph].

[A216-1969] K. Wilson, ' Nonlagrangian models of current algebra', Phys. Rev. 179 (1969) 1499 .

[A217-1954] M. Gell-Mann and F. Low, 'Quantum electrodynamics at small distances', Phys. Rev. 95 (1954) 1300 .

[A218-1973] S. Weinberg, 'New approach to the renormalization group', Phys. Rev. D8 (1973) 3497 .

[A219-1982] O. V. Tarasov, Dubna report JINR P2-82-900 (1982) , unpublished .

[A220-1994] K. G. Chetyrkin, C. A. Dominguez, D. Pirjol and K. Schilcher, 'Mass singularities in light quark correlators : the strange quark case', Phys.Rev.D51 (1995) 5090-5100 , hep-ph/9409371 .

[A221-2006] P. A. Baikov, K. G. Chetyrkin and J. H. Kühn, Phys. Rev. Lett. 96 (2006) 012003 .
P. A. Baikov, K. G. Chetyrkin and J. H. Kühn, 'Massless propagators: Applications in QCD and QED', SFB-CPP-08-06, TTP08-03, Oct 2008. 7pp., Contributed to 8th International Symposium on Radiative

Corrections (RADCOR 2007): Application of Quantum Field Theory to Phenomenology, Florence, Italy, 1-6 Oct 2007, PoS RADCOR2007 (2007) 023 , arXiv:0810.4048 [hep-ph] .

[A222-2008] P. A. Baikov, K. G. Chetyrkin and J. H. Kühn, 'Hadronic Z- and tau-decays in order α_s^4 ', Phys.Rev.Lett.101 (2008) 012002, arXiv:0801.1821v2 [hep-ph] .

[A223-2008] C.A. Dominguez, N.F. Nasrallah, R.H. Röntsch and K. Schilcher, 'Up and down quark masses from Finite Energy QCD sum rules to five loops', UCT-TP-271-08, Jun 2008. 4pp. , Phys.Rev.D79 (2009) 014009 , arXiv:0806.0467 [hep-ph] .

[A224-1979] M. A. Shifman , A. I. Vainshtein and V. I. Zakharov , 'QCD and resonance physics. Theoretical Foundations.' , Nucl. Phys. B147 (1979) 385 , 448.

[A225-2001] J. Gasser and H. Leutwyler, Annals Phys. 158 (1984) 142, H. Leutwyler, 'Nonlattice determinations of the light quark masses', Nucl.Phys.Proc.Suppl.94 (2001) 108, hep-ph/0011049.

[A226-2008] B. Blossier et al., European Twisted Mass Collaboration, 'Light quark masses and pseudoscalar decay constants from N(f)=2 Lattice QCD with twisted mass fermions', DESY-07-148, FTUV-07-2709, IFIC-07-57, MS-TP-07-23, RM3-TH-07-11, ROM2F-2007-16, SFB-CPP-07-58, TUM-HEP-676-07, Sep 2007. 20pp., JHEP 0804 (2008) 020 , arXiv:0709.4574 [hep-lat].

[A227-2008] C. Amsler et al. (Particle Data Group), Physics Letters B667 (2008) 1, and ref. [6-2012].

[A228-2008] T. Becher and M. D. Schwartz, 'A precise determination of alpha(s) from LEP thrust data using effective field theory', FERMILAB-PUB-08-048-T, Mar 2008, 34pp., JHEP 0807 (2008) 034, arXiv:0803.0342 [hep-ph].

[A229-2011] S. Narison, 'Gluon Condensates and $m_b(m_b)$ from QCD-Exponential Moments at Higher Orders', May 2011. 4pp., arXiv:1105.5070 [hep-ph].

[A230-2011] S. Borsanyi, G. Endrodi, Z. Fodor, A. Jakovac, S. D. Katz, S. Krieg, C. Ratti, K. K. Szabo, 'The QCD equation of state with dynamical quarks', WUB-10-12, Jul 2010. 31pp., JHEP 1011 (2010) 077, arXiv:1007.2580 [hep-lat] .

End of references to 'A selection of papers and textbooks for the entire realm of QCD' – labelled 'r-A2x'

References to 'Condensation phenomena and field theory realizations'
– labelled 'r-condx'

[cond-1-1993] W. H. Zurek, 'Cosmic strings in laboratory superfluids and remnants of other phase transitions', Acta Physica Polonica B7 (1993) 1301 ,
URL → http://th-www.if.uj.edu.pl/acta/vol24/pdf/v24p1301.pdf .

[cond-2-1937] P. L. Kapitza, Nature 141 (1937) 74 , JETP 11 (1941) , 1 and 581 .

[cond-3-1941] L. D. Landau, JETP 11 (1941) 592 and L. D. Landau and E. M. Lifshitz, 'Statistical Mechanics', Pergamon Press, New York 1958 .

[cond-4-1998] Complete list works of L D Landau, Physics-Uspekhi 41 (6), June, 1998.

[cond-5-1961] L. P. Pitaevskii, JETP 13 (1961) 451 .

[cond-6-195(6)7] L. D. Landau, 'The theory of a Fermi liquid', JETP 3 (1957) 920 (english translation) , original : JETP 30 (1956) 1058 ; 'Oscillations in a Fermi liquid', JETP 5 (1957) 101 (english translation).

[cond-7-195(6)7] L. N. Cooper, 'Bound Electron Pairs in a Degenerate Fermi Gas', Phys. Rev 104 (1956) 1189 - 1190 ,
J . Bardeen, L. N. Cooper, and J. R. Schrieffer, 'Microscopic Theory of Superconductivity', Phys. Rev. 106 (1957) 162 - 164 ,
J. Bardeen, L. N. Cooper, and J. R. Schrieffer, 'Theory of Superconductivity', Phys. Rev. 108 (1957) 1175 .

[cond-8-1947] N. N. Bogoliubov, J. Phys. (USSR) 11 (1947) 23 .

[cond-8'-1947] N. N. Bogoliubov, 'On the Theory of Superfluidity', J. Phys. XI (1947) 292-301.

[cond-9-195(4)5] N. N. Bogoliubov and D. N. Zubarew, 'The Wave Function of the Lowest State of a System of Interacting Bose Particles', Sov. Phys. JETP 1 (1955) 83-90, and references cited therein.

[cond-10-2002] J.M. Vogels,K. Xu,C. Raman,J.R. Abo-Shaeer and W. Ketterle, 'Experimental observation of the Bogolyubov transformation for a Bose-Einstein condensed gas', Phys.Rev.Lett.88 (2002) 060402, cond-mat/0109205 .

[cond-11-1988] P. Minkowski, 'Color superconducvtivity and the associated phase transition in QCD', in 'Hadron Interactions', Proceedings of the International Symposium on "Hadron Interactions – Theory and Phenomenology", Bechyne, Czechoslovakia, 26. June - 1. July, 1988, J. Fischer, P. Kolář, V. Kundrát eds., Prague 1988 .

CHAIRMAN: P. MINKOWSKI

Scientific Secretaries: M. Ulybyshev, M. Zoccali,

DISCUSSION

- *Kiehn:*

It seems as if you have defined two different values for the gravitational and the inertial mass of the baryons. My question is twofold: first, am I right in saying that in your calculations there are different values for these two, and, if so, what sensitivity would be required to resolve this difference?

- *Minkowski:*

Let me first clarify a misunderstanding. I was defining gravitational and inertial mass of *quarks* (u, d, s) inside *baryons.* On the contrary baryons –in particular stable ones as *p* (*proton*) and *n* (*neutron*), neglecting (electro-)weak interactions – have assigned masses for which the gravitational and inertial masses are exactly equal. Applied to the so singled out *quark* masses this is an extension of the question or problem that R.v.Eötvös had: namely to assign up to an error of one part in 10^9 the mass of the ball that we may call 'test particle'. However for this ball to be attached to a spring, you have to include the spring, which has a mass too. Now, the problem was solved up to one part in 10^9 by R.v.Eötvös by varying the way this is done in order to separate the spring mass, which does not belong to the ball. If you have gauge boson fields providing the energy and playing the role of the spring, this field energy is the integral part of the ball mass, so these two *quantities* – inertial- and gravitational mass of a *quark* inside a baryon will not be equal at all not in one part in 10^9, but in one part in one. This generates the parametrization which I showed you, and you see how the kinetic energy has to behave as a function of all relative velocities. Once you have this idea, there are only two parameters left, as I showed this morning, and you can solve this completely with the condition that the mass squared should obey in order to avoid velocities exceeding c. The inertial mass of the quark at the end of the bond is essentially zero. This has nothing to do with the assigned mass that it gets by being in a *baryon* – or a *nucleus* or a *meson.*

- *Inguglia:*

Can you provide any interpretation of the X_{3872} in your study?

- *Minkowski:*

No. What we are talking about is only a principal solution which is not exact because of existence of spin and some additional forces and which is at most close to measurable spectroscopic masses when the quarks are well approximated by the chiral limit. The X is the worst example for that. You cannot use my theory in that case, but X itself brings new questions that L. Maiani discussed in his lecture .In the case of light quark flavors – u, d, s – we need more spectroscopic information . This involves the light f0 [500, 980, 1370, 1500, 1710 according to the PDG].I must say that many things regarding the f0-states are still controversial, e.g.:the pole mass of f0 (500) is not at 500 MeV but at 1100 MeV, see ref. [3-2011] in my lecture. This is a transposition of your question and should be investigated analytically and with measurements .

- *Inguglia:*

Do you think that we need to invest on a lower (than LHC) energy experiment for high precision measurements?

- *Minkowski:*

Yes .One of such machines is a facility in Mainz. This new machine makes pion beams possible and can measure for instance the electric and magnetic polarisability of the proton. This is a very nice machine where the "crystal ball" detector was transferred from SLAC. Experimenters and theorists in Mainz are trying to use at maximum this machine, which is perfect. The other is FAIR, that will indirectly provide spectroscopy of the low mass hadrons. And even LHC can do a miracle not looking at Higgs but looking at whatever is not Higgs. So LHC can be extremely useful in analysis of low-energy phenomena in parallel with of course the study of the Higgs boson – being of primary importance. There is a non-trivial potential of the LHC with respect to this low energy physics.

V. MUKHANOV

"Quantum Origin of the Universe Structure"

The Slides of the Lecture can be found at:

http://www.ccsem.infn.it/issp2013/

CHAIRMAN: V. MUKHANOV

Scientific Secretaries: P. Dunin-Barkowski, A. Morozov

DISCUSSION

- *Leithes:*

The question is about the prediction of the spectral index of the scalar perturbations you had on your slides versus that generated from the Planck data. The spectral index is calculated from the low L multipoles where the error bars are largest. Are you worried about these uncertainties in the Planck values for the spectral index? The deficit in the power in the low Ls is also difficult to model. I'm not aware of any inflaton potential that can fit the low Ls and the high Ls simultaneously. Could you please comment on that?

- *Mukhanov:*

Spectral index is extracted from the large multipoles. Lack of power at low multipoles is not, in fact, so serious. If you will take into account the cosmic variance (for low multipoles we have just a few representatives within observable horizon) then the probability of this "low power" is about a percent. If one would have a chance to fly to different universes and everywhere would see the same "low power", then I would worry, but because we see only one realization, it's not a big deal, in fact.

- *Leithes:*

So the Planck team overplayed this?

- *Mukhanov:*

They actually confirmed anomalies that WMAP has seen, but they state themselves that all these anomalies are at the marginal level of detection, 2.5 sigma maximum and this is not more than just an indication that in principle it might be correct.

- *Leithes:*

Which inflatonary model have you actually used in your predictions?

- *Mukhanov:*

It was R^2 gravity, but all reasonable models give more or less the same predictions. Irrespective of the model, I can say that if the universe structure originated from quantum fluctuations in the same matter component which was responsible for accelerated expansion, the spectral index should deviate from unity by at least 3%, namely, it should be smaller than 0.97. I recall a story from 2003: when Lyman Page was applying to get money for Atacama Cosmology Telescope, he asked me if he would find 0.99, would I denounce the inflation? I told him that if he would find 0.99 plus/minus 0.01 at 3-sigma level, then we can forget about inflation as predictive theory. People are more sophisticated and if we would not find what we expected, they would even involve more parameters to "explain" the outcome of the experiment, but the real value of these explanations would be zero. Fortunately this did not happen and therefore we can be sure that we really have learned something about nature.

- *Tawfik:*

Assuming that you have a quantum fluctuation in inflation era, and you are arguing that you can explain the late universe, can you explain how this early fluctuations survives the successive phase transitions: electroweak, QCD, hadronization, and you find some implication of it later on?

- *Mukhanov:*

Normally, when we are speaking about the observed fluctuations, in past they were $10^{(-25)}$ cm, but during inflation they were stretched to about $10^{(-3)}$ mm. After this, universe was expanding in the normal way, and the fluctuations we are talking about were after that most of the time larger than the so-called inverse Hubble radius, i.e. the scale of causality. Therefore, just relying on the causality reasoning, one can say that phase transitions cannot influence the fluctuations on large scale, because there is not enough time for that. The fluctuations in which we are interested in become causal again only when the universe is 100000 old. Before that everything was frozen. When inhomogeneities come again inside the horizon, they begin to behave as standing sound waves in baryon radiation plasma. Before re-entering the Hubble scale the period of the waves was larger than the age of the universe and they looked like they were frozen. After the waves begin to evolve, starting from very simple spectrum and using very well understood physics (plasma physics plus gravity), we can calculate the observed picture of the temperature fluctuations, which emerges from initially featureless spectra. As a result we obtain the so-called Doppler peaks etc.

- *Tawfik:*

Some people say that at the Planck scale the Heisenberg uncertainty principle can get some modifications, or the space itself can become discretized. Can this influence the inflation theory you were speaking about?

- *Mukhanov:*

Of course you can assume everything, but, as you see, data tells us that you do not need any modifications. Everything what we know, including Heisenberg uncertainty relation, gravity, namely, Einstein theory (with possible R^2 corrections due to vacuum polarization effects) works perfectly well. Therefore there is no need at the scale where inflation happens (twelve orders of magnitude below the Planck scale) to introduce any modification to Heisenberg uncertainty relation or even general relativity. You just quantize general relativity in the most simple possible way (at linearized level), and this kind of quantum gravity seems to be most useful because it explains the origin of the universe structure. You do not need to care even about non-renormalizability. Everything what concerns the inflation is happening at "safe" energy scales about twelve orders of magnitude below the Planck scale.

- *Tawfik:*

Could you comment on the equation of state of the primordial plasma and of the dark matter and energy?

- *Mukhanov:*

The equation of state we need for inflation exists in nature because as we know from observations today the matter is dominated by dark energy. This dark energy we observe today seems useless for any "practical" purposes. However it is a good indication that such an equation of state can be realized in past when it was extremely useful. Because of dark energy one can produce the universes like ours in one go, starting from quantum fluctuations.

- *De Palma:*

At the beginning of inflation we should have something like quantum state which is a superposition of classical states, describing all these quantum fluctuations. Then, now we have just a classical state. So something like wave function collapse should have happened in between. Could you comment on this?

- *Mukhanov:*

In principle, you can start with any state. Some people say that you have to postulate Bunch-Davies state, or something like that. This is wrong, because one can start with any state, which does not spoil the stage of accelerated inflation. If inflations lasts more than say 75 e-folds, then the only thing relevant for observed today universe are the minimal quantum fluctuations. These fluctuations on the scales smaller than the Hubble scale are just the same as the fluctuations in Minkowski space. When they are stretched at larger scales they preserve amplitude and as a result the occupation numbers on the scales we are interested are huge. Therefore after inflation one gets a superposition of different classical states, that is, something like the Schroedinger cat. The wave function is a superposition of many classically different universes, where for instance, Magellanic clouds can be in completely different positions with respect to us. Because the state of initial quantum fluctuations was translational invariant it will remain translational invariant in homogeneous expanding background and hence there are no preferable points where particular galaxies can be formed according to the wave function describing inhomogeneities. To explain why nevertheless galaxies were formed in the particular positions one has to go to involve some interpretation of quantum mechanics. According to Bohr, one should say that the first astronomer, when he looked at the sky, has made the collapse of the full wave function to its particular component, which describes our universe. This, of course, sounds like nonsense. Another possibility is to think of this wave function as describing many classically different universes, as was suggested by Everett. I cannot add too much to this deep question that goes back to the Einstein-Bohr discussion. The only guy who has made a significant contribution to this discussion, in my opinion, was Everett, although most of the people do not accept his point of view.

- *Melia:*

I was told that the Planck is an ultimate experiment, we will never have to do it again. Will there be future experiments of this type?

- *Mukhanov:*

There is one more prediction of early stage of acceleration, which was not yet confirmed– the existence of long wavelength primordial gravitational waves that were generated on inflation. To see these waves one has to measure the B-mode of polarization of the temperature fluctuations at low multipoles. Planck gave limit on the ratio of tensor to scalar perturbations at the level of 0.11. This allowed us to rule out a lot of models of inflation like φ^2, φ^4, etc. But for the given spectral index, which was found, there is a low bound on the ratio of tensor to scalar perturbations, which is about 0.003. There will be improvement of the Planck accuracy by next year, which perhaps will allow us to go to the level 0.03, but to improve accuracy by order of magnitude one needs have a space mission. If such a mission will be approved, then its results we will have perhaps not early than in 20 years from now.

- *Melia:*

Will we reach a point when we cannot get any more information from the sky?

- *Mukhanov:*

It is always difficult to forecast such things because nobody knows how things will work in future. Fore sure there will be some ground-based and balloon experiments and I believe that within next 10-20 years cosmic microwave background fluctuations will remain an active field of research, but I'm not sure if it will remain a field of great discoveries.

- *Sastre Bruno:*

Would it be possible to do something similar to CMB for neutrinos originating from the nucleosynthesis?

- *Mukhanov:*

No.

- *Sicking:*

In the universe we have baryonic matter, dark matter and a large amount of dark energy. But so far, nobody knows what this dark energy actually is. Do you know of any good theories which can give an insight of what this dark energy can be?

- *Mukhanov:*

There are too many theories of dark energy and there are also theories, which pretend that they can explain present acceleration without dark energy. But I do not know any good one. This is a big problem. Otherwise there would be a natural candidate for this thing. Right now the simplest theory, if one can call it a theory, is just the theory where dark energy is a cosmological constant.

- *'t Hooft:*

How can it be that the age of the Universe can be derived from the cosmic background observations?

- *Mukhanov:*

Because the temperature fluctuations curve depends on the Hubble constant. In principle, this dependence is the weakest one. Normally Hubble constant comes with Ω. But in one place related to location of the peaks this degeneracy is resolved. Now with all these precision measurements, especially in the region of large multipoles, one can get the value of the Hubble constant with accuracy better than from supernova.

- *Zichichi:*

Planck sees what happens now, at this level. From now to then you have 54 powers of ten. And we know that unification energy where α_1, α_2 and α_3 meet is not at the Planck scale but two orders lower. How can it be that without knowing anything from the Big Bang to the Unification you claim that you have such a precision on the age of the Universe?

- *Mukhanov:*

As I said not too much happens with perturbations in the relevant scales before universe becomes about hundred thousands years old simply because of causality. We know that the perturbations had before the scales much larger than the size of the causally connected region. Therefore you do not need really to care what was happening with them in the very early universe because nothing was happening, the perturbations were frozen. They begun to evolve only when the universe was 20 thousand or hundred thousand years old. This is the main reason that makes things independent of all the phase transitions. Of course there is a lot of particle physics after inflation relevant, for instance, for baryon asymmetry etc.

No-Scale Supergravity in the Light of LHC and Planck

Tianjun Li,[1,2] James A. Maxin,[3] Dimitri V. Nanopoulos,[2,4,5,*] and Joel W. Walker[6]

[1] *State Key Laboratory of Theoretical Physics and Kavli Institute for Theoretical Physics China (KITPC), Institute of Theoretical Physics, Chinese Academy of Sciences, Beijing 100190, P. R. China*

[2] *George P. and Cynthia W. Mitchell Institute for Fundamental Physics and Astronomy, Texas A&M University, College Station, TX 77843, USA*

[3] *Department of Physics and Astronomy, Ball State University, Muncie, IN 47306 USA*

[4] *Astroparticle Physics Group, Houston Advanced Research Center (HARC), Mitchell Campus, Woodlands, TX 77381, USA*

[5] *Academy of Athens, Division of Natural Sciences, 28 Panepistimiou Avenue, Athens 10679, Greece*

[6] *Department of Physics, Sam Houston State University, Huntsville, TX 77341, USA*

We review the No-Scale $\mathcal{F}$-$SU(5)$ model with extra TeV-scale vector-like flippon multiplets and associated collider phenomenology relating to the search for supersymmetry at the LHC. The model framework possesses the rather unique capacity to provide a light CP-even Higgs boson mass in the favored 124–126 GeV window while simultaneously retaining a testably light Supersymmetry (SUSY) spectrum. We further elucidate the emerging empirical connection between recent Planck satellite data and No-Scale Supergravity cosmological models that mimic the Starobinsky model of inflation, drawing upon work conducted by the conference presenter (DVN) with John Ellis (King's College London/CERN) and Keith Olive (Minnesota).

*Contribution to the Proceedings of the International School of Subnuclear Physics - 51st, Reflections on the Next Step for LHC, Erice, Sicily, Italy, 24 June - 3 July 2013, based on a talk given by Dimitri V. Nanopoulos.

I. INTRODUCTION

We have developed a framework named No-Scale $\mathcal{F}$-$SU(5)$ [1–9] that is based upon the tripodal foundations of the dynamically established boundary conditions of No-Scale Supergravity, the Flipped $SU(5)$ Grand Unified Theory (GUT), and a pair of TeV-scale hypothetical *"flippon"* vector-like super-multiplets motivated within local F-theory model building. The union of these features has been demonstrated to naturally resolve a number of standing theoretical problems, and to compare favorably with experimental observation of the real world.

The aggressively minimalistic formalism of No-Scale Supergravity [10–14] provides for a deep connection to string theory in the infrared limit, the natural incorporation of general coordinate invariance (general relativity), a mechanism for Supersymmetry (SUSY) breaking which preserves a vanishing cosmological constant at the tree level (facilitating the observed longevity and cosmological flatness of our Universe [10]), natural suppression of CP violation and flavor-changing neutral currents, dynamic stabilization of the compactified spacetime by minimization of the loop-corrected scalar potential, and a dramatically parsimonious reduction in parameterization freedom. The split-unification structure of flipped $SU(5)$ [15–18] provides for fundamental GUT scale Higgs representations (not adjoints), natural doublet-triplet splitting, suppression of dimension-five proton decay [18, 19], and a two-step see-saw mechanism for neutrino masses [20, 21]. Modifications to the one-loop gauge β-function coefficients b_i induced by inclusion of the vector-like flippon multiplets create an essential flattening of the $SU(3)$ Renormalization Group Equation (RGE) running ($b_3 = 0$) [1], which translates into a wide separation between the primary $SU(3)_C \times SU(2)_L$ unification near 10^{16} GeV and the secondary $SU(5) \times U(1)_X$ unification near the Planck mass. The corresponding baseline extension for logarithmic running of the No-Scale boundary conditions, especially that imposed ($B_\mu = 0$) on the soft SUSY breaking term B_μ associated with the Higgs bilinear mass mixing μ, allows sufficient room for natural dynamic evolution into phenomenologically viable values at the electroweak scale. Associated flattening of the color-charged gaugino mass scale generates a stable characteristic mass texture of $M(\widetilde{t}_1) < M(\widetilde{g}) < M(\widetilde{q})$, featuring a light stop and gluino that are lighter than all other squarks [6].

At this vital juncture in the LHC's history, entering an operational pause for refitting to enable collisions at the full design energy of 14 TeV, we take stock of the No-Scale $\mathcal{F}$-$SU(5)$ model's full experimental status and prospects. The first phase of LHC data collection at $\sqrt{s} = 7, 8$ TeV was highly noteworthy for both a discovery (the isolation of a Standard Model (SM) like Higgs boson at around 125 GeV) and a null result (the stubborn persistence of SUSY to unambiguously rise above the noise floor with increasing luminosity and energy). Both circumstances have pressed the standard Constrained Minimal Supersymmetric Standard Model (CMSSM) and Minimal Supergravity (mSUGRA) parameter spaces to an extreme. As is natural, the $\mathcal{F}$-$SU(5)$ model space has likewise diminished as it has been probed by incoming collider data [8], although a wealth of sufficient room to maneuver remains. This result is made all the more remarkable by the fact that the No-Scale $\mathcal{F}$-$SU(5)$ construction takes the form of a Minimal Parameter Model (MPM), with all essential experimental characteristics established solely by the universal gaugino mass boundary $M_{1/2}$, according to which the full SUSY particle spectrum is proportionally rescaled *en masse*. A MPM of similar form has been much studied in the past (See [22] and references therein. For a review, see [23]). In particular, the precise nature of the high-mass cutoff enforced on the model is at the point where the charged stau and neutral lightest supersymmetric particle (LSP) mass degeneracy becomes so tight that the Planck [24] and WMAP [25] cold dark matter (CDM) relic density observations cannot be satisfied. This hard upper boundary on the model's leading dimensionful parameter, $M_{1/2}$, constitutes a top-down theoretical mandate for a comparatively light (and testable) SUSY spectrum which does not excessively stress natural resolution of the gauge hierarchy problem, and is in itself a rather distinctive and unique No-Scale $\mathcal{F}$-$SU(5)$ characteristic.

A recent analysis [26–28] suggests that a cosmological model based upon the No-Scale supergravity sector yields compatibility with the Planck satellite measurements. With convenient superpotential parameter choices, the new cosmological model compatible with Planck data is a No-Scale supergravity realization of the Starobinsky model of inflation [29–31]. We shall elaborate here upon this intriguing connection between the No-Scale $\mathcal{F}$-$SU(5)$ GUT model and a No-Scale Wess-Zumino model of inflation.

II. THE NO-SCALE $\mathcal{F}$-$SU(5)$ MODEL

Supersymmetry naturally solves the gauge hierarchy problem in the SM, and suggests (given R parity conservation) the LSP as a suitable cold dark matter candidate. However, since we do not see mass degeneracy of the superpartners, SUSY must be broken around the TeV scale. In GUTs with gravity mediated supersymmetry breaking, called the supergravity models, we can fully characterize the supersymmetry breaking soft terms by four universal parameters (gaugino mass $M_{1/2}$, scalar mass M_0, trilinear soft term A, and the low energy ratio of Higgs vacuum expectation values (VEVs) $\tan\beta$), plus the sign of the Higgs bilinear mass term μ.

No-Scale Supergravity was proposed [10] to address the cosmological flatness problem, as the subset of supergravity models which satisfy the following three constraints: i) the vacuum energy vanishes automatically due to the suitable Kähler potential; ii) at the minimum of the scalar potential there exist flat directions that leave the gravitino mass $M_{3/2}$ undetermined; iii) the quantity $\mathrm{Str}\mathcal{M}^2$ is zero at the minimum. If the third condition were not true, large one-loop corrections would force $M_{3/2}$ to be either identically zero or of the Planck scale. A simple Kähler potential that satisfies the first two conditions is [10, 13]

$$K = -3\ln(T+\overline{T}-\sum_i \overline{\Phi}_i\Phi_i) , \qquad (1)$$

where T is a modulus field and Φ_i are matter fields, which parameterize the non-compact $SU(N,1)/SU(N)\times U(1)$ coset space. The third condition is model dependent and can always be satisfied in principle [32]. For the simple Kähler potential in Eq. (1) we automatically obtain the No-Scale boundary condition $M_0 = A = B_\mu = 0$ while $M_{1/2}$ is allowed, and indeed required for SUSY breaking. Because the minimum of the electroweak (EW) Higgs potential $(V_{EW})_{min}$ depends on $M_{3/2}$, the gravitino mass is determined by the equation $d(V_{EW})_{min}/dM_{3/2} = 0$. Thus, the supersymmetry breaking scale is determined dynamically. No-Scale supergravity can be realized in the compactification of the weakly coupled heterotic string theory [33] and the compactification of M-theory on S^1/Z_2 at the leading order [34].

In order to achieve true string-scale gauge coupling unification while avoiding the Landau pole problem, we supplement the standard $\mathcal{F}$-lipped $SU(5)\times U(1)_X$ [15–18] SUSY field content with the following TeV-scale vector-like multiplets (flippons) [35]

$$\begin{aligned} &\left(XF_{(\mathbf{10},\mathbf{1})} \equiv (XQ, XD^c, XN^c),\ \overline{XF}_{(\overline{\mathbf{10}},-\mathbf{1})}\right) , \\ &\left(Xl_{(\mathbf{1},-\mathbf{5})},\ \overline{Xl}_{(\mathbf{1},\mathbf{5})} \equiv XE^c\right) , \end{aligned} \tag{2}$$

where XQ, XD^c, XE^c, XN^c have the same quantum numbers as the quark doublet, the right-handed down-type quark, charged lepton, and neutrino, respectively. Such kind of models can be realized in $\mathcal{F}$-ree $\mathcal{F}$-ermionic string constructions [36], and $\mathcal{F}$-theory model building [37, 38]. Thus, they have been dubbed $\mathcal{F}$-$SU(5)$ [37].

III. NO-SCALE SUPERGRAVITY AND THE STAROBINSKY MODEL [26–28]

Recently, an added phenomenological boost has been given to No-Scale Supergravities by detailed measurement of the Cosmic Microwave Background (CMB) perturbations (the structural seeds of galactic supercluster formation residually imprinted upon the faint afterglow of the big bang) from the Planck [39] satellite. This experiment verified a highly statistically significant tilt $n_s < 1$ in the spectrum of scalar perturbations, and set stronger upper limits on the ratio r of tensor (directional) to scalar (isotropic) perturbations. These measurements particularly of n_s, place many leading models of cosmic inflation in jeopardy. For example, single-field models with a monomial potential $\phi^n : n \geq 2$ are now disfavored at the $\sim 95\%$ CL in the case of ϕ^2 models, and at higher CLs for models with $n > 2$. This has revived interest in non-monomial single-field potentials, such as that found in the minimal Wess-Zumino model [40] [1].

A curious scenario suggested by Starobinsky [29] in 1980 is known [30] to match the CMB data effortlessly, yielding a value of $n_s \sim 0.96$ that is in perfect accord with experiment, and a value of $r \sim 0.004$ that is comfortably consistent with the Planck upper limit [39]. This model is a rather ad-hoc modification of Einstein's description of gravity, which combines a quadratic power of the Ricci scalar with the standard linear term. At face value, this model is rather difficult to take seriously, but there is substantial enthusiasm for the observation that this esoteric model is in fact conformally equivalent to the low energy limit of No-Scale supergravity with a non-minimal $N_C \geq 2$ scalar sector [26, 27]. To be specific, the algebraic

[1] Models with similar potentials were proposed long ago [41–43] and more recently in [44]: see [45] for a review.

equations of motion corresponding to a scalar field Φ with a quadratic potential that couples to a conventional Einstein term may be freely substituted back into the action, resulting in the phenomenologically favorable quadratic power of the scalar curvature [46, 47].

In considering the fundamental problem of how cosmological inflation fits into particle physics, a point of view has been taken [26–28] that this union cries out for supersymmetry [48–50], in the sense that it requires an energy scale hierarchically smaller than the Planck scale, thanks to either a mass parameter being $\ll M_P$ and/or a scalar self-coupling being $\ll \mathcal{O}(1)$. Since cosmology necessarily involves consideration of gravity, it is natural to consider inflation in the context of local supersymmetry, i.e., supergravity [51, 52], which points in turn to the superstring as a sole contender for the consistent master embedding of quantum gravity. This preference is complicated, however, by the fact that a generic supergravity theory has supersymmetry-breaking scalar masses of the same order as the gravitino mass, giving rise to the so-called η problem [53, 54] (Also see, for example, Refs. [55, 56]), where the large vacuum energy density during inflation leads to masses for all scalars of order the Hubble parameter [57]. While inflationary models in simple supergravity can be constructed to avoid the η problem [58, 59], these models rely on a seemingly accidental cancellation, invoking extraneous fine tuning in the inflaton mass [60].

For this reason, No-Scale supergravity has long been advocated [10–14] as the unique natural framework for constructing models of inflation [61–65], representing a low energy limit of the superstring. Moreover, this construction yields very successful low energy phenomenology, while invoking a bare minimum (one or zero) of freely adjustable parameters. These proposals have recently been reinvigorated in light of the Planck data, constructing an SU(2,1)/SU(2) $\times$ U(1) No-Scale version of the minimal Wess-Zumino model [26–28] [2]. It was shown that this NSWZ model is consistent with the Planck data for a range of parameters that includes a special case in which it reproduces *exactly* the effective potential and hence the successful predictions of the Starobinsky $R + R^2$ model [26–28].

Starobinsky considered in 1980 [29] a generalization of the Einstein-Hilbert action to contain an R^2 contribution, where R is the scalar curvature:

$$S = \frac{1}{2} \int d^4x \sqrt{-g}(R + \alpha R^2), \tag{3}$$

[2] For an alternative supergravity incarnation of the Wess-Zumino inflationary model, see [66].

where $M \ll M_P$ is some mass scale. As was shown by Stelle in 1978 [46] and by Whitt in 1984 [47], the theory (3) is conformally equivalent to a theory combining canonical gravity with a scalar field φ, described by

$$S = \frac{1}{2} \int d^4x \sqrt{-g} \left[(1 + 2\alpha\varphi) R - \alpha\varphi^2 \right] , \tag{4}$$

as can be seen trivially using the Lagrange equation for φ in (4). Making the Weyl rescaling $\tilde{g}_{\mu\nu} = (1 + 2\alpha\varphi) g_{\mu\nu}$, equation (4) takes the form

$$S = \frac{1}{2} \int d^4x \sqrt{-g} \left[R + \frac{6\alpha^2 \partial^\mu \varphi \partial_\mu \varphi}{(1 + 2\alpha\varphi)^2} - \frac{\alpha\varphi^2}{(1 + 2\alpha\varphi)^2} \right] . \tag{5}$$

Making now the field redefinition $\varphi' = \sqrt{\frac{3}{2}} \ln \left(1 + \frac{\varphi}{3M^2}\right)$ with $\alpha = 1/6M^2$, one obtains a scalar-field action with a canonical kinetic term:

$$S = \frac{1}{2} \int d^4x \sqrt{-\tilde{g}} \left[\tilde{R} + (\partial_\mu \varphi')^2 - \frac{3}{2} M^2 (1 - e^{-\sqrt{2/3}\varphi'})^2 \right] , \tag{6}$$

in which the scalar potential takes the form

$$V = \frac{3}{4} M^2 (1 - e^{-\sqrt{2/3}\varphi'})^2 . \tag{7}$$

The spectrum of cosmological density perturbations found by using (3) for inflation were calculated by Mukhanov and Chibisov in 1981 [30] and by Starobinsky in 1983 [31]. The current data on cosmic microwave background (CMB) fluctuations, in particular those from the Planck satellite [39], are in excellent agreement with the predictions of this $R + R^2$ model.

Some general features of the effective low-energy theory derived from a generic supergravity theory are recalled from Refs. [26–28]. Neglecting gauge interactions, which are inessential for our purposes, any such theory is characterized by a Kähler potential $K(\phi_i, \phi_j^*)$, which is a hermitian function of the chiral fields ϕ_i and their conjugates ϕ_j^*, and a superpotential $W(\phi_i)$, which is a holomorphic function of the ϕ_i, via the combination $G \equiv K + \ln W + \ln W^*$. The effective field theory contains a generalized kinetic energy term

$$\mathcal{L}_{KE} = K^{ij^*} \partial_\mu \phi_i \partial \phi_j^* , \tag{8}$$

where the Kähler metric $K^{ij^*} \equiv \partial^2 K / \partial \phi_i \partial \phi_j^*$, and the effective scalar potential is

$$V = e^G \left[\frac{\partial G}{\partial \phi_i} K_{ij^*} \frac{\partial G}{\partial \phi_j^*} - 3 \right] , \tag{9}$$

where K_{ij^*} is the inverse of the Kähler metric. Inserting into Eq. (9) the Kahler potential of Eq. (1), for N=2 and using the Wess-Zumino Superpotential

$$W = \frac{\widehat{\mu}}{2}\Phi^2 - \frac{\lambda}{3}\Phi^3, \tag{10}$$

with $\widehat{\mu} = \mu(c/3)^{1/2}$, $c = 2 < \text{Re } T >$, and $\lambda = \mu/3$, we get the potential for the real part of the inflaton (see Ref. [26] for details):

$$V = \mu^2 e^{-\sqrt{2/3x}} \sinh^2(x/\sqrt{6}). \tag{11}$$

Clearly, the Starobinsky potential of Eq. (7) is identical with the No-Scale WZ potential of Eq. (10)!

IV. THE WEDGE OF BARE-MINIMAL CONSTRAINTS

In Refs. [5, 9], we presented the wedge of No-Scale $\mathcal{F}$-$SU(5)$ model space that is consistent with a set of "bare minimal" constraints from theory and phenomenology. The constraints included i) consistency with the dynamically established boundary conditions of No-Scale supergravity (most notably the imposition of a vanishing B_μ at the final flipped $SU(5)$ GUT unification near M_{Pl}, enforced as $|B_\mu(M_{\mathcal{F}})| \leq 1$ GeV, about the size of the EW radiative corrections); ii) radiative electroweak symmetry breaking; iii) the centrally observed WMAP7 CDM relic density (and now the Planck relic density $\Omega h^2 = 0.1199 \pm 0.0027$) ; iv) the world average top-quark mass $m_t = 173.3 \pm 1.1$ GeV; v) precision LEP constraints on the light SUSY chargino and neutralino mass content; and vi) production of a lightest CP-even Higgs boson mass of $m_h = 125.5 \pm 1.5$ GeV, accomplished through additional tree level and one-loop contributions to the Higgs boson mass by the flippon supermultiplets [6, 7, 9], supplementing the Minimal Supersymmetric Standard Model (MSSM) Higgs boson mass by just the essential additional 3-5 GeV amount requisite to attain $m_h \sim 125$ GeV, while also preserving a testably light SUSY spectrum that does not reintroduce the gauge hierarchy problem via very heavy scalars that SUSY was originally intended to solve in the first place. This two-dimensional parameterization in the vector-like *flippon* super-multiplet mass scale M_V and the universal gaugino boundary mass scale $M_{1/2}$ was excised from a larger four-dimensional hyper-volume also including the top quark mass m_t and the ratio $\tan\beta$. Surviving points, each capable of maintaining the delicate balance required to satisfy $B_\mu = 0$

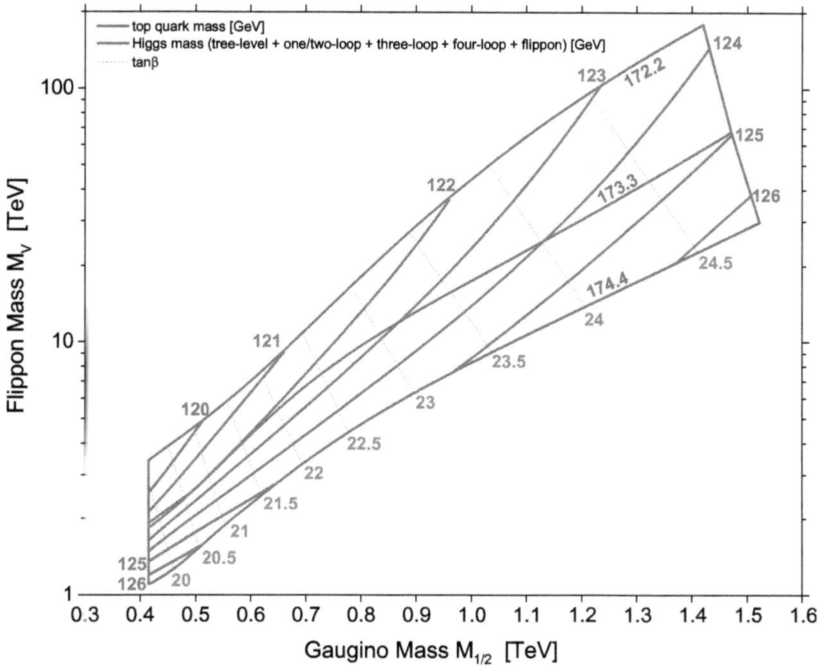

FIG. 1: Constrained model space of No-Scale $\mathcal{F}$-$SU(5)$ as a function of the gaugino mass $M_{1/2}$ and flippon mass M_V. The thick lines demarcate the total Higgs boson mass gradients, including the tree-level plus one/two-loop (as computed by the SuSpect 2.34 codebase), the three-loop plus four-loop contributions, and the flippon contribution. The thin dashed lines represent gradients of tanβ, while the upper and lower exterior boundaries are defined by a top quark mass of $m_t = 173.3 \pm 1.1$ GeV. The left edge is marked by the LEP constraints, while the right edge depicts where the Planck relic density can no longer be maintained due to an LSP and light stau mass difference less than the on-shell tau mass. All model space within these boundaries satisfy the Planck relic density constraint $\Omega h^2 = 0.1199 \pm 0.0027$ and the No-Scale requirement $B_\mu = 0$.

and the CDM relic density observations, were identified from an intensive numerical scan, employing MicrOMEGAs 2.1 [67] to compute SUSY masses, using a proprietary modification of the SuSpect 2.34 [68] codebase to run the *flippon*-enhanced RGEs. The relic density, spin-independent cross-section, and all rare-decay process constraints have been computed with MicrOMEGAs 2.4 [69].

The union of all such points was found to consist of a diagonal wedge (*cf.* Ref. [5, 9]) in the $M_{1/2}$-M_V plane, the width of which (*i.e.* at large $M_{1/2}$ and small M_V or vice-versa) is bounded by the central experimental range of the top quark mass, and the extent of which (*i.e.* at large $M_{1/2} \sim 1500$ GeV and large M_V) is bounded by CDM constraints and the transition to a charged stau LSP. This upper region of the model space corresponds to an exponentially elevated *flippon* mass M_V, which may now extend into the vicinity of 100 TeV. This delineation of the bare-minimally constrained $\mathcal{F}$-$SU(5)$ parameter space, including the correlated values of m_t, $\tan\beta$ and the light CP-even Higgs mass for each model point, is depicted in Figure 1. One obvious concern associated with this circumstance is the appearance of a new intermediate scale of physics, and a potentially new associated hierarchy problem. However, we remark that the vector-like *flippon* multiplets are free to develop their own Dirac mass, and are not in definite *a priori* association with the electroweak scale symmetry breaking; We shall therefore not divert attention here to the mechanism of this mass generation, although plausible candidates do come to mind.

The advent of substantial LHC collision data in the SUSY search rapidly eclipsed the tentative low-mass boundary set by LEP observations. A substantive correlation in the $\mathcal{F}$-$SU(5)$ mass scale favored by low-statistics excesses in a wide range of SUSY search channels, particularly lepton-inclusive searches, at both CMS and ATLAS was remarked upon by our group [8] just below $M_{1/2} \sim 800$ GeV. However, a minority of search channels, particularly lepton-exclusive squark and gluino searches with jets and missing energy [70], were found to yield limits on $M_{1/2}$ that are inconsistent with this fit, and that exert some limited tension against the upper $M_{1/2}$ boundary of the model wedge. This tension is also reflected in one generic limit of a multijet plus a single lepton SUSY search from the CMS Collaboration that places the gluino heavier than about 1.3 TeV [71].

V. CONCLUSIONS

We have considered the experimental status and prospects of the No-Scale $\mathcal{F}$-$SU(5)$ model and No-Scale Supergravity in the light of results from the LHC and Planck Satellite. Given that no conclusive evidence for light Supersymmetry has emerged from the $\sqrt{s} = 7, 8$ TeV collider searches as of yet, we described here the precise nature of the high-mass cutoff enforced on this model at the point where the stau and neutralino mass degeneracy becomes so tight that cold dark matter relic density observations cannot be satisfied. This hard upper boundary on the model's mass scale constitutes a top-down theoretical mandate for a comparatively light (and testable) SUSY spectrum which does not excessively stress natural resolution of the gauge hierarchy problem. The No-Scale $\mathcal{F}$-$SU(5)$ model is consistent with the dynamically established boundary conditions of No-Scale supergravity, radiative electroweak symmetry breaking, the centrally observed Planck cold dark matter relic density, the world average top-quark mass, precision LEP constraints on the light SUSY chargino and neutralino mass content, and production of a 125.5 ± 1.5 GeV lightest CP-even Higgs boson mass.

Building upon ample extant phenomenological motivation for No-Scale $\mathcal{F}$-$SU(5)$, we discussed the potentially significant empirical support recently provided to cosmological models of inflation based upon No-Scale Supergravity by intrinsic Starobinsky-like conformance with the Planck measurements, for a suitable choice of superpotential parameters. The potential deep connection between the No-Scale $\mathcal{F}$-$SU(5)$ and No-Scale Wess-Zumino models is certainly testable in the near future by the upcoming 14 TeV LHC and future higher precision CMB measurements. Whereas the landscape of supersymmetric models is replete with predictions requiring years more of massive data observations and significantly higher LHC beam energies, we emphasize here to the contrary that all the No-Scale $\mathcal{F}$-$SU(5)$ predictions dwell just on the cusp of an experimentally significant discovery. Detection of such a signal of stringy origin by the LHC and complementary experiments measuring the CMB perturbations could reveal not just the flipped nature of the high-energy theory, but also shed light on the geometry of the hidden compactified six-dimensional manifold. Thus, the stakes could not be higher for potential identification of supersymmetric events at the LHC within a No-Scale $\mathcal{F}$-$SU(5)$ construction, and naturally realizable inflation in our $\mathcal{F}$-$SU(5)$ No-Scale $SU(N,1)$ framework, or the revelations more profound.

Acknowledgments

Dimitri Nanopoulos would like to thank Professor A. Zichichi for his warm hospitality at Erice. This research was supported in part by the DOE grant DE-FG03-95-Er-40917 (DVN) and by the Natural Science Foundation of China under grant numbers 10821504, 11075194, 11135003, and 11275246 (TL). We also thank Sam Houston State University for providing high performance computing resources. The conference presenter (DVN) gratefully acknowledges contributions to section III of this proceedings report by John Ellis (CERN) and Keith Olive (Minnesota).

---

[1] T. Li, J. A. Maxin, D. V. Nanopoulos, and J. W. Walker, "The Golden Point of No-Scale and No-Parameter $\mathcal{F}$-$SU(5)$," Phys. Rev. **D83**, 056015 (2011), 1007.5100.

[2] T. Li, J. A. Maxin, D. V. Nanopoulos, and J. W. Walker, "The Golden Strip of Correlated Top Quark, Gaugino, and Vectorlike Mass In No-Scale, No-Parameter $\mathcal{F}$-$SU(5)$," Phys. Lett. **B699**, 164 (2011), 1009.2981.

[3] T. Li, J. A. Maxin, D. V. Nanopoulos, and J. W. Walker, "Super No-Scale $\mathcal{F}$-$SU(5)$: Resolving the Gauge Hierarchy Problem by Dynamic Determination of $M_{1/2}$ and $\tan\beta$," Phys. Lett. B **703**, 469 (2011), 1010.4550.

[4] T. Li, J. A. Maxin, D. V. Nanopoulos, and J. W. Walker, "The Ultrahigh jet multiplicity signal of stringy no-scale $\mathcal{F}$-$SU(5)$ at the $\sqrt{s} = 7$ TeV LHC," Phys.Rev. **D84**, 076003 (2011), 1103.4160.

[5] T. Li, J. A. Maxin, D. V. Nanopoulos, and J. W. Walker, "The Unification of Dynamical Determination and Bare Minimal Phenomenological Constraints in No-Scale $\mathcal{F}$-$SU(5)$," Phys.Rev. **D85**, 056007 (2012), 1105.3988.

[6] T. Li, J. A. Maxin, D. V. Nanopoulos, and J. W. Walker, "A Higgs Mass Shift to 125 GeV and A Multi-Jet Supersymmetry Signal: Miracle of the Flippons at the $\sqrt{s} = 7$ TeV LHC," Phys.Lett. **B710**, 207 (2012), 1112.3024.

[7] T. Li, J. A. Maxin, D. V. Nanopoulos, and J. W. Walker, "A 125.5 GeV Higgs Boson in $\mathcal{F}$-$SU(5)$: Imminently Observable Proton Decay, A 130 GeV Gamma-ray Line, and SUSY Multijets and Light Stops at the LHC8," Eur.Phys.J. **C72**, 2246 (2012), 1208.1999.

[8] T. Li, J. A. Maxin, D. V. Nanopoulos, and J. W. Walker, "Correlated Event Excesses in LHC SUSY Searches at 7 & 8 TeV: New Physics or Conspiring Noise?," Eur.Phys.J. **C73**, 2556 (2013), 1302.6579.

[9] T. Li, J. A. Maxin, D. V. Nanopoulos, and J. W. Walker, "No-Scale $\mathcal{F}$-$SU(5)$ in the Light of LHC, Planck and XENON," Jour.Phys. **G40**, 115002 (2013), 1305.1846.

[10] E. Cremmer, S. Ferrara, C. Kounnas, and D. V. Nanopoulos, "Naturally Vanishing Cosmo-

logical Constant in $N = 1$ Supergravity," Phys. Lett. **B133**, 61 (1983).

[11] J. R. Ellis, A. B. Lahanas, D. V. Nanopoulos, and K. Tamvakis, "No-Scale Supersymmetric Standard Model," Phys. Lett. **B134**, 429 (1984).

[12] J. R. Ellis, C. Kounnas, and D. V. Nanopoulos, "Phenomenological $SU(1,1)$ Supergravity," Nucl. Phys. **B241**, 406 (1984).

[13] J. R. Ellis, C. Kounnas, and D. V. Nanopoulos, "No Scale Supersymmetric Guts," Nucl. Phys. **B247**, 373 (1984).

[14] A. B. Lahanas and D. V. Nanopoulos, "The Road to No Scale Supergravity," Phys. Rept. **145**, 1 (1987).

[15] D. V. Nanopoulos, "F-enomenology," (2002), hep-ph/0211128.

[16] S. M. Barr, "A New Symmetry Breaking Pattern for $SO(10)$ and Proton Decay," Phys. Lett. **B112**, 219 (1982).

[17] J. P. Derendinger, J. E. Kim, and D. V. Nanopoulos, "Anti-$SU(5)$," Phys. Lett. **B139**, 170 (1984).

[18] I. Antoniadis, J. R. Ellis, J. S. Hagelin, and D. V. Nanopoulos, "Supersymmetric Flipped $SU(5)$ Revitalized," Phys. Lett. **B194**, 231 (1987).

[19] R. Harnik, D. T. Larson, H. Murayama, and M. Thormeier, "Probing the Planck scale with proton decay," Nucl.Phys. **B706**, 372 (2005), hep-ph/0404260.

[20] J. R. Ellis, D. V. Nanopoulos, and K. A. Olive, "Flipped heavy neutrinos: From the solar neutrino problem to baryogenesis," Phys.Lett. **B300**, 121 (1993), hep-ph/9211325.

[21] J. R. Ellis, J. L. Lopez, D. V. Nanopoulos, and K. A. Olive, "Flipped angles and phases: A Systematic study," Phys.Lett. **B308**, 70 (1993), hep-ph/9303307.

[22] J. L. Lopez, D. V. Nanopoulos, and A. Zichichi, "A String no scale supergravity model and its experimental consequences," Phys.Rev. **D52**, 4178 (1995), hep-ph/9502414.

[23] J. L. Lopez, D. V. Nanopoulos, and A. Zichichi, *Searching for the Superworld: A Volume in Honor of Antonino Zichichi on the Occasion of the Sixth Centenary Celebrations of the University of Turin, Italy* (World Scientific Series in 20th Century Physics, 2007), vol. 39 Part B., pp. 226–516.

[24] P. Ade et al. (Planck Collaboration), "Planck 2013 results. XVI. Cosmological parameters," (2013), 1303.5076.

[25] E. Komatsu et al. (WMAP), "Seven-Year Wilkinson Microwave Anisotropy Probe (WMAP)

Observations: Cosmological Interpretation," Astrophys.J.Suppl. **192**, 18 (2010), 1001.4538.

[26] J. Ellis, D. V. Nanopoulos, and K. A. Olive, "No-Scale Supergravity Realization of the Starobinsky Model of Inflation," Phys.Rev.Lett. **111**, 111301 (2013), 1305.1247.

[27] J. Ellis, D. V. Nanopoulos, and K. A. Olive, "Starobinsky-like Inflationary Models as Avatars of No-Scale Supergravity," JCAP **1310**, 009 (2013), 1307.3537.

[28] J. Ellis, D. V. Nanopoulos, and K. A. Olive, "A No-Scale Framework for Sub-Planckian Physics," (2013), 1310.4770.

[29] A. A. Starobinsky, "A New Type of Isotropic Cosmological Models Without Singularity," Phys.Lett. **B91**, 99 (1980).

[30] V. F. Mukhanov and G. Chibisov, "Quantum Fluctuation and Nonsingular Universe. (In Russian)," JETP Lett. **33**, 532 (1981).

[31] A. Starobinsky, "The Perturbation Spectrum Evolving from a Nonsingular Initially De-Sitte r Cosmology and the Microwave Background Anisotropy," Sov.Astron.Lett. **9**, 302 (1983).

[32] S. Ferrara, C. Kounnas, and F. Zwirner, "Mass formulae and natural hierarchy in string effective supergravities," Nucl. Phys. **B429**, 589 (1994), hep-th/9405188.

[33] E. Witten, "Dimensional Reduction of Superstring Models," Phys. Lett. **B155**, 151 (1985).

[34] T.-j. Li, J. L. Lopez, and D. V. Nanopoulos, "Compactifications of M theory and their phenomenological consequences," Phys.Rev. **D56**, 2602 (1997), hep-ph/9704247.

[35] J. Jiang, T. Li, and D. V. Nanopoulos, "Testable Flipped $SU(5) \times U(1)_X$ Models," Nucl. Phys. **B772**, 49 (2007), hep-ph/0610054.

[36] J. L. Lopez, D. V. Nanopoulos, and K.-j. Yuan, "The Search for a realistic flipped $SU(5)$ string model," Nucl. Phys. **B399**, 654 (1993), hep-th/9203025.

[37] J. Jiang, T. Li, D. V. Nanopoulos, and D. Xie, "F-$SU(5)$," Phys. Lett. **B677**, 322 (2009).

[38] J. Jiang, T. Li, D. V. Nanopoulos, and D. Xie, "Flipped $SU(5) \times U(1)_X$ Models from F-Theory," Nucl. Phys. **B830**, 195 (2010), 0905.3394.

[39] P. Ade et al. (Planck Collaboration), "Planck 2013 results. XXII. Constraints on inflation," (2013), 1303.5082.

[40] D. Croon, J. Ellis, and N. E. Mavromatos, "Wess-Zumino Inflation in Light of Planck," Physics Letters B **724**, , 165 (2013), 1303.6253.

[41] A. D. Linde, "Primordial Inflation without Primordial Monopoles," Phys.Lett. **B132**, 317 (1983).

[42] A. D. Linde, "," JETP **37**, 724 (1983).

[43] A. Albrecht and R. H. Brandenberger, "On the Realization of New Inflation," Phys.Rev. **D31**, 1225 (1985).

[44] R. Kallosh and A. D. Linde, "Testing String Theory with CMB," JCAP **0704**, 017 (2007), 0704.0647.

[45] K. A. Olive, "Inflation," Phys.Rept. **190**, 307 (1990).

[46] K. S. Stelle, "Classical Gravity with Higher Derivatives," Gen.Rel.Grav. **9**, 353 (1978).

[47] B. Whitt, "Fourth Order Gravity as General Relativity Plus Matter," Phys.Lett. **B145**, 176 (1984).

[48] J. R. Ellis, D. V. Nanopoulos, K. A. Olive, and K. Tamvakis, "Cosmological Inflation Cries Out for Supersymmetry," Phys.Lett. **B118**, 335 (1982).

[49] J. R. Ellis, D. V. Nanopoulos, K. A. Olive, and K. Tamvakis, "Fluctuations in a Supersymmetric Inflationary Universe," Phys.Lett. **B120**, 331 (1983).

[50] J. R. Ellis, D. V. Nanopoulos, K. A. Olive, and K. Tamvakis, "Primordial Supersymmetric Inflation," Nucl.Phys. **B221**, 524 (1983).

[51] D. Z. Freedman, P. van Nieuwenhuizen, and S. Ferrara, "Progress Toward a Theory of Supergravity," Phys.Rev. **D13**, 3214 (1976).

[52] S. Deser and B. Zumino, "Consistent Supergravity," Phys.Lett. **B62**, 335 (1976).

[53] E. J. Copeland, A. R. Liddle, D. H. Lyth, E. D. Stewart, and D. Wands, "False vacuum inflation with Einstein gravity," Phys.Rev. **D49**, 6410 (1994), astro-ph/9401011.

[54] E. D. Stewart, "Inflation, supergravity and superstrings," Phys.Rev. **D51**, 6847 (1995), hep-ph/9405389.

[55] D. H. Lyth and A. Riotto, "Particle physics models of inflation and the cosmological density perturbation," Phys.Rept. **314**, 1 (1999), hep-ph/9807278.

[56] J. Martin, C. Ringeval, and V. Vennin, "Encyclopaedia Inflationaris," (2013), 1303.3787.

[57] A. S. Goncharov, A. D. Linde, and M. I. Vysotsky, "Cosmological Problems for Spontaneously Broken Supergravity," Phys.Lett. **B147**, 279 (1984).

[58] D. V. Nanopoulos, K. A. Olive, M. Srednicki, and K. Tamvakis, "Primordial Inflation in Simple Supergravity," Phys. Lett. **B123**, 41 (1983).

[59] R. Holman, P. Ramond, and G. G. Ross, "Supersymmetric Inflationary Cosmology," Phys.Lett **B137**, 343 (1984).

[60] A. D. Linde and A. Westphal, "Accidental Inflation in String Theory," JCAP **0803**, 005 (2008), 0712.1610.

[61] J. R. Ellis, K. Enqvist, D. V. Nanopoulos, K. A. Olive, and M. Srednicki, "SU(N,1) Inflation," Phys.Lett. **B152**, 175 (1985).

[62] A. S. Goncharov and A. D. Linde, "A Simple Realization of the Inflationary Universe Scenario in SU(1,1) Supergravity," Class.Quant.Grav. **1**, L75 (1984).

[63] P. Binetruy and M. K. Gaillard, "Noncompact Symmetries and Scalar Masses in Superstring - Inspired Models," Phys.Lett. **B195**, 382 (1987).

[64] H. Murayama, H. Suzuki, T. Yanagida, and J. Yokoyama, "Chaotic inflation and baryogenesis in supergravity," Phys.Rev. **D50**, 2356 (1994), hep-ph/9311326.

[65] S. Antusch, M. Bastero-Gil, K. Dutta, S. F. King, and P. M. Kostka, "Chaotic Inflation in Supergravity with Heisenberg Symmetry," Phys.Lett. **B679**, 428 (2009), 0905.0905.

[66] K. Nakayama, F. Takahashi, and T. T. Yanagida, "Polynomial Chaotic Inflation in the Planck Era," Phys.Lett. **B725**, 111 (2013), 1303.7315.

[67] G. Belanger, F. Boudjema, A. Pukhov, and A. Semenov, "Dark matter direct detection rate in a generic model with micrOMEGAs2.1," Comput. Phys. Commun. **180**, 747 (2009), 0803.2360.

[68] A. Djouadi, J.-L. Kneur, and G. Moultaka, "SuSpect: A Fortran code for the supersymmetric and Higgs particle spectrum in the MSSM," Comput. Phys. Commun. **176**, 426 (2007), hep-ph/0211331.

[69] G. Belanger, F. Boudjema, P. Brun, A. Pukhov, S. Rosier-Lees, et al., "Indirect search for dark matter with micrOMEGAs2.4," Comput.Phys.Commun. **182**, 842 (2011), 1004.1092.

[70] "Search for squarks and gluinos with the ATLAS detector using final states with jets and missing transverse momentum at $\sqrt{s} = 8$ TeV," (2012), ATLAS-CONF-2012-109, URL `http://cdsweb.cern.ch`.

[71] "Search for supersymmetry in pp collisions at $\sqrt{s} = 8$ TeV in events with a single lepton, multiple jets and b-tags," (2013), CMS PAS SUS-13-007, URL `http://cdsweb.cern.ch/`.

CHAIRMAN: D. NANOPOULOS

Scientific Secretaries: C. Wymant, A. Rebesh

DISCUSSION

- *Karananas:*

I have one question and one comment to make. The question concerns the class of theories that you presented in the first part of your talk. And you stressed that the cosmological constant vanishes at tree level. I was wondering what happen when quantum corrections are taken into account.

- *Nanopoulos:*

Yes, this is a very good question. That is why I am stressing the fact that this was at the classical level. For what we were doing, and for the moment, we never claim to resolve the cosmological constant problem. The thing that we claim is that we have flat directions and have a potential whose minimum is not close to the order of M_{pl}^4. Now, for the corrections to this picture. Suppose we start from the string and we get from there an effective theory. From this moment, that is your theory, and that is why you impose boundary conditions there. Now, we can do quantum field theory corrections, but you are not supposed to do gravitational corrections there, because this may affect the theory. So, the corrections that are going to be happening in this theory are not going to be catastrophic, as would have been the case if the vacuum were not zero at the classical level. This being said, we never claimed that we've solved the cosmological constant problem. What is kind of interesting is that behind these kinds of theories, including the theory that we found in 1983, there are non-compact symmetries in effect. Several people have calculated corrections to higher order and they have found that, to two loops, there is no modification. Maybe our friend here, Dan, can have a look at it to see how long we can go without any corrections appearing. What I am trying to say is that any time symmetry exists behind a result, then things are a little bit interesting, and it may be easier to understand why we find zero.

- *Karananas:*

Thank you. The comment is about the Wess-Zumino realization of Starobinsky inflation. We have already discussed that but I wanted to share with other people that might have the same interest. You identify the scale μ in the superpotential to the Starobinsky mass m. In a completely analogous way one could identify the mass m there with the effectively inflationary coupling constant of Higgs inflation. And then you would get Higgs inflation as well.

- *Nanopoulos:*

Yes, you are right. I even had a transparency on this, as you said. And you get analogous results, as you see.

- *Karananas:*

The results should be exactly the same, since the inflationary physics is almost identical between these two models.

- *Nanopoulos:*

That is exactly right. And that's what you would find actually, we make it explicitly now. You are right.

- *Hartmann:*

Your model has a local symmetry group SU(5) x U(1). Is it compatible with an additional global, discrete flavour symmetry?

- *Nanopoulos:*

Yes! This local symmetry group is not a unique solution; it a framework in which you can do other things. Also, the see-saw mechanism arises even more naturally here than is usually the case, since the right-handed neutrinos are there automatically, in order to fill out the representations of this larger group.

- *Wymant:*

You mentioned that in the MSSM [Minimal Supersymmetric Standard Model], a Higgs boson mass m_h of around 126 GeV is possible, but only with a large stop mass and/or large stop mixing, whereas in your model the flippons boost m_h and so reduce the need for large stop mass/mixing. However, if the flippons increase m_h, I would expect them to also play a role in the electroweak symmetry breaking conditions, possibly reintroducing unnaturalness despite the smaller stop mass/mixing. Did you check whether this is the case?

- *Nanopoulos:*

I have two students looking at this in detail now. The flippon masses are from a SUSY invariant bilinear term and the corresponding soft masses for the scalar components of the flippon superfields should be compatible with those of the MSSM squarks, depending upon the relevant Yukawa couplings. The flippons allow for smaller stop mass/mixing, and their contribution to the fine-tuning is expected to be smaller than that of the stops; hence, considering only the fine-tuning due to the largest single contributor, the introduction of flippons is a relative improvement.

- *Wymant:*

Do the flippons play any role in the collider phenomenology? Can they be produced with a reasonable cross-section, and do they have any interesting decays?

- *Nanopoulos:*

The flippons are always heavier than the squarks and gluino, and so do not contribute to the dominant collider signature. Their decays are fairly standard, like those of heavy quarks, say.

- *Carlucci:*

Is there any indirect evidence for your theory, in particular in the flavour sector? For example the decay B_s to two muons is very sensitive to the presence of new scalars.

- *Nanopoulos:*

Yes, we have performed a complete analysis of all the relevant constraining observables and we obtain an impressive agreement. In fact, things are better than the Standard Model case in some cases, for example explaining certain positive signals such as dark matter, $(g\text{-}2)\mu$, and even small excesses in SUSY jets+MET searches, rather than merely satisfying non-observation constraints.

- *Downes:*

A quick question about the inflationary sector. The Starobinsky-type potential is [referenced] a lot in string compactification models, for example in large volume scenarios, and people worry a lot about what is called the η problem.

- *Nanopoulos:*

Good, because you see, we don't have the η problem. The potential is very flat here because the universal scalar mass m_0 is zero at the boundary, and that's what plays a very big role.

- *Downes:*

I was under the impression that in a lot of those models, you have this kind of no-scale structure anyway.

- *Nanopoulos:*

Yes, but that is not this type. They use the no-scale structure just to get a post potential, like Kakel-Dist, so it is a little bit different from what we are doing. This is specifically what we call SU(N,1) inflation, where the inflaton is just one of the obvious guys inside the Kähler function.

- *Downes:*

So it's the same and then there's this SU(2,1)XU(1) symmetry. Yes, this thing is pretty.

- *Nanopoulos:*

Yes. Exactly.

Hidden Beauty in $\mathcal{N} = 4$ Supersymmetric Gauge Theory

E. Sokatchev[a,b,c]

[a] Laboratoire d'Annecy-le-Vieux de Physique Théorique LAPTH*
B.P. 110, F-74941 Annecy-le-Vieux, France
[b] Institut Universitaire de France, 103, bd Saint-Michel F-75005 Paris, France
[c]Physics Department, Theory Unit, CERN, CH -1211, Geneva 23, Switzerland

*UMR 5108 associée à l'Université de Savoie

1 Introduction

1.1 Why scattering amplitudes? Hep-PH motivation

The search for the Higgs boson at LHC involves processes of the type

$$\text{gluon} + \text{gluon} \to \text{top quark} + \text{antitop quark} + \text{Higgs}$$

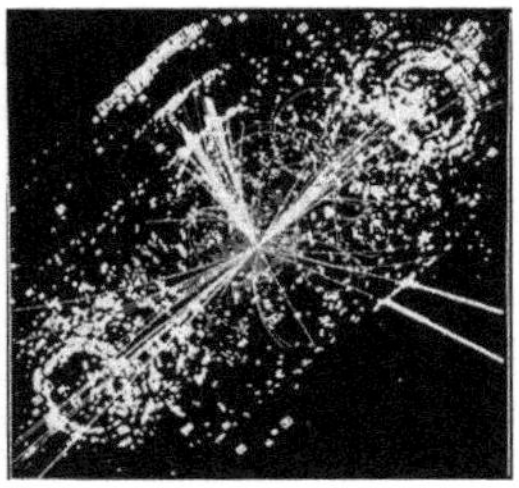
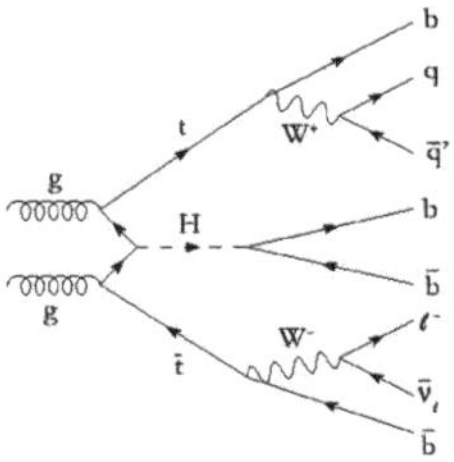

Theory Experiment Feynman diagram

The analysis of the experimental data relies on the following:

- Ability to deal with lots of produced particles in the final state leading to a large background
- The identification of Higgs boson requires a detailed understanding of complicated scattering amplitudes
- The theory should provide a solid basis and efficient calculation tools for a successful physics program at the LHC

1.2 Why scattering amplitudes? Hep-TH motivation

The scattering amplitudes of n particles (gluons, quarks, etc.) are defined as on-shell matrix elements of the S−matrix:

$$\mathcal{A}_n =$$

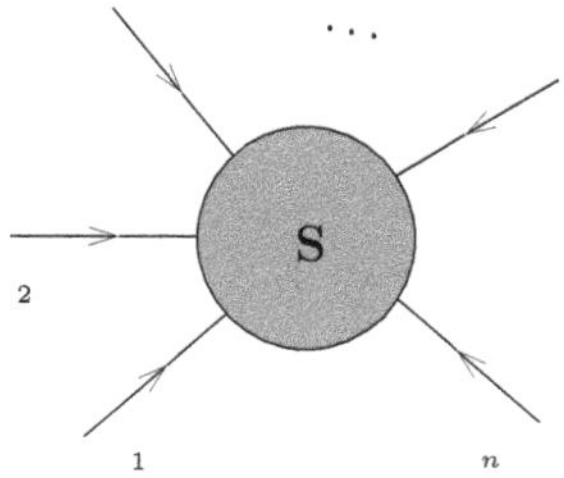

If the external legs of the amplitude are massless (e.g., gluons) the amplitude is infrared (IR) divergent. Numerous studies in QCD have lead to a good understanding of the IR divergent part which has a rather simple structure (exponentiation of the IR poles). However, the finite par of the amplitudes is in general extremely complicated.

In the maximally supersymmetric *planar* gauge theory, or $\mathcal{N} = 4$ super-Yang-Mills (SYM), the scattering amplitudes have a remarkable structure:

- They are simpler than the QCD amplitudes but share many of their properties, like the IR divergent structure.
- There exist all-order conjectures based on integrability and a proposal for describing the theory at strong coupling via the AdS/CFT correspondence.
- They have a new hidden symmetry of dynamical origin – *dual superconformal invariance*. It allows one to essentially predict all tree level amplitudes and the integrands of the loop-level corrections.

Another interesting feature of the scattering amplitudes is their duality with light-like Wilson loops and with the light-cone limit of correlation functions of gauge invariant operators. The origin of this duality is not fully understood. It may be the manifestation of yet another *hidden symmetry*.

1.3 General properties of amplitudes in gauge theories

We consider two types of amplitudes, tree-level and the loop corrections to them. Here are the main features of both types:

1.3.1 Tree amplitudes

- They are well defined in $D = 4$ dimensions, being free from UV and IR divergences and not needing regularization.
- They preserve all *classical* (Lagrangian) symmetries, in particular, the conformal (super)symmetry of the $\mathcal{N} = 4$ SYM theory.
- *Gluon tree amplitudes* are the same in all gauge theories (QCD, ..., $\mathcal{N} = 4$ SYM). Thus, the powerful constructions for such amplitudes in $\mathcal{N} = 4$ SYM can be directly used for phenomenological applications.

1.3.2 Loop amplitudes

- Loop corrections are not universal (gauge theory dependent). In general, their structure is much simpler in $\mathcal{N} = 4$ SYM than in QCD, which makes it possible to make all-order predictions.

- They suffer from IR divergences, hence are not well defined in $D = 4$ dimensions (in the dimensional regularization scheme).
- The classical conformal (super)symmetry of the $\mathcal{N} = 4$ SYM Lagrangian is broken.

1.3.3 Three main questions in this lecture:

- What are the hidden (non-Lagrangian) symmetries of the scattering amplitudes in $\mathcal{N} = 4$ SYM?
- How powerful are these symmetries to completely determine the scattering amplitudes?
- Which dual models describe the all-loop amplitudes?

2 Maximally supersymmetric $\mathcal{N} = 4$ gauge theory

We start by briefly recalling some basic points about the $\mathcal{N} = 4$ SYM theory:

- It is the particle theory with highest possible (super)symmetry (without gravity).
- Its Lagrangian is uniquely specified by the gauge group, e.g., the number of colors N_c for $SU(N_c)$.
- It is a conformally invariant (UV finite) quantum field theory in $D = 4$ dimensions. Its beta function vanishes to all perturbative orders.
- It admits a weak/strong coupling duality (AdS/CFT correspondence). This allows us to use string theory for making non-trivial predictions in gauge theory.

Here is the particle content of the theory:

wavy line	massless spin-1 gluon	(= the same as in QCD)
solid line	4 massless spin-1/2 gluinos	(= cousins of the quarks)
dotted line	6 massless spin-0 scalars	

The interaction between theses particles are described by several vertices:

all of which have the same dimensionless coupling g, being related to each other by supersymmetry.

3 Gluon amplitudes in $\mathcal{N} = 4$ super-Yang-Mills

3.1 General properties

The gluon scattering amplitudes are on-shell matrix elements of the $S-$matrix (see the future on page 1). Each incoming particle is characterized by the following quantum numbers: color index $a_i = 1, \ldots, N_c^2 - 1$, light-like momentum $(p_i^\mu)^2 = 0$ and polarization state (helicity) $h_i = \pm 1$. The color structure of the amplitudes can be very complicated, but it becomes almost trivial in the sector of the so-called color-ordered *planar* ($N_c \to \infty$) gluon amplitudes:

$$\mathcal{A}_n = \mathrm{tr}\left[T^{a_1} T^{a_2} \ldots T^{a_n}\right] A_n^{h_1, h_2, \ldots, h_n}(p_1, p_2, \ldots, p_n) + [\text{Bose symmetry}]$$

The gluon amplitudes are divided into classes according to the helicity assignments of the scattered gluons. In a supersymmetric theory like $\mathcal{N} = 4$ SYM there exists a supersymmetry all-loop Ward identity which forbids the existence of amplitudes where all (or all but one) particles have the same helicity, e.g.

$$A_n^{++\ldots+} = A_n^{-+\ldots+} = 0$$

The non-trivial amplitudes are classified according to their total helicity $h = \sum_1^n h_i$:

$$\mathrm{MHV} = \{A_n^{--+\ldots+}, A_n^{-+-\ldots+}, \ldots\}, \qquad \mathrm{NMHV} = \{A_n^{---+\ldots+}, A_n^{-+--\ldots+}, \ldots\},$$
$$\mathrm{N^2MHV} = \{A_n^{----+\ldots+}, A_n^{-+---\ldots+}, \ldots\}, \quad \ldots, \qquad \overline{\mathrm{MHV}} = \{A_n^{++-\ldots-}, A_n^{+-+\ldots-}, \ldots\}$$

Here MHV stands for 'maximally helicity violating', NMHV for 'next-to-MHV', etc. The PCT conjugate to the MHV amplitude is $\overline{\mathrm{MHV}} = \mathrm{N}^{n-2}\mathrm{MHV}$.

3.2 Hints for hidden symmetry

The gluon amplitudes at tree level are given by a huge number of Feynman graphs with very complicated structure, already for a relatively small number of external legs. The table below shows that the number of diagrams grows factorially with the number of external gluons (or number of loops in the case of loop corrections):

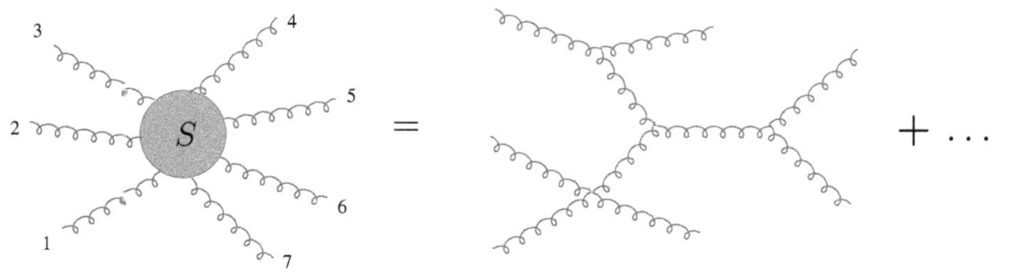

Number of external gluons	4	5	6	7	8	9	10
Number of 'tree' diagrams	4	25	220	2485	34300	559405	10525900

Nevertheless, the final expression (the sum of all Feynman graphs) looks remarkably simple (details to come)

$$A_n^{\text{tree}}(1^-2^-3^+\dots n^+) = \frac{\langle 12\rangle^4}{\langle 12\rangle\langle 23\rangle\dots\langle n1\rangle}\delta^{(4)}\big(\sum_i p_i\big)$$

The question we will try to answer on the next few pages is: *Where does this surprising simplicity come from?*

3.3 Tree MHV amplitudes

The gluon scattering amplitudes describe massless particles with given helicity. This is best done using the so-called spinor-helicity formalism [1]. Here we summarize its basic points:

- Parameterization of the four-momentum $p^\mu = (p_0, p_1, p_2, p_3)$ as a 2×2 matrix

$$\hat{p}^{\alpha\dot{\alpha}} = p_0\sigma_0 + \sum_{i=1}^{3} p_i\sigma_i = \begin{pmatrix} p_0 + p_3 & p_1 + ip_2 \\ p_1 - ip_2 & p_0 - p_3 \end{pmatrix}$$

- On-shell gluon momenta are massless,

$$p_\mu^2 = 0 \qquad \Longrightarrow \qquad \det\|\hat{p}\| = 0 \qquad \Longrightarrow \qquad (\hat{p})^{\alpha\dot{\alpha}} = \lambda^\alpha(p)\tilde{\lambda}^{\dot{\alpha}}(p)$$

  The masslessness condition is solved in terms of

- Commuting complex two-component spinors: $(\hat{p})^{\dot{\alpha}\alpha}\lambda_\alpha = \tilde{\lambda}_{\dot{\alpha}}(\hat{p})^{\dot{\alpha}\alpha} = 0$ defined up to a phase called 'helicity':

$$\lambda^\alpha(p) \quad [\text{helicity } -\tfrac{1}{2}], \qquad \tilde{\lambda}^{\dot{\alpha}}(p) \quad [\text{helicity } +\tfrac{1}{2}]$$

Amplitudes are homogeneous functions of $\lambda_i = \lambda(p_i)$, $\tilde{\lambda}_i = \tilde{\lambda}(p_i)$ with a fixed helicity weight [2]:

$$A_n^{\text{MHV}}(1^-2^-3^+\dots n^+) = \frac{\langle 12\rangle^4}{\langle 12\rangle\langle 23\rangle\dots\langle n1\rangle}\delta^{(4)}\big(\sum_1^n \lambda_i^\alpha\tilde{\lambda}_i^{\dot{\alpha}}\big)\,, \qquad (\text{with } \langle ij\rangle \equiv \lambda_i^\alpha\epsilon_{\alpha\beta}\lambda_j^\beta)$$

Here the delta function encodes the condition of total momentum conservation $\sum_1^n p_i^\mu = 0$ (all particles incoming). The factor in front is made of the Lorentz-invariant contractions $\langle ij\rangle$ of two spinors of the same chirality, each carrying negative helicity weight. Counting the weights at each point, we clearly see the helicity assignment of the MHV amplitude.

The next question we are asking is: *What are the symmetries of this amplitude?*

3.4 Conformal symmetry of the amplitude

$\mathcal{N}=4$ SYM is a quantum field theory with (super)conformal symmetry $SU(2,2|4)$. The tree-level scattering amplitudes are UV and IR finite, they do not require any regularization, therefore the conformal symmetry of the Lagrangian is preserved. Here are a few key points about the conformal symmetry of the amplitudes:

- Conformal symmetry acts locally in x–space, i.e. on the coordinates x_μ of the particles. An example of such a conformal transformation is inversion $x_\mu \to x_\mu/x^2$.
- Conformal symmetry acts *non-locally* in p–space, i.e. on the momenta p^μ of the particles. The reason for this is the Fourier transform $F(p) = \int d^4x\, e^{ipx} f(x)$.
- Realization of conformal symmetry on amplitudes by 2nd-order operators [3]

$$k_{\alpha\dot{\alpha}} = \sum_i \frac{\partial^2}{\partial\lambda_i^\alpha \partial\tilde{\lambda}_i^{\dot{\alpha}}} \qquad \Longrightarrow \qquad k_{\alpha\dot{\alpha}}\mathcal{A}_n^{\rm MHV} = 0 \tag{1}$$

  This reflects the non-local nature of the symmetry in momentum space. It also makes the symmetry non-manifest.
- The conformal symmetry of MHV amplitudes can be extended to the $SU(2,2|4)$ superconformal invariance of the full super-amplitude

$$g\cdot\mathcal{A}_n = 0\,, \qquad g = \{p,\, m,\, d,\, k,\, q,\, \bar{q},\, s,\, \bar{s}\} \in SU(2,2|4) \tag{2}$$

  Here the algebra g includes the generators of translation p, Lorentz m, dilatation d, conformal boosts k, Poincaré $(q,\bar{q})$ and special conformal $(s,\bar{s})$ supersymmetry. This larger symmetry is easy to verify in the case of MHV amplitudes, but is much less trivial for NMHV, N^2MHV,... amplitudes [4].

Conformal (super)symmetry alone is not powerful enough to fix the form of all tree amplitudes, although it imposes some strong constraints. So, our next question is: *What else do we need?*

4 Dual (super)conformal symmetry

4.1 Dual space

Surprisingly, the $\mathcal{N}=4$ SYM (super)amplitudes have a second, *dual (super)conformal symmetry* of dynamical (non-Lagrangian) origin [5], [6]. Here is a simple example illustrating the new symmetry. Consider again the MHV amplitude but this time in the form (we drop the momentum conservation delta function)

$$\left|\hat{A}_n^{\rm MHV}\right|^2 = \frac{(s_{12})^4}{s_{12}s_{23}\dots s_{n1}}\,, \qquad (\text{with} \quad s_{ij} = (p_i + p_j)^2) \tag{3}$$

Due to the mod squared, this expression has vanishing helicity weight and is given in terms of the Mandelstam kinematical invariants s_{ij}. Now, the key trick is to introduce *dual variables* (attention, this not a Fourier transform from $p \to x$!):

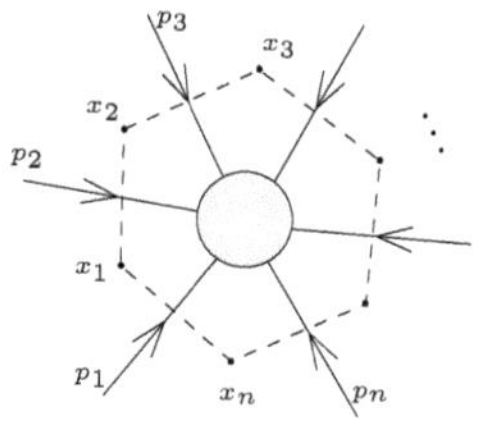

- $p_i = x_i - x_{i+1}\,, \quad x_{n+1} \equiv x_1 \qquad \Rightarrow$ solves $\sum_1^n p_i = 0$
- $p_i^2 = 0 \Longrightarrow (x_i - x_{i+1})^2 = 0$
- $s_{i,i+1} = (x_i - x_{i+2})^2$

The new dual coordinates are simply a means to solve the momentum conservation condition. They can be depicted by a closed polygon with vertices x_i inserted between the pairs of momenta p_i, p_{i+1}, as shown in the figure. The masslessness condition on the momenta implies that the edges of the polygon are light-like. Finally, notice that the Mandelstam invariant $s_{i,i+1}$ for two adjacent momenta is given by the interval between two points in dual space (this is not true for all s_{ij}).

Now, armed with this new tool, let us rewrite the familiar MHV amplitude in dual space:

$$\left|\hat{A}_n^{\rm MHV}\right|^2 = \frac{[(x_1 - x_3)^2]^3}{(x_2 - x_4)^2(x_3 - x_5)^2 \dots (x_n - x_2)^2} \tag{4}$$

It looks like an $n-$point correlation function in $x-$space, BUT here the x's are momenta!

4.2 Dual (super)conformal symmetry in dual space

Conformal symmetry can be regarded as the combination of Poincaré symmetry and conformal inversion. The expression (4) clearly is Poincaré invariant (in the sense of dual space!), but what about inversion? Conformal inversion acts in dual $x-$space as follows:

$$x_i^\mu \to \frac{x_i^\mu}{x_i^2} \qquad \Longrightarrow \qquad s_{i,i+1} \to \left(x_i^2 x_{i+2}^2\right)^{-1} s_{i,i+1} \tag{5}$$

We emphasize the local action on the momenta, unlike the ordinary conformal symmetry of the $\mathcal{N} = 4$ SYM Lagrangian. In fact, the two symmetries are not related to each other.

Now, it becomes obvious that the tree-level MHV (super)amplitude is *covariant* under dual conformal inversion,

$$I\left[\mathcal{A}_n^{\rm MHV}\right] = \frac{x_2^2 x_4^2 x_5^2 \dots x_n^2}{x_1^2 x_3^2} \times \mathcal{A}_n^{\rm MHV}$$

We repeat that this property is new and does not follow from the ordinary conformal symmetry condition (1). Moreover, this new conformal symmetry is manifest, unlike the previous one. This is due to the local action of the generator

$$K_\mu = \sum_{i=1}^{n} (x_i^2 \partial/\partial x_i^\mu - 2x_{i\,\mu} x_i^\nu \partial/\partial x_i^\nu - d x_{i\,\mu}) \tag{6}$$

on the dual coordinates. Unlike the generator $k_{\alpha\dot\alpha}$ in (1), the dual one in (6) is a 1st-order operator.

Dual conformal symmetry can be extended to dual superconformal $\widetilde{SU}(2,2|4)$ symmetry

$$G \cdot \mathcal{A}_n = 0\,, \qquad G = \{P,\, M,\, D,\, K,\, Q,\, \bar{Q},\, S,\, \bar{S}\} \in \widetilde{SU}(2,2|4)$$

Dual superconformal symmetry is a property of *all tree-level (super)amplitudes* (MHV, NMHV, N^{2}MHV,...) in $\mathcal{N} = 4$ SYM theory [6], [7].

4.3 Symmetries of tree amplitudes

- The relationship between conventional and dual superconformal symmetries can be summarized by the following diagram:

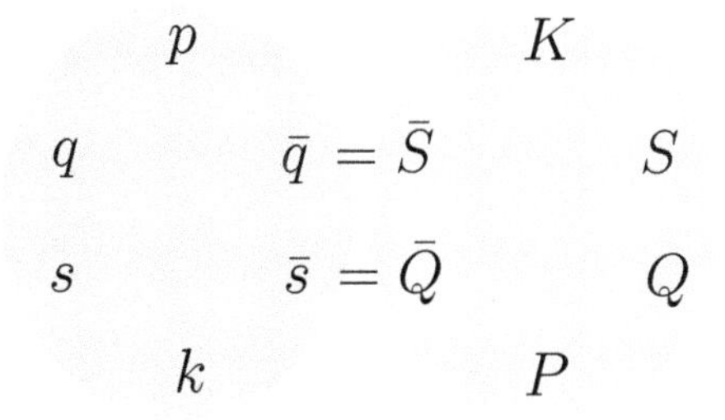

  We see a partial overlap between the two algebras, some of the supersymmetry generators are the same. The rest of the generators differ in the sense explained above (1st versus 2nd-order).

- The same symmetries appear at strong coupling via the AdS/CFT correspondence. They follow from the invariance of the $AdS_5 \times S^5$ sigma model under bosonic [8] and fermionic ([9], [10]) T-duality.

- The closure of the two symmetries is an infinite-dimensional algebra with *Yangian structure* [11]. This is considered as a clear signal of integrability, best seen in the superstring sigma model.

- All tree $\mathcal{N} = 4$ amplitudes are uniquely fixed by:

– supersymmetric BCFW recursion relations [12], [13], [14], [15]

– or equivalently, by symmetries + analytic properties [4], [16], [17], [18]

Our next question is: *What happens to these symmetries at loop level?*

4.4 Planar gluon amplitudes at weak coupling

Loop corrections to the tree amplitudes are characterized by the appearance of (loop) Feynman integrals. Due to the presence of massless particles in them, they have IR divergences. We illustrate the main points on the simplest example of the MHV amplitude.

- All loop corrections to MHV amplitudes are given by a single scalar function [2]

$$A_n^{\text{MHV}}(p_i) = A_n^{(\text{tree})}(p_i) M_n^{\text{MHV}}(\{s_{ij}\}; g)$$

- Example: four-gluon amplitude at one loop

$$A_4/A_4^{(\text{tree})} = 1 + \frac{g^2 N_c}{8\pi^2} I^{(1)}(s,t) + O(g^4)$$

  – Scalar box in dual variables $p_i = x_i - x_{i+1}$ with $p_i^2 = x_{i,i+1}^2 = 0$

$I^{(1)}(s,t) =$ 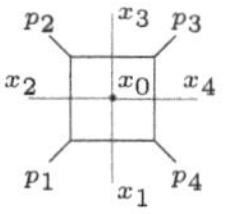

$\sim \int \frac{d^D x_0\, x_{13}^2 x_{24}^2}{x_{10}^2 x_{20}^2 x_{30}^2 x_{40}^2}$

  – Invariant under dual conformal transformations, e.g. $x_i \to x_i/x_i^2$, in *dual* space if $D = 4$

  – This symmetry *is not related* to the conformal symmetry of $\mathcal{N} = 4$ SYM

  – All scalar integrals contributing to A_4 are dual conformal! This has been tested by direct calculations up to five loops [5], [19].

- Dual conformal symmetry is broken by IR divergences. The breakdown is controlled by an *anomalous Ward identity* [20], derived from the duality with light-like Wilson loops ([21], [22], [23]) in dimensional regularization for $D = 4 - 2\epsilon$. It completely fixes the form of the 4- and 5-gluon amplitudes (thus explaining the so-called BDS conjecture [24]), but leaves room for a function of conformal cross-ratios (remainder function [25], [26], [27]).

5 Correlation functions of BPS operators

5.1 Definition and basic properties

The dual conformal symmetry of scattering amplitudes has an equivalent realization as the natural conformal symmetry of correlation functions of gauge invariant operators. Among them, a special role is played by the so-called protected (or short, or 1/2-BPS) $\mathcal{N}=4$ superconformal operators made from the 6 scalars $\phi_{AB} = \frac{1}{2}\epsilon_{ABCD}\bar{\phi}^{CD}$ of the $\mathcal{N}=4$ SYM theory:

$$\mathcal{O}(x) = \mathrm{Tr}(\phi_{12}\phi_{12}), \qquad \tilde{\mathcal{O}}(x) = \mathrm{Tr}(\bar{\phi}^{12}\bar{\phi}^{12})$$

These operators are protected from quantum corrections in the sense that their scaling dimension equals their naive (free theory) dimension. In addition, the two- and three-point correlation functions of such operators also keep their free theory form. The simplest non-trivial correlation function involves four points:

$$G_4 = \langle 0|T(\mathcal{O}(x_1)\tilde{\mathcal{O}}(x_2)\mathcal{O}(x_3)\tilde{\mathcal{O}}(x_4))|0\rangle = \frac{N_c^2}{x_{12}^2 x_{23}^2 x_{34}^2 x_{41}^2}\mathcal{F}(u,v;g)$$

Conformal symmetry fixes its form up to an arbitrary function of the two conformal cross-ratios

$$u = \frac{x_{14}^2 x_{23}^2}{x_{13}^2 x_{24}^2}, \qquad v = \frac{x_{12}^2 x_{34}^2}{x_{13}^2 x_{24}^2}$$

Such four-point functions can provide a lot of data about various operators in the conformal theory. This is done in the singular limit $x_i \to x_j$ corresponding to the operator product expansion (OPE). Here are some key results:

- Extract anomalous dimensions of all twist-two operators [28]
- Confirm maximal transcedentality conjecture to 3 loops [29]

5.2 Correlation functions on the light cone

Besides the standard OPE limit, let us consider a new, light-cone limit in which *all neighboring points* become light-like separated

$$x_{i,i+1}^2 \to 0, \qquad x_i \neq x_{i+1}, \qquad (i=1,\ldots,n) \qquad \Rightarrow \qquad u,\ v \to 0$$

The light-cone limit of G_4 is singular:

(i) For $x_{i,i+1}^2 \to 0$ the correlator develops pole singularities already at tree level

$$G_4^{(\mathrm{tree})} \sim \frac{N_c^2}{x_{12}^2 x_{23}^2 x_{34}^2 x_{41}^2} + \text{subleading terms}$$

but we consider the ratio

$$\mathcal{F}_4 \equiv \lim_{x_{i,i+1}^2 \to 0} G_4(x_i)/G_4^{(\text{tree})}(x_i)$$

(ii) In addition, loop integrals develop logarithmic light-cone singularities $(u, v \to 0)$

$$\mathcal{F}_4 = 1 + \frac{g^2 N_c}{8\pi^2} \frac{i}{\pi^2} \int \frac{d^4x_0 \, x_{13}^2 x_{24}^2}{x_{10}^2 x_{20}^2 x_{30}^2 x_{40}^2} + \ldots = 1 - \frac{g^2 N_c}{8\pi^2} \ln u \ln v + \ldots .$$

(iii) The leading divergent terms can be resummed *to all loops*

$$\mathcal{F}_4 \sim \exp\left(-\frac{1}{2}\Gamma_{\text{cusp}}(g^2) \ln u \ln v\right)$$

The factor Γ_{cusp} is the so-called cusp anomalous dimensions. It appears in many physicals processes. In particular, it governs the leading IR singularity of the on-shell scattering amplitudes.

5.3 Duality correlation functions/scattering amplitudes

In the light-cone limit we observe a surprising duality [30]:

$$\lim_{x_{i,i+1}^2 \to 0} \ln\left(G_n/G_n^{(\text{tree})}\right) = 2 \ln\left(A_n^{\text{MHV}}/A_n^{(\text{tree})}\right) + O(1/N_c^2) \tag{7}$$

It generalizes the light-like Wilson loops/scattering amplitudes duality. Both objects in (7) are divergent and require regularization (UV for G_n, IR for A_n). A convent scheme equally suitable for both purposes is dimensional regularization in $D = 4 - 2\epsilon$ dimensions, with $\epsilon > 0$ for UV and $\epsilon < 0$ for IR divergences. Thus, the duality (7) consists in matching two sets of *divergent integrals*.

The duality can also be formulated in terms of the *finite integrands*:

- Loop corrections via Lagrangian insertions

$$g^{2\ell} \frac{d^\ell}{dg^{2\ell}} G_n = \prod_{i=5}^{4+\ell} \int d^4x_i \, \langle \mathcal{O}(x_1) \ldots \mathcal{O}(x_n) \mathcal{L}_{\mathcal{N}=4}(x_{n+1}) \ldots \mathcal{L}_{\mathcal{N}=4}(x_{n+\ell}) \rangle^{(\text{tree})}$$

- The $\ell-$loop integrand is an $n-$point correlator with ℓ Lagrangian insertions computed at *tree level*:

$$G_{n+\ell}^{(\text{tree})} = \langle \mathcal{O}(1) \ldots \mathcal{O}(n) \mathcal{L}_{\mathcal{N}=4}(n+1) \cdots \mathcal{L}_{\mathcal{N}=4}(n+\ell) \rangle^{(\text{tree})}$$

- Duality with the integrand of the $\ell-$loop amplitude (no regularization needed!):

$$\lim_{x_{i,i+1}^2 \to 0} \ln\left(G_{n+\ell}^{(\text{tree})}/G_n^{(\text{tree})}\right) = 2 \ln\left(Int_\ell[A_n^{\text{MHV}}]/A_n^{(\text{tree})}\right) + O(1/N_c^2)$$

- Matches exactly the integrand of the momentum-twistor construction [17]

5.4 Hidden permutation symmetry of the integrand

The general form of the integrand for $n = 4$ is predicted by $\mathcal{N} = 4$ superconformal symmetry [31]:

$$\langle \mathcal{O}(1)\ldots\mathcal{O}(4)\mathcal{L}(5)\ldots\mathcal{L}(4+\ell)\rangle^{(\text{tree})} \sim \frac{x_{13}^2 x_{24}^2}{x_{12}^2 x_{34}^2} \times \frac{P^{(\ell)}(x_1,\ldots,x_{4+\ell})}{\prod_{1\leq i<j\leq 4+\ell} x_{ij}^2}$$

The numerator $P^{(\ell)}$ is a homogeneous polynomial in x_{ij}^2 of conformal weight $(1-\ell)$ at each point, *invariant under all* $S_{4+\ell}$ *permutations of* x_i.

Here are some simple examples at 1 and 2 loops:

$$P^{(1)}(x_1,\ldots,x_5) = 1\,, \qquad P^{(2)}(x_1,\ldots,x_6) = \frac{1}{48}\sum_{\sigma\in S_6} x_{\sigma(1)\sigma(2)}^2 x_{\sigma(3)\sigma(4)}^2 x_{\sigma(5)\sigma(6)}^2$$

The ℓ–loop integrand is obtained by drawing pictures (graph theory). As an example, consider the five-loop planar integrand:

- We find only 7 planar integrand graph topologies [32]:

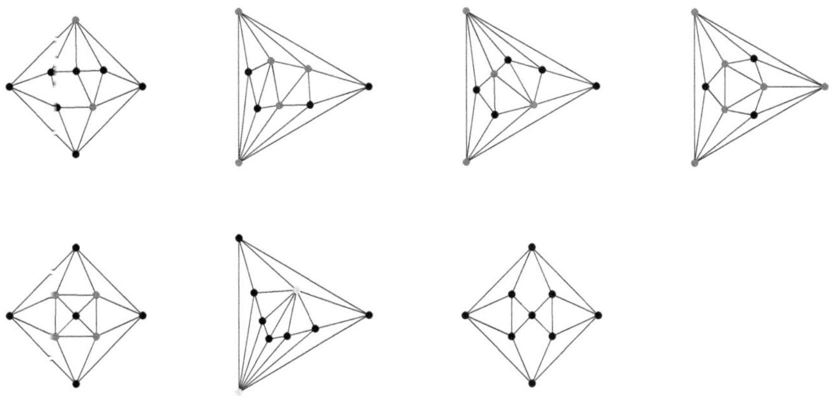

- All coefficients in the planar sector are fixed by a simple log singularity criterion. A few arbitrary constants remain in the non-planar sector.

- This method allowed the direct calculation of the five-loop Konishi anomalous dimension [33].

6 Open questions

- What is the origin of dual (super)conformal symmetry?
- Is it related (equivalent?) to the integrability of the $\mathcal{N} = 4$ SYM Hamiltonian?
- Understand *why* the duality correlators/amplitudes works. Are there some hidden symmetries of the correlator, which fix it to a unique form?
- What happens in the non-planar sector?
- Which of these features survive in gauge theories with less supersymmetry?
- Amplitudes are IR divergent, so not directly related to physical observables. Look for IR-safe observables, e.g. energy-energy correlations.
- In the year 2113: conformal collider physics?

Acknowledgements

I am indebted to my collaborators Fernando Alday, James Drummond, Burkhard Eden, Johannes Henn, Paul Heslop, Gregory Korchemsky, Juan Maldacena, Volodya Smirnov with whom most of these results have been obtain. I am very grateful to Prof. A. Zichichi for the kind invitation to present this lecture at the Erice 2013 International School of Subnuclear Physics.

References

[1] Z. Xu, D. -H. Zhang and L. Chang, Nucl. Phys. B **291** (1987) 392.

[2] S. J. Parke and T. R. Taylor, Phys. Rev. Lett. **56** (1986) 2459.

[3] E. Witten, Commun. Math. Phys. **252** (2004) 189 [hep-th/0312171].

[4] G. P. Korchemsky and E. Sokatchev, Nucl. Phys. B **832** (2010) 1 [arXiv:0906.1737 [hep-th]].

[5] J. M. Drummond, J. Henn, V. A. Smirnov and E. Sokatchev, JHEP **0701** (2007) 064 [hep-th/0607160].

[6] J. M. Drummond, J. Henn, G. P. Korchemsky and E. Sokatchev, Nucl. Phys. B **828** (2010) 317 [arXiv:0807.1095 [hep-th]].

[7] A. Brandhuber, P. Heslop and G. Travaglini, Phys. Rev. D **78** (2008) 125005 [arXiv:0807.4097 [hep-th]].

[8] R. Kallosh and A. A. Tseytlin, JHEP **9810** (1998) 016 [hep-th/9808088].

[9] N. Berkovits and J. Maldacena, JHEP **0809** (2008) 062 [arXiv:0807.3196 [hep-th]].

[10] N. Beisert, R. Ricci, A. A. Tseytlin and M. Wolf, Phys. Rev. D **78** (2008) 126004 [arXiv:0807.3228 [hep-th]].

[11] J. M. Drummond, J. M. Henn and J. Plefka, JHEP **0905** (2009) 046 [arXiv:0902.2987 [hep-th]].

[12] R. Britto, F. Cachazo, B. Feng and E. Witten, Phys. Rev. Lett. **94** (2005) 181602 [hep-th/0501052].

[13] J. M. Drummond and J. M. Henn, JHEP **0904** (2009) 018 [arXiv:0808.2475 [hep-th]].

[14] M. Bianchi, H. Elvang and D. Z. Freedman, JHEP **0809** (2008) 063 [arXiv:0805.0757 [hep-th]].

[15] N. Arkani-Hamed, F. Cachazo and J. Kaplan, JHEP **1009** (2010) 016 [arXiv:0808.1446 [hep-th]].

[16] T. Bargheer, N. Beisert and F. Loebbert, J. Phys. A **44** (2011) 454012 [arXiv:1104.0700 [hep-th]].

[17] N. Arkani-Hamed, J. L. Bourjaily, F. Cachazo, S. Caron-Huot and J. Trnka, JHEP **1101** (2011) 041 [arXiv:1008.2958 [hep-th]].

[18] L. J. Mason and D. Skinner, JHEP **0911** (2009) 045 [arXiv:0909.0250 [hep-th]].

[19] Z. Bern, J. J. M. Carrasco, H. Johansson and R. Roiban, Phys. Rev. Lett. **109** (2012) 241602 [arXiv:1207.6666 [hep-th]].

[20] J. M. Drummond, J. Henn, G. P. Korchemsky and E. Sokatchev, Nucl. Phys. B **826** (2010) 337 [arXiv:0712.1223 [hep-th]].

[21] L. F. Alday and J. M. Maldacena, JHEP **0706** (2007) 064 [arXiv:0705.0303 [hep-th]].

[22] J. M. Drummond, G. P. Korchemsky and E. Sokatchev, Nucl. Phys. B **795** (2008) 385 [arXiv:0707.0243 [hep-th]].

[23] A. Brandhuber, P. Heslop and G. Travaglini, Nucl. Phys. B **794** (2008) 231 [arXiv:0707.1153 [hep-th]].

[24] Z. Bern, L. J. Dixon and V. A. Smirnov, Phys. Rev. D **72** (2005) 085001 [hep-th/0505205].

[25] J. M. Drummond, J. Henn, G. P. Korchemsky and E. Sokatchev, Phys. Lett. B **662** (2008) 456 [arXiv:0712.4138 [hep-th]].

[26] J. M. Drummond, J. Henn, G. P. Korchemsky and E. Sokatchev, Nucl. Phys. B **815** (2009) 142 [arXiv:0803.1466 [hep-th]].

[27] Z. Bern, L. J. Dixon, D. A. Kosower, R. Roiban, M. Spradlin, C. Vergu and A. Volovich, Phys. Rev. D **78** (2008) 045007 [arXiv:0803.1465 [hep-th]].

[28] F. A. Dolan and H. Osborn, Nucl. Phys. B **599** (2001) 459 [hep-th/0011040].

[29] A. V. Kotikov, L. N. Lipatov, A. I. Onishchenko and V. N. Velizhanin, Phys. Lett. B **595** (2004) 521 [Erratum-ibid. B **632** (2006) 754] [hep-th/0404092].

[30] L. F. Alday, B. Eden, G. P. Korchemsky, J. Maldacena and E. Sokatchev, JHEP **1109** (2011) 123 [arXiv:1007.3243 [hep-th]].

[31] B. Eden, P. Heslop, G. P. Korchemsky and E. Sokatchev, Nucl. Phys. B **862** (2012) 193 [arXiv:1108.3557 [hep-th]].

[32] B. Eden, P. Heslop, G. P. Korchemsky and E. Sokatchev, Nucl. Phys. B **862** (2012) 450 [arXiv:1201.5329 [hep-th]].

[33] B. Eden, P. Heslop, G. P. Korchemsky, V. A. Smirnov and E. Sokatchev, Nucl. Phys. B **862** (2012) 123 [arXiv:1202.5733 [hep-th]].

CHAIRMAN: E. SOKATCHEV

Scientific Secretaries: D. Antonenko, I. Kononov

DISCUSSION

- *Morozov:*

My name is Morozov Andrey, I'm from ITEP, so my question is: for N=2 super Yang-Mills there is a relation, called AGT relation which says that the partition functions of this N=2 super Yang-Mills and of the conformal field theory are equal due to both being the parts of some 6-dimensional model. So is there some kind of such relation for N = 4 super Yang-Mills?

- *Sokatchev:*

No that I'm aware of it, but what does this have to do with the subjects of the talk which I give, which is amplitudes and correlation functions. I don't see the connection.

- *Morozov:*

Well, the talk was about N=4 super Yang-Mills, so…

- *Sokatchev:*

Not in general, but an application to scattering amplitudes in this theory and related topics as correlation functions, so, what I can say is that there are many ways you can think of N=4 super Yang-Mills being integrable which means exactly soluble. But then you have to choose your particular object you want to study (in my case it was scattering amplitudes) and try to find out the actual structure of the object which shows some signs of integrability and of the exact solvability which is what I was trying to give indications for. But N=4 super Yang-Mills is a very vast subject.

- *Morozov:*

Thank you!

- *Torres Bobadilla:*

I'm William Torres from Columbia. What does supersymmetry all-loop Ward identity mean?

- *Sokatchev:*

Okay. So the amplitudes I was discussing involve gluons, which can have helicities +/- 1 as a spin-one massless particle.

- *Torres Bobadilla:*

Page 7.

- *Sokatchev:*

Page 7, okay. So you see that the incoming particles can have helicities as indicated here. We can pull out the color factor which in the planar case can be factorized. Then the actual amplitude depends on the particular choice of helicities you put at each leg.

- *Torres Bobadilla:*

This is tree level, right?

- *Sokatchev:*

No-no, this is very general.

- *Torres Bobadilla:*

No-no-no, I mean this equation on $A_n^{++...+}$.

- *Sokatchev:*

No, this is all levels. This is just a general property of the amplitude. So, you could imagine a situation where all gluons have the same helicity or all but one have the same helicity. This is possible. However supersymmetry forbids it and, in particular, you need N=4 to show that these scattering amplitudes cannot exist. So there are various ways to see that these amplitudes vanish. Just apply supersymmetry transformations and you will see that such states do not exist. There is a very elegant way to show it but I could not develop this here. What I was mostly talking about are MHV (maximal helicity violating) amplitudes, but you can combine all kinds of amplitudes in one big object which is called a superamplitude. For that you have to introduce in addition to these twistor variables (λ, α I introduced) some anticommuting, Grassmann variables, or so to speak supertwistor variables. Then you can combine all amplitudes (MHV, NMHV, ...), for all kinds of particles, in one big object, called a superamplitude. And there you will see immediately, by expanding this object in the Grassman (or odd, or anticommuting) variables (supertwistors) that there are no terms with such helicities. So this is what the Ward identity means and it's valid to any order of perturbation theory. That's a very general statement. That's why MHV amplitudes are the simplest object we start with. The others are considerably more complicated but they still have similar features.

- *Torres Bobadilla:*

Thank you!

- *Nikitin:*

Vladimir Nikitin, Moscow State University. What regularization schemes did you use for infrared divergences?

- *Sokatchev:*

It's a very good question; it's a very relevant question. Not only in the particular context, but in a broader context. When we talk about comparing amplitudes, Wilson loops and correlation functions, we should never forget that (apart from tree level, of course, where they are finite) that these objects are divergent for different reasons. So, you need some regularization scheme, you can't live without that. And in some recent theories, well, in some constructions, in particular something which I was showing near the end of the talk one compares not the divergent integrals for the amplitudes or correlation functions or Wilson loops, but rather their integrands which are finite, conformal and very nice. You are may be missing some very important features of the object. There are some things which are not visible at the level of the integrand. You need a regularization, you need to work out the whole integral, that' s the true object you are comparing. So, to come back to your question, well, the most practical scheme which everybody uses, (Zvi et al. use it all the time and they produce important results) is dimensional regularization. In this particular context it's very useful, because at some point we have to compare (well, it's near the end of the talk), we have to compare two completely different objects. One of them is a correlation function, that is put on the light cone. In this way, it develops logarithmic divergences. And, on the other hand, we have the scattering amplitude which has built-in infrared divergences. A priori these things are completely different but, in the dimensional regularization scheme – that's why it is very convenient, you can use the same regulator, epsilon, say. Just change the sign of the regulator, one works for infrared, the other for ultraviolet divergences and, adjusting also the scale of dimensional regularization, it becomes possible to really compare these expressions, i.e. Laurent series in epsilon. Thus, compute correlation functions on this side, and amplitudes on that side, then compare all poles, finite parts, and they turn out to be identical. This is not the only way to regularize, though. There are some more recent regularization schemes, one of them is called the Higgs regularization. It is a way to regularize the Feynman integrals in 4 dimensions by introducing something which looks like a mass, which is just a regulator and it can be done in a way not to loose conformal invariance. Some people have been computing anomalous dimensions using this scheme, it's also very interesting. One of the main problems of using dimensional regularization here (it's very convenient, I mean not techniques, how to use it) is that conformal invariance, this dual conformal invariance I talked about, is broken when you go away from 4 dimensions, it is broken. Then you have to find a way to restore it, which means to look for an anomalous Ward identity. In alternative regularization schemes it is possible to preserve even this dual conformal invariance property in the presence of a regulator. This anomalous conformal Ward identity has been tested in various calculations at a rather high number of loops for 4 points, or many legs at 1 or 2 loops (one loop is trivial, but 2 loops is not). For such calculations dimensional regularization is the most convenient scheme. So, that is the answer to your question. But, one should never forget that this relation is between divergent objects. This relation, here, is between integrands in 4 dimensions, that is, between finite objects, but it only grasps part of the big picture. So, one should be careful!

- *Nikitin:*

Thank you!

-*Krylov:*

Evgeny Krylov, Moscow. Can the full symmetry group of this theory be embedded in some larger group which maybe can give a more general theory?

- *Sokatchev:*

Which symmetry group do you mean? Superconformal symmetry?

- *Krylov:*

No, the full group, superconformal and dual superconformal.

- *Sokatchev:*

Yes, okay. This I mentioned very briefly. It's also something very interesting. What does it mean "embedded"? Actually, it means that the 2 symmetries are realized on some object (say, the amplitude), which is invariant under both. So, it's a logical question to ask what happens if I start commuting the generators of one with the generators of the other? In the beginning of the talk I mentioned that the conventional conformal, (or superconformal) symmetry is realized in terms of second-order operators. This has to do with the fact that space-time conformal symmetry acts on the momenta of the particles non-locally. So, one of the symmetries is realized non-locally, the other one, the dual conformal symmetry acts locally on the momenta. When you start commuting local with non-local you expect that the closure is infinite-dimensional, because every time you commute two of these generators, they produce a new one and so on, and so forth. Even if this by itself is not surprising, what is really nice is that this is a particular algebraic structure, which is called a Yangian and that's typical for integrable systems. Many integrable systems, for instance, the string sigma-model has this Yangian symmetry and that's a sign that N=4 SYM is integrable. By doing this you do not create a new symmetry because saying that the amplitude is invariant under this and under that is all you can say, you cannot gain more consequences. But it is the fact that the true symmetry is infinite-dimensional. You may say: "I can embed these into an infinite-dimensional algebra", or I may say: "These two together generate an infinite-dimensional algebra ". This is considered to be one of the strongest signs that, at least the N=4 gauge theory, is integrable in the sense that the object (in this case scattering amplitude) is exactly soluble. And people are doing a lot of interesting work. I'll take the freedom just to make a comment, I didn't have the time today. Coming back to these divergent amplitudes , at the end of my talk I said that we should be looking for infrared finite quantities. So one of them, which has been studied intensively, is the so-called remainder function, which is something very interesting... Well, maybe I am taking too much time now... Zvi and friends invented the so-called BDS conjecture, which states that it is enough to know the one-loop amplitude, and then you can exponentiate it, and you will find all other loop levels. It turns out to be true for 4 and 5 gluons, and the explanation is given by dual conformal symmetry or rather its anomalous version. But beyond, starting with 6 gluons, conformal symmetry cannot fix the whole object. But now this new object (the remainder function) by itself is finite, because it is sufficient to take the full amplitude and divide it by what the BDS conjecture predicts. You take away all the divergences, they do not depend on the details of the theory. Infrared divergences are universal. So you obtain an object which is finite, and dual conformal symmetry is restored. So people have studying this object, calculated it at rather high loop orders and discovered an interesting integrable structure in it. So now at this moment there is very active research topic, it is believed or practically proved, that this so-called remainder function, which is the essential piece of the amplitude, is exactly solvable. This is due to the dual conformal symmetry and to the Yangian structure here.

- Alba:

I'm Vasyl Alba from ITEP Moscow and Princeton University. I have a couple of questions. How do you use locality and unitarily in N=4 amplitudes? Is it possible to derive locality and unitarity from something else?

- Sokatchev:

It is a fundamental principle in quantum field theory, I don't think that you can derive it. Well, maybe if you put this symmetry as a basic fundamental principle, then unitarity will follow somehow, but I think, this is not the natural way. It doesn't mean that one cannot look for other alternative ways, for instance, this dual conformal symmetry was very useful historically for constructing higher loop 5-gluon amplitudes. Zvi and collaborators were using unitarity to build amplitudes. They have a very efficient mechanism how you can, first, guess the set of integrals. But it turned out later that dual conformal symmetry severely restricts the choice of such integrals, you can predict them. The only thing which is left is to fix the coefficients, which again unitarity does for you. So, knowing the symmetry helps unitarity, but does not replace it.

- Alba:

Do you have integrability in your theory?

- Sokatchev:

This is what I've been talking about...

- Alba:

One of the motivation was that studying N=4 can help us to understand QCD. However in QCD we have neither supersymmetry nor integrability.

- Sokatchev:

That is not true... There is integrability, it has been known for many-many years, the QCD Hamiltonian is integrable, it goes back to the 1990's. It was later taken over in N=4 Yang-Mills where, without surprise, that Hamiltonian of N=4 Yang-Mills is also integrable. But this is the same phenomenon as in QCD, so the integrability is definitely there in all these gauge theories. N=4 has more chances to produce an exact solution, because the integrability is stronger, there is more symmetry. But it is not true, that integrability is important only in this maximally supersymmetric theory.

- Alba:

But what is the reason for all answers becoming much more uglier, than in the massless theory?

- Sokatchev:

There is no conformal symmetry if you have masses. This symmetry is lost, but it does not mean you lose everything. That's one of the main problems, when you try to apply what we have learnt from my lecture to some realistic theory, like the Standard Model. At the end of my talk I appealed to our experimental colleagues, "Please, build a conformal collider for us, then we can test all these theories". It is true, that N=4 is an extreme case without masses, in addition we go to the planar limit. There is one thing, which I want to comment here. Planarity is very important for this to work. Because my basic trick was introducing this dual space. It only works for planar graphs, and does not work for non-planar.

- 't Hooft:

A conformal circular collider will be very easy to make, because large circles are conformally equivalent to small circles.

- Novikov:

I am Oleg Novikov, St. Petersburg State University. Again about locality and unitarity, this stuff is a way to get some new insights in our physics by reformulating the theory in the form where this dual superconformal symmetry is manifest, and locality and unitarity are not. So, my question is do you know any other examples of theories with singular structure. And the second question is do you know if there have been any attempts to play with this new formulation to get some more realistic models.

- Sokatchev:

The answer to both questions is no. I am not aware of anything like this.

- Novikov:

Thank you!

- 't Hooft:

The infrared structure of conformal theories is very counter-intuitive. Therefore, I think, that adding mass terms in the Lagrangian would be a very interesting exercise. So my question is, what happens if I would add any mass term I like, as long as it obeys Lorenz invariance, and respects ordinary gauge invariance. Usually, there will be just a small number of such terms that one can insert, so, you have a very limited number of parameters compared with some more general theory. I can imagine, for instance, that I make approaches to the Standard Model that way.

- Sokatchev:

Let me answer your question by a question to you, because you invented this planar limit in Yang-Mills theory. We know that going to the planar limit, we gain a lot in simplicity, things become much more transparent. To go back is not at all easy, start introducing color parameters, then things become much more complicated, and you may lose some essential features, which is what happens here.

- *'t Hooft:*

The planar limit corresponds to the local gauge group being infinite-dimensional; replacing it by a finite-dimensional group implies that these planar structures behave as interacting particles. The effective field theory that would describe such interactions would not be of the super renormalizable type. This implies that it generates many more infinities that need to be regularized by new counter terms.

- *Sokatchev:*

Trying to introduce masses, maybe it is what we should do. Nobody has tried to do this. Try to introduce very small masses and see the small breaking of this dual conformal symmetry.

- *'t Hooft:*

Regardless how small or big they are, mass terms define the scale. They not only break completely the supersymmetry, but also conformal symmetry is broken only by these mass terms. This must be a big effect on the spectrum of physical states, so it would be worth-while to add such mass terms and just see, what happens.

- *Sokatchev:*

I'm afraid I will lose all features starting with the finiteness of N=4. This is due to supersymmetry.

- *'t Hooft:*

You still keep the ultraviolet finiteness. Now you have no longer a completely continuous and degenerate spectrum of states, so you can talk about unitarity. But in the far ultraviolet all the masses disappear.

- *Sokatchev:*

The answer is nobody has seriously tried to do this, and, of course, it's a true challenge, because if we want to apply all this to some realistic theory, you have to go away from the planar limit and introduce masses.

- *Zichichi:*

I was disturbed when you said that some of the amplitudes were like masses and you said that after a moment of a very complicated computation, you got expressions mass-like. What do you mean by mass-like? Looking like masses? You said this, not me.

- *Sokatchev:*

I've said that people invented the regulator, which looks like a mass, but in 5 dimensions, and extra dimension looks like mass. But in such a way, you do not break the conformal symmetry in 4 dimensions. That's not a true mass.

- *Zichichi:*

Thank you!

- *'t Hooft:*

Thank you very much!

ITP-UU-14/03
SPIN-14/03

Three Erice Lectures

Gerard 't Hooft

Institute for Theoretical Physics
Utrecht University
and
Spinoza Institute
Postbox 80.195
3508 TD Utrecht, the Netherlands
e-mail: g.thooft@uu.nl
internet: http://www.staff.science.uu.nl/~hooft101/

Abstracts

I **The conformal black hole.**
Canonical quantum gravity has a hidden local conformal symmetry which is broken by a BEH mechanism. In this short lecture a summary is given of the arguments that may be used to exploit this symmetry to address the black hole information problem and to cope with the putative "firewall". Non-trivial consequences of this theory are outlined.

II **Classical and quantum theories.**
Usually, "quantization" of a theory means that a classical theory is replaced by a different theory, by replacing Poisson brackets by commutators. In this lecture an other procedure is outlined: some models allow for a description that is quantum and classical at the same time. Such models could point the way towards a promising class of hidden variable models. We briefly discuss what might happen with the well-known no-go theorems in such systems.

III **Closing Lecture.**
The next step for LHC.

January 10, 2014

1 Lecture I, The conformal black hole

The black hole horizon[1] was recognized to present us with a major fundamental problem already some 35 years ago. However, then as much as now, a majority of physicists took the attitude that this problem could be fixed with some rope and cello-tape to make a black hole look like ordinary matter again, and that anyhow superstring theory[2][3] would have the correct answer to this problem built in[4], so that there should be no reason to worry.

In reality the problem is much deeper, forcing us the think more about the nature of space, time and matter themselves. It should be emphasized that the information lingering in or near a black hole horizon cannot be addressed at all using textbook calculational procedures, regardless which of today's quantum gravity theories one adheres to. The problem is also not to identify what is wrong with the text books, and how one might cure the disease.

The problem is to find a solution that allows one to de detailed calculations of the black hole properties, by modifying the existing rules in a fundamental way.[5]

Indeed, the black hole should be handled as an ideal laboratory for thought experiments in quantum gravity, but as of today it seems that quantum black hole physics is in a state of confusion that is very similar to the state elementary particle physics was in before the advent of the Standard Model.

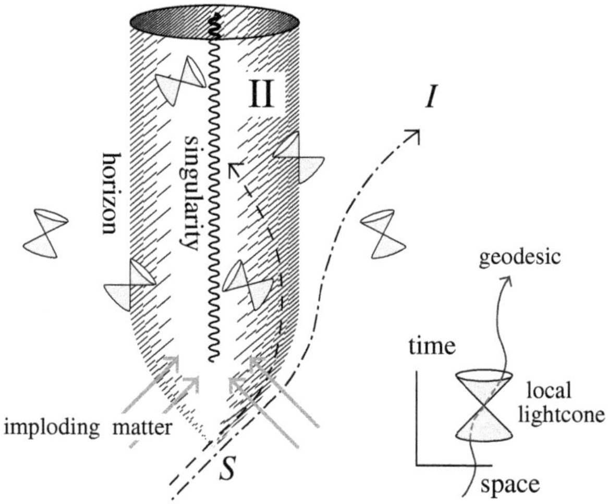

Figure 1: Space-time of a black hole in formation. Time runs vertically. Orientation of local light cones is indicated.

Hawking's important observation[1] was the following: what is perceived as a vacuum to an observer who enters a black hole through its horizon, is a state of entangled particles to an outside observer. The entanglement is between particles visible outside the black hole (indicated by I in Figure 1) and particles that disappeared into it (indicated by II in Figure 1. Since the outside observer does not observe the particles that disappeared, the states (s)he does observe are to be described by a density matrix.

What our modified text book should contain is a prescription as to how this density matrix ought to be replaced by some *pure quantum state*, depending on the material out of which the black hole is formed. The essential ingredient that is still not entirely understood is called *complementarity*[6]

However, whatever that procedure is, it modifies the state described by the observer who went in. To the observer going in, this modified state contains a curtain of particles: a *firewall*[7]. In a meaningful 'textbook procedure', this 'firewall' should not be a replacement of the observations made by an ingoing observer, but a mathematical transformation of that. The description with, and the one without firewall, should be *equivalent*. This is what *black hole complementarity* really is about. It is also a transformation that links the vacuum state with states containing matter, or, a transformation under which the vacuum is not invariant. We have seen such transformations before: the BEH mechanism.[8][1]

Apparently, the general coordinate transformation that links the inside observer to the outside observer is more complicated than previously thought. We now have a proposal:

> There is a "causality principle": two observers linked by black hole complementarity[9] may disagree about the matter content of their world, but they do agree about the causal ordering of space-time points (future, past, or space-like separated)[2]. Therefore, they agree about light cones and light-like geodesics; they may still disagree about distance- and time scales, and about space-like or time-like geodesics. [11]

This situation can easily be implemented in canonical quantum gravity. There, in any given coordinate frame, we have a metric tensor $g_{\mu\nu}(\vec{x},\,t)$ that has 10 independent components (as it is symmetric in its two 4-indices). One of these independent components is the overall factor, to be referred to as the "conformal factor" ω:

$$g_{\mu\nu}(\vec{x},\,t)=\omega^2(\vec{x},\,t)\,\hat{g}_{\mu\nu}(\vec{x},\,t)\ , \tag{1.1}$$

where $\hat{g}_{\mu\nu}$ is defined *modulo* its conformal factor, so that it has only 9 independent components (the square in ω^2 is there just for later convenience). One then observes that the light cones and the light-like geodesics depend on $\hat{g}_{\mu\nu}$ but not on ω. If two observers disagree about $\omega(\vec{x},\,t)$, they still do agree about the causal relations between space-time points.

[1]In gauge theories, also when the BEH mechanism is at work, we have learned to view the vacuum state as being gauge-invariant anyway. What we mean here is simply that the vacuum would be characterized by the expectation value $\langle 0|\varphi(x)|0\rangle=\binom{F}{0}$, which would change after a local gauge transformation.

[2]This is opposed to a theory often phrased from a string theory stand point, which states that space and time are emergent[10]. How any concept of causality, more or less in accordance with special and/or general relativity, is to be formulated in such a theory is not clear to this author.

Indeed, our complementary observers may well agree about this ordering, in spite of the fact that they disagree about the matter content of their universe. Consider a black hole that is slowly decaying due to Hawking radiation. Since the decay is slow in comparison with the events of absorption and emission of particles, we may approximate a decaying black hole very precisely by rewriting its space-time metric as

$$\mathrm{d}s^2 = M(\tilde{t})\Big(-\mathrm{d}t^2(1-2/r) + \frac{\mathrm{d}r^2}{1-2/r} + r^2\mathrm{d}\Omega^2\Big) \,. \tag{1.2}$$

Here, we rescaled the usual r and t coordinates by a factor M, and we take M to depend on time $\tilde{t}$, which may well be taken to be the same as t itself, but one may also choose the retarded or advanced time coordinate; all this makes little difference since the time evolution is slow.

The variable $M(\tilde{t})$ is now seen to be merely a conformal factor in the metric, and the orientation of the light-like geodesics does not depend on it. Now note, that an observer falling into the black hole sees a constant M throughout, while (s)he passes through the horizon, while a distant observer sees that, at the horizon, $M(\tilde{t})$ shrinks to zero. Indeed, this time dependence of $M(\tilde{t})$ generates a Ricci curvature that can well be interpreted as the Ricci curvature caused by the Hawking particles. Or: the curvature caused by the back reaction of the Hawking particles can be switched on or off, depending on who is observing.

In canonical quantum gravity, $\omega(\vec{x},\, t)$ is a dynamical field, with a vacuum expectation value, which can be taken to be 1.

If we now take the Einstein-Hilbert action for gravitation in terms of $\omega^2\, g_{\mu\nu}$ and add to this a matter Lagrangian, which we split into a conformally invariant part and a couple of dimensionful terms, we find:

$$\mathcal{L} = \sqrt{-\hat{g}}\Big(\tfrac{1}{12}\hat{R}\tilde{\omega}^2 + \tfrac{1}{2}\hat{g}^{\mu\nu}\partial_\mu\tilde{\omega}\partial_\nu\tilde{\omega} + \mathcal{L}^{\mathrm{conf}}(\hat{g}_{\mu\nu},\, \varphi^{\mathrm{matter}}) + \tfrac{1}{2}\tilde{\omega}^2\tilde{m}^2\varphi^2 + \cdots \Big) \,, \tag{1.3}$$

where $\mathcal{L}^{\mathrm{conf}}$ is the locally conformally invariant part of the matter Lagrangian, and we made some rescalings:

$$\omega = \sqrt{\tfrac{4}{3}\pi G_N}\, \tilde{\omega} \,, \qquad \tilde{m} = \sqrt{\tfrac{4}{3}\pi G_N}\, m \,, \tag{1.4}$$

where G_N is Newton's constant. Note that, rescaled this way, the mass terms became dimensionless; if all terms that originally had the dimension of mass terms raised to some power, are multiplied by the appropriate power of $\tilde{\omega}$, then this makes everything manifestly conformally invariant. We now note that the vacuum value

$$\langle \emptyset |\, \tilde{\omega}\, | \emptyset \rangle = \sqrt{\frac{3}{4\pi\, G_N}} \,, \tag{1.5}$$

is the only physical effect that breaks the local conformal invariance.

Note, that writing the gravitational Lagrangian as in Eq. (1.3), turns the $\tilde{\omega}$-field into a totally normal looking interacting scalar field, just as the other scalar fields in the

Lagrangian, with one important exception: it appears to have negative metric, since the sign in the action is anomalous. Equivalently, one may say that the contribution of $\tilde{\omega}$ to the total energy-momentum tensor is of the 'wrong' sign.

Since now the original Newton term in the action has been eliminated, the Einstein equation can be written as

$$T_{\mu\nu}^{\rm matter} + T_{\mu\nu}(\tilde{\omega}) = 0 \ , \tag{1.6}$$

and indeed, one can easily verify that, if we return to the original variable $g_{\mu\nu}$, then

$$T_{\mu\nu}(\tilde{\omega}) = \frac{-1}{8\pi\, G_N}\, G_{\mu\nu} \ . \tag{1.7}$$

Now all of this would be merely a re-arrangement of the equations of canonical quantum gravity, which could be interesting mathematics with no physical consequences, if we did not have a very interesting and important twist to add to this story: the $\tilde{\omega}$ field should indeed not be treated differently from the other fields, and this means that our action (1.3) should be kept totally regular as $\tilde{\omega} \to 0$. Now this is the small-distance limit of the theory, where one might expect lots of new physics, but we claim that points in space-time where a field happens to vanish should in no way be special; the functional integral over the matter fields and the $\tilde{\omega}$ field together is a conformally invariant expression and therefore should be treated the same way we always treat conformal theories. True local conformal invariance now dictates that there should be *no local conformal anomalies.* Usually, such anomalies are there[12], so now we postulate that these anomalies cancel out. Only then can we employ the picture of spontaneous local conformal symmetry breaking.

Unfortunately, this theory can as yet not be checked, or rather, attempts to produce physically interesting conformally invariant theories along these lines have not (yet?) been successful. The theory demands that, *if the originally dimensionful terms are added into our conformal theory*, then all Callan-Symanzik β-functions should exactly vanish[11]. This gives us as many equations as there are coupling parameters in the theory, which suggests that maybe constructions can be found where all these parameters are computable,and this emphatically includes the cosmological constant and all masses of the theory. Up till today, no such theories have been found (including supersymmetric systems that would be physically viable).

There are two important remarks to be added: so-far, the author checked explicitly the behavior of the β-functions in a flat $\hat{g}_{\mu\nu}$ background. In reality however, also the equations for $\hat{g}_{\mu\nu}$ must be found (according to the canonical theory, these equations correspond to the demand that *all ten components* of Eq. (1.6) must vanish. This condition may also suffer from anomalies.

Finally, the theory formulated so-far is too simple to be completely valid: it would yield computable coupling strengths, masses and cosmological constant that are all of order one at the Planck scale. This means that we have not yet addressed the *hierarchy problem*[13]: in the real world, quite a few coupling parameters are extremely tiny at the Planck scale, and there should be a natural explanation of that.

2 Lecture II, Classical and quantum theories

An enormous body of literature[14] concerns the question how to interpret the equations of quantum mechanics that describe with extreme precision processes among tiny amounts of matter, typically atoms, molecules and sub-atomic particles. A major fraction of that addresses the question what happens, or what could be happening, when experimenters do measurements in some experimental set-up. It is assumed that they understand what a particle is, and how their detection devices work.

In this lecture it is briefly indicated how one might approach these issues differently. We take as a fact that any physical system can be described in terms of a Hilbert space, and that a process that is being studied can be represented there as a vector in this Hilbert space. What makes standard quantum mechanics so powerful in practice, is that one may introduce any coordinate frame one likes in this Hilbert space, which means that one may choose a basis of elementary states, and consider all possible structures that can arise by producing superpositions of these states.

It is a peculiar feature of quantum mechanics that any superposition of these states appears to be allowed: experimenters can put a particle in an eigenstate of the momentum operator, the position operator or any operator one chooses. Nevertheless, this is manifestly *not* true for "macroscopic" states: a person may stand here or may stand there, but (s)he can never stand in a quantum superposition of such "states". Not all is right with the superposition principle.

What we propose now is to consider or to construct simple models[15]. These models have the property that they can be described perfectly well using Hilbert space, and as such, they must be regarded as "quantum theories". Yet where they differ from conventional quantum theories is, that these models also allow a special basis, the "ontological basis". In this basis, all observable quantities are diagonal operators, at all times, and hence they commute. In the ontological basis, the quantum nature of these models disappears and they become classical. Models of this sort form a large class. In principle, they may include all sorts of interactions, including strictly local interactions, and as such they seem to falsify no-go theorems for local hidden variables. As of today, there is no contradiction because none of my models are sufficiently general to be able to substitute for the Standard Model, so if you firmly believe that hidden variables have been shown to be impossible you can go back to sleep. Alternatively, however, we could try to generalize these models further and try to understand how one might approach the real world along these lines.

2.1 The Dirac notation

We start with a completely general discussion of quantum systems. These systems are characterized by the existence of a Hilbert space. In theories familiar to us, this Hilbert space allows us to use any frame of basis elements that we please. A basis is an (usually infinite) orthonormal set of vectors that completely span this Hilbert space. Let us phrase the resulting description in terms of axioms:

- *States:* the states a physical system can be in form two vector spaces, that of the bras, indicated as $\langle\psi|$, and that of the kets, $|\psi\rangle$. Usually, these spaces are infinite-dimensional. The letter ψ stands short for the description of the physically observable properties of the state.

- *superposition:* In a microscopic system, one may consider the states obtained by superimposing physical states: $|\psi_3\rangle=\alpha|\psi_1\rangle+\beta|\psi_2\rangle$, or $\langle\psi_4|=\gamma\langle\psi_1|+\delta\langle\psi_2|$, where α, β, γ and δ are usually complex. There is an anti-linear mapping $|\psi\rangle\leftrightarrow\langle\psi|$, an inner product $\langle\psi_1|\psi_2\rangle$ exists, and $\langle\psi|\psi\rangle$ is a positive norm. Physical states will be normalized to one.

- *Dynamics:* There is a linear Schrödinger equation:

$$\frac{\mathrm{d}}{\mathrm{d}t}|\psi\rangle=-iH|\psi\rangle\ , \tag{2.1}$$

  where H can be any hermitean operator. Usually, one demands that this operator is bounded from below, or: $H\geq 0$.

Axioms that we shall *not* impose are:

- The probability that one normalized state $|\psi_1\rangle$ is *equal* to another normalized state $|\psi_2\rangle$ is given by $|\langle\psi_1|\psi_2\rangle|^2$.

- The expectation value for an operator X, when measured in any state $|\psi\rangle$, is given by $\langle X\rangle=\langle\psi|X|\psi\rangle$.

This is because we will not admit *any* state $|\psi\rangle$ as a physically acceptable state, even though one can have microscopic states of any form. The macroscopic states, that is, states that include the description of a measuring or detection device, must be elements of the orthonormal ontological basis.

In stead, we do have other axioms that will be used:

- There is an *initial condition:* the universe begins in one, given state $|\psi(0)\rangle$.

- We have *template states* $|\psi_i\rangle$. They may form an over-complete basis of Hilbert space, and their evolution laws may be known to us. An example is Fock space of elementary particles. The probability that the universe is actually in one of these template states is given by $|\langle\psi_i|\psi(0)\rangle|^2$. The expectation value of an operator X is always $\langle\psi(0)|X|\psi(0)\rangle$.

In any other state $|\psi\rangle$, the expectation value of an operator would be

$$\langle\psi|\psi(0)\rangle\langle\psi(0)|X|\psi(0)\rangle\langle\psi(0)|\psi\rangle\ . \tag{2.2}$$

Note, that this coincides with the usual expression only if X is diagonal in the ontological basis of states $|\psi(0)\rangle$. As for the physical world, one would then have to assume that other observable operators X do not exist; those are artifacts of our imagination only.

Conventional quantum mechanics makes use of template states; these are the states that describe single particles such as photons, nucleons and so on. In terms of these template states, $|\psi(0)\rangle$ may be *entangled*:

$$\psi(0) = \alpha|\psi_1\rangle|\psi_2\rangle + \beta|\psi_3\rangle|\psi_4\rangle \ , \tag{2.3}$$

or whatever. We emphasize that our weakened axioms should not imply any observable modifications or limitations in the generally accepted concept of Quantum Mechanics.

But it opens the door to "underlying theories".

2.2 Classical theories that can be mapped onto Dirac quantum systems.

Theories that will be briefly discussed are the following:[3]

Classical theory	Quantum theory
• Discrete periodic system	Finite number of equally spaced energy levels.
• Continuous periodic system	Harmonic oscillator.
• Flat classical sheet	First quantized massless non-interacting chiral fermion.
• Many classical sheets	Second-quantized massless fermions.
• Classical string theory on a lattice (26 or 10 dimensions)	Quantum (super-) string theory on the continuum (26 or 10 dimensions), as yet without interactions.

What these models have in common is that the observables in the classical theory can be identified with a special set of operators in the corresponding quantum system. These operators will be called "beables". Beables $\mathcal{O}_i(t)$, where the index i is a counter, are a set of time-dependent quantum operators that has been selected in such a way that they all commute at all times:

$$[\mathcal{O}_i(t), \mathcal{O}_j(t')] = 0 \quad \text{for all} \ \ t, \ t' \ . \tag{2.4}$$

2.2.1 Discrete periodic system

Our first example is a system that is discrete and periodic. It can be in N distinct positions, called 'states'. The evolution law is simple:

$$(1) \to (2) \to (3) \to \cdots \to (N) \to (1) \ . \tag{2.5}$$

[3] A much larger class of models with interactions was found more recently[16]

Let us elevate these states as basis elements of a (finite dimensional) Hilbert space. The evolution law

$$|1\rangle \to |2\rangle \to \cdots |N\rangle \cdots |1\rangle \ , \tag{2.6}$$

can be described by an evolution operator $U(t)$:

$$|\vec{Q}(t+\delta t) = U(\delta t)|\vec{Q}(t)\rangle \ , \tag{2.7}$$

where δt is the unit time step, so that $U(\delta t)$ takes the form of a matrix:

$$U(\delta t) = \begin{pmatrix} 0 & 0 & \cdots & 1 \\ 1 & 0 & \cdots & 0 \\ 0 & 1 & \cdots & 0 \\ & & \cdots & \end{pmatrix} . \tag{2.8}$$

It is not hard to find a hamiltonian operator H such that $U(\delta t) = \exp(-iH\delta t)$, One can perform a unitary transformation to any other basis, for instance, the basis where $U(t)$ is diagonal. The eigen values of $U(\delta t)$ can be written as $e^{-iE_k\delta t}$, since U is unitary, and because $U(\delta t)^N = U(N\,\delta t) = \mathbb{I}$, we deduce that the N eigenvalues are the N N^{th} roots of 1, so that

$$E_k = \frac{2\pi k}{N\,\delta t} \ , \qquad k = 0, 1, \cdots, N-1 \ . \tag{2.9}$$

The energy eigen states, $|E_k\rangle$, are the finite Fourier transforms of the ontological states $|1\rangle, \cdots, |N\rangle$ of Eq. (2.6). These energy eigen states are of course not the ontological states of the system. The energies are equally spaced, as in an atom with spin $\ell = \frac{1}{2}(N-1)$ and magnetic moment, in a homogeneous magnetic field.

We conclude that the discrete periodic system and the system with equally spaced energy levels are different mathematical descriptions of the same physical system. Besides this, there is an important physical implication: if the quantum system is identified with the ontological system this implies a restriction for the wave function: it is one of the ontological states. At time $t = 0$ the world may have started in one state $|\psi(0)\rangle$, and after that, it never goes into a superposition, but we can use superpositions to do calculations in this model, or to describe statistical ensembles if it is not known in which state our system started.

The philosophy of our approach is that, since $|\psi(0)\rangle$ is not known, it is useful in practice to expand it in terms of whatever template states $|\psi\rangle$ we have available. In terms of these template states, superpositions are ubiquitous.

The beable (2.4) in the present model is the operator n that counts the states in Eqs. (2.5) and (2.6), which are the finite Fourier transforms of the energy eigen states $|E_k\rangle$ of the atom.

2.2.2 Continuous periodic system

Consider the continuum limit of the model in the previous subsection. This is the case where we replace the discrete index $i = 1, \cdots, N$ in Eq. (2.5) by a continuously varying angle $\varphi \in [0,\, 2\pi)$. Let this angle increase with constant velocity v:

$$\frac{\mathrm{d}\varphi}{\mathrm{d}t} = v \,, \qquad |\,\varphi = 2\pi\rangle \;=\; |\,\varphi = 0\,\rangle \,. \tag{2.10}$$

Define on the Hilbert space spanned by the periodic wave functions $\psi(\varphi) = \langle\varphi|\psi\rangle$ an operator $\hat{p} = -i\partial/\partial\varphi$. Then our time evolution is given by

$$\frac{\mathrm{d}}{\mathrm{d}t}|\psi(\varphi)\rangle = -v\frac{\partial}{\partial\varphi}|\psi(\varphi)\rangle \,, \qquad \text{so that} \tag{2.11}$$

$$\frac{\mathrm{d}\hat{\varphi}}{\mathrm{d}t} = -i[\hat{\varphi},\, H] = v \,; \qquad H = v\,\hat{p} \,. \tag{2.12}$$

The energy eigen states are the eigen states of $\hat{p}$, which are the periodic plane waves on $[0,\, 2\pi]$. The period of the motion is $T = 2\pi/\varphi$, so we have

$$e^{-iHT} = \mathbb{I} \,, \quad \rightarrow \quad E_k = \tfrac{2\pi}{T}\,k \,, \qquad k = 0, 1, \cdots, \infty \,. \tag{2.13}$$

This is simply the continuum limit of Eq. (2.9). There is however a caveat: one could argue that the spectrum of states should be represented by all integers, not just the positive ones. Here, we decided to take the continuum limit with a bit more care; since Eq. (2.9) has a lowest value, we keep this same lowest value in the continuum limit, a deliberate choice in our mapping.

Thus we arrived at the same energy spectrum as that of the harmonic oscillator. We claim that, just because this can be regarded as the continuum limit of the trivial finite system of Subsection 2.2.1, the mapping between the harmonic oscillator and the wave functions on the Hilbert space of a continuous periodic system is one-to-one and unitary.

For simplicity, take the case $T = 2\pi$, $v = 1$. Now consider the harmonic oscillator in x space, with $\hat{p}_x = -i\frac{\partial}{\partial x}$. It must have the same periodicity $T = 2\pi$. Consider its hamiltonian therefore to be

$$H = \tfrac{1}{2}(\hat{p}_x^2 + x^2 - 1) \,. \tag{2.14}$$

Here, we subtracted the vacuum energy. If we had not done that, the wave functions would be *anti*-periodic in T, a situation we wish to avoid. Suppose now that we map the energy eigen states $|E_k\rangle$ of the periodic system onto the eigen states $|E_k\rangle$ of the harmonic oscillator. Since we know all the inner products, this leads to a unitary transformation between the states $|\varphi\rangle$ and the states $|x\rangle$. What is this unitary matrix?

It is an elementary calculation, but the ensuing exercise in special function theory is non-trivial. Call the energy eigen states $|n\rangle$, $n = 0, 1, \cdots \infty$. With the proper normalization[4] we have

$$\langle\varphi|n\rangle = \frac{1}{\sqrt{2\pi}}\,e^{in\varphi} \,, \qquad \langle x|n\rangle = \frac{H_n(x)\,e^{-\frac{1}{2}x^2}}{2^{n/2}\pi^{1/4}\sqrt{n!}} \,, \tag{2.15}$$

[4]In earlier papers, the author chose to short-cut these normalization issues by taking $e^{2\pi} = 535.5$ as the base of our exponentials. We decided not to use such conventions here.

where H_n are the Hermite polynomials. This gives our desired inner product:

$$\langle\varphi|x\rangle = \sum_{n=0}^{\infty} \frac{H_n(x)\, e^{-\frac{1}{2}x^2}}{2^{(n+1)/2}\, \pi^{3/4}\sqrt{n!}}\, e^{in\varphi} \ . \tag{2.16}$$

It is difficult to work out this sum analytically; the main problem is the square root that originated from the normalization. It is not a good idea to terminate the sum after 100 or so terms since the expression is very unstable: for sizable values of x the Hermite polynomials become very large and oscillate. However, a very good approximation can be worked out (including the required correction terms). Use the generating function of the Hermite polynomials:

$$\sum_{n=0}^{\infty} H_n(x) s^n/n! = e^{-s^2+2sx} \ , \tag{2.17}$$

We then use the approximation

$$\frac{\sqrt{\pi}}{2^n} n! = \Gamma(\tfrac{1}{2}n + \tfrac{1}{2})\, \Gamma(\tfrac{1}{2}n + 1) \approx \Gamma(\tfrac{1}{2}n + \tfrac{3}{4})^2 \ . \tag{2.18}$$

Substituting this in Eq. (2.16) removes the square root, and one then finds good approximate expressions for the desired inner product (the correction terms could also be computed but turn out to be very small). Thus, we found expressions for $\langle\varphi|x\rangle$ and $\langle\varphi|p\rangle$ which we can plot against φ and x or p, see Fig 2.

In this figure, we clearly see that the sign of x switches as $\varphi = \pm\pi/2$. The most rapid oscillations could not be made visible clearly.

In this model, the beable is the operator φ defined by the matrix (2.16).

2.2.3 "Neutrinos"

In this subsection, we consider non-interacting chiral fermions in 3+1 dimensions. Being chiral, the fields of these fermions have only two spinor components. As such, they resemble neutrinos, in the limit where their masses are neglected. We call these particles "neutrinos" in quotation marks. First consider the first-quantized case: a single "neutrino".

The quantum system is then described by the hamiltonian

$$H = m + \sigma_i\, \hat p_i \ , \qquad \hat p_i = -i\frac{\partial}{\partial x_i} \ , \tag{2.19}$$

$$\sigma_i = \left(\begin{pmatrix} 0 & 1 \\ 1 & 0 \end{pmatrix}, \begin{pmatrix} 0 & -i \\ i & 0 \end{pmatrix}, \begin{pmatrix} 1 & 0 \\ 0 & -1 \end{pmatrix} \right) , \tag{2.20}$$

$$\text{with } \sigma_i\, \sigma_j = i\varepsilon_{ijk}\sigma_k + \delta_{ij} \ , \tag{2.21}$$

but, from now on, we keep $m = 0$. We then define an operator $\vec q$ that specifies the *orientation* of the momentum $\vec p$ without specifying its length, or its sign:

$$p_r \equiv \pm|\vec p| \ , \quad q_i \equiv p_i/p_r \ , \quad |\vec q\,| = 1 \tag{2.22}$$

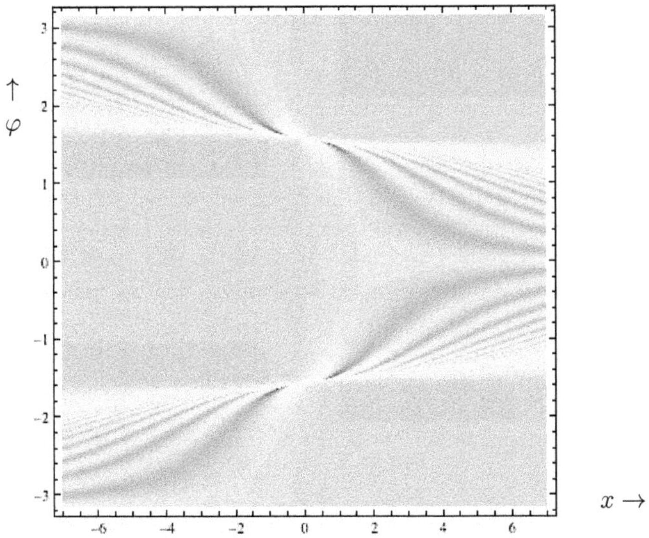

Figure 2: Plot of the real part of the transformation matrix $\varphi|x\rangle$. Horizontally, x runs from -10 to 10, vertically φ runs from $-\pi$ to π .

(which is well-defined in momentum space, and can be extended to position space by the appropriate Fourier transformations). Next, we define the operators

$$s \equiv q_i\sigma_i = \vec{q}\cdot\vec{\sigma} = \pm 1 \ , \qquad r \equiv \tfrac{1}{2}(\vec{q}\cdot\vec{x} + \vec{x}\cdot\vec{q}) \tag{2.23}$$

(because $\vec{x}$ does not commute with $\vec{q}$, the operator r was symmetrized, so as to ensure that it is hermitean).

Now, we claim to have found the beables $\mathcal{O}_i$ of this "neutrino". They are the set

$$\{q_i,\ s,\ r\}\ . \tag{2.24}$$

Proving that these all commute at all times is not hard. First, all q_i and s commute trivially, since the p_i all commute, and the σ_i commute with all p_j . The fact that $[r,\, \vec{q}] = 0$ is most easily seen in momentum space, where $x_i = i\partial/\partial p_i$. Here, r is $1/p_r$ times the dilaton operator in p space, but $\vec{q}$ has fixed norm, so it is invariant under dilatations. So, $\vec{q}$ commutes with all parts of the definition of r . Because of this, r must also commute with s . Furthermore, we have the equations of motion:

$$\frac{\mathrm{d}}{\mathrm{d}t}\vec{q} = 0\ , \qquad \frac{\mathrm{d}}{\mathrm{d}t}s = 0\ , \qquad \frac{\mathrm{d}}{\mathrm{d}t}r = s = \pm 1\ , \tag{2.25}$$

which only involves beables again, and this implies that, in terms of these variables, the entire system evolves classically.

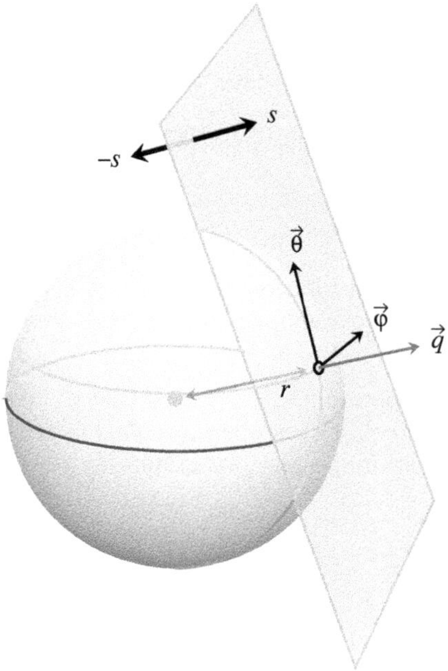

Figure 3: neutrinos are infinite sheets.

The best physical description of this result is that the system behaves as an infinite sheet, propagating with the speed of light in one of the two possible orthogonal directions (indicated by the variable $s = \pm 1$), see Fig. 3

The set of beables, (2.24), is complete. This means that their eigen states span the entire neutrino Hilbert space. This means that a unitary transformation must exist that transforms the eigen states of $\vec{q}$, s, and r onto the Hilbert space of states $|\vec{x}, \sigma_3\rangle$ unambiguously. This mapping, however, is technically rather delicate. I report the result of this technical calculation here, without derivation. One may ask: what is the operator x_i in terms of operators acting on our neutrino sheet?

To answer this question, we need to introduce some more operators for the neutrino sheet. We need operators that flip the sign of s. There are two of these, called s_1 and s_2. We call $s = s_3$. We define them as

$$s_1 = \vec{\theta} \cdot \vec{\sigma} \, , \qquad s_2 = \vec{\varphi} \cdot \vec{\sigma} \, , \qquad s_3 = s = \vec{q} \cdot \vec{\sigma} \, , \tag{2.26}$$

where

$$\vec{q} = \begin{pmatrix} q_1 \\ q_2 \\ q_3 \end{pmatrix} , \quad \vec{\theta} = \frac{1}{\sqrt{q_1^2 + q_2^2}} \begin{pmatrix} q_3 \, q_1 \\ q_3 \, q_2 \\ q_3^2 - 1 \end{pmatrix} , \quad \vec{\varphi} = \frac{1}{\sqrt{q_1^2 + q_2^2}} \begin{pmatrix} -q_2 \\ q_1 \\ 0 \end{pmatrix} , \tag{2.27}$$

which are all normalized to one. They obey the usual spinor algebra:

$$s_i\, s_j = i\varepsilon_{ijk}s_k + \delta_{ij} \ . \tag{2.28}$$

So, using the standard representation (2.20) now for s_1, s_2, and s_3, we have found for the Pauli matrices

$$\sigma_i = \theta_i\, s_1 + \varphi_i\, s_2 + q_i\, s_3 \ , \tag{2.29}$$

which is of course an orthogonal rotation. It was necessary to define s_1 and s_2, and the only option was to identify two vectors $\vec\theta$ and $\vec\varphi$ orthogonal to the vector $\vec q$, which is why we impose Eq. (2.27).

Next, we had to define operators that rotate the sheet. We call their generators $L_i^{\rm ont}$. They do not coincide with the rotation operators L_i in position space, nor the operators J_i for total angular momentum, because we assume L_i only to act on the sheets, not on s_3. We found

$$L_i^{\rm ont} = -i\varepsilon_{ijk}\, q_j\, \frac{\partial}{\partial q_k} \ , \tag{2.30}$$

$$L_i^{\rm ont} = L_i + \tfrac12\Big(\theta_i\, s_1 + \varphi_i\, s_3 - \frac{q_3\,\theta_i}{\sqrt{1-q_3^2}}\, s_3\Big) \ . \tag{2.31}$$

In terms of these operators we van now calculate x_i. Again, without derivation:

$$x_i = q_i(r - \tfrac{i}{p_r}) + \varepsilon_{ijk}\, q_j\, L_k^{\rm ont}/p_r + \frac{1}{2p_r}\Big(-\varphi_i\, s_1 + \theta_i\, s_2 + \frac{q_3}{\sqrt{1-q_3^2}}\,\varphi_i\, s_3\Big) \ . \tag{2.32}$$

Here, the imaginary term is there to cancel an imaginary term ensuing from the non-commutativity of q_i and $L_k^{\rm ont}$ in the next term.

If $\alpha = 1,2$ enumerates the two spinorial eigen states of σ_3, then the conventional basis used to span the Hilbert space is the states $|\vec x,\, \alpha\rangle$. And now one can calculate the matrix that links these states to the eigen states of the beables $\vec q$, r, and s:

$$\langle \vec x,\, \alpha|\vec q,\, r,\, s\rangle = \frac{i}{2\pi}\delta'(r - \vec q\cdot\vec x)\,\chi^s_\alpha(\vec q) \ , \tag{2.33}$$

where δ' is the derivative of the Dirac delta function.[5] This derivative turns out to be an inescapable consequence of a normalization requirement on the corresponding matrix in momentum space, which leads to an extra factor p_r. The spinors χ^1_α and χ^{-1}_α are defined by the eigen-spinor equation

$$(\vec q\cdot\vec\sigma_{\alpha\beta})\,\chi^s_\beta(\vec q) = s\,\chi^s_\alpha(\vec q) \ . \tag{2.34}$$

The derivation of Eqs. (2.32) and (2.33) will be given in a separate publication.

[5]In discussions at Erice the question came up why this expression was not symmetrized in the variables $\vec q$ and $\vec x$. The point, of course, is that here these are c-numbers; $\vec q$ sits in the ket and $\vec x$ is in the bra.

2.2.4 Second quantized models

Second-quantizing fermion models is easy in this formalism. We basically use the same philosophy as Dirac did originally. All we have to do is take an indefinite number ν of neutrino states, declaring the total hamiltonian of the system to be

$$H = \sum_{\nu} H_{\kappa} - C \ , \tag{2.35}$$

where κ enumerates the particles. Being fermions, they must obey a Pauli exclusion principle. The wave function must be antisymmetric under exchange of two particles, so they must all sit in distinct states. The subtraction constant C is chosen such that the energy of the lowest possible energy eigen state, which has all negative energy states filled but all positive energy states empty, is normalized to zero. To avoid ambiguous infinities, one can, temporarily, consider the variables $\vec{q}$ and r to lie in a lattice rather than form a continuum. Every single neutrino sheet still evolves deterministically, so the second-quantized system is also deterministic.

The difficulty in these theories is the masses and the couplings between our "neutrinos". These are not easy to introduce, so as yet, they are omitted.

Finally, we mention (super)string theory. This example was treated in last year's Erice lecture. It is an amazing feature of this theory that the bulk of a string behaves completely deterministically on its world sheet. In units where $\alpha' = (2\pi)^{-2}$, one can separate the integer parts and the fractional parts of the (super) string 's space-time coordinates. It then turns out that the integer parts of these coordinates are beables: strings behave as deterministic models living on a space-time lattice. The fermions on a superstring world sheet can then be handled as Boolean variables ± 1 traveling in one or in two directions on this string.

However, the question whether this has a chance of remaining so when strings are allowed to interact has not yet been answered. It was noted that a deterministic superstring interaction can be postulated but we do not know whether, and if so in what way, this interaction is related to the standard string interaction descriptions.

2.3 On hidden variables in quantum mechanics

The fact that the quantum models described here turn classical when written in an especially chosen basis for Hilbert space, suggests that we might search for more such models. If anything comes up resembling the Standard Model, one might think that this could be the ultimate explanation why our world seems to be quantum mechanical: in the standard description we are using the wrong basis to describe Hilbert space. It would be easy to explain *why* we are using the wrong basis: if the fundamental time scale of our deterministic system is at the Planck time, then all energy eigen states are superpositions of huge numbers of states, because we normally limit ourselves to energies much lower than the Planck energy, when considering quantum phenomena. The states we do use, and the sectors of Hilbert space that we are familiar with, are limited to energies that do not exceed a few TeV, which is much smaller than the Planck energy, and therefore, all states that we ever encountered in quantum systems, are superpositions of huge numbers of ontological states. One may suspect that this is why we cannot recognize the ontological states in any experiment.

This theory looks very elegant at first sight. Some issues that are often hotly debated in discussions about the foundations of quantum mechanics, would find quite natural explanations. One such issue is the question of the "collapse of the wave function", or the question how a measurement can lead to a wave function that is constrained by the outcome of the experiment[17]. This is also the question why Schrödinger's cat always comes out either dead or alive but never in a superposition of two of such states in Schrödinger's famous thought experiment.

In our theory, the answer to all these questions is that the wave function is always in an ontological state; in the basis of the beables, the "true" state of our world is never a superimposed state. Consider now a macroscopic phenomenon such as a planet being in a certain position, a person standing somewhere, or the needle on a scale in a measuring device. In this theory it is totally natural to assume that these states can be distinguished one from another by checking the states of the beables. This allows us to add an extra axiom concerning macroscopic states[18]:

Macroscopic states are always described by observables that are diagonal in the basis of the beables.

There are macroscopic states where a cat is dead and states where a cat is alive, but the superimposed state is not ontological. What this really means is that the superposition axiom does not hold for macroscopic states.

In a similar fashion, one can understand Born's rule for the probability of an event in terms of the norm-squared of a wave function[18]. Probabilities in our theory are determined by the frequency of the occurrence of a phenomenon under initial conditions that otherwise look alike. One indeed finds that such probabilities end up as the norm squared of the wave functions used.

However, in most discussions on the foundations of quantum mechanics, one finds as a common denominator the statement that this would mean that in such theories quantum mechanical phenomena would be mimicked by classical stochastic variables. *And hidden variables,* most certainly "local" hidden variables, *are forbidden by no-go theorems.*

The most famous example of such no-go theorems was provided by J.S. Bell[19], when he phrased what is now called the Bell inequalities[20]. Consider an experiment where quantum-entangled particles are produced, for instance the famous Einstein-Podolsky-Rosen thought experiment, or one of the modern versions of that with entanglement in the polarization of photons or electrons. The particles are detected and measured by two observers, A (Alice) and B (Bob), who are far separated from one another. They both measure whether a photon passes through a polarization filter, and the rotation angle of both filters used are given by vectors $\vec{a}$ and $\vec{b}$.

It is now essential that A and B can each choose the polarization of their filters $\vec{a}$ and $\vec{b}$ any way they like. After the experiment is done they compare the tables of notes they made of their observations. The correlations they find cannot possibly be explained in terms of information that might have been carried along by these entangled particles; one has to use pristine quantum mechanics.

The experiment, and various alternative experiments, have been actually done and to no-one's surprise, the outcome agrees with quantum mechanics[21]. Explanation of the outcome in terms of hidden variables appears to be impossible. For our theory this means that we have to identify the cause of this apparent disagreement. What is wrong with the models described in the previous sections? Is there anything wrong with Bell's theorem? Somewhere along the lines of the arguments a weak spot must be found.

Most theorists argue that Bell's inequalities, as well as other arguments, simply prove that the models we presented above can never be extended to include systems such as the Standard Model. The real world cannot have a basis of beables. On the other hand, there seems to be no impediment against including interactions in our models; classical systems can be chosen as complex as we want, while still local. We can also *always* define the classical states of a model to form the basis elements that span a Hilbert space. Thus, the question remains: what is wrong with that?

2.3.1 Superdeterminism

There are weak spots in the standard arguments. We must point out that it was assumed that Alice and Bob have the 'free will' to choose the polarization directions $\vec{a}$ and $\vec{b}$ of their filters, any time they want. But what is free will? Is it not obviously so that decisions made both by Alice and by Bob, are dictated by laws of physics that determine what goes on in their heads? This is called "super determinism". Usually, this is countered by the statement that, yes, their decisions are determined by laws of physics, but no, these forces are far too complex to be included as parts of the arguments; if these phenomena would have to be included then one would have to assume that the entangled particles emitted by the source, long before Alice and Bob decided how to place their filters, emitted their photons in such a way that the two decisions were already anticipated. The two photons would have had to "calculate in advance" what would go on in Alice's and Bob's heads. This can't be right. This is called "conspiracy", and, according to many researchers, conspiracy is absurd and disgusting.

Is that an air-tight argument? What does 'disgusting' mean in mathematical terms? Suppose we applied it to number theory. Consider a 'ridiculously large' prime number P. Then consider an even larger number, $Q = 2^{P-1} - 1$. What are the odds that such a large number Q is dividable by P? Is the assumption that the remainder be always zero not 'absurd'? It is, until the simple proof is found that indeed the remainder is zero. The proof requires counting all the way to P. No-one can count that far, one would need more than the life-time of the universe. Yet the statement is true. Something can only be absurd in the eyes of the beholder.

We can also argue as follows.Take a time-reversible deterministic model. Take the state found just after Alice and Bob made their observations. Then extrapolate backwards in time. Two deterministically polarized photons return to the atom that was supposed to emit these in an entangled state. The contradiction seems to disappear, except that it may be difficult to extend the extrapolation to time before the atom emitted these particles. But then it is evident that correlations in the initial state have to be taken into account. Usually, these correlations are taken to be 'random', but of course, they are not. If the universe started in a single ontological state $|\psi(0)\rangle$, then, in terms of the template states $|\psi\rangle$, all emerging states of the universe at later times will exhibit highly complex and non-trivial correlations, but nevertheless, the real 'wave function of the universe', $|\psi(0)\rangle$, can only evolve into ontological states.

Let us point out that, according to quantum field theories, the entire scattering matrix can be derived from the correlation functions of the vacuum fluctuations. Thus, if we know the vacuum fluctuations with infinite precision, we can predict the future. What this means is that there could well exist 'conspiracy' in the correlations of the initial state. This conspiracy could well be all we need to explain away Bell's theorem.

It is difficult to foresee where this discussion will end. In the mean time, it is of interest to investigate and extend the models exhibited here further.

3 Lecture III, Concluding remarks

What should the next step for the LHC be? In spite of the fact that the Large Hadron Collider could operate only at half its design energy up till today, it already produced spectacular results. With coordinated efforts, convincing evidence was found for the existence of a particle that agrees remarkably well with the predictions for the Higgs particle[22]. Besides that, new bounds have been set for the masses of supersymmetric particles. Certainly any fears that high energy physics will soon be a dying branch of science have as yet been found ungrounded. Yet we do see changes in the domains of biggest activities in science. Astrophysics, astro particle physics, and in particular cosmology and even quantum cosmology are gaining in strength. in accordance with several recent findings, of LHC as well as cosmological observations, our view of the world of elementary particles and forces is changing. The Standard Model turns out to function much better than previously thought possible. We did succeed in filling out more details. In particular the neutrino sector is being clarified further, and it did not lead to great surprises: neutrinos move in accordance with special relativity just like everything else. Also the Standard Model of cosmology is being further corroborated.

Indeed, cosmology is gaining ground, but particle physicists are not giving up lightly. Plans are well on their way to design the successors of LHC: the linear collider, CLIC and a new circular accelerator, the super LHC.

LHC is giving theoreticians a lot to think about. We are aware of the fact that experimental physicists are only too eager to do experiments proving theory to be "wrong", Surprisingly now, what has been achieved so-far is that the Standard Model has been mostly confirmed. Is this a disappointment? Not really: most theoreticians have been predicting that in the LHC domain the SM should collapse, yet it did not, so indeed, experimenters did prove all those theoreticians wrong! No SUSY in sight, no strong violations of unitarity indicative for new particles. What is happening?

Perhaps we must ask for even more accuracy. The Standard Model becomes technically unnatural[23] at energies much higher than a TeV, so something must happen, we should not give up. There are many questions now to be considered. Theoretical calculation techniques are becoming more and more powerful, as shown by the incredible calculations of the infinities in quantum gravity models at very high orders, done by Zvi Bern and his group[24]. It will become indeed very interesting to compare precision calculations with precision experiments.

Every now and then I do address one warning to those theoreticians who think that simple unified models can be devised that should "explain everything". My point is that such models cannot be extremely simple. We have to allow our Future Standard Model to explain to us why the Universe is so tremendously complex. We might attribute the complexity of the universe to the fact that constants of nature come in ratios that correspond to extraordinary large numbers such as 10^{122} (the inverse cosmological constant in Planck units). But where does such a hierarchy of scales come from? This is still a great mystery. In my first lecture I explained that a model might be found in which all coupling strengths, including the masses and the cosmoloical constant, are fixed and computable; but they all come out to be of order 1 at the Planck scale. That obviously cannot be right, but how does one construct models that generate large but finite numbers by themselves? This is something we have to learn to understand.

In my theory I emphasize the role of local conformal symmetry in quantum gravity. In a sense, this symmetry is spontaneously brolen as in the B.E.H. mechanism[8]. Local conformal symmetry could explain the apparent lack of structure at extremely high energies, as well as a similar lack of structure found in the cosmic background radiation. But we are not there yet, something is still wrong in our way of thinking.

I did indicate in my second lecture how one might think of this. All today's theories are based on quantum mechanics. Quantum mechanics is the universal frame for everything. I have questions here, and I phrased them in terms of simple but suggestive models. What these models have in common is that they do not claim quantum mechanics to be wrong or even incomplete, but rather that a different language can be designed that makes us look at quantum mechanics differently. The way these models are phrased now suggests that we are trying to identify "hidden variables". Much work has been done to establish that such hidden variables do not exist, but the theories usually considered may be based too much on pre-conceived notions of what a hidden variable theory should look like. Researchers investigating quantum mechanics are very close to secure quantum mechanics from any hidden variable scenario. It is therfore important to investigate which tiny cracks have not yet been closed. That's where theoreticians should search further.

Thanks to innumerable very important experiments in the domain of particle physics and cosmology, we do possess of a quite impressive body of data as to what the most fundamental particle interactions seem to look like. They put us in a very interesting and promising position to continue searching for that one model that can account for Nature's book keeping system at the Planck scale in a consistent manner. How long it will take to find such models nobody can tell at present.

References

[1] S.W. Hawking, *Particle Creation by Black Holes*, Commun. Math. Phys. **43** (1975) 199, Err. ibid. **46** (1976) 206; J.B. Hartle and S.W. Hawking, *Path Integral derivation of black hole radiance*, Phys. Rev. **D13** (1976) 2188

[2] M.B. Green, J.H. Schwarz and E. Witten, *Superstring Theory*, Cambridge Univ. Press, 1987, ISBN 0 521 32384 3.

[3] J. Polchinski, *String Theory*, Cambridge Monographs on Mathematical Physics, P.V. Landshoff et al eds., Vol. I, An Introduction to the Bosonic String, 1998, ISBN 0 521 63303 6; Vol. II, Superstring Theory and Beyond, 1998, ISBN 0 521 63304 4.

[4] K.Skenderis, *Black holes and branes in string theory*, Utrecht introductory lecture notes, ArXiv:hep-th/9901050v3 Feb. 1999.

[5] G. 't Hooft, *On the quantum structure of a black hole.* Nucl. Phys. **B256** (1985) 727; *The black hole interpretation of string theory.* Nucl. Phys. **B335** (1990) 138; *The scattering matrix approach for the quantum black hole: an overview.* J. Mod. Phys. **A11** (1996) 4623, gr-qc/9607022.

[6] L. Susskind, L. Thorlacius and J. Uglum, *The Stretched horizon and black hole complementarity*, Phys. Rev. **D 48** 3743 (1993) [arXiv:hep-th/9306069].

[7] S.M. Mathur and D. Turton, *Comments on black holes I: The possibility of complementarity*, arXiv:1208.2005[hep-th].

[8] F. Englert and R. Brout, *Phys. Rev. Lett.* **13** (1964) 321; P.W. Higgs, *Phys. Lett* **12** (1964) 132; *Phys. Rev. Lett.* **13** (1964) 508; *Phys. Rev.* **145** (1966) 1156.
G.S. Guralnik, C.R. Hagen and T.W.B. Kibble, *Phys. Rev. Lett.* **13** (1964) 585; T.W.B. Kibble, *Phys. Rev.* **155** (1967) 1554.

[9] G. 't Hooft, *The Holographic Principle*, hep-th/0003004; L. Susskind, J. Math. Phys. **36** 6377 (1995), arXiv:hep-th/9409089; L. Susskind, L. Thorlacius, and J. Uglum, Phys, Rev. **D48** (1993), 3743.

[10] E.P. Verlinde, *On the Origin of Gravity and the Laws of Newton*, JHEP **1104** (2011) 029, DOI: 10.1007/JHEP04(2011)029, e-Print: arXiv:1001.0785 [hep-th].

[11] G. 't Hooft, *Probing the small distance structure of canonical quantum gravity using the conformal group*, ITP-UU-10/30; SPIN-10/25; arXiv:1009.0669[gr-qc]; *The Conformal Constraint in Canonical Quantum Gravity*, ITP-UU-10/40, Spin-10/33, arXiv:1011.0061[gr-qc]; *A class of elementary particle models without any adjustable real parameters*, Found Physics **41** 1829, ITP-UU-11/14, Spin-11/08, arXiv:1104.4543.

[12] D.M. Capper and M.J. Duff, *Conformal Anomalies and the Renormalizability Problem in Quantum Gravity*, Phys. Lett. **53A** 361 (1975); M.J. Duff, *Twenty Years of the*

Weyl Anomaly, Talk given at the Salamfest, ICTP, Trieste, March 1993, arXiv:hep-th/9308075;
E.S. Fradkin and A.A. Tseytlin, *Conformal Anomaly in Weyl Theory and Anomaly Free Superconformal Theories*, Phys. Lett. **134B** 187 (1984).

[13] E. Gildener, *Gauge Symmetry Hierarchies*, Phys. Rev. **D14** (1976) 1667
S. Weinberg, *Gauge Hierarchies* ,Phys. Lett. **B82** (1979) 387
F. Jegerlehner, *The hierarchy problem of the electroweak Standard Model revisited*, DESY-13-093, HU-EP-13-25, DESY 13-093, DESY 13-093 e-Print: arXiv:1305.6652 [hep-ph].

[14] W.E. Lamb, Jr., *The Interpretation of Quantum Mechanics*, Collected papers, J. Mehra ed., Rinton Press, Princeton, 2001, ISBN 1-58949-005-3;
Quo Vadis Quantum Mechanics?, The Frontiers Collection, A. Elitzur et al, Eds, Springer, Berlin, Heidelberg, New York, ISBN 3-540-22188-3.
The Quantum Structure of Space and Time, Proceedings of the 23rd Solvay Conference on Physics, D. Gross et al, eds., Brussels, 2005.

[15] G. 't Hooft, *Determinism and Dissipation in Quantum Gravity*, in Basics and Highlights in Fundamental Physics, The Subnuclear series, Vol. **37**, World Scientific, 2001, A. Zichichi, ed., p. 397, SPIN-2000/07, hep-th/0003005; *Relating the quantum mechanics of discrete systems to standard canonical quantum mechanics*, ITP-UU-12/14; SPIN-12/12, arXiv:1204.4926; *Duality between a deterministic cellular automaton and a bosonic quantum field theory in 1+1 dimensions*, ITP-UU-12/18; SPIN-12/16, arXiv.org/abs/1205.4107; *Discreteness and Determinism in Superstrings*, ITP-UU-12/25; SPIN-12/23, arXiv:1207.3612v2[hep-th].

[16] G. 't Hooft, *Hamiltonian formalism for integer-valued variables and integer time steps and a possible application in quantum physics*, Report-no: ITP-UU-13/29, SPIN-13/21 (2013), http://arxiv.org/abs/1312.1229.

[17] A. Albrecht, *Following a 'collapsing' wave function*, Phys.Rev. **D48** (1993) 3768, FERMILAB-PUB-92-318-A, IMPERIAL-TP-92-93-03 DOI: 10.1103/PhysRevD.48.3768;
A. Bassi and G. Ghirardi, *A General argument against the universal validity of the superposition principle*, Phys. Lett. **A275** (2000) 373, DOI: 10.1016/S0375-9601(00)00612-5, e-Print: quant-ph/0009020.

[18] G. 't Hooft, *How a wave function can collapse without violating Schrödinger's equation, and how to understand Born's rule*, ITP-UU-11/43, SPIN-11/34, arXiv:1112.1811[quant-ph].

[19] J. S. Bell, "On the Einstein Podolsky Rosen Paradox", *Physica* **1** (1964) 195.

[20] M.P. Seevinck, *Parts and Wholes. An Inquiry into Quantum and Classical Correlations*, Doctoral Dissertation, Utrecht University, October 2008.

[21] A. Aspect, J. Dalibard and G. Roger, *Experimental test of Bell's inequalities using time varying analyzers* , Phys. Rev. Lett. **49** (1982) 1804.

[22] The CMS Collaboration, *Observation of a new boson at a mass of 125 GeV with the CMS experiment at the LHC*, Phys. Lett. **B716** (2012) 30, CMS-HIG-12-028, CERN-PH-EP-2012-220, e-Print: arXiv:1207.7235 [hep-ex] ;
The ATLAS Collaboration, *Observation of a new particle in the search for the Standard Model Higgs boson with the ATLAS detector at the LHC*, Phys. Lett. **B716** (2012) 1, CERN-PH-EP-2012-218, arXiv:1207.7214 [hep-ex].

[23] G. 't Hooft, Lectures given at the Cargèse Summer Institute, 1979, in 'Recent developments in gauge theories'. Ed. by G. 't Hooft et al., (Plenum Press, New York NY 1980): *LECTURE III: Naturalness, chiral symmetry, and spontaneous chiral symmetry breaking.* Reprinted in: Dynamical Symmetry Breaking, A Collection of Reprints. Ed. by E. Farhi et al. (World Scientific, Singapore). Cambridge 1982, 345.

[24] Z. Bern et al, *Simplifying Multiloop Integrands and Ultraviolet Divergences of Gauge Theory and Gravity Amplitudes*, Phys. Rev. **D85** (2012) 105014, SLAC-PUB-14529, e-Print: arXiv:1201.5366 [hep-th]; Z. Bern et al, *he Five-Loop Four-Point Amplitude of N=4 super-Yang-Mills Theory*, Phys. Rev. Lett. **109** (2012) 241602, UCLA-12-TEP-105, SU-ITP-12-22, SACLAY-IPHT-T12-056, e-Print: arXiv:1207.6666 [hep-th]. See also lectures held at this School.

CHAIRMAN: G. t' HOOFT

Scientific Secretaries: A. Sastre Bruno, R. Zubov

DISCUSSION I

- Mojaza:

I would like to know about instanton corrections to the scalar mesons. How do these corrections behave in other strongly coupled theories?

- 't Hooft:

I don't know if Prof. Maiani is going to talk about that. We did work on this topic and the observation is that the instanton vertex is an effective vertex which is particularly significant in the three flavor case, where three left handed quark flavors go to three right handed quark flavor. All together, this is is a 6 quarks vertex. It can be used to connect a 2 quark state to a 4 quark object, so that these form mixed states. Without the instanton effect, a scalar quark-antiquark meson cannot easily mix with a meson that contains two quarks and two antiquarks. But this is exactly what the instanton vertex does: it mediates between those different states.

- M. Csanad:

There has been some experimental evidence on the creation of low mass eta prime mesons in a hot and dense medium. Do you think this can be a hint of a possible partial chiral symmetry restoration at high temperatures and what is the role of instantons there?

- 't Hooft:

Indeed in a quark-gluon plasma, i.e., unconfined phase probably also chiral symmetry is no longer spontaneously broken, and in the symmetric phase of chiral symmetry the quarks in principle are massless apart from the current mass; the strange quarks still keep their current mass but that is a lesser problem. So now of course you have four combinations of massless quarks: the 3 pions plus an eta type, that is, the eta before it started to mix with the strange-antistrange object to form the familiar eta. So now if that mixing gets undone then you might get very light eta fields. I must say that this is going rather deeply into the quark-gluon plasma business and QCD with which I'm less familiar so I'm not sure I'm giving you a right answer now. Also I don't know how strong the evidence for light etas is, because the instanton, well, I would expect it still to have a sizeable effects even in any phase that QCD might be in. An important aspect of the instanton is its long distance feature. It goes away rapidly if you go to short distance scales. So if you go to very high energy, so that the dominant wavelength is short, then maybe the instanton might be less effective. And that could be the explanation why instantons are less effective in a very hot quark-gluon plasma.

- Someone:

The electric field in instantons are screened in the plasma, right? So they got short ranged. So then it is less effective.

-'t Hooft:

Yes, that could be a topological reason for the instanton to be removed out of the plasma and then of course this mixing will be undone. That could be the answer.

- *Csanad:*

And can a scenario be likely where the eta prime mass decreases strongly, but the eta mass and the pion mass don't change too much? Because that is what can be seen, but not very strongly.

- *'t Hooft:*

I wouldn't be able to say whether this will be an instanton effect, because that happens in the SU(3) singlet channel and the strong mixing of different quarks goes accordig to SU(3) flavor. That could be an effect of instantons. I'm not sure I know the right answer to all these questions. *Note added in proof:* if the eta-prime mass decreases rapidly this could be an indication for a strong reduction of the influence of the instantons.

- *Melia:*

If you had computers with an infinite amount of computing power, so that you can do any calculation on the lattice. What other questions do you think we would be asking about the theory of QCD?

- *'t Hooft:*

Lattice QCD is in principle the answer. If you took a lattice with a very densely grained lattice then you could simulate QCD quite accurately if you had such a fantastic computer. My belief is that the answer to your question is yes, but it is not obvious without a mathematical improvement to solve problems like the mass gap. Physically we believe there is a mass gap and we believe we understand QCD so well to have a way to compute everything you like. I think your question can be split in two questions: First: can we make such calculations in principle, with a complete control of the errors? And secondly, are the computers powerful enough to do calculations? But both questions are connected, if you know the mathematics very well, then you can improve your numerical calculations.

- *Melia:*

But, if you can do it perfectly, with arbitrary high precision numerically, then?

- *'t Hooft:*

I am sure the people who do these calculations are going to discover a lot of things but I think that is not the end. The situation is similar to QED. In QED you can do perturbation theory at any order; we just stop because the are too many diagrams. But nevertheless QED is not a perfect theory, you have the Landau ghost, and this means that you cannot compute diagrams with 137 loops or more, which is why for mathematicians QED doesn't exist. In QED, we do not have this problem in practice because we stop at 5 or 6 loops. In QCD the situation is similar, but much more serious: from a pragmatical point of view we can use QCD to do precise calculations, but we can ask if the theory is mathematically defined in a meaningful way.

- *Sun:*

In your slides you write down the solution for the one-instanton configuration. I'm wondering: is there any systematic way to write down the solutions for many instanton configurations so far? What the difficulty of that?

- *'t Hooft:*

It is a subject of a quite considerable amount of mathematical literature. It's not so easy. I did happen to tumble into a solution of arbitrary many instantons, but not the most general one, mine was a very special solution. Solutions with any number of instantons are exact. We know that there are much more solutions and in effect mathematicians know a lot about these solutions. I'm not sure whether they have written them out, all of them explicitly. I won't be surprised if no-one will ever succeed in doing that, but I haven't followed all these details. The problem gets much more severe if you take an instanton and one anti-instanton together. Then in principle that's unstable, so that there doesn't need to be any solution; maybe you can have metastable solutions or accidental solutions for that. So then the question becomes again very hard. As I said the mathematical literature is full of this and not only about the instantons I described; there are higher dimensional objects and magnetic monopoles. There is the other question: are there exact solutions for magnetic monopoles particularly in the Bogomol'nyi case where you have first order equations for magnetic monopoles? Then again you can write down the N monopole configuration. Formally these things can be solved but to understand explicitly it always becomes very difficult to write down explicit solutions; mathematicians have made huge efforts in their literature on these issues.

- *Korthas-Altes:*

The mathematicians have reduced the problem of finding instantons to an algebraic problem, which is hellishly difficult to solve but anyway it's progress. And thanks to that the calorons that you mentioned this morning were found.

- *Sastre:*

Are there non-selfdual solutions in Yang-Mills?

- *'t Hooft:*

In principle you can think of an instanton-antiinstanton. The instanton is self-dual and anti-instanton is antiself-dual but both together are not a solution, in general. Maybe accidental solutions exist, but I am not sure of that. Probably such solutions don't exist because there is attraction between them so they are unstable. In Minkowski space-time there are lots of solutions, but not so many in closed form since the equations are non-linear.

- *Zichichi:*

I would like to report on the experimental needs for instantons to be there, and the point is, when the story started namely with the fact that pseudo-scalar and vector mixing are different from zero and not equal. This story was deduced from bubble chamber physics. And then we need to go to non-bubble chamber devices. So the instanton became needed indeed as the prediction for theta pseudoescalars different from theta vectors. The eta and eta prime are very near to SU(3) flavor axis. The only theoretical support for that comes from instantons.

- *'t Hooft:*

But you know that theoreticians introduced instantons in QCD for many other reasons. So the prediction that instantons have to exist came from various different arguments.

- *Zichichi:*

We cannot forget than the eta prime was taken to be a tensor meson. The proof that eta prime was a pseudoscalar was done by non-bubble chamber physics.

- *Alba:*

Suppose we have a theory with an instanton solution. What happen with the instanton when we move along the renormalization flow for this theory?

- *'t Hooft:*

The instanton is a scale invariant solution but upon renormalization the theory is no longer scale invariant. So the instanton will actually be unstable against scaling, such that it would not like to collapse to a point because the smaller you make it, the smaller the coupling constant, and the bigger the action. Therefore, the lowest action instantons will be very large ones, which means that the instantons tend to go to the infrared. Actually, in Euclidean space you might not be interested in completely exact solutions but in approximate solutions, such as instanton plus antiinstanton solutions. They also play a role in the theory because you can have many of our effective vertexes. So all you need is approximate solutions and they exist. What you have to do in practice is, to keep the scale parameter rho as an independent parameter, and you have to do the integral over all values of rho. That integral was a flat integral before renormalization, but then the renormalization flow would be divergent because it would be an integral of rho from zero to infinity, and that would be a divergent integral. Now renormalization changes all that, it makes things scale dependent and then instantons are not likely to become very small anymore so as the rho integration is cut off at small distances. Now the rho integration might also require a cutoff at large distances due to confinement. So its size will be limited in practice. And because you also have an instanton-antiinstanton gas, instantons will meet there friends and then they can't become bigger than that. So in fact the multi instanton solution might be a way to introduce a cutoff for the single instanton expressions. This becomes a dynamically very complicated question, which really is the question of solving the entire non-perturbative QCD theory, and that of course is notoriously difficult. Another way to proceed is the lattice theory. Instantons are there in the lattice. When you do the lattice integration you find that some of those configurations you are looking at contain instantons and anti-instantons. So the ultimate way of solving a theory is not by looking at instantons but looking at the completed generic gauge field configurations. I think the answer to your question will be that the problem is merging with all the other problems of non-perturbative QCD.

- *Ivanytskyi:*

Since instantons are non-perturbative solutions it seems that they can shed light on such non-perturbative phenomena as the deconfinement phase transition. However as I know they doesn't contribute much to the partition function of QCD at finite temperature. Can you comment this?

- *'t Hooft:*

Well I don't quite agree. You can consider the instanton-antiinstanton gas, so you could think of approximating a solution in QCD by having a lot of instantons and antiinstantons there and by prescribing how they interact. And then you have to solve the Van der Waals problem of a gas of interacting "molecules". Now you just call them instantons, and they interact with each other in a very complicated way. You can make mean field approximations and such things, use methods from statistical physics to find approximate solutions there. And that all is non-perturbative because the instanton is really something very different from perturbation theory. Combining that with the perturbative oscillations around single instanton configurations you could arrive at a superior approximation technique to solve all of QCD. And then, once you do that, you might be able to hit upon phase transitions just as the Van der Waals gas has phase transitions: it can be liquid, solid and so on. But what I'm now suggesting would be a very very difficult program to follow in practice.

CHAIRMAN: G. 't HOOFT

Scientific Secretaries: L. Borsten, S. Nagy

DISCUSSION II

- Backhouse:

Quantum theories have superposition, in a classical theory you can be in only one state, so you have your harmonic oscillator, you go to this other basis which is a classical basis, and in general, there would be superposition in that basis as well but because you chose special initial conditions that you fix it to a single state there? Is that correct?

-'t Hooft:

In principle one is allowed to use any basis one likes, but there is one basis which we refer to as the ontological basis, that describes, between quotation marks, things that are really happening. So when I map the neutrinos on sheets, I imagine the sheets are the things that are really happening in this world, once you know where one sheet is you can follow it in space and time everywhere and it will just behave totally classically. So in this way then, a quantum theory may have turned into a classical theory.

- Backhouse:

But in this ontological basis you don't have superposition, you just have one state?

- 't Hooft:

Well, you might discover that you don't know what the exact state of nature is, because our universe is too complex to be able to follow every single atom. But then you want to talk of probabilities instead. Classical probabilities, but it wouldn't do any harm to describe those as if they were a wavefunction, because then you can say: all I really need for the wavefunction is its amplitude, I'm not interested in its phase at all - the phase is unphysical - but even though it's unphysical, it may be mathematically interesting to have a phase, just to decompose your extremely complicated calculations into templates, which are simple. Then you introduce superposition as a mathematical machine, an artifact, which is not physical but helps you do the calculations. Sometimes, this simplifies things so much, that you can't even imagine a world without those apparent superpositions. So, for mathematical reasons, sometimes you go to another basis where superpositions happen all over the place. That is my interpretation of what elementary particles are: elementary particles are templates, they make our life simple, we now understand what's going on, but in terms of the elementary particle basis, I think I see superpositions, that's because I haven't yet had the idea of going to the "ontological basis", where superpositions are meaningless. The wave functions in that basis are totally meaningless, but as soon as you go to another basis, those phases mean something.

- Backhouse:

So, you have a quantum harmonic oscillator and you make what we would normally call a measurement there, and you collapse it to one of our quantum states, that you introduce a lot of superposition in the ontological states. Does that mean your theory doesn't have measurement, do you have some measurement problem?

- 't Hooft:

It turns out that in this way of looking at things, measurement becomes something very simple. It says: look at the ontological state nature is in. Now, we don't know what that state is, and it turns out the template states, the states of the elementary particles that we use to describe Fock space today, the ontological state of the universe is very very complicated. If you knew it, the measurement would be very simple. You say: which of these things are switched on and which are not. Where are the ones, where are zeroes. One means it's happening, zero means it's not happening. And a measurement means you find an inner product between that and a template state. You have some idea that there may be a particle moving there or maybe not. Well, I take the inner product of Nature's state with that template state; it tells you there's a particle and if the inner product turns out not to be zero, then you're describing a situation where the probability of a particle being there has that value. But that is, of course, quantum mechanics. So you use quantum mechanics as a tool, and that is the bottom line of what I am trying to say, that quantum mechanics is a machinery, a very very smart machine to compute all the probabilities that you are encountering in doing physics.

- *Backhouse:*

If you could find through a more realistic system the commuting basis, would that be a good calculational trick, would that help us to calculate things?

- *'t Hooft:*

Well, our problem today is that we want to figure out what Nature's laws are and we still haven't found out what the ontological basis is of Nature and it will be quite a while until we figure that out. As long as we don't, then when we measure things, we encounter superpositions all the time.

- *Backhouse:*

I'm sure a lot of people don't like your ontology, but they may be able to use it as a natural tool, to calculate other things.

- *'t Hooft:*

That's right, so think of the Ising model as a gigantic game of chess, too complicated to play just analytically, so you want to use statistical methods, and in statistical physics, the outcome of that calculation was of extreme interest: give a classical answer to a classical question but use the quantum machinery on its way. And that may be helpful in other branches of physics. Adding a note in proof: ontology might not become a handy tool to calculate things, but rather a handy tool to interpret what is going on when you investigate experimental observations in a quantum system, even if you are not convinced that my views are totally correct. So I prefer the converse of what you say: the ontological system is technically so complicated that we use quantum mechanics as a mathematical tool to compute the probabilities.

- *Mojaza:*

How does the uncertainty principle come into play in this game? Is it still a fundamental part or does it emerge at the quantum level description? How should I understand it, if there is a classical underlying description?

-*'t Hooft:*

Well, the uncertainty relation only comes about when you consider two non-commuting operators and you think that they both describe reality, then that is a conflicting situation. In the real world that's not true. Only one, or probably none of them, some complicated combination of them, describes reality. As soon as you split things up in terms of non-commuting operators, what happens is your own responsibility. It doesn't quite work, and that's why you can't get certainty as an answer. So that's, to my mind, the origin on the uncertainty relations, that the particle doesn't have momentum and position simultaneously defined. If you try to do that, you end up with foggy answers, you're using operators which don't describe the real world and therefore you are not getting an accurate answer to your questions. That is in a very rough way the meaning of the uncertainty relations as they are in quantum mechanics. I try actually to change as little as possible of what quantum mechanics really is, because quantum mechanics is a theory that works so perfectly that we don't want to disturb it. Ultimately, the Schrödinger equation is a true equation, so I don't want to change that. But when you introduce non-commuting variables you know that they cannot all describe reality at the same time.

- *Downes:*

How do we understand fermions in terms of this ontological basis?

- *'t Hooft:*

I had no time to discuss fermions in my lecture. Fermions are very simple quantities in this picture: they are just pluses and minuses or zeroes and ones. In very many cases in the classical world, there are boolean variables, just flip-flop variables, which take only two values and you can describe them classically, by saying I have an operator sigma which is zero or one or minus one or plus one, however you define it, and these things propagate in time; they play a role in the evolution of everything else, so they are important entities in nature, but it turns out that to work with them classically leads also to complications technically. If, however, you replace them with Grassmann variables, anti-commuting variables, then it makes things a lot easier because you get linear equations. So, using operators that replace ones by zeroes and add minus signs there in such a way that you get anti-commutation rules is tremendously efficient as a mathematical tool so fermions are mathematical tools to describe boolean variables. I have many examples, the fact that neutrinos are fermions is just one of them, those minus signs you introduce do not have a direct physical meaning in the ontological basis but as soon as you want to make transformations, you want to go from one basis of elements to another, these transformations become easy, because the fermionic fields are linear and you can use those linear fields to rapidly find some more convenient basis. But then, as soon as you have done that, all these minus signs you get with anti-commuting fermions become significant in the calculation; you can't argue them away and that's why we now think we have fermionic particles, where these minus signs are important; two fermions cannot be at the same point; at one point you only have zero or one, you don't have a two or a minus one; so the mapping of fermions onto a classical system is very beautiful, using the flip-flop variables.

- *Maksym:*

Quantum logic operates with much quirkier rules than boolean logic (corresponding to classical physics) - it has much wider possibilities. How to interpret that? What about your formalism?

- *'t Hooft:*

I'm somewhat simplifying the whole notion of quantum logic. Quantum logic is a mathematical device. The suspicion is that once you dig all the way to the Plank scale, you no longer encounter quantum logic, there the logic is completely classical, Bayesian logic, so that it's yes or no or maybe, but you don't encounter superpositions, so you don't have qubits, but only ordinary bits, fermions at that level are ordinary bits, not qubits; again, qubits are a mathematical tool - as soon as fermions are replaced by superpositions of states, you'll find it mathematically useful to introduce qubits. The original theory is assumed to be at the Plank scale; the Plank scale is very far away from ordinary physics, so the suspicion that things are different there should not affect ordinary physics. It does affect of course our ability to make theories at the Plank scale, where my belief is that quantum logic will no longer be applicable, just classical logic. On our way from the Plank scale to the SM scale (remember, that's 16 orders of magnitude, which is an enormously big distance), you'll find that quantum logic is more suitable to describe what you think is happening; just because you're integrating over billions and billions of bits of information; the result of that integration is something that looks at first sight very simple, but what you pay for that is that your logic no longer comes out as simple as the laws at the Plank scale, it is more complex and it's what we call quantum logic.

- Borsten:

You appealed to the idea of "conspiracy" to circumvent Bell's theorem. Can the same principle be used to avoid the Kochen-Specker theorem (i.e. classical theories are non-contextual)?

- 't Hooft:

The powerful Kochen-Specker type theorems all refer to noncommuting operators: they start with the idea that you can measure any operator whether or not it commutes with other operators and you can think of nature being in an eigenstate of one such operator. Yet two noncommuting operators are not simultaneously diagonalizable: nature cannot be in an eigenstate of these two operators at the same time. Nevertheless, we use such operators all the time – the bottom line of my theory is that this cannot be the fundamental description of the real world. Our use of such operators is valid for the standard model etc, but not so at the Plank scale. To understand physics at the Plank scale one should only use commuting operators, which may be diagonalised all at the same time – in which case the contradictions disappear. Because today we use noncommuting operators we have the Kochen-Specker type theorems; theorems about noncommuting operators which are mathematical facts that cannot be disputed. These facts are reproduced by experiment, so one should not try replace or contradict quantum mechanics, instead my proposed theory seeks to explain the origin of quantum mechanics.

- Kiehn:

You introduced very nice examples of exact correspondences between classical and quantum systems. Are there systems for which no such correspondence between a classical and quantum description exists?

- 't Hooft:

It depends on your demands. For example, if you insist on locality then it is harder to find classical models that agree with certain quantum systems. If these demands are relaxed then any quantum system can be made classical, although in the extreme cases they may be very complicated. Hence, formally, for any quantum system one can find a classical equivalent. However, most of these mappings will not obey my very stringent requirements that they be physically meaningful and logical. In many cases one might reject the mapping from quantum to classical on the basis that it is too contrived and physically unappealing.

-Borsten:

Could you comment on the relation (if any) to de Broglie-Bohm theory, which also has a classical ontology and, in instances, a preferred set of observables? Are there any tests to distinguish your proposal from quantum mechanics?

- G. 't Hooft:

It is very important that I do not want to change quantum mechanics – I want to understand where it comes from, rather than changing it. So, if done correctly, there should be no way to distinguish my theory from a genuinely quantum mechanical one. Except my theory should allow me to find underlying theories. For instance, we are searching a theory that works well at the Plank scale but we are unhappy with today's understanding of quantum gravity. The motivation for my work is to find better theories at the Plank scale and those should be experimentally testable. However, first one has to understand better what quantum mechanics is and in doing so I will not make any predictions which deviate from standard quantum mechanics. There may be one exception though – the quantum computer. My formalism is in contradiction with the claim that one can make a quantum computer that can perform calculations beyond the capability of any classical computer. If Nature is classical at the Plank scale then one could build a Plank scale classical computer that would be able to do anything a quantum one can. Since in principle a quantum computer can solve non-polynomial problems, in order for a classical computer to outperform it, it would have to scale exponentially, quickly taking it beyond the Plank scale. Hence, my proposal predicts that such a quantum computer is not actually possible.

- *Maksym:*

Does your argument imply that there exists a classical polynomial factorizing algorithm, which could compete with Shor's algorithm?

- *'t Hooft:*

No, I believe it is more likely that if one where to try building a quantum computer to carry out Shor's algorithm it would fail due to decoherence. A quantum computer requires a very strong degeneracy between qubits, such that energy levels must strictly be zero and there should be no interactions between the qubits. On the other hand, one must read out what the qubits say, so there you need very strong interactions. Those are conflicting demands, they cannot both be met at the same time, and therefore we expect decoherence and errors. These errors will accumulate as the number of qubits grows to the required scales and I predict the error correcting codes will fail. This appears to be corroborated by some of the engineers working on decoherence – it becomes increasingly difficult to avoid decoherence as the size of the computer grows.

- *Backhouse:*

The difficulty of prime factorization is not proved. Moreover we do not know if P is NP. So, you are not necessarily making a prediction about quantum computers; it could instead be a prediction about the existence of better classical algorithms.

- *'t Hooft:*

Yes, it could be that some problems that are currently thought to be non-polynomial in fact are polynomial. That would be another way out. However, I believe that decoherence is more of a problem than is normally thought. If it happens that all problems are polynomial that would be a somewhat stranger way out from a mathematical perspective, but you are right, it would be a possibility.

S. BERTOLUCCI

"Highlights from ATLAS – ALICE - CMS"

The Slides of the Lecture can be found at:

http://www.ccsem.infn.it/issp2013/

CHAIRMAN: S. BERTOLUCCI

Scientific Secretaries: M.V.Carlucci, M.Colocci

DISCUSSION

- Kiehn:

My question is a very general one. You showed multiple projects far beyond the LHC, ILC or an even bigger ring at CERN. LHC already took a while to be built, the next machine would probably account for the whole lifetime of a scientist, and anything beyond that would be even bigger. My fear is that at some point these projects become unfeasible, and I fear that it could already take decades for the next big project. What are your thoughts on that?

-Bertolucci:

I think that we should be worried about what you say, it is a very serious question. Personally I do not believe that the next project is unfeasible, but it is clear that the ones after the next might be unfeasible, if we dot invent new technologies. We are studying ways in which we can make more compact machines, new ways in which we can get higher energies (if it will be needed, because maybe in a while we convince ourselves that next scale is the Planck mass and this would make it impossible to study at an accelerator!). For instance, one of the things that I have mentioned in my transparencies is that we are studying at CERN plasma wakefield acceleration, in order to move away from RFs, which in its most sophisticated application can give you 100 MeV per meter. If instead of 100 MeV per meter you could get several GeV per meter, then you could imagine that very compact/very high energy colliders could be built.The other question you are raising, the large amount of time needed to build frontier accelerators, can only be solved by avoiding to put large project in series, but try to overlap them, organizing the programs in a global way. It is evident that CERN, in this moment, cannot afford to start immediately to build the next big machine, because this would impede the exploitation of LHC. But if for instance Japan proposes to lead the construction of the ILC, which is a mature machine and a mature technology, we should support the project from the other regions, because it will contribute considerably to the advancement of physics and at the same time it will avoid a large gap between projects. Another important question that we have to address is that there is more and more clear evidence that the answer to our question is becoming very common at different scales, from the large scale, astrophysics, cosmology, to the small scale, high energy physics, and again more collaboration between these communities is needed in order to have more diversity in the field, addressing the same set of fundamental questions.

- Wymant:

You mentioned, about extrapolating Higgs coupling, precisions in the future a factor of 2 improvement in theory, in the context of about 3000 fb^{-1}, if I remember correctly. Where is this factor 2 coming from, is it based on some historical precedent, or just a challenge to the theorists, and have you had any reaction from, say, the Higgs Cross Section Working Group about what they think on a factor 2 improvement?

- Bertolucci:

I am an experimentalist, so this was my optimistic assumption. The theory groups, including the Higgs x-section group, say that it is not unrealistic to expect such an improvement in the next 5-10 years. It will be a tough challenge, requiring large programs of higher order calculations to be executed. This will also be helped by the data collected at LHC, (for instance by improving the knowledge of PDF's). But even judging from my direct experience, the precision of theory from the beginning of CDF at the Tevatron, and the last measurements at the Tevatron and the LHC, have improved by large factors, an impressive achievement in a relatively short time span! Therefore, I think that the assumption is realistic. What is probably a bit on the optimistic side is thinking that the systematics of the experiments will scale down like one over the square root of the statistics, considering that HL-LHC, due to its ambitious luminosity, will provide a very challenging experimental environment (with an event pileup of 140!). Retaining the same data quality as now will be a tough challenge!

- *Leontsinis:*

I would like to ask what you think about the future, what is going to happen, your personal thinking, about three matters. First of all, if the machine is going to run with 25 or 50 ns, second if the next machine will be ILC in Japan or TLEP or LEP3 at CERN, and finally if a muon collider is completely out of the question for the next 20 years.

- *Bertolucci:*

The machine will start running probably for the setup at 50 ns, just to check that everything is under control, before the start of the physics run, where we plan to go for 25 ns. Concerning the 25 ns we know already that the parameters of the injected beams are good enough. What we still do not know is how severe will be the "electron cloud" problem.When you have a lot of current in a machine, you have a phenomenon which is called "electron cloud": essentially you are extracting electrons from the beam-pipe wall, and these electrons get accelerated by the field induced by the beam on the pipe surface: these electrons hit the beam pipe and extract more electrons, forming a cloud which spoils the quality of the beam and the vacuum in the machine. Usually the "cleaning" of the machine is done with the beam itself, with the so-called "scrubbing", which consists in using a high current beam at lower energies to maximize electron extraction.What we did in the past at the beginning of each running period was scrubbing the machine at 25 ns bunch spacing, to have the highest induced fields. Now we cannot any longer use this advantage, because next time we cannot scrub the machine at 12.5 ns, so we have just to scrub, clean the machine and use the machine at the same bunch spacing, and we do not know how effective it will be. Second question: what first, ILC, TLEP? You know, there are "religions" about that, so there are people who like linear machines, people who like circular machines. Both machines have plus and minuses. TLEP, would require a 80 - 100 km tunnel in order to get at 350 GeV in the center of mass (just above the t-tbar threshold) at the cost of a massive amount of power ($\geq$ 60 MW) spent in synchrotron radiation. Increasing the c.m energy let's say a factor of 2, would increase SR power by a factor of 16, which is clearly a no-go. On the other hand, it would have the advantage of a better luminosity, also because you could use in parallel more than one experimental region. The ILC has the advantage that probably it can be done reasonably soon, so it could run in overlap with LHC. Since now I am a person who has to do management, there is also another advantage: ILC would bring into particle physics 6-7 billions of dollars which otherwise will not belong to the field, and this is not a small thing. So in principle I think that the only reasonable machine in this moment is ILC, because it could come reasonably early, possibly overlapping the LHC program, facilitating an early assessment of the physics case for the next large scale accelerator. Concerning the muon collider, already a low energy muon collider would be a perfect Higgs factory because, given a decent luminosity (~ $10^{33}cm^{-2}s^{-1}$), would produce Higgs bosons in the s-channel, and would be very small in size. The problem is that nowadays nobody knows how to build it, not a minor detail. In order to do that, we have to be able to produce a large number of muons (by throwing a few MW proton beam on a target), cool them down in such a way that they can be injected in a machine, accelerate them fast before they decay and keep sustaining this cycle for years! Will it be impossible forever? Probably not: there are ideas how to do that, there are running projects addressing the various steps (again you see how important is accelerator R&D), and the effort is justified, because a circular muon collider would be probably the only way to reach very high energies in a lepton collider and the technology would also allow to have very intense monochromatic neutrino beams. But we are still very far from that machine.

- *Hartmann:*

I have a strange question. There are two main detectors at the LHC, ATLAS and CMS. You showed that there are some inconsistencies. Has there ever been a talking about having a third main detector as a sort of consistency check? If one makes some mistakes, how would you know which one? It may be interesting to have that in a future ring, if there will be a future ring.

- *Bertolucci:*

You may think that in a circular machine you could put as many interaction regions as your budget will allow. This is essentially not true, because any interaction region (the place in which the highly focused beam interacts) will add to a parameter, the beam-beam linear tune-shift, which is a measure of the effect of the mutual disturbance of the two beams. Now for each machine there is a budget on the amount of tune shift you can afford, so at some point more interaction regions would result in less luminosity for each one of them. There are of course more practical and technical reasons (beyond money) that forbid the proliferation of interaction regions. Coming to ATLAS and CMS, so far there has been no real inconsistency, we might argue about some tensions, which are still well consistent within statistics. In general these magnificent detectors have enough internal redundancy to check consistency of the their results internally, and having two of them, with a different design philosophy and jet a very similar sensitivity, is more than enough in my opinion. Moreover, since the beginning of the data taking, we have set up a procedure in which the two experiments inform each other (and us), before declaring anything big. At a future ring there will be certainly more than one experiment (do not forget that LHC has seven experiments, running at four interaction points), but their number will be determined by their physics program. Two general purpose detectors will probably be enough also for the future machine.

- *Ruthmann:*

I have a question, I would like to know, for linear collider experiment like ILC, whether it exists a lower energy threshold, and so what are typical injection energies to such linear colliders.

- *Bertolucci:*

The ILC scheme foresees injection in the main LINAC at 5 GeV, which is the energy of the dumping rings needed to provide beams of appropriate emittance. The final energy out of the main LINACs can be tuned from that value up to the maximum value (125 GeV in the initial realization, then 250, then 500). Of course, the higher the energy, the higher the luminosity, due to the Lorenz factor. But still the ILC would be capable of luminosities at the Z pole producing a giga sample of Z, which would allow a number of very precise test of the SM.

-*De Palo:*

Just a curiosity. How much precisely do you need to know your beam energy in this kind of experiments, because you say that you will start running at 6.5 TeV and then with energy slowly rising to 7 TeV, so is this 0.5 TeV beam energy difference significant? How do you treat the data in this case?

-*Bertolucci:*

The absolute beam energy at LHC is determined with a precision of 0.05% (~ 3.5 Gev for a 7 TeV beam). The absolute energy determination is important for some specialized experiments, like for instance the measurement of the total proton-proton cross section, which is based on the optical theorem. Besides that, the very precise knowledge of the proton energy is somehow irrelevant, because we are interested in the collision of its constituents (quarks and gluons), which depends on the fraction of the total momentum carried by the interacting partons, whose statistical distribution is described by the Parton Distribution Functions (PDFs). PDFs are very steep functions of this fraction, so even a modest increase of the proton momentum, will result in a very large increase of the probability to find a parton at a given energy. Increasing the energy by 1 TeV will therefore extend considerably the discovery potential of LHC. Conversely, at a lepton collider, precise knowledge of the beam energy can be used to constrain the kinematics of the event and to perform high precision measurements (e.g. the Z mass and width). This precious feature of the lepton colliders will be progressively lost at future machines like the ILC, due to the emission of high energy photons at very small angles, due to the interactions of the particles with the strong e.m. field of the other beam ("beamstrahlung").

LNGS: Past, Present and Future

Fernando Ferroni*
Sapienza Universita' di Roma & INFN
P.za A Moro 2, Roma, I-00184 –Italy

1) A laboratory for the cosmic silence

In 1979 a vision of Professor A. Zichichi started to take real form by the submission to the italian parliament of the project of excavating in the hearth of Gran Sasso massif a huge underground laboratory consisting of three experimental halls and connecting galleries.
The laboratory would have been accessed by the highway that connects L'Aquila to Teramo province. With an extraordinary foreseeing the halls were oriented toward CERN. In 1989 the first experiment, MACRO [1], started data taking searching for monopoles.

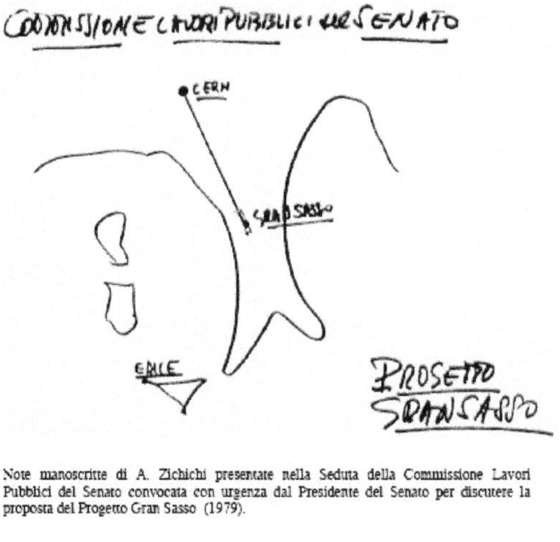

Note manoscritte di A. Zichichi presentate nella Seduta della Commissione Lavori Pubblici del Senato convocata con urgenza dal Presidente del Senato per discutere la proposta del Progetto Gran Sasso (1979).

Fig. 1: The original drawing presented to the Italian parliament showing the possibility of shooting a CERN beam to the proposed Gran Sasso laboratory.

It is remarkable that the motivation given at the time are still actual today.
-Nucleon stability
-Neutrino astrophysics
-New cosmic phenomenology
-Neutrino oscillations
-Biologically active matter
-Ground stability.

***email fernando.ferroni@roma1.infn.it**

In short, the laboratory offers three halls (size 100 x 20 x 20 m3) at a depth of 3800 m.w.e. The residual muon flux is 3 x 10-4 per square meter per second. The access is horizontal and big preconstructed detector can be transported by trucks.

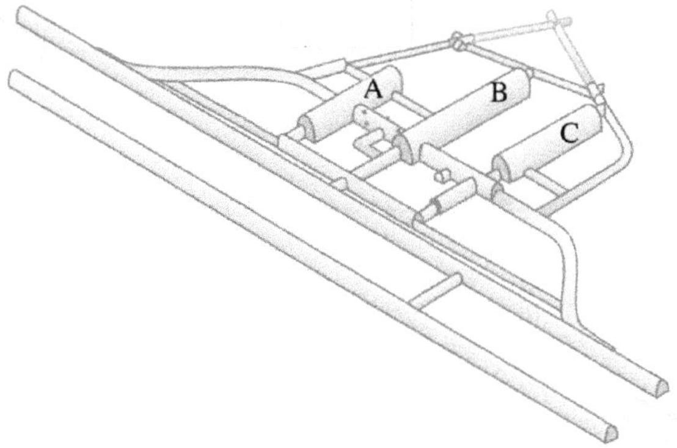

Fig.2 : Schematics of LNGS

2) A busy laboratory

Gran Sasso laboratory is used by a large community from all over the world. The almost 900 users represent about 30 countries. The largest communities are from Italy, Germany, Russia and USA.
A snapshot of the experiments [2] active at this time includes:
DAMA-Libra, Cobra, VIP, CUORE, CRESST, LUNA, GERDA, LVD, XENON100, ERMES, GIGS, ICARUS, OPERA, Borexino, Dark Side 50 and I have likely missed something.

The focal points of the recent activity are found in Neutrino Physics (experiments with CNGS beam, double beta decay experiments, astrophysics neutrinos) and Dark Matter searches. An additional experimental line of nuclear astrophysics is performed by using the LUNA little accelerator.
The strongest point of the lab is to offer a first class location for ultrare event searches.

Let us see the main result that have been recently produced in the Lab.

2.1 Neutrino Physics with CNGS

2.1.1 OPERA

OPERA [3] is a magnetic spectrometer with very good tracking power. Its goal is to observe interactions (charged currents) induced by tau neutrinos that were born as muon neutrinos at CERN.

The target is made by a huge number (in excess of 120000) of lead/photographic emulsion that allows to visually confirm the decay of the tau lepton created in the interaction.
Three events [4] have been positively identified giving the first direct evidence of the transformation mu/tau neutrino.

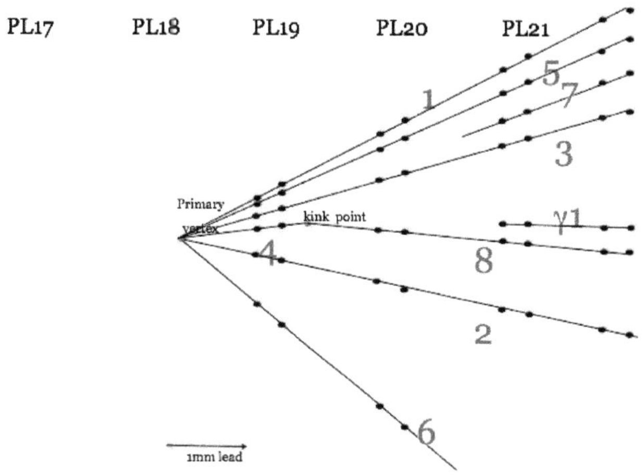

Fig. 3: One of the event identified as a charged current interaction of a tau neutrino.

2.1.2 ICARUS

ICARUS [5] , a massive (600 ton) LAr time projection chamber is a powerful demonstrator [6] of the superiority of this technique for electronic reconstruction of neutrino interactions. Its success will justify the large investments on much larger scale LAr detectors for future long baseline experiments aiming at measuring CP-violation in the neutrino sector

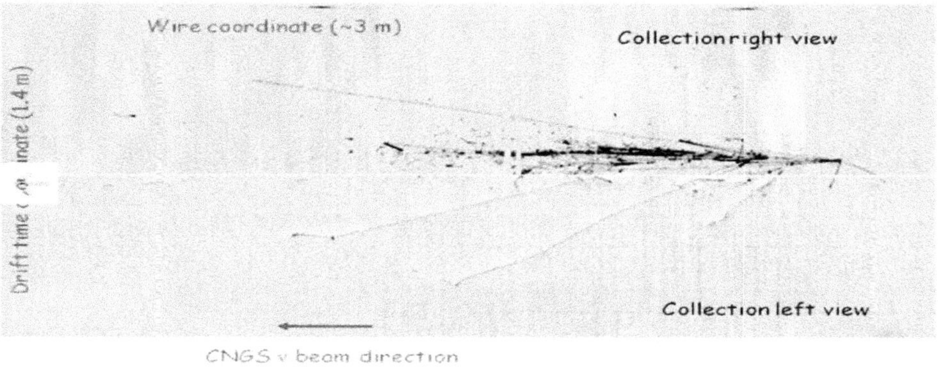

Fig. 4: An impressive demonstration of the capability of the LAr in reconstructing a neutrino event from the CNGS.

2.2 Solar and Terrestrial Neutrinos

Borexino [7], a sphere filled with few hundred tons of ultrapure liquid scintillator is operating with the task of observing as many interactions as possible of neutrinos produced by the different fusion processes happening in the core of the Sun.

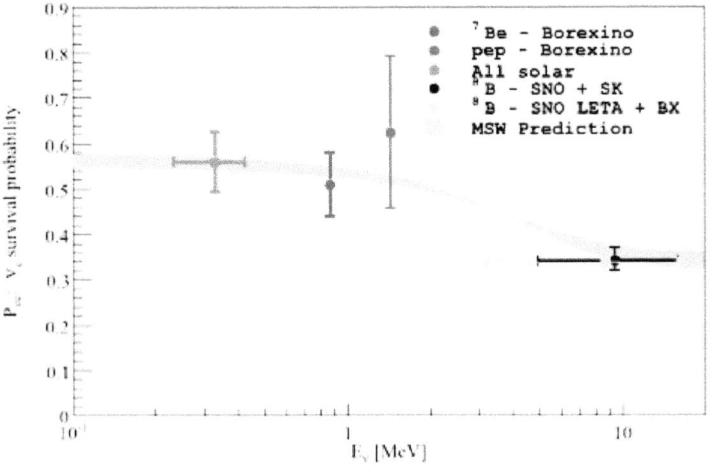

Fig. 4: Observed neutrino flux compared with the oscillation model (MSW) prediction.

As seen in Fig.4 Borexino measurements [8] span over almost the entire phase confirming inside the same experiment the MSW [9] model of oscillations in matter.
The extremely ultrapure scintillator further allows the experiment to work with a very low threshold and be therefore sensitive to the neutrinos [10] emitted by U and Th decay in the Earth crust and mantel.

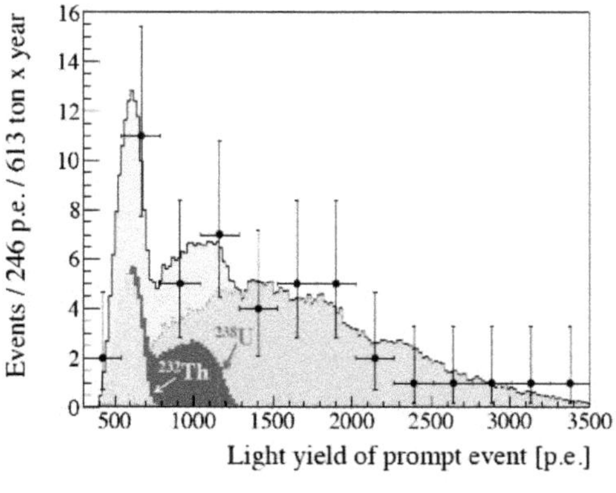

Fig. 5: Spectrum observed in Borexino that demonstrate the observation of geo neutrinos.

This measurement is extremely useful to understand how much of the Earth heath is due to radioactive decays.

An evolution of this experiment has been recently approved. Thanks to the ERC European Union program a campaign of measurement making use of artificial neutrino generators will be possible. Two sources [11] will be operated to provide neutrinos from 51Cr and antineutrinos from 144Ce. This program aims to cover an important fraction of the phase space that would be the realm of sterile neutrinos, the family that does not belong to Standard Model and whose existence has been claimed by a couple of experiments without a solid confirmation.

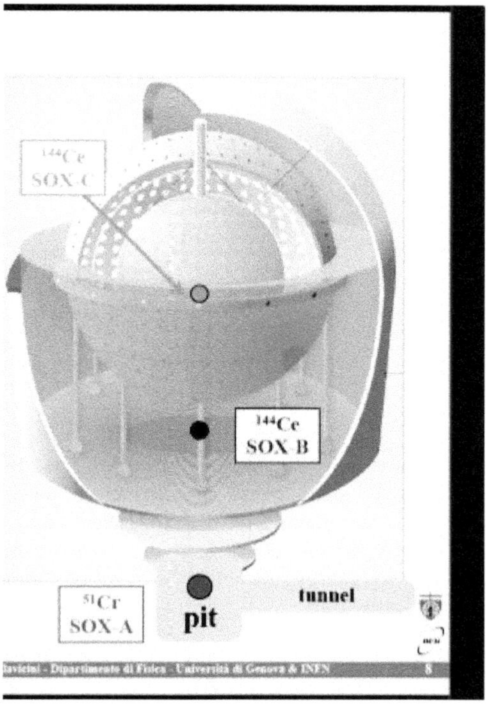

Fig. 6: Configurations foreseen for the different sources in the future SOX (and CeSOX) experiment making use of Borexino.

2.3 Neutrinoless Double Beta Decay searches

Fundamental question about neutrinos are still in front of us after many decades. Are neutrinos Majorana particles or Dirac ones ? Is global lepton number violated or not ?
The only known way to try to give an experimental answer to this very important question is to observe events due to neutrinoless double beta decay. Easy to say, difficult to do.
The challenge is best explained by the picture of what you might expect in terms of the quantity that determines the effective cross section of the process you would like to observe and the formula that predicts the sensitivity of your experiment.

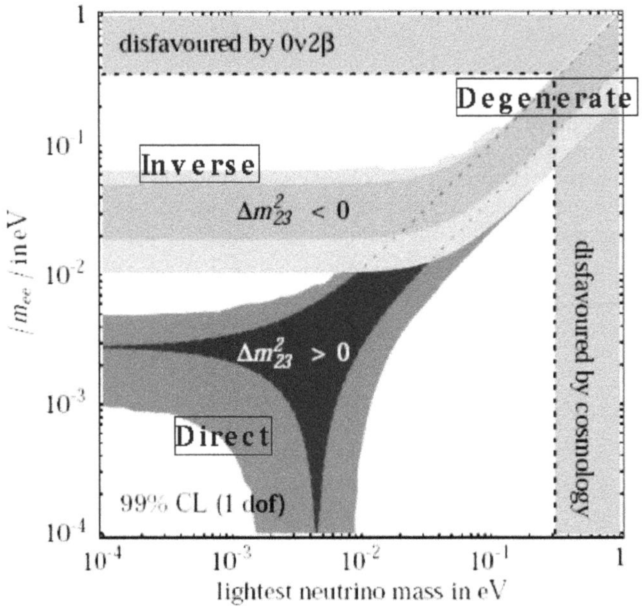

Fig. 7: The effective neutrino mass as seen by the neutrinoless double beta decay experiment as a function of the lightest neutrino mass. Green region shows what is expected in case of inverted hierarchy , red region for direct hierarchy.

The difficulty of detecting the process is best explained by the formula that describe the lifetime: $1/t = G(Q,Z)|M_{nucl}| < M_{ee} >^2$. A combination of small phase space, nuclear matrix element and smallness of the neutrino masses points to experiment with required sensitivity of about 10^{27} years . Experimentalists at Gran Sasso have chosen to bring to the extreme the technique that allows to combine an exceptionally good energy resolution with a very tiny background (dominated by radioactive decay products) ideally zero.

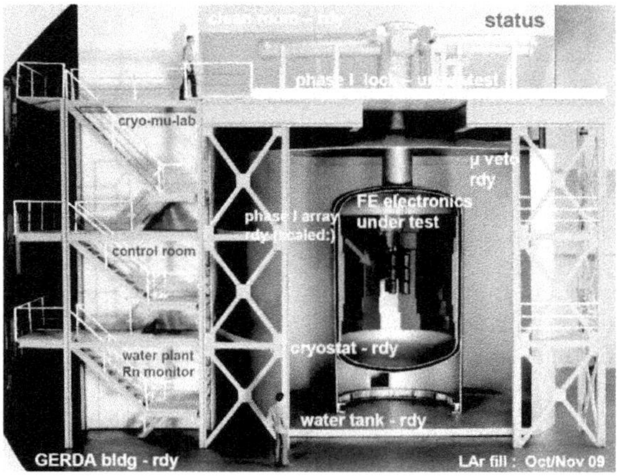

Fig. 8 : The Germanium diode of GERDA experiment immersed in LAr inside a steel cryostat

GERDA [12] experiment, based on germanium crystals, is running since quite a while and has already reached a sensitivity [13] that seriously challenge the dubious result from a part of the HdM group.

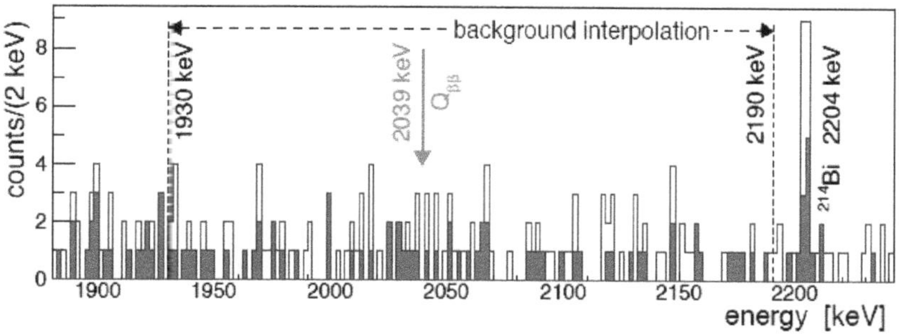

Fig. 9 : The result of the first run of GERDA. No signal is observed at the transition energy.

CUORE [14], the gigantic successor of Cuoricino[15], is under construction and will start to take data in 2015.

Fig. 10 : A few of the 19 towers of tellurium oxide crystals that will constitute the CUORE detector.

The advantage of CUORE will be in the highest transition energy of ^{130}Te (also by far the best possible isotope in term of natural abundance with a respectable 34%) with respect to Ge therefore suffering a substantially lower background from radiative photons. The weak point is in the limited ability of rejecting degraded alpha particle energy deposit that can mimick the signal. A vibrant program for exploiting the background rejection power of scintillating crystals bolometrically read out is therefore going on. Promising results are obtained with crystals of zinc selenide [16] and zinc molibdate [17]. The difficulty with this materials is the need of a many hundred of tons of isotopic enrichment capacity for which today there is neither industrial availability nor a workable economics.

2.4 Supernovae neutrino searches

In the process of a supernova formation there is a huge quantity of neutrinos that would be flying away in all directions in space. Profiting by the exceptional condition of low background present in the underground laboratory one can hope to be able to detect the signal induced by the arrival of those neutrinos. A lot of information could then be gained by the intensity, the energy spectrum, the duration of this pulse. A modular detector, LVD [18], based on 1 Ton of liquid scintillator is waiting in the depth of the laboratory for this signal coming from an explosion happening in our galaxy. The expected rate is low (one per a few dozen of years) and challenges the patience of scientists that are however waiting with faith the event to occur. LVD, that is running since more than 20 years, looks at events that would happen inside 20kpc radius.
LVD has been very beneficial to help in disproving the OPERA claim on superluminal neutrinos. By making use of the horizontal muons that cross both OPERA and LVD it has been shown [19] that the time shift between the two detectors changed by the famous value of 60 nsec sometime in 2008. And it went back to what it was after the misbehaving cable was found and fixed in OPERA.

2.5 Dark matter searches

More than three quarters of the entire matter of the universe are dark. A form of matter that neither emits nor absorbs electromagnetic radiation.. We are aware of its presence by several effects, the rotation speed of the galaxies, the lensing effect, the collision of galaxies, big bang nucleosynthesis, however the particles of which it is composed are particularly reluctant to manifest either in the product of the high energy collision at LHC or by hitting the nuclei of our underground detectors. The most accredited hypothesis on what dark matter could be, advocate the Weakly Interacting Massive Particles (WIMPs). They represent a class of dark matter particles that froze out of thermal equilibrium in the early universe with a relic density that matches observation. This field of research is fastly evolving. Contradictory results exists and a lot of experiment using very different techniques are running or are in preparation. The landscape is still foggy but everybody feels that we are exploring a new land.
Gran Sasso is participating to the competition with several experiments based on very different techniques : room temperature scintillators (Dama/Libra) [20], bolometric crystals (CRESST) [21] and noble cryogenic liquids (XENON [22], Dark-Side [23]). We miss semi-conductor based detectors.

2.5.1 Dama/Libra experiment

This experiment is running since 15 years operating an ultrapure set of NaI(Tl) crystals. The principle of detection is based on probing the modulation [24] that should be induced by the motion of the Sun (and its planetary system) through the cloud of dark matter in direction of the Cygnus constellation combined with the rotation of the Earth around it. The effect should be sizeable in terms of cross section and with predictable shape and phase.

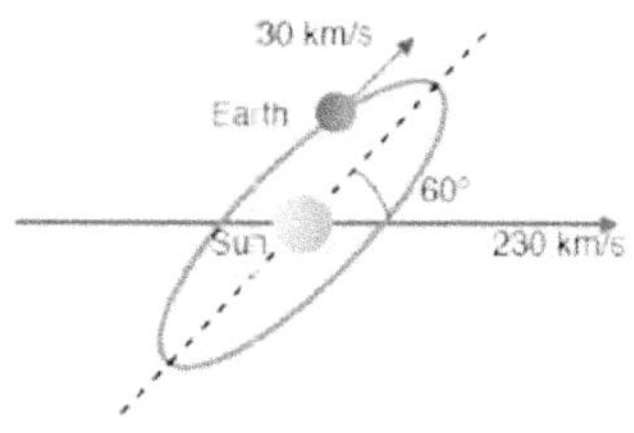

Fig. 10 : Sketch of the source of annual modulation for dark matter

Indeed the experiment shows an indisputable result [25] in terms of modulation with a phase that is very compatible with what expected from the model. The result can be interpreted as elastic scattering of a 10 GeV WIMP.

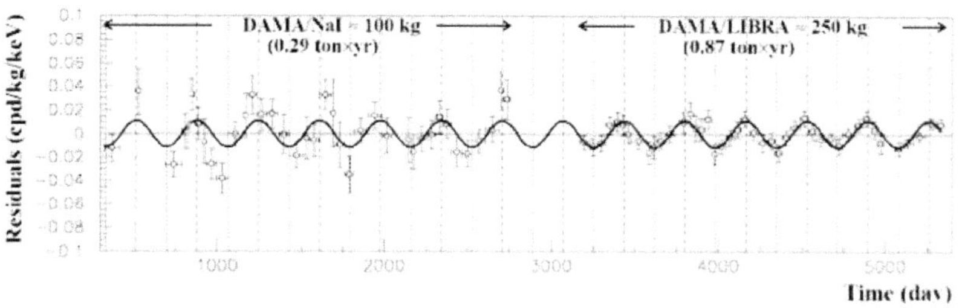

Fig. 11: The results of a more than ten year running of the Dama/Libra experiment. A clear annual modulation is observed in the data

The only problem is that the experiment is still waiting from the support of another positive evidence from an independent one. It has to be said that Dama/Libra by far has accumulated more data than anybody else. More than 1Ton * year.

2.5.2 CRESST

This experiment is using the technique of scintillating bolometers by using CdWO4 crystals whose signal is recorded both in the phonon channel, that measures precisely the deposited energy, and in the light channel that helps in discriminating electromagnetic interactions by nuclear recoils. A sort of positive evidence [26] for some excess of events that might be attributed to dark matter interactions is observed. Statistical significance though is not strong and the region does not coincide with the one of Dama/Libra.

2.5.3 The XENON family

One of the technique that is followed for detecting dark matter is based on the observation that the elastic scattering of a WIMP can efficiently transfer kinetic energy to a nucleus allowing to measuer its recoil in a sensitive, low-radioactivity detector based on a noble liquid. LXe is a possible choice and indeed in Gran Sasso one detector (XENON10 [27] has been operated and it has demonstrated the potentiality of the technique. A bigger successor (XENON100) is still running and (at the time of this school) has put the best limits [28] to the existence of this elusive form of matter.

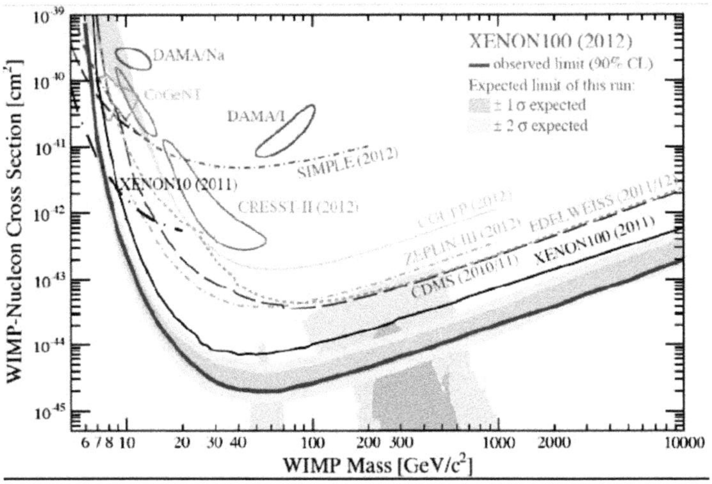

Fig. 12 : The conflicting results of all the experiment searching for dark matter at the time of this school.

It has to be said that a clear tension exists among the results of these many experiments.
A race is now existing with a competing, very similar experiment, LUX [29], that has shown a superior performance for the low energy part of the spectrum. XENON1T [30], now in construction at LNGS is the response of the collaboration to challenge posed by the competition. This new detector will have a mass 10 times higher than the previous one and aims to a background 100 times less. The predicted sensitivity for the WIMP-nucleon cross section is in the range of $2 * 10^{-47}$ cm $^{-2}$.

2.5.4 The liquid argon competition

Liquid Xenon is not the sole option for building a detector based on noble liquids. An appealing alternative is given by liquid argon. A prototype of a future, huge, LAr detector for the dark matter detections is now functioning at Gran Sasso. It is the Dark-Side50 experiment [31]. This liquid argon dual-phase TPC is nestled within 30-tons of liquid scintillator neutron veto counter. The liquid scintillator neutron veto counter is hosted inside a 1,000-tons active water Cerenkov muon veto.

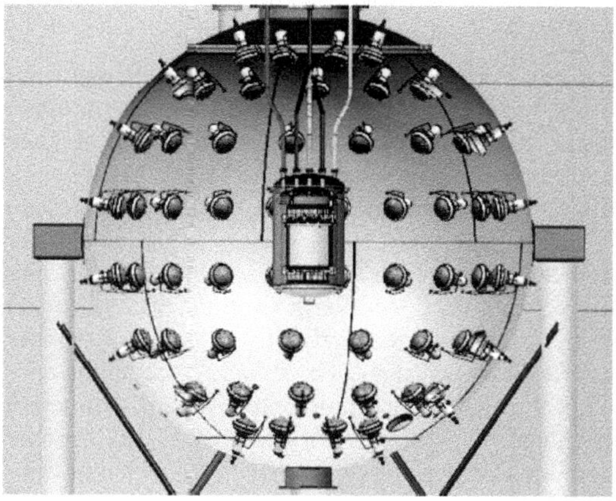

Fig. 13: Schematic of Darkside-50 liquid argon dual-phase TPC.

If successful the following step would be a 5000 Kg detector with the same reach of XENON1T.

3. *Nuclear Astrophysics*

3.1 LUNA

Since long a little accelerator (400 kV) is operated by the LUNA collaboration [32] with the purpose to study thermonuclear reactions down to the energy of the nucleosynthesis inside stars. Many significant results [33] on the reactions that drive our Sun evolution have been obtained. A new project has been recently aiming to expand the physics scope by using a new single ended positive ion accelerator (a 3.5 MeV machine) hydrogen and helium. The physics becomes more relevant at higher temperatures than hydrogen burning. Two beam lines are foreseen for solid and gas target experiments. The LUNA-MV project is financed by a special grant for Italian ministry of research.

4. *Life science*

4.1 Cosmic Silence

Following a campaign of studies in-vitro [34] on a kind of cell isolated from a transgenic mouse (pkZ1) the proposed (and approved) experiment aims to study the influence of the dose absorbed by mice kept inside the underground laboratory when compared with the same animals placed outside. Essentially testing the 'Linear No-Threshold' model [35].

5. Conclusions

Gran Sasso laboratory has played a fundamental role in developing the sector of physics that is now known as 'astroparticle physics'. It is now hosting experiments that address the hottest questions in particle physics such as the nature of neutrinos or of the elusive dark matter. The big experiments that received the CERN-to-Gran Sasso neutrino beam have given the final proof of the transmutation of this particle while propagating and also have set stringent limits to existence of new types of neutral leptons.

The sun thermonuclear reactions have been probed more completely than anywhere else by a single experiment and at the same time even neutrinos produced in the Earth were detected. The same experiment is now ready to accept (anti)neutrinos from radioactive decays of chosen isotopes to restrict even more the phase space for the so-called sterile neutrinos. Or perhaps discover them.

A fascinating aspect of the research performed in this deep underground laboratory is that here we study both the mechanism of formation of stars (their infancy) and their death when they become a supernova. Going in the depth of Earth to explore the Cosmos.

Intuition and a strong willing, the ingredients that allowed A. Zichichi to make a dream come true and donate to Italy a world class laboratory. The same element that shall guide all the scientists in their daily research.

References

[1] M. Ambrosio et al., Nucl. Instrum. Meth. A486 (2002) 663; M.Ambrosio et al., Eur. Phys. J. C25(2002) 611
[2]www.lngs.infn.it/lngs_infn/contents/lngs_it/research/experiments_scientific_info/library_publications/preprint_reports/report2012.htm
[3] R. Acquafredda et al., JINST 4 (2009) P04018
[4] N. Agafonova et al, JHEP 1311 (2013) 036
[5] S. Amerio et al., Nucl. Instrum. Meth. A527 (2004) 82
[6] C. Rubbia eta al., JINST 6 (2011) P07011
[7] G. Alimonti et al., Astropart. Phys. 16 (2002) 205
[8] G. Bellini et al., arXiv:1308.0443 (2013)
[9] L. Wolfenstein, Phys. Rev. D17 (1978) 2369; S.P. Mikheyev and A. Yu. Smirnov, Prog. Part. Phys. 23 (1989) 41
[10] G. Bellini et al., Phys. Lett. B722 (2013) 295
[11] G. Bellini et al., JHEP 1308 (2013) 038
[12] K. H. Ackermann et al., Eur. Phys. J. C73 (2013) 2330
[13] M. Agostini et al., Phys. Rev. Lett. 111 (2013) 122503
[14] R. Ardito et al., hep-ex/0501010 (2005).
[15]E. Andreotti at al., Astropart. Phys. 34 (2011) 822
[16] J. W. Beeman et al., Adv. High. Energy Phys. 2013 (2013) 237973; J.W. Beeman et al., JINST 8 (2013) P05021
[17] D.M. Chernyak et al., Nucl. Instrum. Meth. A729 (2013) 856; L. Cardani et al., arXiv: 1312.4680
[18] C. Alberini et al., Nuovo Cim. (1986) 237
[20] P. Belli et al , Phys. Rept. 307 (1998) 269
[21] M. Buehler et al., Nucl. Instrum. Meth. A370 (1996) 237
[22] E. Aprile et al., New Astron. Rev. 49 (2005) 289
[23] T. Alexander et al., JINST 8 (2013) C11021
[24] K. Freese, M. Lisanti and C. Savage, Rev. Mod. Phys. 85 (2013) 1561
[25] R. Bernabei at al., Eur. Phys. J. C73 (2013) 2648
[26] F. Petricca et al., J. Phys. Conf. Ser. 375 (2012) 012003
[27] J. Angle et. al, Phys. Rev. Lett. 100 (2008) 021303.
[28] E. Aprile et al., Phys. Rev. Lett. 109 (2012) 181301
[20] D. S. Akerib et al., Nucl. Instrum. Meth. A704 (2013) 111
[30] E. Aprile, arXiv:1206.6288 (2012)
[31] T. Alexander et al., JINST 8 (2013) C11021
[32] C. Broggini and the LUNA collaboration, AIP Conf. Proc. 853 (2006) 358
[33] A. Guglielmetti, PoS (ENAS 6) (2012) 057
[34] F. Antonelli et al., Radiat. Res. 164 (2005) 514
[35] J. W. Gofman, Radiation and Human Health, Sierra Books Inc. (1981).

CHAIRMAN: F. FERRONI

Scientific Secretaries: E. Messina, M. Vignati

DISCUSSION

- Kiehn:

In one of the slides the possibility of making an experiment on a very fundamental principle such as the Pauli principle has been outlooked. I ask you how would such an experiment look like.

- Ferroni:

The experiment, which is called VIP, has been proposed by Sergio Bertolucci. I kindly invite the audience to ask directly to him, since he is giving later a lecture, in order to get a satisfying answer.

- Antonenko:

I would like to have a deeper insight on the possibility of applying the facility of the Gran Sasso laboratory to biology and medicine.

- Ferroni:

In the past some culture of cells had been kept for a long time at a very low radiation level. The outputs were then processed in hospitals and biology laboratories. Since the results of these experiments were encouraging, the study of the impact of radiations on little genetically modified animals (such as mice) had also been proposed. The main issue when studying behavior at 'zero' dose is due to the presence of a relevant underlying background radiation in animal bodies. We suspect that the analysis is very difficult to be performed. This experiment, which has not yet been done, has received the endorsement of the Gran Sasso community.

- Ruthmann:

Are you aware of projects to build the next iteration of DAMA, somewhere else in the world, to try to establish the signal?

- Ferroni:

It is a very nice question. We know the problems of repeating that experiment. The first problem is that DAMA has a huge amount of mass (200 kg essentially of NaI). This NaI has been made long ago at Crismatech, I believe, which is a company making crystals, under very very special conditions since this is the most radiopure NaI ever produced at this level of contamination. Potassium is the real killer because of the 3 keV line at low energy. Nobody is able to reproduce these things today. Of course you can invest money, you can wait and you may succeed: this is the first problem. The point is that NaI in itself is not very expensive. Few 100 kg of NaI you buy for a normal PET detector, it does not cost much. But here there is a serious investment to carry on and the success is not guaranteed. My preferred option, but I do not know if it would be possible, is to take the experiment and bring it to the south hemisphere. At least there you eliminate or you check one of the things: the season essentially is an effect that changes because winter becomes summer. So you have something that changes, the average increase of temperature, the different humidity, whatever has an annual modulation would change. Certainly it is extremely difficult but in principle it can be done. To my knowledge, there are no laboratories equipped in the south hemisphere to be ready to accept such an experiment. I know there is an experiment done in Korea, not with NaI but with CsI. CsI cannot be as radiopure as NaI, because Cs of course has the isotope which emits photons, but however it has an interesting threshold and could run for a long time. There is now an

R&D activity which is going on in China with CsI crystals. If you want to compare the signal of a neutron using ionization and the one of an electron or a photon, the signal in CsI at room temperature looks pretty much the same in its time shape. But if you start to change the temperature and go down, the signal of the neutron shrinks and shrinks and the signal of the electron stays the same. Of course I say neutron because the assumption is that neutron is scattering almost like the WIMP. So if you change the shape you would be able to reject most of the background induced for example by the potassium line, which manifestly is a photon. The maximum separation is obtained at a temperature which is, just to give you an idea, the one of liquid Xenon. Now, they claim a rejection factor of the order of 10^7, with this technique. Of course they are at the level (I have visited the lab) of a crystal like this – some cm cube in a little cryostat. We have opened a collaboration with them. We offered our expertise from the people that for example participated in making the process of radiopure crystals of the CUORE experiment (TeO_2). So we have all the expertise in controlling how to build these things in the best possible way. We have offered our help to the Chinese with the idea that they will end up with 1 ton, 40 crystals 25kg each, of these things to be put in Gran Sasso. This will be to my knowledge a very interesting powerful check of the result. Of course here the difference would be that you can tell whether the interaction is coming from a neutron-like particle or an electromagnetic one, which I think is already a big step forward. Sandro you wanted to add something, possibly?

- *Bettini:*

If you want really to test Dama, I think that you should do that with NaI. The NaI is not extremely difficult to do. It is a matter of time, it is not very expensive. There is another small laboratory in Europe that is Canfranc. There we have developed that. The issue is potassium, the potassium for Dama was 100 ppb in the powder, and 20 ppb in the crystals. Now we produced powder with 90 ppb (that is ok), but the crystal is 40 ppb, so we are not very far from doing that. Once the procedure is understood, to produce the crystals is not very expensive. 250kg we do not know the price, but it is a few 100.000 Euro, not very expensive. The point is that if you take one of these crystals before testing, the cosmogenics go away (6 months).

- *Ferroni:*

Dama is going on since 10 years, first. Second, in principle in a normal world if you repeat precisely the same experiment, you would get the same result. You should add something. For example are you able to go to a lower threshold? This would be an important issue: here the threshold is around 2 keV, and, you know, whenever an experiment observes an effect at the threshold, there is a potential problem. So you can go to a very much lower threshold: this is an issue. But otherwise in principle doing the same experiment in the same condition in the same lab, because Canfranc is not better nor is it worse, I do not think you learn much. You should add something that allows you to disentangle one of the possible effects.

- *Ruthmann:*

The iteration of Xenon are going on and on, I wasn't aware of the costs which somehow limit the next iteration of DAMA:

- *Ferroni:*

1-ton of LXe is doable, not a huge quantity. Certainly at a certain moment you get a problem with the cryostat. Because you go larger and larger, you have to drift the signal, you have to purify the things with a faster rate. So there are limits, but I do not think that 1 ton is the real limit. The real problem is when you look at the sensitivity of Xenon, projected to 1 ton. At a certain moment, in a mass region from 10 GeV to 1000 GeV, you have reached the limit you believe WIMPs can do. I mean looking at gaining another factor 10 or 100, you loose the faith of the fact that Dark Matter is a WIMP. At a certain point you don't believe the prejudice that Dark Matter is made of WIMPs, you abandon this model.

- *Noferini:*

What if the mass is 10 TeV?

- *Ferroni:*

If the mass is 10 TeV, this is not the right measurement method to go essentially. Because the flux becomes so poor. Certainly it costs less than doing the superLHC at 100 TeV center of mass, but that thing is moving up (indicates at the cross section sensitivity plot). You could do 10 tons. But you know, experiments are somewhat guided but some prejudice. At a certain point, when you don't believe that that is the right way, people start to move out. My suggestion is that we should take more attention to axions, for example, as a possible candidate for dark matter. We do not invest enough for this specific field of research, because we thought that WIMPs were the paradigmatic solution.

- *'t Hooft:*

We know roughly the velocity, because that's the galactic curve (100 km/s) that's independent of the mass of the particle. So 10 TeV can give the same effect of 1 TeV.

- *Ferroni:*

Yes. But you have less particles. The loss of sensitivity is due to the fact that the flux reduces, you have the same amount of mass distributed in less particles, essentially.

- *Bettini:*

It is due to the kinematics. The energy it gives to the Xenon, depends on the two masses. If they are equal the transfer of energy is maximum. If the mass of the WIMP is bigger it gives less energy to Xe.

- *Ferroni:*

Also you have less particles. The Dark Matter is a certain amount of mass, is not more or less. You divide by the number of elements it is composed of. You have 10 times less particles.

- *Biancofiore:*

What is the origin of the modulation of the DAMA spectrum, if it is a signal?

- *Ferroni:*

Dark Matter is everywhere right? And the Earth is going around, so the cross section depends on the relative velocity of your detector with respect to the Dark Matter. The Earth rotates around the sun, which is moving in one direction, you pick up the particle that hit against you and the other, and you just go in subtracting the thing. It's not a tiny effect, because the speed we are talking about is ~100 km/s (not the speed of light). So we are talking of an important effect, because in winter or in summer you have a different effect. Perfectly a sinus or a cosinus, you follow it. Unfortunately there are many things that have an annual modulation. Also the flux of neutrons. Of course the experiment tried to convince everybody in the world that they have eliminated all the possible sources of spurious modulation, still people are not convinced. That's life.

- *Morozov:*

There is a highway tunnel nearby. Shouldn't it affect the precision of the experiment?

- *Ferroni:*

You mean: the passing cars make some movement of the ground? I do not understand. I think this would affect the only experiment that for sure I have not mentioned, which is Ginger, an interferometer to measure the gravitational effects. You know that in the next couple of years, one satellite will fly, which is the LISA path finder, for demonstrating that a gravitational wave detection could be done in space. The idea is that you put 3 satellites with a laser going around and you just test the effect of gravitational wave that moves them with respect to the perfect point of the orbit. Nothing to do with Gran Sasso, but once pieces of equipment of this LISA path finder were to be tested at Gran Sasso. The question that the scientific committee asked, "come on, you have an highway close by, how wouldn't this affect the frequency" (practically in the same range of frequencies interesting for these things, Hz or subHz). For such an experiment I do not really see why shacking the detector would do it, I do not think you could simulate the hit of one neutron or nucleus just by shacking the detector, neither for the solid nor gaseous or the liquid. I do not think that any detector is affected in practice by the nearby highway. Actually there is only traffic on Saturday and Sunday, when people go skiing, the other day of the week you can forget to see a car inside (not true but almost).

Highlights from ALICE

Paolo Giubellino
INFN Torino and CERN Geneve
E-mail: `paolo.giubellino@CERN.ch`

The LHC has delivered for the first time collisions of Nuclei in November 2010, at an energy of 2.76 TeV per nucleon pair, which represents a jump of more than an order of magnitude over the highest energy nuclear collisions ever studied before. The high energy, the quality of the state-of-the art detectors, and the readiness of the experimental collaborations at the LHC have allowed a rich harvest of important scientific results in a very short time. In this lecture a short overview will be given of how the results from the LHC, and in particular from the ALICE experiment, have provided new insight on the properties of matter under extreme conditions of temperature and pressure, analogous to the conditions present in the early phases of the evolution of the Universe. First results from the proton-nucleus runs will also be presented.

1.Why high-energy Heavy-Ion Collisions

Collisions between very high-energy nuclei, normally referred to as Heavy-Ions since they are atoms stripped of their electrons to allow acceleration, produce each a huge number of particles, over ten thousand in a single event at the LHC. A typical event is shown in Fig. 1. The study of such extremely complex events poses a formidable challenge both to the detections systems and to the software and analysis methods to be used to treat and eventually understand the data. So, it is legitimate to ask first of all the question of why experimenters are willing to face such extreme problems, what are the underlying questions which we aim to answer.

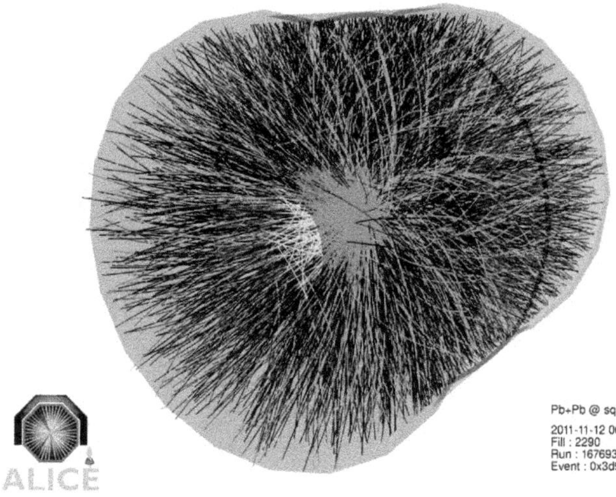

Fig. 1 A collision of two Lead Nuclei as seen by the ALICE detector at the LHC. (CERN-EX-1111290 04)

The strong force, which holds together the nucleons, protons and neutrons, in nuclei and the quarks within the nucleons, is described to great precision by a very successful theory, called Quantum Chromo Dynamics (QCD), in which the quarks possess a color charge and interact exchanging force carriers, the gluons, which are also colored. A fundamental feature of the interaction is the fact that it grows with the distance between the quarks, much like a spring connecting two balls. Therefore, any attempt to isolate a quark are in vain, since the energy stored in the spring will eventually be more than the mass of a quart-antiquark pair: the spring will break generating a pair, and the quarks will not be isolated. This property is called "confinement" and has major consequences. In particular, most of the mass of the hadrons is to be attributed to this feature: more technically, one can say that the mass is acquired dynamically because of the interaction between the quarks, whose Higgs mass would account for just a small fraction, about 1%, of the mass of the light hadrons as shown in Fig. 2. Therefore, since the mass of the matter around us is essentially the mass of the nuclei of the atoms, it can be said that, while the Higgs mechanism is at the base of the masses of the quarks, strong interaction is at the origin of most of the mass of ordinary matter.

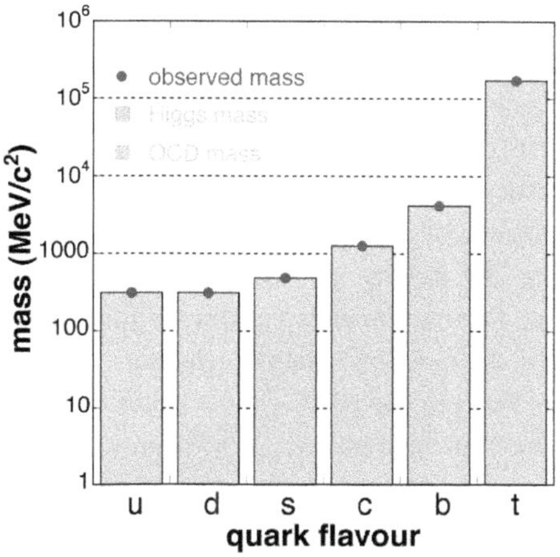

Fig. 2 QCD and Higgs mass components for the light and heay quarks from Ref [1]

The method of colliding high-energy nuclei to heat and compress hadronic matter dates from the early eighties, but the path had been indicated already several years before. It was in 1975 that T.D. Lee pronounced the famous sentence "it would be interesting to explore new phenomena by distributing a high amount of energy or high nuclear density over a relatively large volume", and the idea of deconfinement had been around for some time. Already in 1965 Hagedornhad observed that the mass spectrum of hadronic states led naturally to the concept of a "limiting" or "critical" temperature beyond which the concept of hadron would cease to make sense. In 1973 the development of QCD as the theory describing strong interactions introduced the concept of asymptotic freedom [3][4]: at small distances quarks and gluons would be loosely interacting, or quasi-free.

In 1975, N. Cabibbo and G. Parisi [5]proposed for the first time the existence of a ``different phase of the vacuum in which quarks are not confined", and drew the first schematic diagram of hadronic matter with temperature and baryonic density as axes. Two regions were defined: at low-T and low baryonic density, ordinary hadrons, at high-T, high-μ_B a deconfined state: a new field of research was born! In the same year, J.C. Collins and M.J. Perry[6] developed the idea that matter under extreme density conditions, as in neutron star cores and in the early universe, is made of quarks instead of hadrons, which overlap and lose their identity. Their argument was based on the fact that quarks interact weakly when they are close together, and therefore at high densities could be described as a gas of free massless quarks. Finally, in 1980 E. Shuryak introduced [7]the term quark gluon plasma (QGP) to describe the state of nuclear matter in which quarks are deconfined, stressing the analogy between a classical plasma made by ionized atoms and the state made of colored strongly interacting objects.

In a nucleus-nucleus collision the energy of the incoming projectiles is dissipated in the relatively large volume defined by the overlap region of the two colliding nuclei, thus creating the conditions for a phase transition to deconfined quark matter by heating.

So, by colliding nuclei at high energies we compress and heat a system with a large number of quarks, to form a "bubble" of deconfined strongly-interacting matter, which in a very short time, of the order of 10^{-23}s, expands and cools until ordinary hadrons reconstitute, just as they did in the evolution of primordial Universe, some 10 millionth of a second after the Big Bang.

Over the last three decades, a vigorous experimental program, carried both in Europe and in the United States, has dramatically increased our understanding of nuclear matter under extreme conditions of temperature and density, allowing a real exploration of the phase diagram of strongly interacting matter. Our current understanding is graphically summarized in Fig. 3, with the trajectories allowed by the most important accelerators in operation or under construction: the two colliders LHC (in Europe) and RHIC (in the USA) and the fixed-target high-luminosity facility FAIR under construction in Germany.

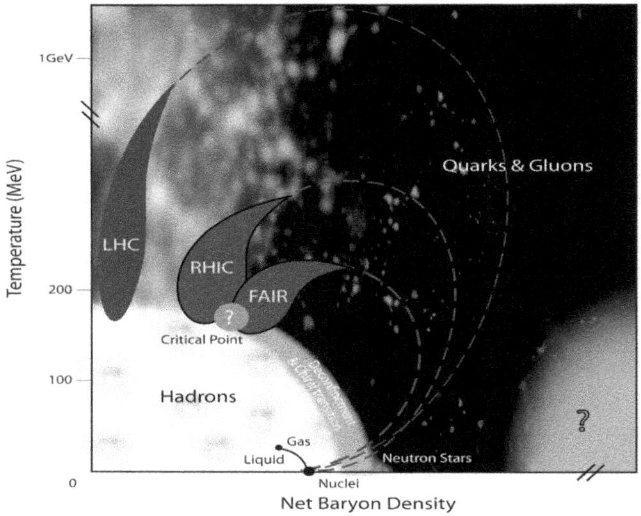

Fig.3 The phase diagram of nuclear matter, from Ref[1]

Together, these programs aim at providing answers to questions such as:
What are the properties of matter at the highest temperatures and densities?
What is the QCD equation of state? How can we test it?
What are the dominant microscopic mechanisms of QCD non-equilibrium dynamics and thermalization?
How does hadronization proceed dynamically? How is it changed in dense QCD matter?
How can the QCD phase diagram be efficiently explored?

Clearly, a comparison of the experimental results with theory is not a trivial task. Since the transition to the deconfined phase is outside the limits of validity of the perturbative QCD, a number of different approaches has been developed to derive the equation of state and the transition temperature. Only at the LHC some aspects of the interactions will be accessible with perturbative methods. Effective lagrangians have been extensively used, and also potential models, and even calculations based on the AdS/CFT correspondence have been providing a fresh look and new insight. Still, the most important and widely used are QCD calculations on a space--time lattice. They generally predict a phase transition to occur, and are able to determine its temperature. Current results indicate a critical temperature aroundT_c~170 MeV, or about 10^{12} K, 100000 times higher than at the center of the Sun, corresponding to a critical energy density of ε_c~ 700 MeV/fm^3

In the following, I will give a brief overview of the main experimental results obtained so far at the LHC, just little over one year after the very first collision of Lead Nuclei was recorded in the experiments. Three experiments joined in this pioneering work: ALICE[9], a dedicated experiment specifically designed to study nuclear collisions, and both ATLAS [10]and CMS[11], the two general-purpose experiments of the LHC, primarily designed for the study of high-p_T processes in pp collisions. I will focus mostly on the ALICE results.

While I will not describe the experimental apparata, I would like to remind here that ALICE has a very different optimization compared to the other LHC experiments. For the understanding of nuclear collisions it is essential that as many of the produced particle as possible are measured and identified, for the widest possible range of transverse momenta (typically from 0.1 to 100 GeV/*c*) and for all particle types: electrons, muons, photons and hadrons. So, the two guiding principles in the ALICE design have been robust tracking, able to handle the very large number of tracks which characterize Heavy Ion Collisions, and Particle Identification capability. The first is provided by very high granularity tracking detectors providing 3-Dimensional space points (several hundred millions of points) operating in a moderate magnetic field and with very low material. To identify leptons, photons and hadrons, including short-lived particles, such as hyperons, D and B mesons, ALICE features essentially all known techniques: dE/dx, Cherenkov & Transition radiation detectors, Time Of Flight, calorimeters, muon filter and topology (secondary vertices, kinks).

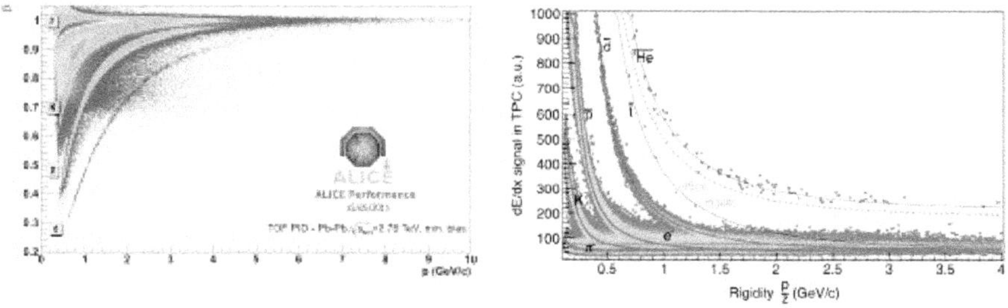

Fig. 4: PID in ALICE with TOF (left panel) and TPC (right panel) from ref.[12]

As an example of the ALICE PID capability, Fig. 4 shows the of the performance of the Time Of Flight (TOF) and of the Time Projection Chamber (TPC)'s energy loss measurement. Both are performing better than originally specified, a remarkable result in detector systems of such complexity observing events with thousands of tracks! The TOF has provided a time resolution below 100 picoseconds, and the TPC an energy loss measurement with a precision better than 6%.

In figure 5 (left panel) the momentum coverage of the various ALICE PID detectors is shown, demonstrating the extent of the coverage achieved. In addition, short-lived particles are identified through their decay topologies: in the right panel of figure 5 the impact parameter resolution expected in ALICE is shown, which is adequate for the measurement of D and B meson decays. K and Λ decays can be efficiently reconstructed for transverse momenta up to more than 10 GeV/*c*. The very fine granularity of the tracking detectors, providing a large number of space points per track, provides high efficiency tracking for charged particles even at the highest particle multiplicities foreseen at the LHCwith Pb beams, with a lower limit in momentum lower than 200 MeV/*c*. The very long track length measured allows ALICE to achieve an excellent momentum resolution ($\delta p/p < 5\%$ at 100 GeV/*c*). Several smaller detectors (ZDC, PMD, FMD, T0 andV0) used for global event characterization and triggering are located at small angles. The combination of the central and forward detectors provides coverage for the measurement of charged multiplicity which extends from -3.4 to +5 in pseudorapidity. An Electromagnetic Calorimeter, EMCal, covers about a third of the azimuth and provides improved energy resolution and triggering for jets. A high-resolution crystal photon spectrometer, PHOS, is placed at a very large radial distance from the interaction point (4.6 m). The muon spectrometer covers the forward rapidities, between-2.5 and -4, and features a complex arrangement of absorbers, starting very close to the interaction point (90cm) to minimize muon background from weak hadronic decays, and fourteen planes of tracking and triggering chambers, using 2-D pad readout to handle the high multiplicities.

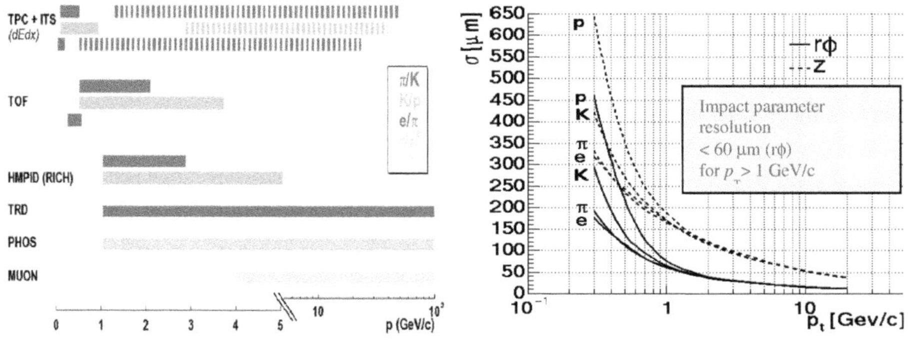

Figure 5: Left: particle identification in ALICE: momentum range and detectors used. Topological identification, not included in this plot, can reach over 10 GeV/*c* for K and Λ. Right: ALICE impact parameter resolution as a function of transverse momentum.

The hardware trigger in ALICE combines the input from detectors with fast trigger capability (T0, V0, ZDC, SPD, TOF, PHOS, EMCal, Muons, ACORDE). It operates at several levels (pretrigger, L0, L1 and L2) to satisfy the individual timing requirements of the different detectors, as the ALICE electronics is in general not pipelined. In addition, a software based High-Level Trigger (HLT) using a farm of up to 1000 multiprocessor PCs performs a preprocessing of the data allowing for a compression of the data volume by a factor of 3.5. It also allows real time data quality monitoring.

2. ALICE results

The first LHC heavy-ion results were already published during the 2010 lead–lead run, lasting for a month. In fact, there were three papers submitted almost simultaneously: the multiplicity measurement [13] and the measurement of v_2 coefficient of the azimuthal anisotropy [14] by ALICE, and by the ATLAS collaboration a paper on jet-energy imbalance [15]; these appeared together in one issue of Phys. Rev. Lett. One of the most spectacular findings at RHIC was that the matter generated in heavy-ion collisions flows like a liquid with very low viscosity, almost at the limit of what is allowed for any material in nature. This tells us that the constituents of this QGP are quite different from freely interacting quarks and gluons, and led to the definition of the matter under study as sQGP, for strongly interacting Quark Gluon Plasma, a liquid very different from the expected gas of quasi-free quarks and gluons. The first LHC azimuthal anisotropy measurement [14] confirms the RHIC results: elliptic flow of particles with the same transverse momenta is almost identical at the two energies. The nearly-perfect fluid has been found to be opaque to even the most energetic partons (quarks and gluons), which appear as jets of particles from the collisions, an effect known as jet quenching. This is an interpretation of the reported strong jet-energy imbalance [15]. However, the physical mechanisms underlying these phenomena are not well understood. Another manifestation of jet quenching manifests is a reduced yield of high-transverse-momentum-particle in central collisions compared to that expected from the measurements in proton–proton reactions. To express such reduction a properly normalized ratio of yields in heavy-ion and in proton–proton interactions – the nuclear modification factor R_{AA} – is used. Since proton-proton collisions are rescaled according to the number of binary collisions, R_{AA} should be well below one at low transverse momentum, where particle production is rather proportional to the number of participant nucleons, and grow toward one for increasing transverse momenta. The suppression pattern observed by ALICE [16] gives factor about 7 (= $1/R_{AA}$) lower charged-particle production in lead–lead collisions at transverse momentum around 6GeV/c, a striking evidence of the energy loss in the extreme density matter.

The size of the pion-emitting source in central lead-ion collisions is deduced from the shape of the Bose-Einstein peak in the two-pion correlation functions [17]. The collective flow makes the size of the system appear smaller with increasing momentum of the pair. This behaviour is clearly visible for the radii measured in the ALICE experiment. The results for measurements of the radius of the pion source in three dimensions indicate a short duration for the emission, hence an “explosive” emission. Time when the emission reaches its maximum is 10–11fm/c, significantly longer than it is at RHIC, see fig. 6 (left side). Moreover, the product of the three components at low pair-momentum, an estimate of the homogeneity volume of the system at decoupling, is twice as large as at RHIC, see fig. 6 (right side).

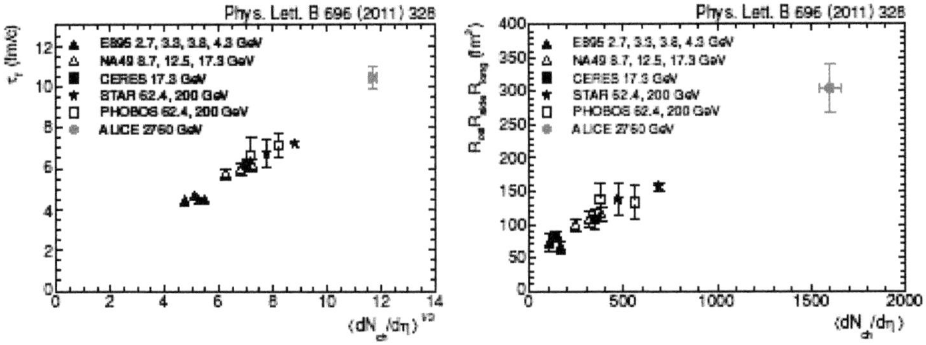

Figure 6. Bose–Einstein pion-interferometry results from different experiments [17]. Estimate of lifetime of emitting source as a function of cube root of particle density (left side). Volume of homogeneity region as a function of particle density (right side).

One of the crucial measurements for the characterization of the fireball produced in heavy-ion collisions are the spectra of identified hadrons, which encode the collective expansion velocity developed in the QGP and in hadronic stages. Moreover, the overall abundances of identified hadrons are believed to be fixed at hadronization, thus they indicate the temperature when chemical composition was established. The measured spectra [18] show an increase of about 10% in the radial-flow velocity when compared to RHIC results. At present, however, the yield ratios observed by ALICE seem to challenge both previous experiments and theory. While the K/π, Ξ/π and Ω/π ratios are compatible with the expectations from the thermal model with a temperature for the "chemical freeze-out" of about 165MeV, as in previous observations, the p/π ratio points to a significantly lower temperature. On the experimental side, there are indications of a similar effect at lower energies, which call for further investigations. On the theoretical side, a number of different possibilities are being investigated, none of them conclusive at the moment.

As mentioned above, the first results on elliptic flow were published [14] during the initial Pb–Pb run. Flow is a fundamental observable since it carries information on the equation of state and the transport properties of matter created in a heavy-ion collision. This observable relates final state anisotropies with features of the initial one, thus allowing study of the medium response and characteristics. The azimuthal anisotropy in particle production is a clear experimental signature of collective flow. It is caused by multiple interactions between the constituents of the created matter and the initial asymmetries in the spatial geometry of a non-central collision. The amount of elliptic flow measured by the v_2 coefficient, integrated over transverse momentum, increases by 30% compared to RHIC energies. However, this increase is entirely a consequence of the growth in transverse momenta. The overall azimuthal distribution of particles emitted in a heavy-ion collision can be described with a Fourier expansion, whose coefficients relate to different aspects of the initial state. ALICE measured also the higher harmonic coefficients of azimuthal distribution [19]; they exhibit much shallower centrality dependence than v_2, and the symmetry planes of v_2 and v_3 coefficients are uncorrelated (see fig. 7, left side). These observations point to fluctuations in the initial geometry as the origin of higher-order azimuthal asymmetries. The flow coefficients can also be studied with a different approach, using two-particle correlations in azimuthal angle [20]. A Fourier decomposition of the correlation function gives squares of different v coefficients, and describes well the structures seen in azimuthal correlations, see fig. 7 (right side), such as the "long-range ridge" (a small angle correlation between particles distanced in longitudinal direction) and the "Mach cone" (a correlation on two sides of back-to-back direction). The collective response to initial spatial anisotropy that causes elliptic flow economically explains these puzzling features, once event-by-event initial-state density fluctuations are considered. The ALICE experiment has analyzed azimuthal asymmetry also separately for different particle species, which constrains models incorporating a realistic initial state and hydrodynamic evolution.

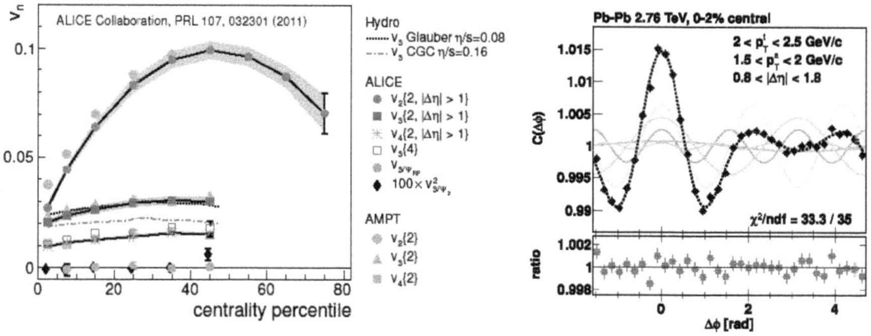

Figure 7. Measurements of azimuthal anisotropy in lead–lead collisions at LHC. Dependence of different v_n coefficients (see legend) on collision centrality [19] (left). Fourier decomposition of two-particle azimuthal correlation function in 2% most central collisions [20] (right).

The two-particle azimuthal correlations at higher transverse momenta measure particle-yield modifications in jet-like structures [21]. When compared to the expectation from proton–proton interactions, the observed yield around high-transverse-momentum particle is slightly higher, while the yield on the opposite side is reduced. This measurement assists the understanding of jet-quenching phenomena. The ALICE collaboration continues to contribute to this topic by studying reconstructed jets, underlying event fluctuations, modifications of jet-fragmentation function, jet-particle composition and providing detailed studies of the dependence of the energy loss on the parton traversing the QGP.

Heavy-flavour particles are recognized to be effective probes of a very dense and hot medium formed in nucleus–nucleus collisions; they are expected to be sensitive to its energy density, through the mechanism of in-medium energy loss. At LHC energies, heavy-flavour particles are copiously produced and thus provide ideal tools for QGP studies. The nuclear modification factor R_{AA} is well established as a sensitive observable for the study of the interaction of hard partons with the medium. Parton energy-loss is caused by the strong interaction, hence the amount of energy loss depends on colour charge of parton. Therefore, quarks are predicted to lose less energy than gluons. In addition, heavy quarks (up to some high momentum), which are slower than light, cannot emit gluons within the so-called "dead-cone" around their trajectory; this effect is expected to reduce the energy loss of heavy quarks with respect to light ones. Thus, a pattern of gradually decreasing suppression (i.e. increasing R_{AA}) should emerge when going from the light-flavour hadrons (e.g. pions), which mainly come from gluons, to the heavier D and B mesons: $R_{AA}(\pi) < R_{AA}(D) < R_{AA}(B)$. The measurement and comparison of these different probes provides, therefore, a test of the colour-charge and mass dependence of parton energy-loss. The ALICE collaboration has measured the production of the charmed mesons D^0 and D^+, detecting their hadronic decays in lead–lead collisions [22]. In central collisions a large suppression with respect to expectations at large transverse momentum was found, indicating that charm quarks undergo a strong energy loss in the hot and dense state of strongly-interacting matter formed at the LHC. This is the first time that D meson suppression has been measured directly in central nucleus–nucleus collisions. The results show a suppression by a factor 4–5, almost as large as for charged pions, above 5 GeV/c (see fig. 8). At lower momenta, there is an indication of smaller suppression for D than for π mesons.

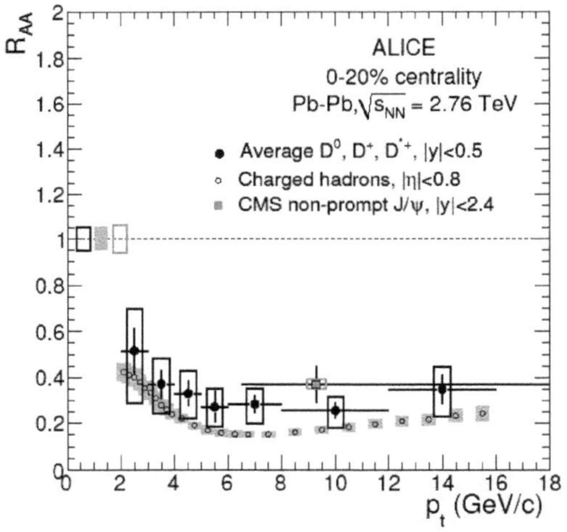

Figure 8. Charmed-meson nuclear modification factor as a function of transverse momentum compared to that of charged hadrons and of non-prompt (mostly from B-decays) J/ψ mesons (as measured by CMS).

Suppression of charmonium production was for a long time considered as one of the main probes for a deconfined medium. At large enough temperatures bound charm–anticharm states are supposed to be dissolved due to Debye screening. However, at LHC energies, new mechanisms of charmonium production in the QGP could occur because of a large number of charm quarks. Around one hundred charm–anticharm pairs are expected to be produced in a central lead–lead collision. Several dynamical transport models predict that charm and anticharm quarks could combine at later stages of the interaction, leading to an enhancement of charmonium production in the most central collisions. ALICE detects charmonium down to very low transverse momentum in two different regions: the central barrel in the dielectron channel and the forward muon arm in the dimuon channel. The detection at low transverse momentum is crucial because the recombination of the charm and anticharm quarks is expected to be the main production mechanism for charmonium below 3GeV/c. The different regions allow for the study of QGP with different charm densities. ALICE has studied the nuclear modification factor R_{AA} for J/ψ mesons as a function of collision centrality [23]. The results indicate that the J/ψ nuclear modification factor shows little dependence on centrality (see fig. 9), a trend that is different from that observed at lower energies. For central and mid-central collisions the J/ψ R_{AA} is larger at the LHC than that measured at RHIC. In a complementary study, the CMS collaborations at the LHC have measured a smaller value for the J/ψ nuclear modification factor at transverse momenta above 6.5 GeV/c. These observations hint at the recombination of charm and anticharm quarks in the QGP as the main mechanism for J/ψ production in central lead–lead collisions at LHC energies.

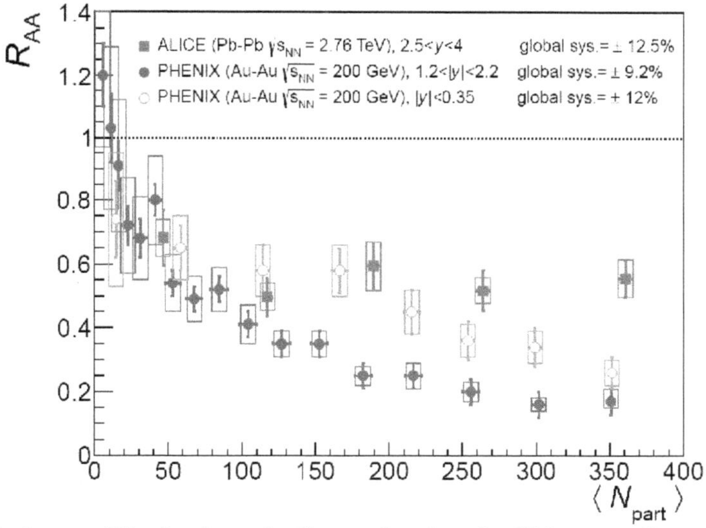

Figure 9. Nuclear modification factor for J/ψ as a function of collision centrality, expressed as number of participants. ALICE data are compared to those from the PHENIX experiment at RHIC.

It is clear from these first results that a coherent picture of the characteristics of the extreme density matter of which our Universe was made of in the first microseconds of its existence. Even more, the tool chosen to study it, heavy-ion collisions at the world's most powerful accelerator, has proven to be the right one, allowing for a continuous improvement of our understanding. The results shown here are but a small subset of the wealth of results obtained so far, and a was a mere hint to the ones coming. It is indeed a very exciting time for the physics of the Quark-Gluon Plasma.

3. First results from the pA run

When studying the difference between pp and PbPb collisions it is essential to understand the initial wave function of the colliding nuclei, and thus disentangle 'initial state effects" from those due to the hot and dense matter created in PbPb collisions. Understanding the structure of the initial state is very interesting in its own right, because at the LHC energy, experiments probe the structure of the nucleus in a novel and unexplored QCD regime of very low values of the longitudinal parton-momentum fraction (x <10–3). In this kinematic region, the extremely high gluon density is expected to saturate by means of strongly non-linear coherent processes, leading theorists to predict the existence of yet another pre-collision state of matter – the so-called colour glass condensate (CGC).Moreover, pPb collisions at the LHC open a new window for the study of high density strongly interacting matter, allowing a study as a function of the volume of the system.

The analysis of the pPb pilot run of 2012 and full run of 2013 are now in full swing, and only a few highlights will be mentioned here.

The results from the basic control measurements, the charged-particle multiplicity density and transverse-momentum spectrum, were in agreement with expectations from theoretical models that incorporate present knowledge of the function describing the longitudinal distribution of partons inside the nucleus [24]. In addition, when compared with the results obtained for pp collisions at 2.76 and 7 TeV, on one hand the charged-particle multiplicity in pPb collisions was found to scale roughly with the mean number of nucleons participating in the collision, as expected for particle production through soft processes. On the other hand, the particle spectrum at large transverse momenta – beyond 3–4 GeV/c – was found to scale with the mean number of binary nucleon–nucleon collisions, as expected for particle production through hard processes. Also charmed mesons showed no significant deviation from binary nucleon–nucleon scaling, as shown in Fig. 10. The above observations confirm that the suppression of high transverse-momenta hadrons and jets observed in central PbPb collisions can be attributed to a final-state effect, namely, the energy loss experienced by partons traversing the medium created in these collisions.

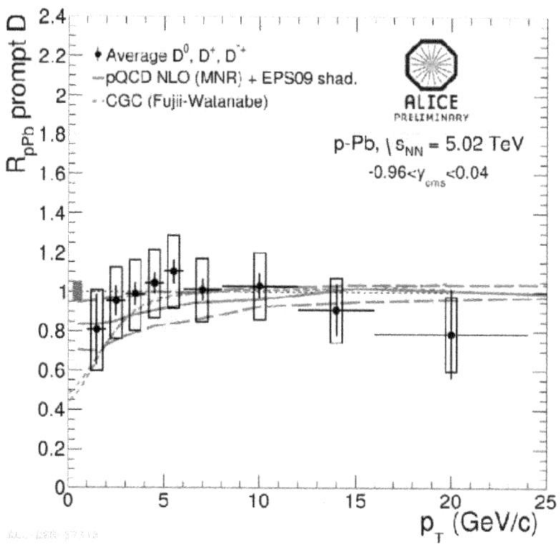

Figure 10. Ratio of charmed mesons production in pPb collisions to the one measured in pp scaled by the number of collisions R_{pPb}. $R_{pPb}$is plotted as a function of pT for prompt D mesons (average of D0, D+ and D*+) in p-Pb collisions at √sNN=5.02 TeV. Results are compared to model calculations of initial-state effects.

The study of two-particle correlations in high-multiplicity pPb events led instead to a surprising result: a near-side, long-range (elongated in pseudorapidity) correlation, forming a ridge-like structure observed in high-multiplicity pp collisions, accompanied by an essentially symmetrical away-side ridge opposite in azimuth ,was observed [25]. This double ridge was revealed after the short-range correlations arising from jet fragmentation and resonance decays were suppressed by subtracting the correlation distribution measured for low-multiplicity events from the one for high-multiplicity events. Similar long-range structures in heavy-ion collisions have been attributed to the collective flow of particles emitted from a thermalized system undergoing a collective hydrodynamic expansion. The origin and nature of this structure are currently subject of debate and research. In the subsequent months, and more evidence of features which in heavy-ion collisions have been attributed to collective behavior have emerged, in particular a characteristic mass scaling in the measurement of v_2 harmonic coefficients of the azimuthal anisotropy, and in the baryon-to-meson ratios. So the questions of how collectivity and even hydrodynamic behavious could take place in such a small system, and of the compatibility with the energy loss results are currently hotly debated. The whole community interested in the study of strongly interacting matter at extreme conditions is eagerly waiting for the further developments of these intriguing findings.

4. The future of ALICE

The LHC operates in a quite different mode than previous CERN accelerators. During running, only a very short stop of operation takes place around the end of the year, while most of the time is devoted to quasi-continuous running. After O(3) years of operation, a long shutdown of more than a year allows for major maintenance and improvements of both the accelerator and the detectors.

Data taking for ALICE in the so-called RUN1 has included, apart from data taking with pp collisions, three period of data taking with nuclear beams. In 2010 the ALICE detector collected with a minimum-bias trigger about 30 million of lead–lead collisions at centre-of-mass energy of 2.76TeV per nucleon pair. In 2011 there was a short period of proton–proton running at the same energy at the start of the LHC operation, to gather comparison data; these were used for normalization in practically all measurements described above. The rest of the proton running proceeded at (a standard) energy of 7TeV. At the end of the year, the LHC switched again to the heavy-ion mode. This time the instant luminosity grew to above $10^{26}cm^{-2}s^{-1}$, and was above the design value for energy 2.76TeV per nucleon pair. ALICE used triggers for different types of lead–lead collisions: central, semi-central, dimuons, photons, jets, ultraperipheral, and others, and collected about 100 million events, inspecting over $0.1nb^{-1}$ of integrated luminosity. At the beginning of 2013, the coming Heavy-Ion period, has been dedicated to proton–lead collisions, to improve the baseline comparison, taking into account modifications of structure functions in nuclei. ALICE collected $0.03nb^{-1}$ of luminosity in the four weeks of pA running, half of which in pA mode and half in Ap mode, to take into account the asymmetry of the apparatus.

Since then the LHC operation has been paused for the upgrade necessary to increase the collision energy. The Long Shutdown (LS1), will last until end of 2014. During this period, ALICE will carry a very intense program of detector consolidation, installation of new detectors (DCAL, a second arm of the ALICE electromagnetic calorimeter, the last five supermodules of TRD and a fouthsupermodule of PHOS).

During RUN2, after the restart in 2015, the ALICE collaboration aims to complete its approved programme, collecting $1nb^{-1}$ of heavy-ion collisions at the higher collision energy (5.5TeV per nucleon pair in the centre of mass, being the design value). The intention is to achieve a significant part of this agenda before the second long shutdown for the preparation of the LHC luminosity increase, planned for 2018.

A major upgrade of the ALICE detector allowing for high-luminosity heavy-ion running after this period, to reach a tenfold increase in integrated luminosity in RUN3 and RUN4 (extending to the mid 2020's), is now being discussed. The Letter of Intent defining the scientific goals has been approved in late 2012, and the Technical Design Reports discussing the technical implementation of the upgrade will be discussed in 2013 and 2014. The extended physics program justifying the LHC operation in heavy-ion mode beyond 2020, which would imply collecting over $10nb^{-1}$ of data, has been included among the top priority goals of the European Strategy for High Energy Physics.

After having confirmed the main discoveries obtained at RHIC, ALICE has entered an exciting phase of new measurements, allowing a much broader and deeper study of the QGP. At the same time, ALICE is already preparing, on the basis of what has been learnt so far, a next step in more detailed characterization of the extreme state of matter produced at LHC; new, unexpected discoveries may come!

References

[1] NUPECC Long Range Plan 2010: Perspectives of Nuclear Physics in Europe, published by the European Science Foundation
[2] R. Hagedorn, NuovoCim. Suppl. 3 (1965) 147-186
[3] D.J. Gross and F. Wilczek Phys. Rev. D 8(1973) 3633—3652.
[4] H. D. Politzer Phys. Rev. Lett. 30 (1973) 1346-1349.
[5] N. Cabibbo and G. Parisi Phys. Lett. B 59 (1975) 67-69
[6] J.C. Collins and M.C. Perry~M.C. Phys. Rev. Lett.34(1975) 1352—1356.
[7] E.V. Shuryak Phys. Rept. 61 (1980) 71—158
[8] B. Muller and J. Nagle, Ann. Rev. Nucl. and Part. Phys. **56** (2006) 93

[9] ALICE Collaboration, Aamodt K, et al., JINST 3:S08002 (2008).

[10] ATLAS Collaboration, Aad G, et al., JINST 3:S08003 (2008).

[11] CMS Collaboration, Adolphi R, et al., JINST 3:S08004 (2008).

[12] A. Kalweit for the ALICE Collaboration, J.Phys.G G38 (2011) 124073.

[13] K. Aamodt et al. (ALICE collaboration), Phys. Rev. Lett. **105** 252301 (2010).

[14] K. Aamodt et al. (ALICE collaboration), Phys. Rev. Lett. **105** 252302 (2010).

[15] G. Aad et al (ATLAS collaboration), Phys. Rev. Lett. **105** 252303 (2010).

[16] K. Aamodt et al. (ALICE collaboration), Phys. Lett. **B696** 30 (2011).

[17] K. Aamodt et al. (ALICE collaboration), Phys. Lett. **B696** 328 (2011).

[18] M. Florisfor the ALICE collaboration, J. Phys. **G38** 124025 (2011).

[19] K. Aamodt et al. (ALICE collaboration), Phys. Rev. Lett. **107** 032301 (2011).

[20] K. Aamodt et al. (ALICE collaboration), Phys. Lett. **B708** 249 (2012).

[21] K. Aamodt et al. (ALICE collaboration), Phys. Rev. Lett. **108** 092301 (2011).

[22] B. Abelev et al. (ALICE collaboration), JHEP **1209** 112 **(2012)**

[23] B. Abelev et al. (ALICE collaboration), Phys.Rev.Lett. **109** 072301(2012)

[24] B. Abelev et al. (ALICE collaboration), Phys.Rev.Lett. **110**082302 (2013)

[25] B. Abelev et al. (ALICE collaboration), PhysicsLetters **B 719** 29(2013)

Planck Highlights

Alain Riazuelo

1 Introduction

The Cosmic Microwave Background (hereafter CMB) can be defined in simple terms as the afterglow of the Big Bang. The early Universe was very dense, very hot and therefore filled by intense radiation. This radiation was diluted and cooled down as the Universe expanded, but has survived till then. In the early phases of the Universe, the high density of matter associated to the fact that it is ionized made it completely opaque to radiation. As a consequence the free mean path of photons was very short. This situation changed drastically when the Universe reached the age of ~380 000 years, the so-called recombination era. Its temperature dropped below 3 000 Kelvins, which corresponds to the threshold where free electrons can combine with hydrogen nuclei. At that temperature, opacity is mostly determined by Thomson scattering of photons on free electrons, so that the disappearance of the latter made rather suddenly the free mean path of the former become larger than the size of the observable Universe. As a consequence, the CMB photons we observe today have travelled mostly unaffected since this epoch and give us a very clean picture of the state of the Universe at that time. More importantly, some events that occured significantly earlier than 380 000 years after the Big Bang, and hence at much higher energies have left an imprint on it, which allows us to recover some information on such events. For these reasons, since around two decades, the CMB has become the most useful tool to study cosmology. The reason for this to have become possible comes from a series of fortunate coincidences that allows us to extract a very large number of information from a single object.

2 A short introduction to CMB physics

Even today, the CMB photons are overdominant as compared to those produced by any other astrophysical processes. The electromagnetic energy that is stored within the CMB corresponds to around 95% of the electromagnetic energy of the Universe, the five remaining percents being mostly split in more or less equal parts between starlight (in the visible part of the spectrum) and the thermal radiation of interstellar dust (in the far infrared domain). All the other processes at other wavelengths (radio, X-ray, gamma) bring an astrophysically interesting but numerically negligible contribution to this total, as shown in Fig. 1. Since the Universe is homogeneous and isotropic on large scales, the CMB is also expected to show a very high degree of isotropy. However, structure formation is a rather slow process and therefore necessitates that even at the time of CMB emission, the matter distribution was not exactly homogeneous. This inhomogeneity in the matter distribution should be somehow reflected in the radiation distribution, which is expected to show some degrees of inhomogeneity, either in its photon number or temperature. Observing these inhomogeneities is at the core of modern studies of CMB. The observability of these inhomogeneities comes from a very fortunate coincidence: not only the CMB dominates within the microwave part of the spectrum over any other sources, but it dominates in a limited frequency interval by more than 10^5. Therefore, even if the homogeneities we are interested in are rather small (10^{-5} in relative amplitude) they are still dominant over the other sources (hereafter named as "foregrounds" since they originate from later epochs and therefore from nearer to us than the CMB). This situation is shown on Figure 2 below.

Where does this 10^{-5} amplitude come from? Before recombination, Thomson scattering between photons and electrons, combined with Coulomb interaction between electrons and atomic

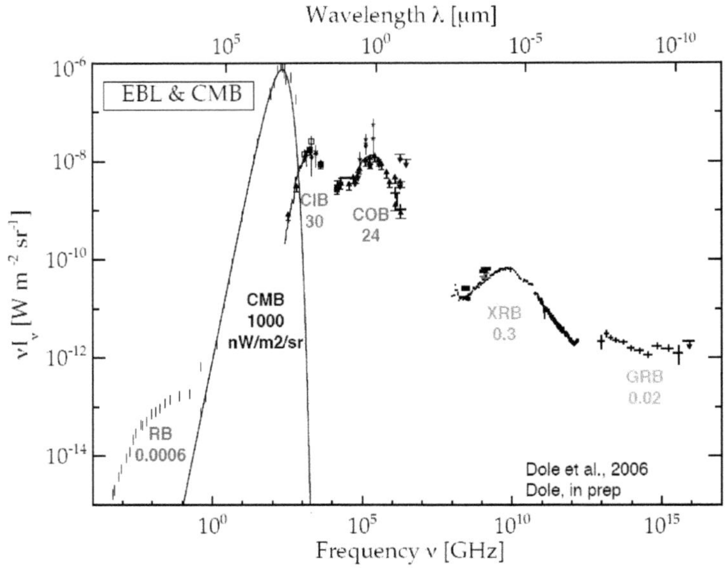

Figure 1: The overall integrated electromagnetic spectrum of the Universe, as seen by a fictious observer far from any source of radiation. The CMB completely dominates any other sources, the others being at most 30 and 40 times smaller (for the Cosmic Infrared Background and the Cosmic Optical Background, respectively). However, what will make the CMB really interesting to study is the relative amplitude of its fluctuations with respect to the other sources as shown in the next Figure.

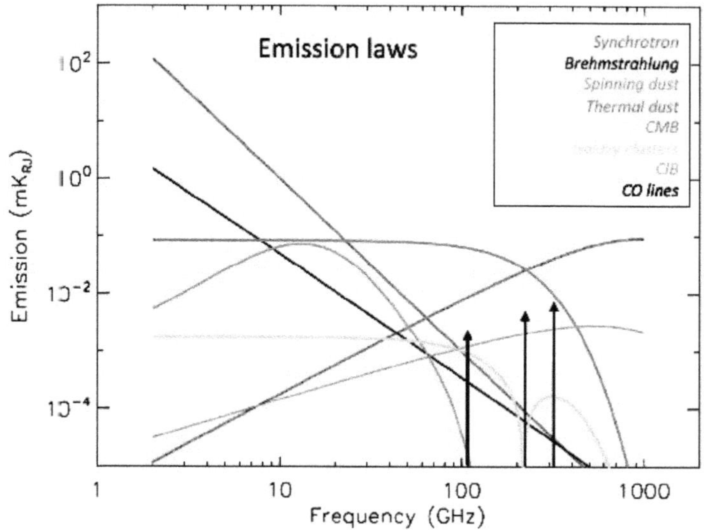

Figure 2: The relative amplitude of CMB fluctuations as compared to various foreground emissions. The amplitude is expressed in term of antenna temperature (therefore a constant value for CMB at low frequency since it is a thermal radiation) as a function of frequency. Some foregrounds dominate at low frequencies such as synchrotron and free-free emission, whereas some others dominate at higher frequencies (thermal emission from dust and the Cosmic Infrared Background). However, and somehow by chance, CMB fluctuations are still dominant in a narrow range of frequencies. In order to perform the most accurate extraction of the CMB map, one observes the sky in a variety of channels, from 30 to 857 GHz, most of which are dominated by foregrounds. This allows to obtain a very accurate map of these, which can then be substracted in the frequencies where the CMB dominates in order to obtain the cleanest CMB map.

nuclei make ordinary matter and raditation behave as a single fluid. Therefore the temperature fluctuations in the photon fluid reflect the density fluctuations in the baryon fluid. After recombination, baryons and photons cease to interact, and baryon then behave as an almost pressureless fluid.

It is well-known that some gravitational instability occurs on large scales when one considers the propagation of sound waves within a medium, the critical wavelength associated to this process being the so-called Jean's length, λ_J, whose value is given by $\lambda_J = c_s\sqrt{\frac{\pi}{G\mu}}$, where c_s is the sound speed within the medium and μ is mass density. A simple interpretation of this result is that, up to some numerical constants, a spherical object of average density μ will have equal content of (positive) thermal energy and (negative) gravitational binding energy if its radius is of order λ_J. Above this radius, it will collapse under its own weight because pressure force are not able to stabilize it. In a non expanding medium, the instability associated to the Jeans length makes overdensities grow exponentially with time, but in an expanding Univers, overdensities can still grow (but only under some conditions), and only linearly with respect to the scale factor a, i.e. the overall relative distance increase between objects because of the expansion of the Universe. Today's CMB temperature is 2.73 K only, a factor 1000 smaller than the temperature at CMB emission[1]. Therefore, one expects that today's fluctuations in the baryon fluid will be 1000 times larger than at recombination. However, when confronting this ansatz to the observations, we find that this is not the case: many lengthscales today have reached the nonlinear regime (i.e., density contrast larger than 1) and therefore, following the above reasonning, were expected to be at least at the 10^{-3} level at recombination, which is not the case since the observed fluctuations are at level 10^{-5}. This result proves that the smallness of the CMB fluctuations necessitates the adjunction of some new physics, the most convincing candidate of which being dark matter. The reason for this is that since (by definition) dark matter does not interact with radiation, its pressure is negligible and consequently it is much more rapidly subject to the Jeans instability than ordinary matter and can have much larger fluctuations than the observed 10^{-5} level of the photon + baryon fluid at recombination. Since the presence of dark matter is, independently, very strongly suggested by the analysis of the galactic rotation curves and even more by the dynamics of the galaxy clusters[2], the existence of dark matter arises as a very natural solution to many astrophyscial problems.

Once this background is settled, let us go back to the physics of CMB before recombination. The combined baryon + radiation fluid has some energy density that is dominated either by the radiation part (at earliest epochs) or matter (at later epochs), as well as some pressure (only contributed by radiation but transferred to baryons as well through Thomson scattering and Coulomb ilteractions). Any inhomogeneities in such a fluid will lead to the propagation of density waves. Overdense regions also contain more photons and therefore are hotter, whereas underdense regions are cooler, which leads to observable patterns in the CMB map. To this simple framework, one has to add the influence of gravitational effects on the wave propagation, which range from the expansion of the universe itself to the presence of dark matter which evolves rather independently from matter and radiation but whose inhomogeneities influence the latter through their gravitational potential. Since the fluctuations are small, all this can be computed by using a linearized version of the General Relativty equations around a homogeneous and isotropic solution, a task

[1]The CMB was emitted when free electrons combined with hydrogen and helium nuclei. A naive estimate of the corresponding temperature would therefore be $T_{\rm rec} \sim 13.6\ {\rm eV}/k_{\rm B}$ ($k_{\rm B}$ being the Bolzmann constant), however, the Universe as a whole is a system with a very large photon-to-baryon ration, which is larger than 10^9. Consequently, what matters at recombination is not the average energy of photons, but the average enery of the tiny (10^{-9}) fraction of photons which still have a chance to interact significantly with atomic nuclei even though the temperature is lower than $13.6\ {\rm eV}/k_{\rm B}$. On can easily show that when this energy is around 13.6 eV, then the temperature is only 3000 K.

[2]In both cases, Kepler third law links the velocity of stars (within galaxies) of galaxies (within clusters) to the total mass of their host. This gravitational mass can then be compared to the baryonic mass deduced from the radiation it emits, once one can modelize the mass-to-light ratio of the system. Whatever structure one observes, there is always an important gap between the luminous, baryonic mass and the gravitational one, the former being systematically lower to the latter. Even if one assumes that some baryons do not radiate much, it does not seem possible to explain the observations without invoking the existence of some new form of matter which does not show any amount of non gravitational interaction with the rest of the Universe (see, e.g., [1].

that is not very complicated technically, but that depends on a moderately large parameter space that has to be explored through Monte-Carlo simulations in order to find the set of parameters that fits the data best.

What is the CMB angular spectrum supposed to look like? Since we are in an expanding universe, physical lengths (and, hence, wavelengths) grow with time. Moreover, there is an important timescale associated with the expansion. It is the Hubble radius, that is the inverse of the expansion rate, which is (in general) proportional to the age of the Universe at the epoch of interest, as well as the size of the observable Universe. General Relativity tells us that the expansion rate is determined by the energy content of the Universe which during most of its history is dominated by radiation (at early times) or matter (at late times). In these two situations, one can show that physical lengths grow more slowly than the Hubble radius. This means that waves are, at sufficiently early time, larger than the then observable Universe before becoming smaller than it. Cosmologist often summarize this by saying that perturbations "enter into the Hubble radius" when their (growing) wavelength becomes smaller than the (growing) Hubble radius. As long as this has not occured, we are in a situation where the perturbations evolve on a timescale larger than the age of the Universe, and therefore have not yet begun to evolve in practice. One therefore may say that at sufficiently early times, any cosmological perturbation is frozen once it has been formed by some yet-to-determine process. Once density waves begin to evolve, they are expected to oscillate with a frequency that is all the more large that the wavelength is small. Therefore, different wavelengths will show a progressive phase shift with time. Finally, since CMB shows a picture of the Universe at recombination, any angular scale in the CMB map can be converted into a physical scale at recombination, and propagating sound waves should translate into an oscillating angular power spectrum, the detailed features of which depends on the detailed physics that is at work there. An outlook of this, as well as a sketch of the approximate dependence of the spectrum with respect to some cosmological parameters is given in Fig. 3.

3 Temperature spectrum and cosmological parameter estimation

After a long and tedious processing of the raw data (which, of course included the foreground emissions), the final processed CMB fluctuation map was obtained from the Planck data. It is shown in Fig. 4 The angular power spectrum can also be extracted from the data (altough in a more subtle way than extracting it from the map itself). The result is shown in Fig. 5. From the temperature data alone, it is possible to constrain the cosmological parameters. The main difficulty we have to face is that we do not know for sure which underlying model has to be confronted with the data. The first step is therefore to determine which minimal model can possibly fit the data. Since the advent of the WMAP data ten years ago, we know that we need at least six parameters in order to interpret the CMB data. These six parameters are:

- **Baryon density.** This is the most obvious parameter. At present, no particle physics theory is able to predict the baryon asymetry which settled the baryon-to-photon ratio at around $n_{\rm b}/n_\gamma \sim 6 \times 10^{-10}$, so that we need to extract this parameter from the data. In practice, we extract the present day average baryon density $\rho_{\rm b}^0$, and, following the usual conventions, it is expressed in term of the baryon density parameter times the squared normalized Hubble constant, i.e. the quantity $\omega_{\rm b} \equiv \Omega_{\rm b} h^2$ where h corresponds to the normalized Hubble constant, i.e. $h = H_0/(100 \text{ km s}^{-1} \text{Mpc}^{-1})$. The conversion factor is then $\rho_{\rm b}^0/(1 \text{ GeV m}^{-3}) \simeq 10.48\omega_{\rm b}$.
- **Dark matter density.** The exact nature of dark matter (axions, WIMPs, or whatever) does not matter from a cosmological point of view. The only property which plays a role in structure formation is the epoch at which it begins to behave as a pressureless fluid. In case this happens when any scale of cosmological interest has not yet begun to evolve, one talks about cold dark matter (in the other case, the conventional name is "hot" or "warm" dark

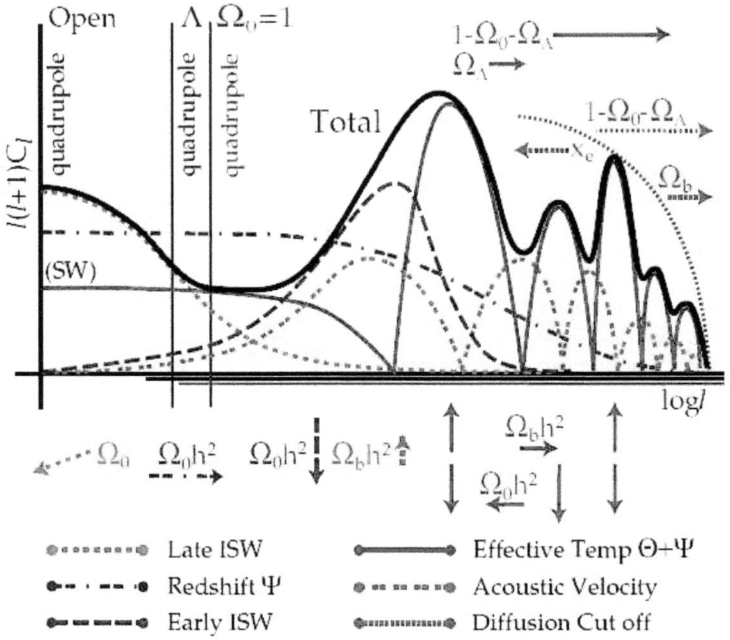

Figure 3: A sketch of what the CMB angular spectrum is expected to look like, and its dependence on some cosmological parameters. To each angular scale corresponds (roughly) a physical scale whose evolution mostly depends on the ratio of its period to the age of the Universe at recombination. The series of peaks we see here corresponds to mode which did not have time to evolve (left of the first peak), which experienced half of an oscillation (first peak; from overdensity to underdensity, or vice-versa), one full oscillation (second peak) and so on. The relative amplitude of the peaks mostly depends on the baryon density ($\Omega_b h^2$) and at a lesser extent the total (dark + baryonic) matter density ($\Omega_0 h^2$). Some damping occurs because of viscosity and heat conduction at small scales (right of the spectrum) and its amplitude is determined by the baryon density as well. Finally, the overall physical to angular scale conversion factor mostly depends on the history of the expansion of the universe and therefore dark energy as well as its geometry (Ω_Λ and $1 - \Omega_0 - \Omega_\Lambda$, respectively). For more details, see Ref. [2] from which this picture was borrowed.

Figure 4: The Planck CMB temperature fluctuation map (left). The Planck resolution is of course much better than the pixel size of the picture shown here, so that there is in fact more information in the real map. As a comparison, the right picture shows the increase in resolution between the historical COBE satellite map (1992), the WMAP data (2003-2009) and Planck. In addition to the obviously much better resolution, the increase in sensitivity and foreground removal is also crucial.

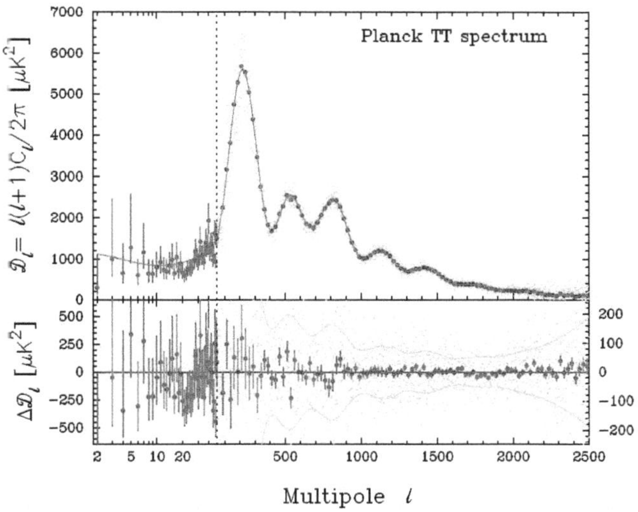

Figure 5: The corresponding Planck CMB temperature fluctuation spectrum. The angular two point correlation function of the map $C(\theta)$ is expanded in term of Legendre polynomial, $C(\theta) = \sum_\ell C_\ell P_\ell(\cos\theta)$, the coefficient of which are shown here (grey dots) up to some factor. The spread in individual measured C_ℓ can be reduced by binning the data (blue circles). The red curve shows the best fit to the data for the six parameter cosmological model described in the text. The break in the curve comes from the fact that the graph shifts from logarithmic to linear horizontal scale at $\ell = 45$. The bottom part of the graph shows the residuals once the best fit is sbstracted from the data. Only the small scale ($\ell > 45$) data are used for parameter extraction as explained later in the text. Figure taken from Ref. [3].

matter depending on the relativistic to non relativistic epoch). Just as for baryonic matter, its present day energy density $\rho_{\rm m}^0$ is parametrized by $\omega_{\rm m}h^2$, with the same conversion factor as above.

- **Dark energy.** In addition to normal and dark matter, the study of the history of the expansion of the Universe in the recent billions of years reveals that an extra component of negative pressure has to be present in the Universe. The reason is that we measure an acceleration of the scale factor with time, i.e. of the distance between any pair of far away galaxies. General Relativity predicts that the second derivative of the scale factor a has to obey an equation like $\ddot{a} \propto -(3P+\rho)$, where P and ρ respectively stand for the total pressure and energy density of the matter-energy content of the Universe. A positive $\ddot{a}$ therefore implies a negative total pressure ($P < -\rho/3$, actually), something which cannot be achieved with radiation, neutrinos, baryons or even dark matter. The extra component responsible for this negative pressure is conventionally named "dark energy". It could be vacuum energy (constant energy density with exactly opposite pressure) or something more complicated.

- **Reionization.** Today, no star whose composition can be determined through spectroscopy shows a total absence of metal, i.e., anything else that hydrogen and helium. This is at odds with the expected content of the first stars which originated from matter that had been unprocessed since the Big Bang Nucleosynthesis and that should therefore be devoid of any metal. This is not a surprise, however, since it has been known for long that zero metallicity stars are far less opaque to radiation and therefore much more short lived than their present-day counterparts. Even though these short lived objects have never been seen, their presence is known indirectly by the fact that spectroscopy reveals that most matter is in ionized form one billion years after the Big Bang, a result that can be explained only if one assumes that some intense UV radiation has been emitted before then, just as one expects to be the case from these primordial stars which, being less opaque, could burn their fuel at a much higher rate and hence had a much larger surface temperature. If matter was reionized sufficiently early, which therefore amounts to say that the first generation of stars formed quite early as well, then the free electron density increase can possibly rescatter once or twice a fraction of the CMB photons and alter the primordial CMB map.

- **Initial power spectrum of density fluctuations (2 parameters).** Cosmological perturbations naturally evolve from a set of initial conditions, but cannot arise from nothing. For example, Poissonian fluctuations in the baryon or photon number are far, far too small to explain the amplitude of the observed temperature fluctuations at recombination. We therefore need to a starting point for the conventional evolution of the cosmological perturbations that already contains fluctuations, and we therefore need to assume some yet unknown phenomenon that produced them. Observations suggest however that, at least approximatively, this "initial" spectrum is close to be a power law, that need (at least) two parameters to be described: an amplitude at some fixed scale and a spectral index. These are the parameters that can be extracted from the data. Then, having a class of early Universe scenarios that could produce cosmological perturbation during a very early epoch, these parameters can possibly be related to the true physical parameters of the underlying theory that was at work then.

A fit to the Planck data was made in the framework of this six parameter model, the results of which are given in Fig. 6.

Then, some extension of this parameter space was attempted by extending the model in various ways such as:

- **Non zero curvature.** Any early Universe scenario has to address the question of why is the Universe so homogeneous and isotropic. Whatever the detailed explanation, it usually predicts that the spacial curvature of the Universe (i.e. the inverse square of the scale beyond which the Euclidean laws of egometry cease to be valid) had to be negligible. This can be checked within the data, mostly for consistency purpose.

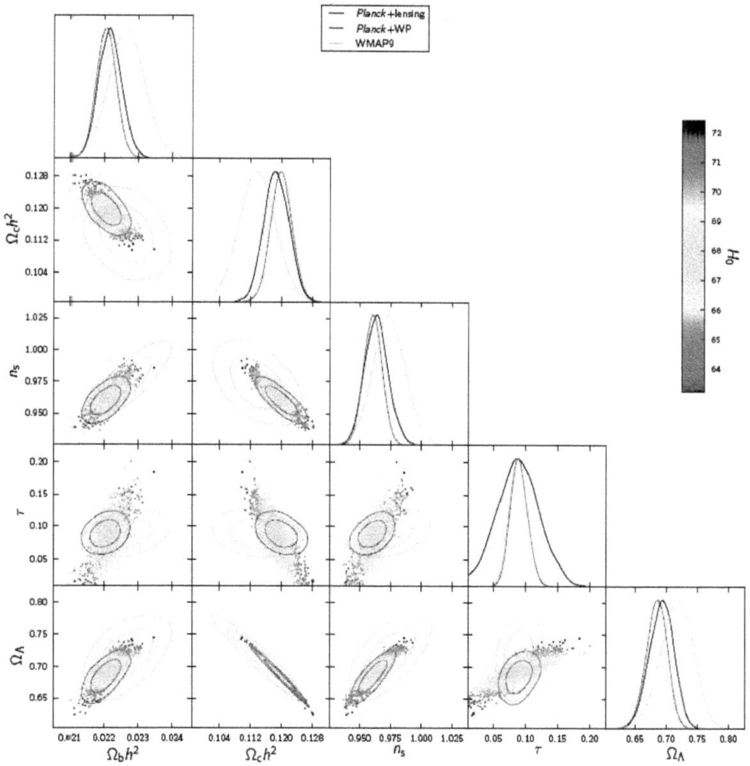

Figure 6: The one- and two-dimensional likelihoods for five of the six parameters of the so-called "concordance model" outlined by the WMAP data and more accurately constrained by the Planck data. From top to bottom are shown the estimation for the baryon density ($\Omega_b h^2$), the dark matter density ($\Omega_m h^2$), the initial spectrum spectral index (n_S), the reionization optical depth (τ) and the dark energy fraction (Ω_Λ). The one dimensional likelihood are on the diagonal and the two-dimensional likelihood under it. The constraints from Planck data alone (including lensing) are in black, those from Planck (without lensing) + WMAP polarization in red. The previous WMAP data are in grey. In most case, the Planck alone constraints are significantly better than the WMAP data, at the exception of the reionization constraints, an unsurprising result since reionization is mostly constrained by the large scale polarization data which are not yet public for Planck. The expansion rate of the Universe, the so-called Hubble constant is not estimated directly from the process, but comes trivially from the formula $H \propto \sqrt{(\Omega_b h^2 + \Omega_c h^2)/(1 - \Omega_\Lambda)}$. The corresponding value of the Hubble constant is shown in the two-dimensional likelihood plots through color codes. Adding some external constraints on it can possibly shift the best fit. The two dimensional plots show some of the degeneracies between parameters. Some parameters are not very well estimated because of these, whereas some combinations of them are exquisitely well constrained. The only parameter that is not shown is the amplitude of the perturbations at some fixed scale, which is not really important in practice. Figure taken from Ref. [3].

- **Non trivial dark energy properties.** Dark energy is usually supposed to behave as a cosmological constant, or, equivalently, as vacuum energy within constant energy density and opposite pressure. However, more complicated dark energy models can be considered, where dark energy is made of some scalar field with negative but dynamical pressure, which is not exactly opposite to the energy density.
- **More complicated description of initial conditions.** It is clear that an amplitude (at some scale) and a spectral index are the minimum ingredients to the description of initial conditions, but more realistic models usually predict deviation from this simple law. Most notably, the spectral index might depend on the scale, and in addition to density perturbations, there may exist primordial gravitational waves which may leave an imprint on the CMB.
- **Variation of fundamental constants.** Laboratory experiments can put very stringent constraints on the variation of fundamental constants, but only during a limited time span (rarely more than one year – at most a decade). On the contrary, astrophysical data can scan larger timescales, although with less precision in the constraints (or model-dependent constraints). Recombination era is mostly determined by Thomson scattering which depends on the fine structure constant. Also, the efficiency of the baryon drag by photons depends on the electron-to-proton mass ratio. Moreover, any epoch of structure formation depends on the formulation of the gravitation theory, which can be more complicated that General Relativity. All these possibilities can, at some extent, be tested or constrained by the CMB data.

The initial six parameter model was not really in difficulty when confronted to the new Planck data, so that adding one extra parameter did not improve anything. Only upper limits one these extra parameters could be put with Planck, however, this situation might evolve when the polarization data are included in the analysis next year.

4 Some crosschecks

The Planck data very strongly strengthen the already popular six parameter "concordance model". However, the now standard model can be even more strengthened by including other data coming from the Planck satellite.

4.1 Polarization

It is known that Thomson scattering generates some polarization. If one considers the scattering of a plane wave, then the outgoing radiation is linearly polarized perpendicularily to the scattering plane. Consequently, if one considers scattering at right angle from two orthogonal direction (horizontal and vertical, see Fig. 7), then if there is more radiation coming along the two horizontal directions than along the two vertical directions, then the scattering radiation seen by the observer will acquire a net vertical linear polarization. In other word a quadrupole anisotropy in the photon distribution function at scattering translates into polarized scattered radiation. Now, if one starts from only density fluctuations in the photon distribution, then it is pretty obvious that density gradients will generate dipolar fluctuations in the distribution function and that subsequently dipole gradients will generate quadrupolar fluctuations (and so on). This shows that in addition to the temperature anisotropy, one expects to have polarization anisotropies which will be partially correlated with the temperature ones. In the end, one should observe two extra spectra that have to be confronted to the cosmological models one wants to test: the polarization angular spectrum and the cross correlation between temperature and polarization. At present, the polarization data are not ready, but the Planck collaboration have made public the present state of the data processing, which is given in Fig. 8. The blue points show the likely polarization data (on small scales, since the large scale polarization data are more foregound dependent and more difficult

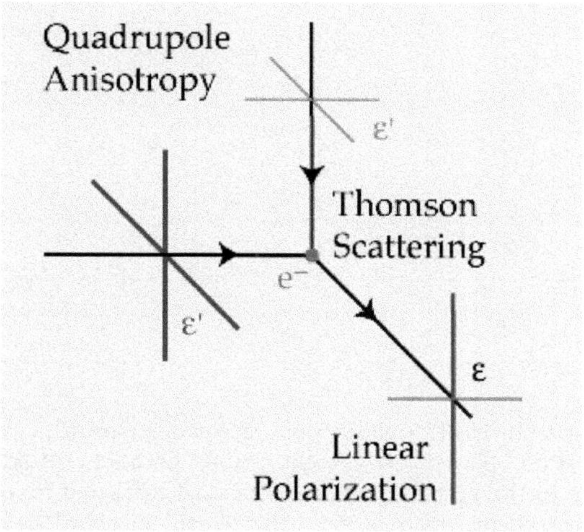

Figure 7: A simple sketch showing how an anisotropic radiation becomes linearly polarized because of Thomson scatteing. Considering radiation that is only scattered at 90 degrees, if the incoming radiation along the horizontal direction of graph has a different flux than the one that travels along the vertical direction before scattering, then the observed radiation gets more linear polarization in one direction than in the others abnd therefore exhibits some net linear polarization. The polarization can be mapped throughout the microwave sky and its power spectrum can be computed, just as for temperature. The cross-correlation between temperature and polarization can also be measured.

to process), and the red curve show the best fit obtained *from the temperature data only*. The impressive agreement between the two is a very convincing proof that no serious problem is likely to occur in the interpretation of the data. However, it has to be noted that it is very difficult to find distinct models that share the same temperature spectra but different polarization spectra, so that the agreement shown on Fig. 8 is mostly a check that the foreground removal procedure (and hence much of the foreground modelling) is very satisfactory.

4.2 Gravitational lensing

Probably the most impressinve crosscheck that can be done in the Planck data comes from gravitational lensing. i.e. the deflection of light produced by the foreground matter distribution fluctuations. The main effect of gravitational lensing on CMB (at the scale observed by Planck) is to distort the "true" map of the CMB fluctuations.This distortion does not induce loss of information as in the case of turbulence, because the distortion does not increase when one goes to smaller scales. However, an initially Gaussian distribution will exhibit non Gaussian features because of this distortion, a very fortunate thing since those can be identified through the connected four-point correlation function of the CMB fluctuations. Although the lensing signature had already been found at more than 5σ in previous CMB observations starting from the Atacama Cosmology Telescope in 2011 [4], it is the first time that the data are sufficiently precise so that the observed CMB map can be delensed: the lensing is detected here at more than 25σ. This makes possible to deduced the distortion field produced by lensing and hence map the projected gravitational potential of all the structures between the last scattering surface and us. In practice, this reconstruction is difficult to do very accurately, and the gravitational potential maps that are shown in Fig. 10 have in fact a rather low signal-to-noise ratio (around 0.6 per individual mode), however, what one is in fact interested in is the correlation functions associated to the map, and

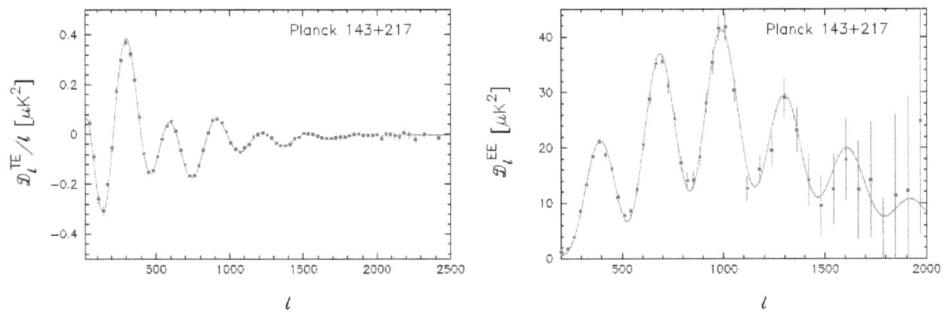

Figure 8: Present (March 2013) polarization anisotropy spectrum and the temperature-polarization cross spectrum. The dots represents still-to-be improved data points, whereas the solid curve corresponds to the best-fit cosmological model extracted from the temperature data only. The agreement is excellent, but at present, this mostly tells that the polarization foreground removal technique is accurate. How much the polarization data will improve the parameter estimation remains undetermined at the moment.

it is possible to construct the projected potential two-point correlation function. Such a function can then be compared to the one that is predicted by the best-fit model of the temperature data. This is shown in Fig. 9, and even if noisy, the agreement between the data and the predicted curve is excellent. More stringent cross checks can be done with lensing as the projected gravitational potential maps are expected to show correlation with galaxy catalogs in the regions where they overlap. This is indeed what is found between 7 and 20σ level when considering the NVSS quasar catalogue (at 20σ), the SDSS luminous red galaxies catalogue (at 10σ), and the MaxBCG SDSS cluster catalogue and the WISE infrared satellite catalogue (at 7σ).

4.3 Baryon content and nucleosynthesis

One of the most impressive result of CMB study is that it allows to recover the baryonic content of the Universe independently of the predictions of the Big Bang Nucleosynthesis (BBN). During the Big Bang, weak equilibrium breaks down as the Universe cools down. This fixes the amount of available neutrons when the temperature then drops to 0.1 MeV so that nucleosynthesis can begin. With the CMB, we see a coupled baryon + photon fluid whose hydrodynamical properties allow to infer the relative amount of baryons with respect to photons. In other words, cosmology allows to measure the baryon content of the Universe either by doing some nuclear physics at $T \sim 0.1$ MeV or by studying the propagation of sound waves at a temperature of a few thousand Kelvins only. These measurements are made by completely different observations, instruments, teams, etc. The systematics, uncertainties are also completely unrelated, but still both methods give compatible results, with a nucleus-to-photon ratio of order of 6×10^{-10} and an uncertainty that is now of only a few percents.

However, the CMB parameter extraction necessitates to know the separate amounts of hydrogen and helium since recombination history is a three-step process, with a first electron combining to helium nuclei, then the second one and soon after hydrogen recombination. Usually parameter estimation is made by fitting the total baryon content of the Universe to CMB data assuming that the hydrogen-to-helium ratio is given by the predictions of standard nucleosynthesis as a function of the total baryon density. But the Planck CMB data are so precise that the *separate* imprint of hydrogen and helium can now be recovered directly from the data, so that the hydrogen-to-helium ratio can be estimated independently of the BBN prediction. And, again, we recover a perfect agreement between CMB and nucleosynthesis predictions.

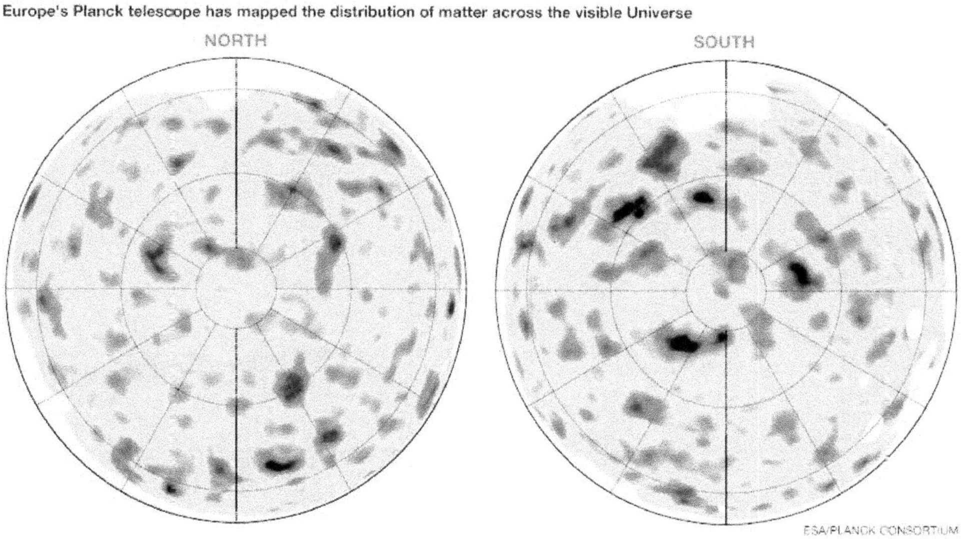

Figure 9: From the non Gaussinaities induced by gravitational lensing, one can attempt to reconstruct the projected gravitational potential on the celestial sphere, as shown here for the north and south Galactic pole regions. The lensing reconstruction suffers from large uncertainties, so that the maps shown here are compatible with the observed lensing signal, but very noisy. The SNR for individual mode is 0.6, so that there is in fact more noise than signal here.

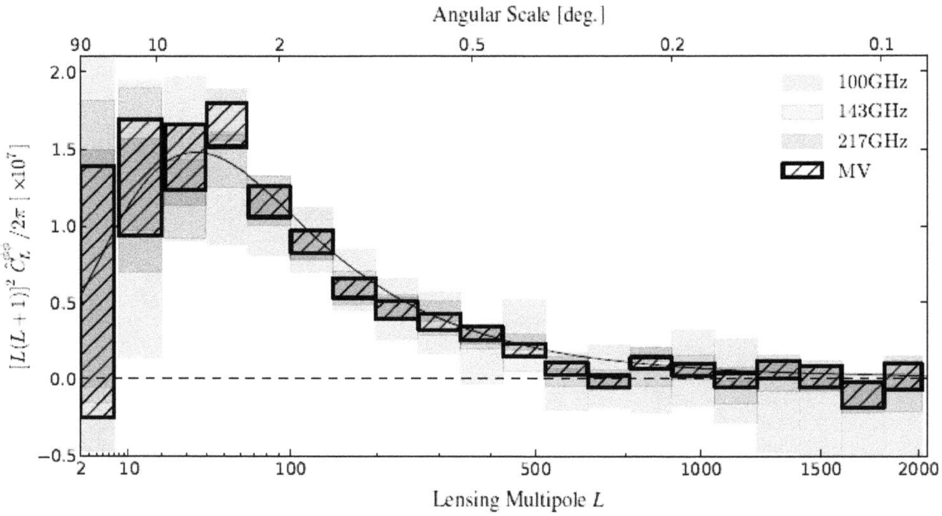

Figure 10: From the (noisy) reconstructed projected gravitational potential, one can reconsctruct the associated two-point correlation function. The procedure can be done in several ways (using different frequency maps, or a combination of them. The hashed bins correspond the best combination. The solid curve corresponds to the expectation of the best-fit model found without using the lensing data, so that the agreement between the curve and the bins is in fact a very powerful crosscheck of both the relevance of the cosmological model and the data processing pipeline.

4.4 Curvature

In the six parameter model estimation, we have explicitely assumed that the spacelike sections of the Universe were Euclidean, i.e., "flat" according to cosmologists' ambiguous vocabulary. This is justified by the fact that most early universe models that generate large scale coherent fluctuations such as the ones that are observed also contribute to erase any departure from Euclidean spacelike section, and this is indeed one of the first prediction that was made from the inflationary paradigm [5], but this has to be checked at some point. From the temperature spectrum alone this difficult as spatial curvature almost only affects the angular scale under which the CMB features are seen, just as dark energy does, so that the two parameters are degenerate and only a combination of the two can be extracted from the data. In order to remove this degeneracy, one has to combine the temperature spectrum with some other data such as the Hubble parameter constraints of the supernova data, however both are shown to be in (very) slight tension with the CMB data. It is therefore better to perform some internal checks of the flatness of space, which can be (again!) done with lensing. Indeed the amplitude of the distortion map is directly linked to and angular scale at which a given physical length appears when seen from some redshift, and this angular size vs. redshift relation depends on the dark energy + curvature content of the universe, which allows to break the above mentioned degeneracy.

From the Friedmann equation, one can define the density parameter Ω_X of any form of matter X by its ratio with respect to the critical density, $\rho_{\rm crit} = 3H^2c^2/8\pi G$. Then, the departure from 1 of the sum of all the Ω's, $\Omega_K \equiv \sum_X \Omega_x - 1$ can be put in correspondance with the local curvature radius $R_{\rm curv}$ through the formula $R_{\rm curv} = c/H\sqrt{|\Omega_K|}$. The Planck 1σ constraint on Ω_K here transform from $\Omega_K = -0.042^{+0.027}_{-0.018}$ without lensing data to $\Omega_K = -0.0096^{+0.010}_{-0.0082}$ including them. Taking a realistic value of 65 $\rm km\,s^{-1}\,Mpc^{-1}$ for the Hubble parameter, the curvature radius is bound to be larger than more than 100 billion lightyears. Restraining ourselves from considering Planck data alone, the curvature can be further constrained by adding the baryon acoustic oscilation (BAO; they correspond to featured that are observed in the present three-dimensional distribution of galaxies) data, which also allow to extract some even more accurate information on the angular size vs. redshift relation at lower redshift. When doing so, the 2σ (instead of previously 1σ) constraint on Ω_K reaches $\Omega_K = -0.001^{+0.0062}_{-0.0065}$, so that the curvature radius is then bound to be at least 3.7 times larger than the radius of the observable Universe at the 2σ level.

5 Search for neutrino properties

One of the hottest topic in CMB study was what could be deduced about the energy density of neutrinos. The motivation for ths originates from the so-called reactor anomaly, which is a series of observations which suggests a third neutrino oscillation that is not compatible with he two other, well-established, oscillations, the atmospheric neutrino oscillation and the Solar neutrino oscillation. The inconsistency of the reactor anomaly comes from the fact that the squared mass difference it predicts is neither compatible with the sum or the diffrence of the two other squared mass diffences found in the two well-established oscillations, hence cannot corespond to a situation with only the three known neutrinos. The reactor anomaly remains somewhat controversial and needs future experiments to be settled in either way. Before these, cosmology can give some answers by measuring independently the ultra relativistic matter energy density at recombination. In standard cosmology, there are only three species of neutrinos. As long as the temperature is above 1 MeV, weak equilibrium is established and both neutrinos and photons share the same temperature. Below this temperature, the coupling does no longer occur and neutrinos and photons evolve independently from each others. The neutrinos temperature keeps the $1/a$ scaling due to the expansion of the Universe, just as the photons do, except during the electron positron annihilation that occur around 0.5 MeV where some energy injection occurs in the photon fluid. In order to estimate this increase of temperature with respect to the usual $1/a$ behaviour, we remind that since there are (almost) an equal number of electrons and positrons, their chemical potentiel is

0, so that according to the first law of thermodynamics, the photon fluid evolves according to $\mathrm{d}U = -P\mathrm{d}V + T\mathrm{d}S$, just as for the photon fluid. However the "conservation law" for the common stress energy tensor of the photons + electrons + positrons fluid implies that for this system, one has $\dot{\rho} = -3H(P+\rho)$, where ρ is the energy density, which corresponds to the thermodynamic constraint $\mathrm{d}U = -P\mathrm{d}V$ once we note that $U = \rho V$ and that $3H$ is the relative time variation of a comoving volume, i.e. $3H = \dot{V}/V$. Consequently, the energy injection of the electron positron annihilation into the photon fluid is adiabatic, and the entropy S is conserved, or, equivalently, the entropy density $s \equiv S/V$ scales as $1/a^3 \propto T_\nu^3$, so that the quantity s/T_ν^3 is constant. Now, since U, V and S are extensive variables, the law $\mathrm{d}U = -P\mathrm{d}V + T\mathrm{d}S$ translates into $U = -PV + TS$ or, when dividing by V, $s = (\rho + P)/T$. As well known, the energy density for a relativistic fluid writes $\rho = 2g\sigma T^4/c$, where g is the number of fermionic or bosonic degrees of freedom and σ the Stefan-Boltzmann constant. Long before the electron positron annihilation, the quantity s/T_ν^3 is therefore given by $s/T_\nu^3 = (g_{e^+e^-} + g_\gamma)2\sigma T_\gamma^3/cT_\nu^3 = (g_{e^+e^-} + g_\gamma)2\sigma/c$. Long after the annihilation, only the photon fluid exists, so that one has $s/T_\nu^3 = g_\gamma 2\sigma T_\gamma^3/cT_\nu^3$, so that the temperature ratio between photons and neutrinos is given by

$$\left(\frac{T_\gamma}{T_\nu}\right)^3 = \frac{g_{e^+e^-} + g_\gamma}{g_\gamma} = \frac{\frac{7}{8} \times 2 \times 2 + 2}{2} = \frac{11}{4}. \tag{1}$$

The energy density ratio between photons and neutrinos is therefore

$$\frac{\rho_\gamma}{\rho_\nu} = \frac{g_\gamma}{g_\nu}\left(\frac{T_\gamma}{T_\nu}\right)^4 = \frac{2}{3 \times \frac{7}{8} \times 1 \times 2}\left(\frac{11}{4}\right)^{\frac{4}{3}} = \frac{22}{21}\left(\frac{11}{4}\right)^{\frac{1}{3}} \simeq \frac{1}{0.68132...}. \tag{2}$$

In practice however, the electron positron occurs too close to weak equilibrium breaking so that part of the e^+e^- annihilations end up in injecting energy in the neutrino fluid, which, for cosmological purposes, amount to claim that the neutrino to photon temperature is still given by the $(4/11)^{\frac{1}{3}}$ factor, but that the "number" of neutrinos species is larger than 3. In the cosmological literature, one thus parametrizes the neutrino energy density by the "effective number" N_{eff} given by

$$\frac{\rho_\nu}{\rho_\gamma} = \frac{N_{\text{eff}}}{3} \times 0.68132... \tag{3}$$

Detailed weak interaction calculations then show that in the framework of the standard model, one has $N_{\text{eff}} \simeq 3.046$ [6].

Finally, CMB data are weakly dependent on neutrino number or masses. Since the well established squared mass differences that have been found are fairly small, it is safe to consider them as equal for CMB studies, so that in what follows, we shall focus on three equal mass neutrinos (the Standard Model ones), plus possibly something else. This explains the choice of the four analyses that were performed beyond the standard three massless (or negligible mass) neutrino hypothesis:

1. An arbitrary N_{eff} number of massless neutrinos
2. Three neutrinos ($N_{\text{eff}} = 3.046$) of arbitrary equal mass
3. The mixture of these two cases, with an arbitrary number of equal mass neutrinos
4. $N_{\text{eff}} = 3.046$ massless neutrinos, plus another set of massive neutrinos of arbitrary (non integer N_{eff}) amount.

The effect of neutrinos on CMB spectrum is well-known and has deserved significant scrutiny in the recent year because of the insights that CMB could bring to the reacor anomaly puzzle. What CMB constrains well is the matter to radiation ratio at recombination. Therefore, adding extra radiation necessitates to have more matter at recombination, and from Friedmann equations, higher Hubble parameter. This reduces the age of the universe at recombination, which mostly affects the small scale integrated effect of photon + baryon fluid evolution, therefore it is important to add the small scale CMB data beyond Planck in order to address this issue as accurately as possible. Also, polarization helps in reducing the degeneracies between the effect of $N_{\rm eff}$ and other cosmoogical parameters, therefore the minimum combination of data that has to be consider is Planck + WMAP polarization + other ground based CMB observations at higher angular resolultion (hereafter Planck+WP+HighL).

The first case already leads to mixed results. If one takes the Planck data only, then the likelihood function peaks somewhere between $N_{\rm eff} = 3$ and $N_{\rm eff} = 4$, with the 2σ limits at

$$N_{\rm eff} = 3.36^{+0.68}_{-0.64} \qquad \text{Planck} + \text{WP} + \text{HighL}$$

This constraint puts the standard $N_{\rm eff} = 3.046$ at around 1σ to the preferred value, but the $N_{\rm eff} = 4$ is disfavoured at almost 2σ. Constraining better the Hubble parameter should allow to decrease these bounds. Adding the BAO data allows to reduce this uncertainty as the baryon acoustic oscillations are sensitive to some angular scale that depend of the late time value of the Hubble constant. Since BAO data are consistent with Planck data, one expects them of contribute to narrow the constraints on $N_{\rm eff}$, and this is indeed the case, as one obtains $N_{\rm eff} = 3.30^{+0.54}_{-0.51}$, which prefers a little bit more significantly the $N_{\rm eff} = 3.046$ case over the $N_{\rm eff} = 4$ one, which is now exluded at more than 2σ. However this conclusion differs significantly if one instead adds the external Hubble parameter measurements. The reason is that those prefer higher value of this parameter, they accomodate with a higher total radiation + matter content at recombination with a fixed ratio between the two. One obtains here $N_{\rm eff} = 3.62^{+0.50}_{-0.48}$, which now prefer the $N_{\rm eff} = 4$ case over the $N_{\rm eff} = 3.046$ one. Adding the BAO constraints to the Hubble parameter puts the two competing cases on almost equal grounds, with $N_{\rm eff} = 3.52^{+0.48}_{-0.45}$.

Considering now the combined mass of the neutrinos in a $N_{\rm eff} = 3.046$ situation (case 2 above), one observes that the constraints somewhat depend on the datasets one considers, although in a far less dramatic way. By taking the already mention Planck+WP+highL combination, one obtains the constraint

$$\sum_{m_\nu} < 0.66 \text{ eV} \qquad \text{Planck} + \text{WP} + \text{HighL}$$

at the 2σ level. Such constraint are found by using various other combinations, that are more on less stringent, ranging from 0.23 eV if one includes the BAO measurements to more than 1 eV. Although impressive, these results remain inconclusive in order to get some insight on the neutrino masses, since this figure is more than one order of magnitude above the neutrino oscillation constraints (which say that the total neutrino mass has to be larger than 0.06 eV).

Combining the two former assumptions (case 3 above) allow to put simultaneous constraints on $N_{\rm eff}$ and $\sum m_\nu$. The effect of massive neutrinos (which behave as non relativistic matter at late times) is quite different from that of increasing the neutrino number (which only increases the radiation content at early times), so that we do not expect both effects to be degenerate and hence that the simultaneous contraints will differ much from the seperate one. Indeed, by taking the base combination of data, the simultaneous constraints are

$$\left.\begin{array}{l} N_{\rm eff} = 3.29^{+0.67}_{-0.64} \\ \sum m_\nu < 0.60 \text{ eV} \end{array}\right\} \quad \text{Planck} + \text{WP} + \text{HighL}$$

Adding the BAO data slightly improve the constraint to $N_{\rm eff} = 3.32^{+0.54}_{-0.52}$ and $\sum m_\nu < 0.28$ eV at 2σ.

Considering now a massive sterile neutrino of arbitrary abundance in addition to the three ordinary ones (for which one takes $\sum m_\nu = 0.06$ eV in order to realistically account for the oscillation data), one has to distinguish between the actual sterile neutrino mass and what can be called its effective mass. The sterile neutrino becomes non relativistic at the time its temperature $T_{\rm s}$ drops below its mass. But since we deal with a sterile neutrino that might be have decoupled since a long time from the rest of the known particles, $T_{\rm s}$ can possibly be different (and lower) from that of the known neutrinos, just as the normal neutrino have a lower temperature than the CMB. If this sterile neutrino has a thermal distribution, we therefore define the effective sterile neutrino mass $m_{\rm s}^{\rm eff}$ as a function of its real mass in the thermal hypothesis, $m_{\rm s}^{\rm ther}$ by $m_{\rm s}^{\rm eff} = m_{\rm s}^{\rm ther}(T_{\rm s}/T_\nu)^3$.

In practice, it is the effective mass that is contrained, the real mass being somewhat arbitrary, at least as long it is smaller than 10 eV since above this mass, a particle behaves like ordinary cold dark matter and therefore cannot be distinguished from it. The constraint on obtains here is therefore split into two domains. One first allows low $N_{\text{eff}} - 3$ with almost arbitrary $m_{\text{s}}^{\text{eff}}$ since such particles would behave (from CMB parameter extraction point of view) like cold dark matter. Second, and more interestingly, one can put the set of constraint

$$\left.\begin{array}{l} N_{\text{eff}} < 3.91 \\ m_{\text{s}}^{\text{eff}} < 0.59 \text{ eV} \end{array}\right\} \quad \text{Planck} + \text{WP} + \text{HighL}$$

and, as earlier, adding the BAO data only slightly improves the constraint to $N_{\text{eff}} < 3.80$ and $m_{\text{s}}^{\text{eff}} < 0.42$ eV at 2σ. In this respect the most naive, straightforward interpretation of the reactor anomaly would probably be $N_{\text{eff}} \sim 4$ and $m_{\text{s}}^{\text{eff}} = m_{\text{s}}^{\text{ther}} \sim 1$ eV, which is significantly away from the 2σ region.

6 What can Planck could say about inflation

Presumably, the most interesting part of the Planck results, at least from a particle physicist point of view, deals with the contraints on the initial spectrum of cosmological perturbations, from which one can infer constraints on the physical process that produced them. At present, the most favored model for generating the first seeds of cosmological perturbations is (by far) inflation. It relies on the hypothesis that one (or more) scalar fields drove the expansion of the early Universe some time during which the field kinetic term was negligible as compared to the potential term so that it behaved close to a cosmological constant, resulting by an almost exponential expansion phase. One can show that this process naturally explains the homogeneity of the Universe, but also that it freezes toward a classical level the quantum fluctuations of the field itself, which are therefore the seeds of the cosmological pertubations that evolve later on. From this rather general setup, one can build a very large number of inflationary scenario, by varying the field potential and/or by adding some extra fields. If one focuses on single field inflation, then the physical parameter is the inflation potential $V(\phi)$, and the three main predictions of inflation (which only marginally constrain the shape of $V(\phi)$ are:

- Negligibly small spacelike curvature;
- Almost "scale invariant" spectrum of the fluctuations, i.e., with the astrophycists' definition, the spectral index of the fluctuations, n_{s} is expected to be very close but smaller than 1
- (Almost) Gaussian fluctuations.

To this, one may add that one may have primordial gravitational waves, but at a level lower than the density fluctuations. The Gaussianity of the fluctuation is strongly related to single field inflation. With Planck sensitivity, non Gaussianities coming from multi-field inflation were expected to be detectable. Finally, the order of magnitude of the energy density during inflation is related to the amplitude of the density perturbation, and since the latter are of order of 10^{-5}, the energy scale is very likely to be within the range 10^{14}–10^{16} GeV (the precise value depending on the amount of gravitational waves, i.e. the value of the r parameter — see below.

The results from the Planck data analysis confirmed these three predictions. As already said, no spacelike curvature is hinted by the data, which makes the inflationary ansatz relevant to the CMB interpretation. In this framework, no non Gaussianities are seen in the data, which strongly favors single field inflation among the complete inflationary zoology. Finally, and for the first time, deviation from the $n_{\text{s}} = 1$ value is detected for the first time at 5σ level. This conclusion remains robust even is we allow some non zero amount of gravitational waves, and the constraints in the n_{s}–r plane (where r characterizes the gravitational waves-to density fluctuation power spectrum ratio at some scale) are summarized in Fig. 11. In this graph, the predictions associated to some specific inflationary scenario can be drawn, and their consistency with respect

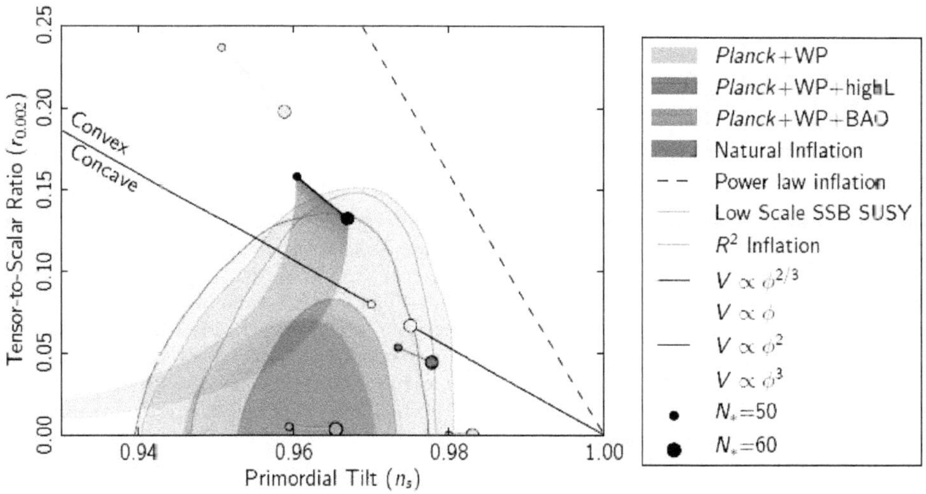

Figure 11: A sketch of the constraint derived from Planck plus other data on the two most important inflationary parameters, the density perturbations spectral index n_s and the gravitational wave to density perturbation power spectrum ratio r. In the six parameter concordance model does not include r as a parameter, and adding it results in interesting upper limits on it. On the contrary, the concordance model already detects a 5σ deviation of n_s from 1, a result that persists when one includes the r parameter in the data. As explained in the text, a given inflationary model can be put somewhere in the $n_s - r$ plane, the exact position depending on the epoch of end of inflation, a given model therefore lying along some short segment. Some simple inflationary models are ruled out, such as the $V \propto \phi^3$ potential (several σ's) and the $V \propto \phi^4$ one (not shown as it is outside the graph). Only models with a sufficently low r are of interest now. Note that in practice, this procedure does not recover/constrain the full inflationary potential $V(\phi)$, but rather its shape on the limited range of ϕ which correspond to the values of the inflaton field when the cosmological perturbations we observe were generated.

to the Planck data can be checked. For example, an important result is that a power law $V \propto \phi^n$ potential for the inflationary field is excluded at 2σ for $n = 2$ and many more sigmas for larger values of n. Actually, locally concave rather than convex potentials are preferred by the data, which means that the potential is more complicated than a simple power law form since it must have some minimum and be locally convex for some value of ϕ (different from the one it was when it produced the fluctations we oserve in the CMB sky). Whether or not it is possible to perform a full reconstruction of the inflationary potential is something that cannot be decided today. In any case, even if the Planck data are not precise enough to pinpoint a very limited region of the n_S–r plane, they are already able to say something about the inflationary potential, and therefore to say something about some process that probably occurred when the temperature of the Universe was around 10^{14} or even 10^{16} GeV. More details on the selection that Planck performs among inflationary models can be found in Ref. [7].

7 Various other results

7.1 Did Planck do some indirect detection of dark matter?

A poorly advertised result of the Planck mission is the discovery/confirmation of the so-called "microwave haze". This radiation was hinted from the WMAP observations back to 2004 [9] and was then received with skepticism[3]. It consists in a synchrotron-like emission that is seen on low frequencies (mostly 30 and 44 GHz), but whose spectrum is harder[4] than that of a typical synchrotron radition, and softer than free-free emission. The WMAP claim became more seriously considered in 2009 when the Fermi satellite discovered the so-called "Fermi-bubbles", which are a large area of high energy (GeV range) emission arising around the Galactic center. The microwave haze could then be naturally explained as the synchrotron radiation emitted by charged particles associated with the high energy photons detected by Fermi, however the association remained to be proven, and this is what Planck could do, by benefiting from its better sensitivity and foreground control [10]. The detailed analysis of the Planck data did justify some of the hidden assumption made in the WMAP analysis, and exclude some of possible the caveats that had been put forward against the existence of the microwave haze. The importance of the haze-Fermi bubble association from a particle physics point of view arises from the fact that it could be the consequence of dark matter coannihilation in the Galactic center region. In their WIMP interpretation, dark matter particles are likely to coannihilate, produing high energy photons, as well as particle-antiparticle pairs which will then travel through our Galaxy magnetic field and then radiate synchrotron radiation. As an exemple, TeV range dark matter particle coannihilation could possibly produce the desired Fermi radiation as well as the observed microwave haze, provided, for the latter, that the Galactic magnetic field has a value of order 5 μG [5]. However, the dark matter coannihilation hypothesis is only a possibility among others as more conventional astrophysical processes could produce it [11]. These include enhanced star formation/supernova rate in the Galactic center, second order Fermi acceleration, galactic wind and active galatic nucleus-like activity from the cental black hole. The Fermi emission remains somewhat puzzling at this stage. First, the emission flux is very flat in the sense that it is very uniform along the whole area of the sky where it is seen, and it has also a sharp cutoff at its edges. This is a surprising result, because if the GeV photons arise from a volume emission, then there should be a brightening in the center and a limb darkening, the latter not being present in the data. If, on the contrary, they are due to a surface emission (related to some shock-wave, for example), then one should see some limb-brightening, something which is not seen either. From what we understand of the different astrophysical explanations, some seem already ruled out because they should be accompanied by various other counterparts that are not seen: for example, galactic winds should produce an Hα emission which

[3] An instructive personal account of this can be found in a blog post [8] by the author of Ref. [9]

[4] i.e., the ratio of the former to the latter increases with frequency.

[5] We apologize for spreading the astrophyscist obsolete habit of using Gauss and other weird esu units instead of Teslas and the rest...

is not seen here. Still, none of the proposed explanation seems able to explained the flatnes of the emission combined with the sharpness of its cutoff. For example, AGN hypothesis should produce volume emissivity and therefore limb darkening. This leaves, at the moment of writing of the se proceedings, the possibility that the dark matter cannihilation is valid, and in any case the Planck data regarding the synchrotron counterpart will play a role in setting the answer.

7.2 Negative results of various exotic searches

We gave a detailed account of what Planck could say about an extension of the six parameter concordance model in the neutrino sector, but this is far from being the only extension that has been considered and for which one obtained upper limits.

For example, the detection of cosmic strings in the Planck temperature map has been considered. Cosmic strings are objects belonging to the general class of topological defects, that can be formed during a symmetry breaking phase transition. String-like objects are known to exist in condensed matter, such as the magnetic flux tube in a superconductor of the vorticity tubes in a superfluid. Regarding the cosmological context, if during some phase transition one breaks an U(1) symmetry, then one can form extended string-like objects, which are either closed loops or infinite strings. Under quite general conditions, these objects can intercommute when they cross, that is, they can exchange their ends (see Figure 12). This important property makes that the number of long strings (those that are either infinite or which form closed loops larger than the size of the observable Universe) per unit of Hubble volume remains more or less constant with time, so that these object constantly generate density fluctuations of order of the mass per unit of length, $G\mu$, whose scale is given by the symmetry breaking scale by $G\mu \sim (E_{\rm SB}/E_{\rm Pl})^2$. Such regime is the so-called scaling regime and is known to be established both during the matter dominated and the radiation dominated eras. It has also been known for a long time that cosmic strings cannot generate a wiggly anisotropy spectrum such as the one that is observed. The reason is that since generate continuously density perturbations, what one sees can be interpreted as the incoherent sum of wiggly spectra which are progressively shifted with respect from each others [12]. The wiggly shape of the observed CMB spectrum therefore rules out cosmic strings as the main contributor to cosmological perturbations, however, it might happen that a symmetry breaking occurs sufficiently soon after an inflationary phase, so that a subdominant string network leaves a small but observable imprint in the CMB anisotropies. In such case, as it is fairly complicated to compute the two-point correlation function produced by a string network, it is better to look for some simple characteristic signature left by sch object. One of these is due to the fact that spacetime around a string is conical, i.e., it is locally Minkowskian, but there is a deficit angle when one turns around a string. The perimeter of a circle of radius R is not $2\pi R$, but $2\pi(1-4G\mu)R$. Consequently, two photons propagating parallel to each other and passing on different side of the string will be focused after passing close to the string, as shown in Fig. 13. This string wake effect will translate into a redshift difference of light rays passing on different side of a string as seen from an observer looking at the string along a direction that is orthogonal to its motion. One therefore expects to see curves of temperature discontinuities in the CMB sky, the so-called Kaiser-Stebbins effect [13], which is proportional to $8\pi G\mu$. The total length of the discontinuities does not depend on $G\mu$ since the scaling regime does not depend on it, but the amplitude of the discontinuities does. The Planck data do not show any hint of discontinuities (or smoothed version of them because of the finite resolution of the telescope), nor any other signature associated with cosmic strings[6], which translates into an upper bound on $G\mu$ of 1.5×10^{-7} [14].

7.3 The intriguing large scale anomalies

Although the data impressively fit the six parameter concordance model, some analyses also reveal some intriguing features in the data.

[6]Actually, the cosmic strings anisotropy spectrum peaks at rather high multipoles, so that it is the direct impact of cosmic strings on the C_ℓ spectrum that constrains them better than the specific non Gaussian signatures, see [14] for the detailed constraints depending on the methods that are used.

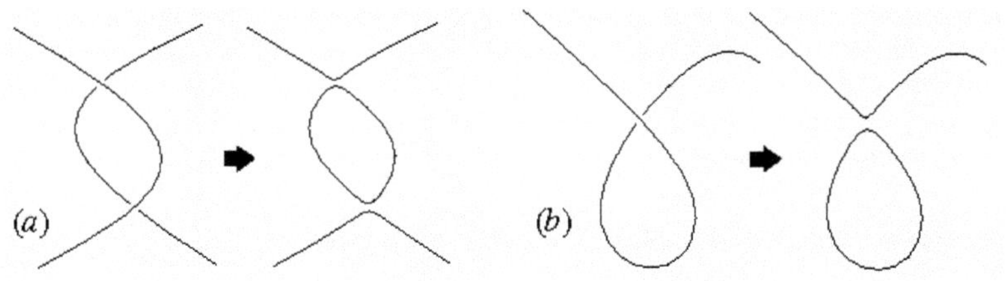

Figure 12: The process of intercommuting (a), possibly leading to the formation of cosmic string loops (b). Picture taken from [15].

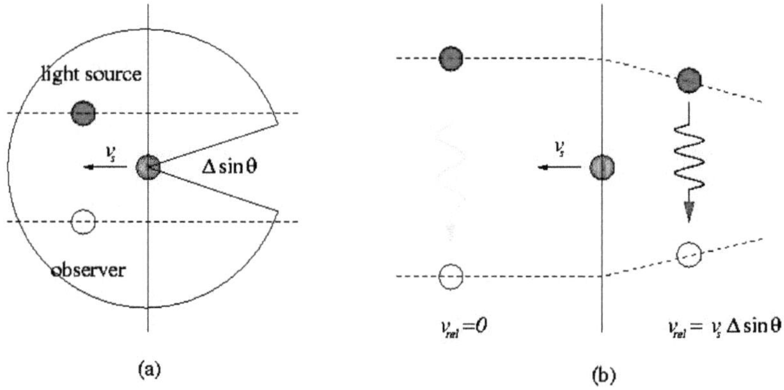

Figure 13: A sketch of the Kaiser-Stebbins effect, i.e. the fact that initially parallel light rays will be focused in the wake of a cosmic string as a consequence of the conical space that exists around these objects. If, in addition, the cosmic string is moving perpendicularily to the observer, a temperature discontinuity will be observed around the string, a feature that can make its detection possible. The amplitude of the discontinuity is proportional to the deficit angle $\Delta \sin \theta$, which in turn is proportional to the cosmic string mass per unit of length μ expressed in Planck units. Picture taken form [16].

One of the most obvious is that there is a slight lack of power at large angular scales when considering the best fit to the data (see left part of Fig.5). Most of the $\ell < 50$ multipoles are below the best fit curve, which is rather unusual from a statistical point of view. Since this feature is rather weak (a less than 3σ effect), it is tempting to discard it completely, but when trying to extend the six parameter model to something else, the inclusion or not of the low ℓ multipoles somewhat affect the outcome of the analysis. Basically, new physics appears to be a more favored possibility when one includes the large angular scales, although with a weak statistical significance. Also, this feature adds to various "anomalies" that were already hinted by the WMAP data and confirmed by Planck. For example, the $\ell = 3$ multipole is unusually flat and well aligned with the quadrupole. There seems also to be many phase correlations between the different multipoles, whereas there are expected to be uncorrelated in a statistically isotropic Universe. A comprehensive list of these "anomalies" is given in Ref. [17].

Even more unexpected is the so-called power asymmetry. It consists in the fact that two opposite hemispheres on the sky have statistically significant different overall amplitude spectra in the sense that we can model the observed temperature map by the phenomenological formula $\Delta T/T(\hat{n}) = s(\hat{n})(1 + A\hat{p} \cdot \hat{n})$, where $s(\hat{n})$ is a statistical isotropic map, A some amplitude and $\hat{p}$ a direction. If one includes a sufficiently large amount of scales, then there seems to be a more than 3σ detection of a nonzero value of A. This fact was hinted by the WMAP results, and proven to be foreground unrelated by Planck. At present, it seems that no simple inflationary model is able to explain this feature.

None of the effect mentionned here are very significant (rarely above 3σ) but their accumulation definitely makes them intriguing. One could argue that we may naturally inclined to pay attention to the rare anomalies that exist in the data and give them more relevance than they deserve, but one has also to consider that these features are real and hint for some incompleteness of our description of the Universe. A fascinating possibility would be that during inflation, large scales experience less homogeneization than small scales, so that the features that are seen only on large scales could give some hints about the pre-inflationary universe. Although some attempts have aleardy been done, it is too early to decide whether these feature will prove being able to change our view about the early Universe.

8 Reflection on the future of CMB studies

The Planck temperature anisotropy data can hardly be improved by future missions because the main limitation we face is the so-called cosmic variance, i.e., the fact that we observe a finite volume of the Universe, which therefore limits, from a purely statistical point of view, the robustness of our results. Regarding polarization, the situation can be very significantly improved since we are (very) far from being in reach of the ultimate sensitivity achievable, since we are here limited by the diffraction limit and photon noise. Some reflection have begun since a few years regarding a possible successor to Planck. A recently proposed project (from European side) was called PRISM (Polarized Radiation Imaging and Spectroscopy Mission) [18]. It was based on a larger telescope (3.5 m instead of 1.5 m of equivalent diameter) and more importantly, a much larger number of detectors (7600 vs. 74 for Planck). This huge leap in sensitivity could allow a number of new studies to be done. For example, the CMB frequency spectrum could be measured to such a precision that some tiny deviations from thermalization due to exotic phenomena as early as $z \sim 10^6$ could be detected. Among these, let us mention dark matter annihilation, dark matter decay, or primordial black hole evaporation.

A considerable amount of technical challenges had to be addressed in order to make this project possible, since, for example, the amount of data collected by the instrument would have been too large to be downloaded to Earth so that a heavy amount of real time processing had to be done onboard. All this strongly impacted the overall cost of the mission and put it in the category of the Large ESA mission (overall cost around 1 billion euros) whose first available launching slots would not occur before 2027. Very recently, ESA chose to attribute the two presently planned L-class mission slots to other fields (spaceborne gravitational waves observatory and X-ray telescope), so that the PRISM project will probably have to be rescaled down to a medium class mission, but still keeping the large increase in detector number of the first L-class proposal.

9 Conclusion

The results presented here come from the first release of the Planck data. They confirm the overwhelming consistency of the concordance model, but point out to some anomalies whose interpretation (and existence) are unclear. The situation regarding these anomalies should improve in 2014 with the release of the polarisation data. Whether or not they will disappear or point toward the necessity to improve the concordance model is impossible to tell now.

References

[1] K. Freese, B. Fields, D. Graff, Proceedings of the 19th Texas Symposium on Relativistic Astrophysics and Cosmology (CD-ROM), arXiv:astro-ph/9904401 .

[2] W. Hu, Lect. Notes Phys., **470**, 207 (1996).

[3] P. A. R. Ade, *et al.*, arXiv:1303.5076 .

[4] S. Das, *et al.*, *Physical Review Letters*, **207**, 021301 (2011).

[5] A. H. Guth, *Physical Review D*, **23**, 347–356 (1981).

[6] G. Mangano, *et al.*, *Nuclear Physics*, **B729** 221–234 (2005).

[7] J. Martin, *et al.*, arXiv:1312.3529.

[8] `http://blogs.discovermagazine.com/cosmicvariance/2012/09/04/ /guest-post-doug-finkbeiner-on-fermi-bubbles-and-microwave-haze/`

[9] D. P. Finkbeiner, *Astrophysical Journal*, **614**, 186–193 (2004).

[10] P. A. R. Ade, *et al.*, *Astronomy and Astrophysics*, **554**, A139 (2012).

[11] G. Dobler, *Astrophysical Journal*, **750**, 17, 2012.

[12] A. Riazuelo, & N. Deruelle, *Annalen der Physik*, **9**, 288–298 (2000).

[13] N. Kaiser, & A. Stebbins, *Nature*, **310**, 391–393 (1984).

[14] P. A. R. Ade, *et al.*, arXiv:1303.5085 .

[15] `http://www.damtp.cam.ac.uk/research/gr/public/cs_home.html`

[16] `http://aether.lbl.gov/eunhwa_webpage_2/coverpage.html`

[17] , P. A. R. Ade, *et al.*, arXiv:1303.5083 .

[18] `http://www.prism-mission.org/`

CHAIRMAN: A. RIAZUELO

Scientific Secretaries: A. Laszlo, R. Aggleton

- *M. Csanad:*

On one of the last slides the galactic haze was mentioned and its origin was suggested being the dark matter halo. Could you go into details, in particular, what other sources could cause it?

- *Riazuelo:*

In 2004 WMAP we had detected some low frequency anomalous emission softer then free-free emission, harder than synchrotron emission and we didn't know what it is. There were many potential explanations at that time and people did not worry much. Then, in 2009 there was the Fermi bubble announcement.of the Fermi Telescope. It seems that if you look at the gamma-ray sky, the location of this soft emission matches almost the Fermi bubbles. So it seems that we see a synchrotron counterpart of a high energy emission. It is expected that if you have high energy electrons, this will radiate synchrotron radiation, because of the galactic magnetic field. And in order to explain this as the counterpart of the Fermi bubbles, one needs a 5 microGauss galactic magnetic field, which is OK. Planck could confirm the WMAP detection because of the much better foreground control, one could eliminate the more or less implicit assumptions of WMAP analysis. The WMAP team of course could prove the consistency of their analysis but could not rule out the possibility that it was an artefact. With Planck we made sure that this haze is there, and we are almost sure that this is counterpart of the Fermi bubble. Now the question is if there are any conventional astrophysical explanations to these bubbles which were not expected at the time they were discovered. There are lot of papers about this, but in my opinion many of these are biased toward or against one of these hypotheses. What is strange that on the Fermi map there is a very flat emission with a sudden cutoff at the edge. This is a bit strange as if it would be some volume emission, one would expect that it is brighter at the center. If it were a surface emission like of a shock wave, it would then be the contrary. In the Fermi data we see something which is hard to associate either with surface or volume emissivity. Therefore we do not really know the cause. With this respect the Planck data do not bring much, as they seem to be the cause [but rather the consequence] of this high energy emission. So, for the high energy emission is that we have to look for an explanation. Among all the astrophysical explanations there are some which are already ruled out because they predict in addition some other emissions (Lyman-alpha etc) which are not there. In this respect, the dark matter explanation is not ruled out, even if it does not seem to explain this rather unusual luminosity profile [better than other interpretation]. So, the situation is that it maybe has to do with the dark matter.

- *Lange:*

Could you explain the meaning of the primordial tilt parameter n_s and how it is connected to inflation and how it is measured in the temperature fluctuation spectrum?

- *Riazuelo:*

Basically what we have is that we start from some initial fluctuations, which is more or less flat, and that is to be modulated by physics. So if you are able to remove these contributions, you can recover the initial spectrum. Now the problem is that there are some degeneracies between the initial shape and some subtle effects that are here. E.g. there is the reionization of the universe, which gives a slight boost [on large angular scales], which mimicks a tilt that is slightly reduced. So, if you want to reconstruct the parameters of the initial condition, you must have some very precise data. If you have only the temperature data, this is the situation with the tilt: there is some degeneracy of the reionization and the primordial tilt. So, the more precise your data are you more remove these degeneracies and with WMAP we could not claim that this primordial tilt was smaller than 1 at 5 sigmas, here we can. With the polarization we can do much better in fact, because when you look at the polarization spectrum you expect you have a bump here which is due to reionization, so polarization gives you a much more precise measurement on the reionization and therefore breaks the degeneracy. Today it seems that if you don't assume some weird physics, then primordial tilt is smaller than 1 by 5 sigmas. Of course if you add some extra physics with some extra parameters even coming from the inflation side, you may relax that sort of bound. In any case, the constrain on your estimated parameters depend on the number of parameters you allow in your model. The physical meaning of n_s is just the spectrum of the fluctuations. You assume some powerlaw on the density fluctuations and if n_s is smaller than one you have more fluctuations on the large scales. So this is just the relative amount of fluctuations on the small scales and large scales. The problem is that you cannot go to arbitrarily large scales because the observable universe has finite size and it is also problematic to go very small scales because of many astrophysical effects which contribute to the spectrum there. From time to time there are some claims from comparing power spectrum at the large scales and small scales, but there can be lot of complicated astrophysics behind. With Planck there is about 2.5 orders of magnitude of scale which is spanned, which is marginally enough to do that sort of thing. Should we be able to measure much smaller scales, we might be able to much better constrain this.

- Jenni:

Can you please comment on the number of neutrinos?

- Riazuelo:

We observe two type of neutrino oscillations: solar and atmospheric neutrinos. And on smaller range we have the so called reactor anomaly which may be explained by the depletion of the expected neutrino flux. And it could be explained with a fourth type of neutrinos, because the mass-square difference does not match the others. So, there were people considering extra type of neutrinos and observing their implications on the CMB. There are several ways to analyze this and there are several sets of data you can take. This plot shows what is the situation if there is an extra zero or negligible mass neutrino and you vary their number. And here you have the likelihood depending on which data you take. If you take the Baryon Acoustic Oscillation data [in addition to the Planck data] data you narrow the constraint: this disfavors 4 species of neutrinos. If you take into account the Hubble constant measurement which is independent of Planck, then you shift the likelihood to a larger number of neutrinos. This is very inconclusive in a sense, because there is no strong exclusion of either 3 or 4 species, but the result depends crucially on which dataset you take. This [Hubble constant] dataset seems not to be compatible with Planck, and if you take them for sure, together they don't fit well to the six parameter models and therefore may be a hint for new physics. Other thing what people considered to take 3 neutrinos fixed but look at their mass. The mass of the neutrino has some effect on the structure formation at small angular scales. We have only upper limits on the masses, which again depend on which set of data we take. By combining the two constraints (more than three neutrinos, all of equal mass) does not change the picture.
What makes more sense in my opinion is to consider 3 massless neutrinos along with a massive sterile one. It does not show hint of a large mass difference or a large number of total neutrinos. At present all this is inconclusive [because neutrino have a weak effect of CMB anisotropy], but with polarization data you will break some of the degeneracy which we have here. In the far future [10 years] we will have some lensing measurements from ESA (Euclid project), which can measure better at smaller scales, and tighten the constraints. At present we do not see anything, and certainly we do not favor a fourth generation of neutrinos.

- *Leontsinis:*

Are there any plans for other searches, such as gravitational waves or dark matter, using PRISM?

- *Riazuelo:*

Next step is to include polarisation. Polarisation is some arrowless vector field, and can be decomposed into 2 parts: a gradient part, the "E" mode; and a curl part, the "B" mode. The main difference is that the latter is not invariant under parity transformation whereas the former is, and theory tells us that density waves only produce these E modes, whereas gravitational waves produce both. So if you want a better constraint on primordial gravitational waves, it's not very interesting to look at temperature and anisotropy spectrum, unless the gravitational wave contribution to the large scale fluctuations is important. But it is better to look for these B modes. There is some intrinsic limit on what you can detect for gravitational waves, because unfortunately as I showed on the plot [goes to slides], the density perturbation contributes to E modes, but lensing distorts the field and transforms part of E mode signal to B model. So there is some contamination of B modes which comes from density waves. So the minimal amount of gravitational waves you can constrain in CMB comes from this. The next step is to go towards this limit. Planck will not be able to scan this, be the possible successor to Planck will try. PRISM is here, and they think it will be only about a factor of 3 above the ideal case. It is probably the best that can be thought of. Of course, before this there will be constraints form Planck with respect to the polarization, so this will help reduce some of the degeneracy between parameters. Detailing this is complicated, but we can carry out better tests due to polarisation.

- *Morozov:*

1) You showed Planck scans the sky going around. It scans equatorial regions more than polar regions. How does this method affect the results and precision of the results in different regions of the sky?

2) Did Planck produce any new data about the letters 'SH' on the CMB map?

- *Riazuelo:*

1) It does not affect it much. This is the way the sky is scanned. As you said, there are more scanning points here [points to poles] than the rest. It is very difficult to do otherwise. This important thing is that the ecliptic plane is scanned with at least the amount that was obtained. The most important thing is to check calibration of the experiment to check there is no drift in the electronics [the 1/f noise] that would change the signal strength, so that you have to observe the same region all along the full mission. Nevertheless, people think that some other observation strategy would have been better because we also have to observe planets [i.e. bright pointlike sources] many times. This is important to get an idea of the beam size, the point spread function of each detectors. For the temperature and isotropy the chosen scanning strategy may not be optimal, but it is not the most annoying thing in the whole process. For polarization the situation is different. It is much more difficult when you have two detectors with polarisers which have different orientations to check the cross-calibration between the detectors. And being able to stop the rotation of the satellite to do some long observation of a region of the sky could have been useful for polarization. Since the data for polarization is not out yet, we do not know if this choice was a bad one or not.

2) I did not talk about some of the anomalies observed in data. I did not talk about the dangers of an a posteriori analysis. In the WMAP data you can see a S and H imprinted on the CMB map. This is the danger to do an a posteriori analysis, for example what is the probability to have Stephen Hawking's initials in the sky? A priori, it is very small, but after we notice that fluctuations draw by chance these letters in the sky, the odds to find them are of course much larger! By adding some smaller scale fluctuations [with the Planck data] you tend to erase this pattern. It is like the

face on Mars [that was advertized by the Viking mission] in 1976. Higher resolution satellite images showed it disappeared. It is the same thing here [the S and the H almost disappeared]. Some anomalies have survived to the Planck data. This is why when we do the analysis, we drop all the large angular scale part, with strange features here [points to map]. We don't really know what to do about this. Since we expect some correlation between temperature and polarisation, it depends on whether anomalies are seen in the polarization. If they are, people will be more concerned. They are not under the carpet, but they are not yet emphasized.

- De Palma:

In the future, how much can the measurement of the CMB improve? Is there some intrinsic limit?

- Riazuelo:

It has to do with the next step, the PRISM mission. Here we are exactly in the position where a photon is going to hit the primary mirror, and what you see here is the focal plane. Obviously, the focal plane is rather empty. Only 1% of photons which hit the primary mirror will end up in the detectors. It's like having photographic plates instead of a CCD. The next step is to have a bolometer array instead of these individual horns. People are working hard to make this possible. PRISM announced that they will try and have 100 more bolometers than on Planck. How realistic this is, and on what timescale, it is not know, but we can definitely improve the number of photons collected by having a more packed configuration of detectors [with respect to what it found on Planck focal plane]. If you try and measure temperature fluctuation, this is not necessary, since the precision of the data we have is already limited by cosmic variance. We observe only 1 universe and this is the main limitation for the temperature spectrum. If you are looking at the polarization spectrum, then the signal is very low and you need more detectors. This is what people are aiming for. It is not an issue if you want to do it from the ground [but then you have to deal with the atmospheric emission]. But if you want to do it in space, we have a problem with the amount of data you transmit to Earth. So you will have to do some onboard processing, and this is quite complicated. Planck didn't have enough power to do this. Even though PRISM is bigger it still faces the same problem. But this is a very preliminary design, and the launch slot is in 15 years [at least]. It is obvious that given the poor coverage of the focal plane on Planck we can do better. How much better, and at what cost, I do not know.

- Lukács:

What does Planck data tell us about cosmic strings? The best fit several years ago was that cosmic strings were more likely to exist than not, and if yes, their tension was close to that of GUT strings.

- Riazuelo:

First, when you have cosmic strings you generate cosmological perturbation at every epoch. Thus the nice oscillatory features you see are completely destroyed. So 15 years ago, when people did not see any structure in this spectrum, the fact CMB fluctuations were due to topological defect in general or inflationary-like perturbation was an open question. Planck data tell us that the dominant contribution was due to inflationary-like perturbation, which were generated very early in the universe. So how can you constrain the amount of cosmic strings from these maps? There are several features. If you look at the spacetime around a cosmic string, it is locally flat, but there is a deficit angle – space is somehow conical around it. And if you have 2 photons travelling parallel towards the cosmic string, they will be focused after having gone past the string. If you are at rest, and a cosmic string passes in front of you, you will observe a temperature discontinuity. People have been looking of such discontinuity in the map, and these are not found. This gives an upper limit on temperature discontinuity, giving a limit on mass per unit length of a cosmic string. One

Planck paper [arXiv:1303.5085] addresses this. You may also have some non-Gaussian signatures from them, which are not seen, giving an upper limit. I have no idea in terms of model-building how it constrains a cosmic string model. If you have a model that predicts a cosmic string with a mass per unit length of 10^{-10} in Planck units you have no chance of seeing this in the CMB.

Latest Results from BNL and RHIC-2013

M. J. Tannenbaum *
Physics Department, 510c,
Brookhaven National Laboratory,
Upton, NY 11973-5000, USA
mjt@bnl.gov

November 21, 2013

1 Introduction

Quite different from CERN, Fermilab and SLAC, Brookhaven National Laboratory (BNL) is a multipurpose laboratory, founded in 1947 to promote basic research in the physical, chemical, biological and engineering aspects of the atomic sciences and for the purpose of the design, construction and operation of large scientific machines that individual institutions could not afford to develop on their own. The present mission has expanded to both basic and applied research at the frontiers of science, including nuclear and high-energy physics; physics and chemistry of materials; nanoscience; energy and environmental research; national security and nonproliferation; neurosciences; structural biology; computational sciences; and to provide "cutting edge" research facilities for these purposes. BNL is located on Long Island roughly 100km east of New York City. In fact the Relativistic Heavy Ion Collider (RHIC) at BNL can be seen from outer space since it is not buried in a tunnel but is in an enclosure on the surface, which is covered by earth for shielding (Fig. 1a).

RHIC is composed of two independent rings, of circumference 3.8 km, containing a total of 1,740 superconducting magnets. RHIC can collide any species with any other species and since beginning operation in the year 2000 has provided collisions at 13 different values of nucleon-nucleon c.m. energy, $\sqrt{s_{NN}}$, and nine different species combinations (counting differently

*Research supported by U.S. Department of Energy, DE-AC02-98CH10886.

RHIC Run	Year	Species	c.m. Energy	Ldt
Run-1	2000	Au+Au	130 GeV	1 μb-1
Run-2	2001-2	Au+Au	200 GeV	24 μb-1
		Au+Au	19 GeV	
		p+p	200 Gev	150 nb-1
Run-3	2002/3	d+Au	200 GeV	2.74 nb-1
		p+p (L)	200 GeV	350 nb-1
Run-4	2003/4	Au+Au	200 GeV	241 μb-1
		Au+Au	62.4 GeV	9 μb-1
Run-5	2005	Cu+Cu	200 GeV	3 nb-1
		Cu+Cu	62.4 GeV	0.19 nb-1
		Cu+Cu	22.4 GeV	2.7 μb-1
		p+p (L)	200 GeV	3.8 pb-1
Run-6	2006	p+p (T+L)	200 GeV	10.7 pb-1
		p+p	62.4 GeV	100 nb-1
Run-7	2007	Au+Au	200 GeV	813 μb-1
Run-8	2007/2008	d+Au	200 GeV	80 nb-1
		p+p (T)	200 GeV	5.2 pb-1
		Au+Au	9.2 GeV	
Run-9	2009	p+p (L)	200 GeV	16 pb-1
		p+p (L)	500 GeV	14 pb-1
Run-10	2010	Au+Au	200 GeV	1.3 nb-1
		Au+Au	62.4 GeV	100 μb-1
		Au+Au	39 GeV	40 μb-1
		Au+Au	7.7 GeV	260 mb-1
Run-11	2011	p+p (L)	500 GeV	27 pb-1
		Au+Au	200 GeV	915 μb-1
		Au+Au	27 GeV	5.2 μb-1
		Au+Au	19.6 GeV	13.7 M events
Run-12	2012	p+p (L)	200 GeV	9.2 pb-1
		p+p (L)	510 GeV	30 pb-1
		U+U	193 GeV	171 μb-1
		Cu+Au	200 GeV	4.96 nb-1
Run-13	2013	p+p (L)	510 GeV	156 pb-1

Figure 1: a)(left) Aerial view of BNL. In addition to RHIC, at the top, there are two synchrotron light sources visible, the rings at the bottom. The dome in the center is the decommissioned High Flux Beam Reactor (1965-1996); and the chimney is the last remnant of the original (1950-1968) Graphite Research Reactor. b) Year, species, $\sqrt{s_{NN}}$, integrated luminosity and proton polarization (**L**ongitudinal, **T**ransverse), for RHIC runs.

polarized protons as different species). The performance history of RHIC with A+A and polarized p-p collisions is shown in Fig. 1b.

2 News from BNL since ISSP2012

In keeping with the spirit of the title, I review the latest news from BNL before going to the RHIC physics results. In September 2012, the panel appointed by the U. S. Department of Energy (DoE), led by Prof. Robert Tribble, to decide whether RHIC, Jefferson Laboratory in Newport News, VA, or the Facility for Rare Isotope Beams (FRIB) project at Michigan State University should be closed (due to insufficient funding) concluded with a good outcome. They recommended that a 2% real increase per year in the DoE nuclear physics budget would save all three labs. This was included in the President's 2014 budget.[1]

There were several important personnel changes at BNL. On January 8, 2013, Berndt Mueller, a leading Nuclear Physics Theorist and Administrator from Duke University became the new Associate Laboratory Director responsible for the Nuclear and Particle Physics programs at BNL. In March

[1]However, the Fiscal Year 2014 started on October 1, 2013, with the U. S. Government shut down for 16 days due to the lack of an approved budget, so ...

2013, Doon Gibbs, who had been the Deputy Laboratory Director, became the new Laboratory Director, while the previous Laboratory Director, Sam Aronson, became the Director of the RIKEN-Brookhaven Research Center at BNL, replacing Nick Samios. About the same time, a new world-class interdisciplinary Science Laboratory building was completed with 60 standard laboratories and 4 specialized laboratories: a humidity controlled dry room, two ultra-low vibration laboratories, and a laboratory which connects the Molecular Beam Epitaxy (MBE) machine, which builds custom designed thin films one atomic layer at a time, to one of the ultra-low vibration labs via a vacuum-locked system which allows the MBE-created samples to be transported for analysis without exposing them to air. Then, a week later, the U. S. Department of Energy, the owner of BNL, announced that they would initiate a competetion for a new contractor to Manage & Operate BNL for the DoE, with evident dissatisfaction in the performance of the present contractor who had managed BNL since 1998, having replaced the original founders of the laboratory, a consortium of nine major northeastern research universities, who had operated the laboratory from 1947-1998.

In June, the 15 m diameter precision storage ring from the BNL muon $g-2$ experiment [1] began a circuitous (5000 km) very delicate cross-contry trip to Fermilab (only 1440 km in a straight line) involving custom built trucks and a specially prepared barge which brought the magnet down the East Coast, around the tip of Florida and up the Mississippi River to Illinois. This new muon $g-2$ experiment at Fermilab would be the fifth such experiment, which was pioneered at CERN.

By some incredible coincidence, a few weeks before this ISSP2013 school, the June 2013 CERN Courier reprinted an article from 1970 with the title "Preparing for a third '$g-2$'." This brought back good memories to me because I had worked on the second $g-2$ experiment when I was a post-doc at CERN in 1965-66 [2]; but, of course, Prof. Zichichi worked on the ground-breaking original $g-2$ experiment at CERN in 1959-1961 [3]. Figure 2a shows the original $g-2$ team seated on top of their 6 m magnet, including a $\sim$ 30 year old Nino. Fig. 2b was taken in 1963 on the BNL-AGS floor during my thesis experiment where the three Professors were discussing what the $\sim$ 24 year old graduate-student should do next. My thesis was muon-proton elastic scattering [4] to find out "Why does the muon weigh heavy?" Even today, with Prof. Higgs in the audience, we still don't know! Other experiments at that time did better: the "two-neutrino experiment" [5] was in the beam to the left (Nobel Prize); while on the right, over the AGS machine, in "inner Mongolia", CP violation [6] was discovered (Nobel Prize). Those were the days; but even more excitement lay ahead.

Figure 2: a)(left) First $g-2$ experiment, CERN, c. 1959. Left-to-Right: Sens, Charpak, Muller, Farley, Zichichi. b) (right) $\mu - p$ experiment BNL, c. 1963: L-R: Tinlot, Cool, Tannenbaum, Lederman.

3 ICHEP1972: Hard-Scattering, Quarks, and QCD

At the International Conference on High Energy Physics (ICHEP) in 1972, there were three momentous developments that inform our work today:

- The discovery in p-p collisions at the CERN ISR of production of particles with large transverse momentum (p_T) which proved that the partons of Deeply Inelastic Scattering (DIS) interacted with each other much more strongly than electromagnetically.
- Measurements of DIS in neutrino scattering presented by Don Perkins who proclaimed that "In terms of constituent models, the fractionally charged (Gell-Mann/Zweig) quark model is the only one which fits both the electron and neutrino data."
- The origin of QCD in the presentation by Harald Fritzsch and Murray Gell-Mann with the title "Current Algebra: Quarks and What Else?"

Figure 3a shows the first observation of scattering at large p_T [7]. The π^0 spectrum breaks away from the e^{-6p_T} dependence known since cosmic ray measurements, with a power-law spectrum that flattens as the c.m energy, $\sqrt{s}$, is increased. Excellent cooperation of experimentalists and theorists showed in 1978 that these data could be explained by QCD [8, 9] if the quarks in a nucleon had "intrinsic" transverse momentum, $k_T \approx 1$ GeV/c (Fig. 3b). Although these QCD calculations in agreement with the high p_T single particle spectra were published in 1978, most experimentalists in the

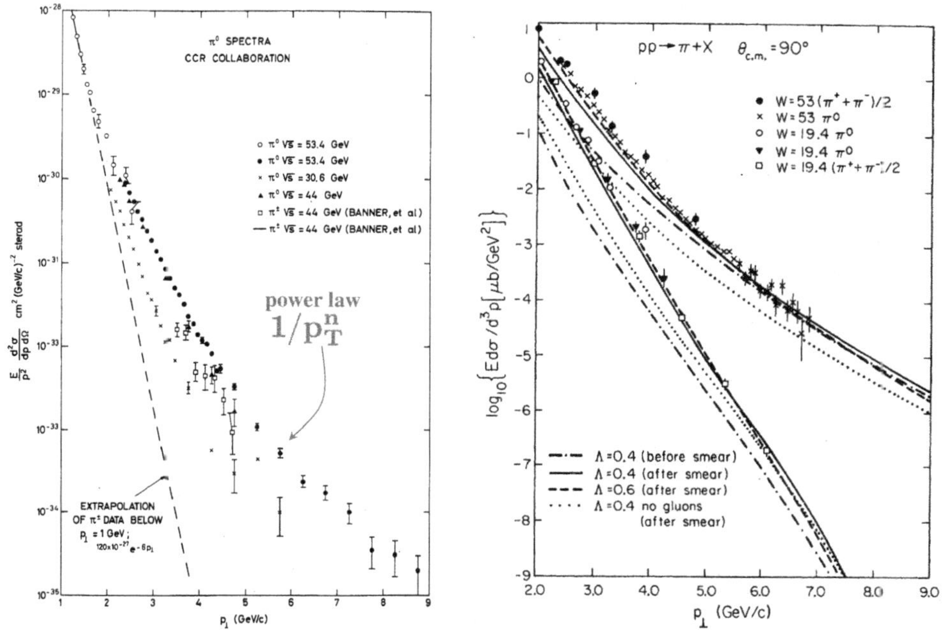

Figure 3: a)(left) Plot of invariant single π^0 cross section vs. p_T for several $\sqrt{s}$ from CCR at the CERN-ISR [7]. b)(right) Feynman, Field, and Fox [9] QCD calculation of mid-rapidity high-p_T π spectra at $\sqrt{s} = 19.4$ and 53 GeV, with and without k_T smearing, for two values of $\Lambda_{\rm QCD}$.

U. S., notably at the first Snowmass conference in July 1982, were skeptical because of evidence against jets presented at the ICHEP1980 by a CERN experiment, NA5 [10].

Bjorken had proposed in 1973 [11] that jets from the fragmentation of high p_T scattered partons should be observed using "4π" hadron calorimeters. The first large aperture measurement was by NA5 [10] (Fig. 4a) who showed a transverse energy (E_T) spectrum at ICHEP1980, where the sum:

$$E_T = \sum_i E_i \sin\theta_i \tag{1}$$

is taken over all particles emitted into a fixed solid angle for each event. In Fig. 4a [10], the solid angle varies from full azimuthal acceptance, $\Delta\phi = 2\pi$, in the c.m. rapidity range $-0.88 < y < 0.76$, to smaller azimuthal regions as shown on the figure. The striking results, which contradicted a previous claim from Fermilab [12], were: i) no jets were seen in the full azimuth data; ii) all the data were far above the QCD predictions; iii) the large E_T observed was the result of "a large number of particles with a rather small transverse momentum".

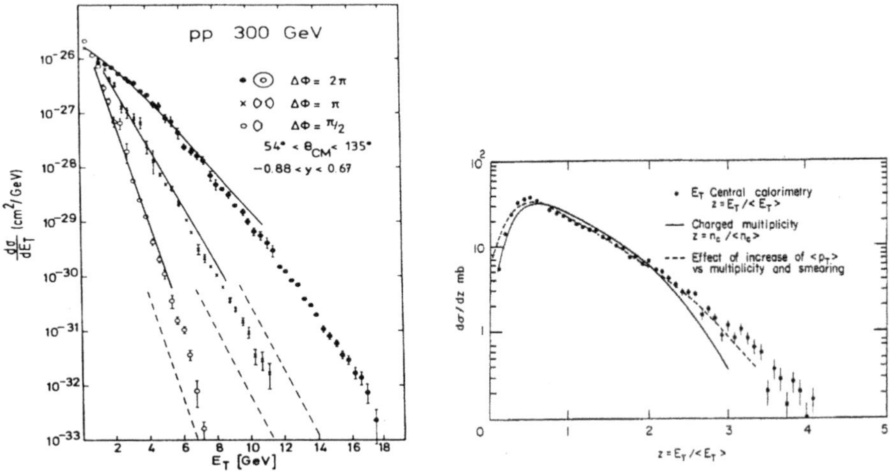

Figure 4: a)(left) E_T distributions [10] in the solid angles indicated. Predictions from soft (low p_T) multiparticle production and QCD hard-scattering are shown by solid and dashed curves respectively. b) (right) E_T distribution [13] for $|\eta| \leq 3$ from p̄-p collisions at $\sqrt{s} = 540$ GeV in the KNO type variable, $z = E_T / \langle E_T \rangle$, used for multiplicity [14].

As we shall see below, E_T distributions are very important in Relativistic Heavy Ion (RHI) physics since they can be used to characterize and study the nuclear geometry of an A+B reaction on an event-by-event basis. The strong relation between E_T and multiplicity distributions and the absence of jets in these distributions was emphasized in a talk by UA1 at ICHEP 1982 (Fig. 4b) [13]. Ironically, this talk immediately followed a talk by UA2 which provided the first evidence for a di-jet from hard-scattering at a level 5-6 orders of magnitude down in the E_T distribution at $\sqrt{s} = 540$ GeV.

There are many additional important results in high-energy physics from this period that are relevant to both QCD and RHI physics, which must be skipped in this brief introduction. I have covered some of these results in previous ISSP lectures and proceedings; but this year, I wrote a book with Jan Rak [15] which covers this information in detail, which is the real introduction to what follows.

4 QGP Physics—RHIC Highlights 2013

The Quark Gluon Plasma (QGP) was discovered at RHIC, and announced on April 19, 2005, with properties of a hot, dense, inviscid liquid, rather than the gas of free gluons and quarks expected.[2] Although soft ($p_T \lesssim 2$ GeV/c) physics dominates particle production in p-p, p+A and A+A collisions and presumably in the thermalized QGP, with $T \approx 300 - 600$ MeV as estimated from "thermal photons" measured in the range $1 \leq p_T \leq 3$ GeV/c [16], the use of hard-scattering as an in-situ probe of the medium in RHI collisions was an important and productive innovation at RHIC. The effect of the medium on outgoing hard-scattered partons produced by the

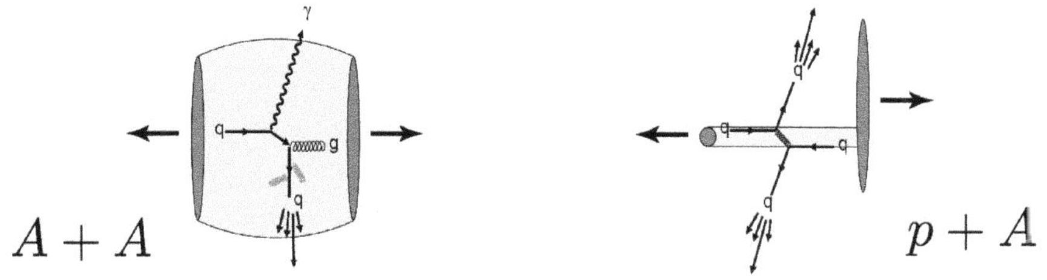

initial A+A collision is determined by comparison to measurements in p+A (or d+A) collisions, where no (or negligible) medium is produced. The impact parameter of the nuclear collision, the distance between the centers of the colliding nuclei, is measured by the upper percentile (e.g. top 5%) of an E_T or multiplicity distribution since the multiplicity increases with increasing overlap of the nuclei (centrality). The highlights this year include hard-scattering and soft physics results in Au+Au together with some important new results in d+Au.

The suppression of high p_T particles in Au+Au collisions at RHIC is reviewed in Fig. 5. For π^0 (Fig. 5a) [17] the hard-scattering in p-p collisions is indicated by the power law behavior p_T^{-n} for the invariant cross section, $Ed^3\sigma/dp^3$, with $n = 8.1 \pm 0.05$ for $p_T \geq 3$ GeV/c at $\sqrt{s_{NN}} = 200$ GeV. The Au+Au data at a given p_T can be characterized either as shifted lower in p_T by δp_T from the point-like scaled p-p data at $p'_T = p_T + \delta p_T$, or shifted down in magnitude, i.e. suppressed. In Fig. 5b, the suppression of the many identified particles measured by PHENIX at RHIC is presented as the Nuclear Modification Factor, $R_{AA}(p_T)$, the ratio of the yield of e.g. π per central Au+Au collision (upper 10%-ile of observed multiplicity) to the point-like-scaled p-p cross section, where $\langle T_{AA}\rangle$ is the average overlap integral of the nuclear thickness functions:

$$R_{AA}(p_T) = \frac{d^2N^{\pi}_{AA}/dp_T dy N_{AA}}{\langle T_{AA}\rangle\, d^2\sigma^{\pi}_{pp}/dp_T dy} \quad . \tag{2}$$

[2]This is discussed in more detail in my proceedings from the ISSP2011, also available as arXiv:1201.5900.

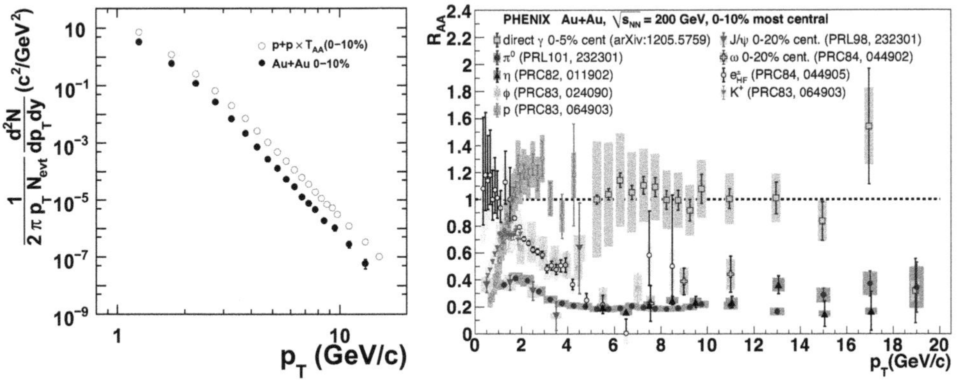

Figure 5: a) (left) Log-log plot of invariant yield of π^0 at $\sqrt{s_{NN}} = 200$ GeV as a function of transverse momentum p_T in p-p collisions multiplied by $\langle T_{AA} \rangle$ for Au+Au central (0–10%) collisions compared to the Au+Au measurement [17]. b) (right) $R_{AA}(p_T)$ for all identified particles so far measured by PHENIX in Au+Au central collisions at $\sqrt{s_{NN}} = 200$ GeV.

The striking differences of $R_{AA}(p_T)$ in central Au+Au collisions for the many particles measured by PHENIX (Fig. 5b) illustrates the importance of particle identification for understanding the physics of the medium produced at RHIC. Most notable are: the equal suppression of π^0 and η mesons by a constant factor of 5 ($R_{AA} = 0.2$) for $4 \leq p_T \leq 15$ GeV/c, with suggestion of an increase in R_{AA} for $p_T > 15$ GeV/c; the equality of suppression of π^0 and direct-single $e^{\pm}$ (from heavy quark (c, b) decay) at $p_T \gtrsim 5$ GeV/c; the enhancement of the protons for $2 < p_T < 4$ GeV/c (baryon anomaly). For $p_T \gtrsim 4$ GeV/c, the hard-scattering region, the fact that all measured hadrons are suppressed, but direct-γ are not suppressed, indicates that suppression is a medium effect on outgoing color-charged partons likely due to energy loss by coherent Landau-Pomeranchuk-Migdal radiation of gluons, predicted in pQCD [18], which is sensitive to properties of the medium.

Comparisons of the suppression of non-identified hadrons in $\sqrt{s_{NN}} = 2.76$ TeV Pb+Pb collisions at LHC to the RHIC Au+Au π^0 data are also very interesting. This is shown both in terms of the suppression, $R_{AA}(p_T)$ (Fig. 6a), and the fractional shift in the p_T spectrum $\delta p_T/p'_T$ (Fig. 6b). Interestingly, despite more than a factor of 20 higher c.m. energy, the ALICE R_{AA} data from LHC [20] are nearly identical to the RHIC measurement [19] for $5 < p_T < 20$ GeV/c. Since the exponent of the power-law at LHC ($n \approx 6$) is flatter than at RHIC ($n \approx 8$), a $\sim 40\%$ larger shift $\delta p_T/p'_T$ in the spectrum from p-p to A+A is required at LHC (Fig. 6b) to get the same

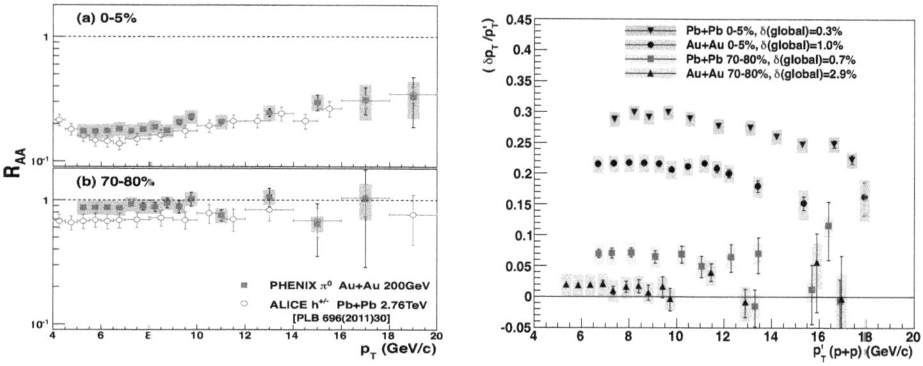

Figure 6: a) (left) R_{AA} of π^0 in $\sqrt{s_{NN}} = 200$ GeV central (0-5%) and peripheral (70-80%) Au+Au collisions [19] at RHIC compared to non-identified charged hadron ($h^{\pm}$) R_{AA} in $\sqrt{s_{NN}} = 2.76$ TeV Pb+Pb collisions at LHC. b) (right) Fractional shift of p_T spectrum $\delta p_T/p'_T$ vs. p'_T (p-p) calculated by PHENIX [19] for RHIC and LHC.

R_{AA}, which likely indicates $\sim$ 40% larger fractional energy loss at LHC in this p_T range due to the probably hotter and denser medium. These measurements can be combined with the previous measurements at RHIC for $\sqrt{s_{NN}} = 39$ and 62.4 GeV [21] (Fig. 7) to reveal a systematic increase of $\delta p_T/p'_T$ in central A+A collisions at $p'_T = 7$ GeV/c, going from 5% to 30% over the c.m. energy range $\sqrt{s_{NN}} = 39$ GeV to 2.76 TeV.

This year, the major event was the p+Pb run at LHC which also spurred new or improved d+Au results from RHIC. Apart from one hard-scattering

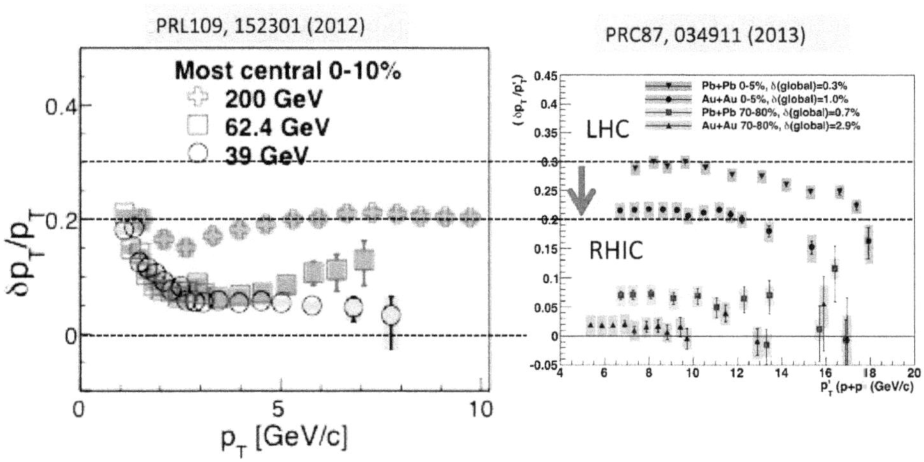

Figure 7: Fractional shift of p_T spectrum $\delta p_T/p'_T$ in central A+A collisions from $\sqrt{s_{NN}} = 39$ GeV to 2.76 TeV

result to be presented first, all the results involve the predominant soft physics of multiplicity and E_T distributions as well as flow. A new measurement of identified hadron production in both Au+Au d+Au at $\sqrt{s_{NN}} = 200$ GeV [22] gives some insight into the baryon anomaly. Figure 8a shows R_{AA}

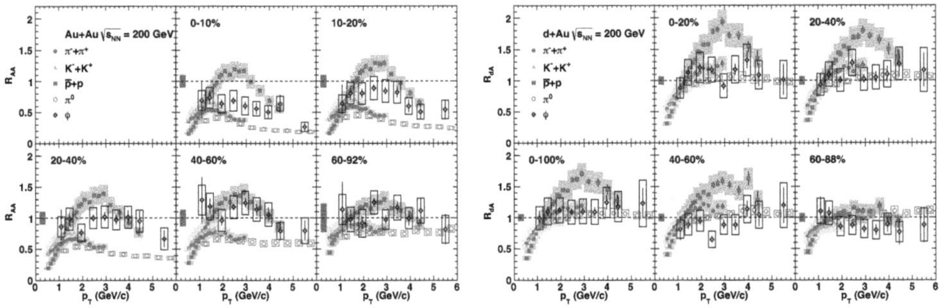

Figure 8: Measurements of R_{AA} of identified particles as a function p_T and centrality at $\sqrt{s_{NN}} = 200$ GeV [22]: a)(left) Au+Au; b) (right) d+Au.

in Au+Au for protons and mesons in the range $0.5 < p_T < 6.0$ GeV/c, where, in central collisions (0-10%), all the mesons are suppressed for $p_T > 2$ GeV/c while the protons are enhanced for $2 < p_T < 4$ GeV/c and then become suppressed at larger p_T. The d+Au results in Fig. 8b show no effect for the mesons, $R_{AA} \approx 1$ out to $p_T = 6$ GeV/c; while the protons show a huge enhancement (Cronin effect) in all centralities except for the most peripheral (60-88%). This suggests the need for a common explanation of the proton enhancement in both Au+Au and d+Au collisions, which is lacking at present.

Returning to the soft physics of multiplicity and E_T distributions, the PHOBOS experiment at RHIC, with a large pseudo-rapidity acceptance $-5 < \eta < +5$ over the full azimuth had presented an instructive measurement of the charged multiplicity density, $dN_{\rm ch}/d\eta$ from the first d+Au run in 2003 (Fig. 9a) [23]. For peripheral collisions, the distribution is symmetric around mid-rapidity, as in p-p collisions. However, with increasing centrality, the multiplicity increases over the whole η range but with a larger increase at negative η (the Au rapidity) such that the peak of the distribution steadily shifts in the direction of the Au nucleus.

These features which are similar to what was first observed in fixed target p+A experiments at $\sqrt{s_{NN}} \sim 19.4$ GeV could be explained (c. 1976) by a simple model, the Wounded Nucleon Model (WNM) [24]. From relativity and quantum mechanics the only thing that can happen to a relativistic nucleon when it interacts with another nucleon in a nucleus is to become an

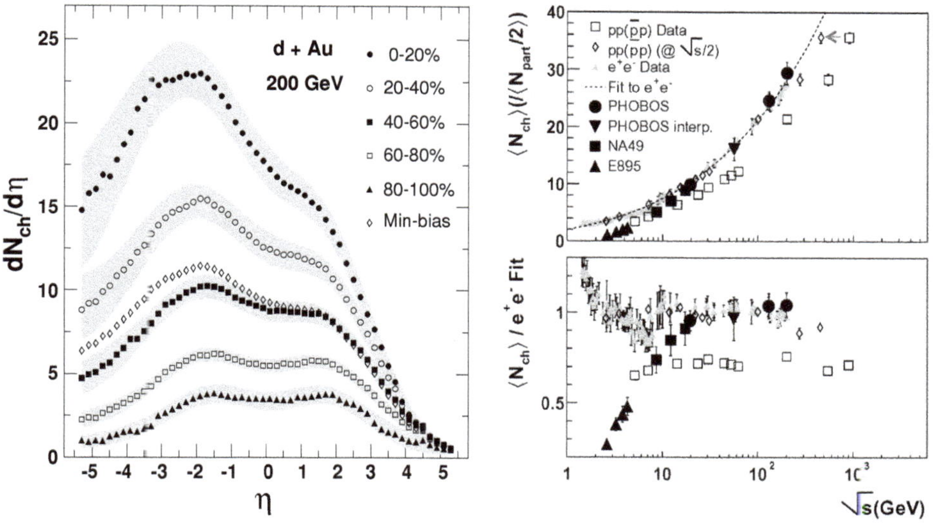

Figure 9: a) (left) Charged particle multiplicity density in rapidity, $dN_{\rm ch}/d\eta$, as a function of centrality in d+Au collisions at $\sqrt{s_{NN}} = 200$ GeV [23]. b) (right) Total charged multiplicity per nucleon pair in p-p and A+A collisions as a function of c.m. energy $\sqrt{s}$ compared to $e^+ + e^-$ collisions [25].

excited nucleon with the same energy but reduced longitudinal momentum (rapidity). It remains in that state inside the nucleus because the uncertainty principle and time dilation prevent it from fragmenting into particles until it is well outside the nucleus. If one makes the further assumptions that an excited nucleon interacts with the same cross section as an unexcited nucleon and that the successive collisions of the excited nucleon do not affect the excited state or its eventual fragmentation products, this leads to the conclusion that the elementary process for particle production in nuclear collisions is the excited nucleon, and to the prediction that the multiplicity in nuclear interactions should be proportional to the total number of projectile and target participants (Wounded Nucleons) [24], rather than to the total number of collisions.

Another interesting effect observed by PHOBOS [25] is that the "leading particle effect" in p-p collisions, as discovered by Zichichi and collaborators [26], in which the total multiplicity at c.m. energy $\sqrt{s_{\rm pp}}$ is equal to that in e^+e^- collisions at $\sqrt{s_{\rm ee}} = \sqrt{s_{\rm pp}}/2$ (the "effective energy") because the leading protons carry away half the p-p c.m. energy, is absent in A+A collisions (Fig. 9b). This observation seems to contradict the WNM, in which the key assumption is that what counts is whether or not a nucleon was struck, not how many times it was struck.

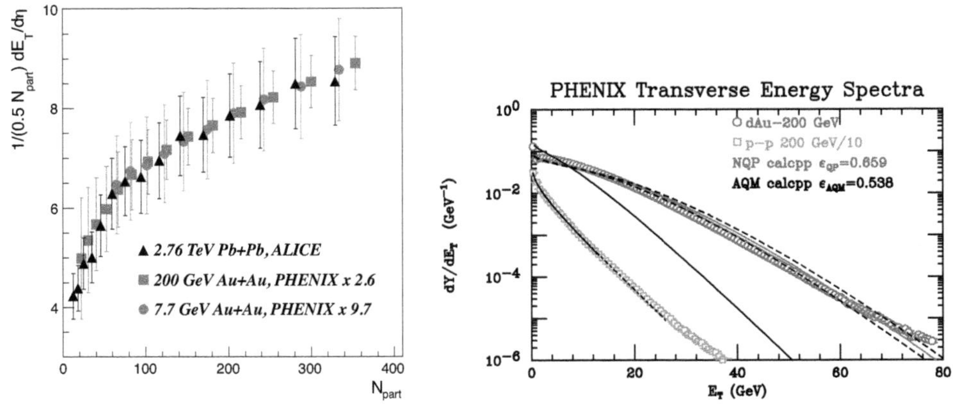

Figure 10: a) (left) $dE_T^{AA}/d\eta/(0.5\,\langle N_{\rm part}\rangle)$ vs. $\langle N_{\rm part}\rangle$ in Au+Au and Pb+Pb collisions from $\sqrt{s_{NN}} = 0.0077$ to 2.76 TeV. b) (right) PHENIX preliminary measurement of E_T distributions for p+p and d+Au at $\sqrt{s_{NN}} =$ 200 GeV with calculations of the d+Au spectrum based on the AQM (color-strings) and the number of constituent-quark participants (NQP).

In fact, the WNM fails badly at mid-rapidity for both $dN_{ch}/d\eta$ and $dE_T/d\eta$ as shown by a plot of $dE_T/d\eta/(N_{\rm part}/2)$ vs. $N_{\rm part}$ from PHENIX which should be constant if the WNM were true (Fig. 10a). The fact that the scaled evolution with centrality is the same from $\sqrt{s_{NN}} = 7.7$ GeV to 2.76 TeV indicates that the dominant effect is the nuclear geometry of the A+A collision. It has been shown that the evolution in Fig. 10a can be explained by a nuclear geometry based on the number of constituent-quark participants, the NQP model [27, 28]. Thus the shape of the data in Fig. 10a is simply the number of constituent-quark participants/nucleon participant, $N_{\rm qp}/N_{\rm part}$.

For symmetric systems such as Au+Au, the NQP model is identical to another model from the 1970's, the Additive Quark Model (AQM) [29]. The AQM is actually a model of particle production by color-strings in which only one color-string can be attached to a constituent-quark participant. Thus, for asymmetric systems such as d+Au, the maximum number of color-strings is limited to the number of constituent-quarks in the lighter nucleus, or six for d+Au, while the NQP allows all the quark participants in both nuclei to emit particles. Fig. 10b shows that the NQP model gives the correct E_T distribution in d+Au, while the AQM has a factor of 1.7 less E_T emission due to the restriction on the number of effective constituent-quarks in the larger nucleus. The positions of the three-constituent quarks are generated about the position of each nucleon, in a standard Glauber calculation, according to the measured charge distribution of the proton, which gives a physical basis for the "proton size fluctuations" recently discussed at LHC [30].

Constituent-quark-participants might also explain the increase of the "effective energy" in A+A collisions compared to p-p collisions as discussed above (Fig. 9b). The $\langle N_{\rm qp}/N_{\rm part}\rangle$ is 1.5 for a p-p collision but rises to 2.3-2.7 for more central (0-50%) A+A collisions. Thus the "effective energy" for particle production increases due to an increase in the number of (constituent-quark) participants, not because of additional collisions of a given participant. This preserves the assumption in these "extreme-independent" participant models that successive collisions of a participant do not increase its particle emission.

The most surprising soft-physics result in p+Pb and d+Au physics this year concerns what looks very much like collective flow observed in these small systems, where no (or negligible) medium or collective effect was expected. Fig. 11a shows a LEGO plot of $\Delta\eta$, $\Delta\phi$, the difference in polar and

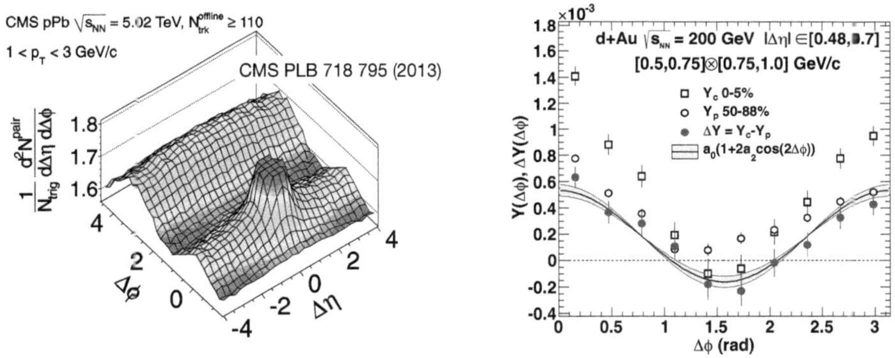

Figure 11: a) (left) CMS ridge in two-particle correlations [31]. b) (right) PHENIX two-particle azimuthal correlations in d+Au at RHIC [32].

azimuthal angles from correlations of two particles with $1 < p_T < 3$ GeV/c in p+Pb by CMS at $\sqrt{s_{NN}} = 5.02$ TeV [31]. A clear $1 + 2v_2 \cos 2\Delta\phi$ modulation of the distribution independent of $\Delta\eta$ is observed, called the 'ridge' in Au+Au collisions where the modulations v_2, v_3, $\ldots v_n$ are attributed to collective flow of the QGP medium. At RHIC, PHENIX confirmed this result in d+Au collisions at $\sqrt{s_{NN}} = 200$ GeV (Fig. 11b) [32]. In order to remove any v_2 effect due to two-particle correlations of hard-scattering, lower p_T triggers were used as well as cuts in $\Delta\eta$ to remove the same-side peak. Also, since there is no suppression of hard-scattering in p+Pb or d+Au collisions (recall Fig. 8), the conditional two-particle yield from di-jets is independent of centrality. Thus, any residual hard-scattering effect was removed by subtracting the peripheral (50-88%) from the central (0-5%) measurement which revealed the beautiful $\cos 2\Delta\phi$ curve characteristic of elliptical flow shown in Fig. 11b.

Figure 12a compares the v_2 measurements vs p_T from d+Au at $\sqrt{s_{NN}} =$ 200 GeV and p+Pb at 5.02 TeV. The larger values from the d+Au results are thought to be due to the larger eccentricity (ε) of the two-nucleon deuteron compared to the single nucleon proton. In fact, the values of v_2/ε from d+Au and p+Pb are consistent with the dependence of v_2/ε on $dN_{\rm ch}/d\eta$ (Fig. 12b) observed in Au+Au and Pb+Pb collisions, which was taken as proof of collective flow from hydrodynamics.

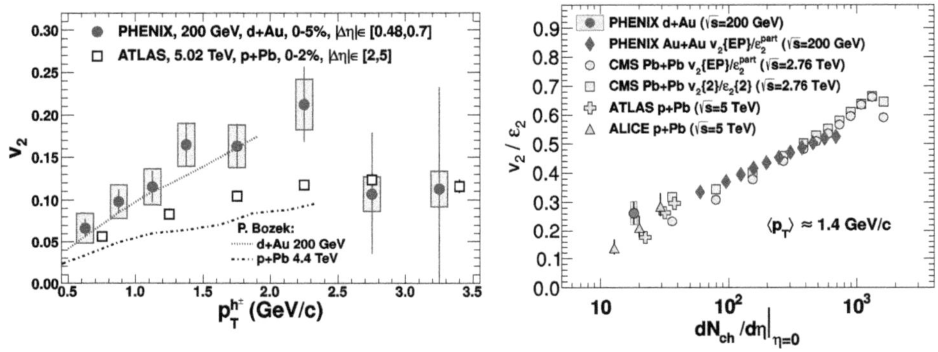

Figure 12: a) (left) Comparison of v_2 vs. p_T in d+Au at RHIC to p+Pb at LHC [32]. b) (right) Compilation of v_2/ε vs. $dN_{\rm ch}/d\eta$ at $p_T = 1.4$ GeV/c in d+Au, p+Pb, Au+Au and Pb+Pb collisions at RHIC and LHC [32].

These new results again underscore the importance of p-p and p+A comparison data to understand the observations in RHI collisions, where the detailed physics of the QGP is far from understood. To this end, PHENIX has proposed a new more conventional collider detector, s(uper)PHENIX, based on a thin-coil superconducting solenoid, to concentrate on hard-scattering and jets. This would replace the very successful but 15 year old special purpose small aperture two-arm spectrometer designed to measure $J/\Psi \to e^+ + e^-$ down to zero p_T at mid-rapidity, the original expected signal for deconfinement, as well as identified particles such as single $e^\pm$ from heavy quark decay, π^0, η and other hadrons (recall Fig. 5b) that could cause background to the J/Ψ but which turned out to be valuable probes of the QGP. This year sPHENIX got a big boost by acquiring the (made in Italy) BABAR solenoid magnet from SLAC which became available when the B-factory in Italy was unfortunately cancelled. Conceptual design of the new experiment is well underway, with mid-rapidity, forward and eRHIC capability [33] (Fig. 13). New collaborators would be most welcome.

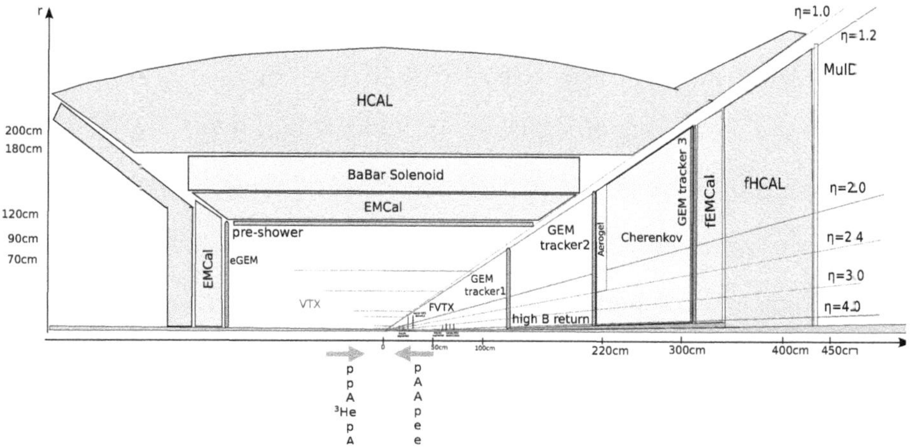

Figure 13: a) sPHENIX concept with forward detector [33].

References

[1] G. W. Bennett, *et al.*, Phys. Rev. D**73** (2006) 072003.

[2] F. J. M. Farley, J. Bailey, R. C. A. Brown, M. Giesch. H. Jostlein, S. van der Meer, E. Picasso and M. Tannenbaum, Nuovo Cimento **45A** (1966) 281-286.

[3] G. Charpak, F. J. M. Farley, R. L. Garwin, T. Muller, J. C. Sens and A. Zichichi, Nuovo Cimento **37** (1965) 1241-1363.

[4] R. W. Ellsworth, A. C. Melissinos, J. H. Tinlot, H. von Briesen, Jr., T. Yamanouchi, L. M. Lederman, M. J. Tannenbaum, R. L. Cool and A. Maschke, Phys. Rev. **165** (1968) 1449-1465.

[5] G. Danby, J-M. Gaillard, K. Goulianos, L. M. Lederman, N. Mistry, M. Schwartz and J. Steinberger, Phys. Rev. Lett. **9** (1962) 36-44.

[6] J. H. Christenson, J. W. Cronin, V. L. Fitch and R. Turlay, Phys. Rev. Lett. **13** (1964) 138-140.

[7] R. L. Cool, *et al.*, in *Proc. XVI Int. Conf. High Energy Physics*, Eds. J. D. Jackson and A. Roberts (NAL, Batavia, IL, 1973), Vol 3, p. 317.

[8] J. F. Owens and J. D. Kimel, Phys. Rev. D**18** (1978) 3313-3319.

[9] R. P. Feynman, R. D. Field, and G. C. Fox, Phys. Rev. D**18** (1978) 3320-3343.

[10] C. De Marzo, *et al.*, Phys. Lett. B**112** (1982) 173-177.

[11] J. D. Bjorken, Phys. Rev. D**8** (1973) 4098-4106.

[12] C. Bromberg, *et al.*, Nucl. Phys. B**134** (1978) 189-241.

[13] G. Arnison, *et al.* (UA1), CERN-EP-82-122 (unpublished).

[14] G. Arnison, *et al.* (UA1), Phys. Lett. B**107** (1981) 320-324.

[15] Jan Rak and Michael J. Tannenbaum, *High p_T Physics in the Heavy Ion Era*, (Cambridge Univ. Press, Cambridge & New York, 2013).

[16] A. Adare, *et al.* (PHENIX), Phys. Rev. Lett. **104** (2010) 132301.

[17] S. S. Adler, *et al.* (PHENIX Collab.), Phys. Rev. C**76** (2007) 034904.

[18] See R. Baier, D. Schiff and B. G. Zakharov, Ann. Rev. Nucl. Part. Sci. **50** (2000) 37–69, and references therein.

[19] A. Adare, *et al.* (PHENIX), Phys. Rev. C**87** (2013) 034911.

[20] K. Aamodt, *et al.* (ALICE), Phys. Lett. B**696** (2011) 30-39.

[21] A. Adare, *et al.* (PHENIX), Phys. Rev. Lett. **109** (2012) 152301.

[22] A. Adare, *et al.* (PHENIX), Phys. Rev. C**88** (2013) 024906.

[23] B. B. Back, *et al.* (PHOBOS), Phys. Rev. C**72** (2005) 031901(R).

[24] A. Białas, A. Błeszynski and W. Czyż, Nucl. Phys. B**111** (1976) 461.

[25] B. B. Back, *et al.* (PHOBOS), Phys. Rev. C**74** (2006) 021901(R).

[26] M. Basile, ... A. Zichichi, *et al.*, Phys. Lett. B**95** (1980) 311-312.

[27] S. Eremin and S. Voloshin, Phys. Rev. C**67** (2003) 064905.

[28] R. Nouicer, Eur. Phys. J. C**49** (2007) 281-286.

[29] A. Białas, W. Czyż and L. Lesniak, Phys. Rev. D**25** (1982) 2328.

[30] ATLAS Collaboration, ATLAS-CONF-2013-096 (September, 2013).

[31] S. Chatrchyan, *et al.* (CMS), Phys. Lett. B**718** (2013) 795.

[32] A. Adare, *et al.* (PHENIX), Phys. Rev. Lett. **111** (2013) 212301, and references therein.

[33] `http://www.phenix.bnl.gov/plans.html`

CHAIRMAN: M.J. TANNENBAUM

Scientific Secretaries: V. Begun, J. Lange

DISCUSSION

- Tawfik:

It is good to know that PHENIX is planning an upgrade to sPHENIX or ePHENIX the same as STAR. I understood from your lecture that you are also planning heavy-ion collisions. Is there any ingredient you are going to add to the STAR program?

- Tannenbaum:

Are we going to add to the STAR program? They don't have a hadron calorimeter. Nor can they ever have one because of the iron yoke that they built. They would have to rebuild that whole magnet.

- Tawfik:

Fine. The other question is regarding the thermal photons. Are you planning to measure it at different center of mass energies?

- Tannenbaum:

Measure what?

- Tawfik:

Thermal photons, thermal photons, photons.

- Tannenbaum:

Direct-photons?

- Tawfik:

Direct-photons.

- Tannenbaum:

We try to measure it whenever we can; but it's a hard measurement because the rate is low. Actually, we are much more interested in direct-photon-hadron correlations to measure what`s happening with an away jet and particles losing energy, because it is really a precision measurement. In fact, we demonstrated, as I showed last year, that we do get the correct fragmentation function measuring direct-photon-hadron correlations in p-p collisions. In fact a new paper for gold-gold just came out today or very soon. It was recently accepted for publication but I already discussed it last year. We measured the fact that in direct-photon-hadron correlations in gold-gold collisions you can see the lost energy on the away side and you can see that it spreads out, like CMS originally showed. We plan to do a lot more gold-gold running measuring direct-photons, but we probably would not go so much lower in energy. I do not see the reason why, because you need a lot of luminosity and typically once you get down below 62.4 GeV c.m. energy the luminosity is bad and you would not get very high in pT with direct-photons.

References:
p-p: PHENIX Collaboration, Phys. Rev. D **82**,072001 (2010)
Au-Au: PHENIX Collaboration, Phys. Rev. Lett. **111**,032301 (19 July 2013)

- *Tawfik:*

Are you also planning to do a beam-scan program to study collisions at different center of mass energies?

- *Tannenbaum:*

Of course, that's the nice thing about a collider. The two experiments have to agree: actually the management has to get the two experiments to select what is the most important according what committees feel; but then both experiments make the measurements. Whatever STAR makes – we make; STAR has some advantages, and we have some advantages. For high pT production, STAR recently presented some measurements, which I showed here last year, but PHENIX had published such measurements a long time ago.

- *Tannenbaum:*

I didn't understand that Tawfik's original question was for thermal photons at low pT, not direct-photons at large pT. For thermal photons (1<pT<3 GeV), the answer is that there are 62.4 GeV data but the PHENIX measurements involve using either internal conversions to eliminate the pizero background, which costs a factor of 1000 in rate, or by external conversion for which the reduction in rate is a factor of 10 to 200 depending on the configuration but which only reduces the hadron contamination. We have not yet published an external conversion result at 200 GeV. When this is complete we will try it with the existing 62.4 GeV data.

- *Zennamo:*

I had a question about the two-particle correlation functions. CMS had observed a ridge in the p-p collisions and I was wondering if RHIC would be able to do anything with that or if there was any insight that could be offered into that at all?

- *Tannenbaum:*

Very interesting. I actually wrote a paper on that with Richard Weiner. So the answer is: are there longitudinal correlations in p-p collisions? Most certainly; but this has been known for years and is the basis for color string models. The issue is whether there is hydrodynamic flow in p-p collisions; or is there a quark gluon plasma (QGP) in p-p collisions? Richard Weiner believes that there is a QGP in p-p collisions and that hydro works. I am agnostic on whether hydro works or not; but I said that I don't believe that there is a QGP in p-p collisions. In this article (which I don't have the reference at hand but just look up Weiner and Tannenbaum, there is only one such paper), I said that in order for me to believe that a QGP is produced in p-p collisions I would have to see the two key signatures. The main signature is that pizeros are suppressed. What I proposed is that you take low multiplicity, or inclusive, p-p collisions and you measure the gamma/pizero ratio. Then you go to high multiplicity and you again measure the gamma/pizero ratio. The gammas will not be absorbed by the medium but if there is a QGP then the pizeros will. The reason why you need to make two measurements is that energy conservation comes into play--you expect that at high multiplicity there is not enough energy available to make high pT particles. The "ridge" effects seen in d+Au are not so surprising. In p-p collisions, the big test, according to Edward Shuryak, is to look for radial flow. What happens in radial flow is that if you have some medium which is undergoing collective expansion, and if a proton which was produced with a certain velocity, beta, is flowing in the medium, it picks up the medium collective velocity, so you have the Lorentz combination of the two velocities and you would see that. The issue is whether you can see radial flow in p-p collisions. I recently learned something interesting on this subject. I got very upset when the CMS paper on the ``ridge" in p-p collisions was published in 2010. Why? Because they claimed in a press release that it was the quark gluon plasma like was seen at RHIC. I looked at this press release (and I am really against physics by press release, as you might have heard in my lectures here two years ago). I looked up this press release and it was not a CERN press release. In fact, it wasn't even a press release, it was a CMS description of their experiment on their web site, where they claimed it. This year, we just had a meeting at MIT about a month ago celebrating Wit Busza. Now, CMS has nice results on flow in d+Pb which I do believe. The interesting thing was that when I asked ``What about this high multiplicity p-p data?" How does that work? Gunther Roland answered, ``Oh that's all full of multi jets." So I do believe in longitudinal correlations in p-p collisions; but once you trigger on something that has 30 jets in it, I don't think that you're ever going to sort that out; but whether there are longitudinal correlations, I certainly do believe that.

References:
CMS Collaboration, JHEP 1009:091, 2010, arXiv:1009.4122 [hep-ex]
M.J. Tannenbaum, R.M. Weiner, arXiv:1010.0964 [hep-ph]

- *Sicking:*

In the interpretation of heavy-ion collisions, we use Glauber calculations, e.g. for the nuclear modification factor R_{AA}. In Prof. Rith's talk, we saw that the EMC effect can be explained, for example, by pions and quark pairs in addition to the nucleons in the nuclei. Have the Glauber calculations been updated for this? Should Glauber calculations be reconsidered taking this into account?

- *Tannenbaum:*

The answer is a) we don't actually need Glauber calculations for hard-scattering because the cross section for a point-like process in A+B collisions is A times B times the cross section in p-p collisions (where A and B are the number of nucleons in the colliding nuclei) and it doesn't matter what the nucleons are doing as long as they are stuck together. That is called minimum bias. Most of our results work in min-bias, the same with flow. It's very easy to make Glauber calculations. In fact, in my book it is all well described; it's really simple and there are papers on this. What we do is the following. We know the nuclear Woods-Saxon distributions, and some have a hard core. Then we generate nuclei with these distributions and sometimes you don't want to space the nucleons too close, so you leave space according to the hard core. This way you generate all the nucleons in the nucleus. And then? What are you saying? What else should we do?

- *Sicking:*

My point is that A and B are then A plus something and B plus something, so the numbers are changed.

- *Tannenbaum:*

Honestly, it's very simple to do these Monte Carlo Glauber calculations. We could put in pions or quark pairs or correlations between two nucleons (in fact we do that with the hard core); but I don't know that there's any hard evidence for that from EMC. I'm still trying to understand what kind of a correlation the nucleons have with this new explanation of the EMC effect that Rith mentioned. It's also very easy to put these constituent quarks in. Some of the theorists say that they do a Glauber calculation and they add quantum fluctuations. What we do is that we first generate the positions of all the nucleons. Then, since Hofstadter has measured a nice exponential distribution of charge in a nucleon, we generate three constituent quarks according to that distribution, so the quarks are no longer at the position of their nucleon, they have moved around. This then gives a physical basis for what the theorists put in ad hoc. In summary, the answer to your question is that if something like what you originally asked were proved, it is trivial to modify a Monte Carlo Glauber Calculation. I could do it this afternoon.

Alessandro Bettini
University of Padua. Department of Physics; INFN. Padua division (Italy)
Laboratorio Subterráneo de Canfranc (Spain)
bettini@pd.infn.it
bettini@lsc-canfranc.es

Status of the Three Neutrinos

1. The three families

The elementary particles come in three "families" with the same structure and distinguished by the masses of their elements. Each of them contains two quarks and two leptons. The leptons are: the electron and its neutrino (e^-, ν_e), having electron lepton flavour, the muon and its neutrino (μ, ν_μ), with muon flavour, and the tau and its neutrino (τ, ν_τ), with tau flavour.
The Standard Model assumes that neutrinos of definite flavour to be stationary states, namely eigenstates of the mass; it assumes also that all their masses to be zero. Both assumptions are contradicted by the experiment.
The SM assumes that neutrinos and antineutrinos are different particles, distinguished by the lepton number, but the assumption is not experimentally proven.
The SM assumes that neutrino flavours are three. This assumption has been experimentally proven with high precision by measuring the line shape of the Z at LEP. Indeed, the width of the resonance increases and its peak height decreases with increasing number of neutrinos into which the Z can decay. The LEP experiments have established $N_\nu = 2.984 \pm 0.008$. Additional information comes from cosmology, which gives two pieces of evidence. The first one is on the "effective neutrino density", namely the density of non-interacting radiation in the early universe, neutrinos or else. The corresponding effective neutrino number expected in the SM is $N_{\nu eff}$=3.0395[1]. The Planck mission recently published its first year data[2] on the Cosmic Microwave Background (CMB) temperature fluctuations and polarisation. A fit including data from other experiments on CMB at higher angular resolution as well on the Baryon Acoustic Oscillation gives $N_{\nu eff} = 3.30^{+0.54}_{-0.51}$. This value is fully compatible with 3, even if 4 is not excluded. The second piece of information are upper limits on the sum of all neutrino masses, including hypothetical ones, called sterile, not coupled to the Z. The limits depend on several assumptions but may safely be taken between 300 and 600 meV.
Claims for sterile neutrinos have been advanced both in ν_e or $\bar{\nu}_e$ disappearance and in ν_μ to ν_e oscillation data, without however being supported by substantial evidence, as we shall now summarise.
For electron neutrino disappearance, the so-called "Gallium anomaly" is the following. Both Gallium-based solar neutrino radiochemical experiments, GALLEX/GNO and SAGE, performed two runs each with radioactive neutrino sources for checking the overall detection performance. The average of the four ratios between the measured interaction rate and the expected ones is R=0.86±0.05. However, the expected values depend on the cross section of the detection process, namely $\nu_e + {}^{71}Ga \rightarrow {}^{71}Ge + e^-$. This includes non-measured contributes that have been calculated, with a not completely known theoretical uncertainty. The latter is not included in the above-mentioned error bar. This is why neither experiment used R for calibration and why no anomaly can be reliably stated.
For electron antineutrino disappearance, the so-called "reactor anomaly" is the following. In 2011 Th. A. Mueller et al.[3] re-evaluated the neutrino flux from power reactors, finding it higher than the previously accepted value by about 3%. Soon afterwards, G. Mention et al.[4],

considered the available measurements of the neutrino flux from several experiments at different distances, shorter than 100 m, from a number of reactors, finding that in average they account for only 0.943±0.023 of the expectations with the new flux calculation. The effect is only 2.5 σ, but was claimed as an anomaly. Notice however, that the fission beta spectra involve some 6000 transitions, one third of which are forbidden. The latter introduce large uncertainties in the predicted antineutrino flux, which remain also in the Mueller at al. re-evaluation and that is equal or larger than the claimed uncertainty and effect. Moreover, in 2013, when all mixing angles were known including θ_{13}, C. Zhang et al.[5] could fit the known distance dependence of the antineutrino flux to the measured values. The extrapolation to zero distance gives 0.959 ± 0.009 (exp) ± 0.027(syst) of the Mueller re-evaluation. This is a 1.4 σ effect, which is much smaller than the systematic uncertainty of the latter. In conclusion, there is no sound reactor anomaly.

Anomalies have been reported also in appearance experiments, which, notice, look at physical processes different from the disappearance ones. In 1995 the LSND experiment observed 3.8 σ excess in $\bar{\nu}_\mu$ to $\bar{\nu}_e$ appearance. If interpreted as an oscillation it would correspond to Δm^2>0.2 eV^2. However, the effect was based only on a background model, a very unsafe procedure. In 2001, the KARMEN experiment excluded a large fraction of the LSND claimed parameter space. MiniBooNE at FermiLab started as ν_μ to ν_e appearance experiment, which should be equal to $\bar{\nu}_\mu$ to $\bar{\nu}_e$ for *CPT*. The experiment worked at higher energy but in the same range of *L/E*, which is the relevant quantity, as LSND. In 2009 the experiment published a result, in which no signal was present in the LSND claimed signal region. There was however excess at lower energies, where the background model is not reliable, and, in any case, not compatible with oscillations. In 2010 MiniBooNE presented results on $\bar{\nu}_\mu$ to $\bar{\nu}_e$claiming a signal in oscillation region having observed an excess over calculated background of 20.9±14.0 events. Notice that the statistical significance is only 1.5 σ and that if real when combined with the neutrino result of the same experiment would imply a large *CPT* violation. Finally, at the Neutrino conference at Kyoto in 2012, MiniBooNE reported that higher statistics anti-neutrino data were "now much more consistent with neutrino data. It is not yet known whether the excesses [those at low energy] are due to oscillations, some unrecognized neutral currents γ background, or something else"[6]. The lesson is that background models are always uncertain and consequently claims of discoveries based only or mainly on an observed difference between data and expectations from such models are not reliable.

Additional information comes from the searches of anomalies in the ν_μ to ν_e appearance by ICARUS[7] and OPERA[8] at the CERN to Gran Sasso beam (CNGS). These lead to the exclusion of a large fraction of the parameter region claimed by MiniBooNE[9]. The surviving one would imply a "new" neutrino mass between 0.6 and 1 eV, in contradiction with the above quoted limit from cosmology.

2. Lepton identification

Let us see how the neutrino states of definite flavour are defined. We call ν_e the neutral particle produced with a positron, for example in β^+ decay and $\bar{\nu}_e$the neutral particle produced with an electron, e. g. in β^- decay, and similarly for ν_μ and $\bar{\nu}_\mu$ and for ν_τ and $\bar{\nu}_\tau$. Neutrinos cannot be directly detected. The charged lepton produced by the neutrino interaction in the detector identifies the neutrino flavour. Notice that this fact implies the question: is the flavour at detection equal to that at production? The answer is YES, if proper time elapsed between production and detection is sufficiently small, NO if it is large, as we shall quantify later.

From the experimental point of view, notice that leptons are usually much rare than the hadrons and consequently that the identification of the charged lepton flavour without ambiguities is mandatory.
The main categories of detectors are the trackers, which show, within a resolution dependent on the technique, the images of the tracks of the charged particles, and the calorimeters that absorb and measure all the energy of the particle.
Electrons and pions produce very equal tracks, but can be distinguished by the different shape of the shower they produce. An electron penetrating in a dense medium will soon radiate a photon, which in turn will convert in electron-positron pair, which will radiate more photons, etc. A pion will behave similarly, but producing a shower containing also hadrons on top of electrons and photons. The difference in shape between the two showers can be enhanced and detected with proper techniques, as we shall see, particularly in their initial phases. A high discriminating power is needed to reliably identify the electrons that are produced very rarely compared to the pions.
Muons can be distinguished from hadrons and electrons due to their much higher penetrating power. Differently from pions, muons go through metres of material without interacting, leaving a long, straight track.
Tau leptons have a lifetime much shorter than the more common hadrons, about 0.3 ps; at a few GeV energy, which are typical in the experiments under discussion, the lengths of their tracks between production and decay points, are of the order of hundreds micrometres.
In 1962 M. Schwartz, L. Lederman, J. Steinberger et al.[10] built the first neutrino beam by colliding the circulating proton beam on an internal target at the BNL AGS accelerator and absorbing all the emerging hadrons and muons with a sufficiently thick Fe absorber. Only neutrinos, those produced in the pion decays together with a muon, could reach the detector. The latter was a set of Al spark chambers providing both the target mass and the ability to distinguish muons from electrons. Only the former were observed, proving that two neutrino flavours exist.
In the same years, A. Zichichi[11] developed the PAPLEP (Proton-AntiProton into LEpton Pairs) two-arm spectrometer at the CERN Proton Synchrotron (PS). It was the beginning of the search for the 3rd sequential lepton family, imagined as a replica of the first two, made of a "Heavy Lepton" (HL) and its neutrino (ν_{HL}). The idea was to look for lepton pairs of opposite sign. These can be produced directly or result from the decays of a HL anti-HL pair, produced by the reaction $\overline{p}+p \to HL^+ + HL^-$, shortly followed by the decays $HL^+ \to e^+\nu\overline{\nu}$ or $HL^+ \to \mu^+\nu\overline{\nu}$ and $HL^- \to e^+\nu\overline{\nu}$ or $HL^+ \to \mu^-\nu\overline{\nu}$. The presence of the intermediate Heavy Leptons can be inferred when one of these decays is into an electron the other into a muon, while in the direct production the leptons have the same flavour. A further signature, due to the presence of undetected neutrinos, is the "acoplanarity", meaning that the plane defined by the momenta of the detected leptons does not contain the direction of the beam.
The detector elements closest to the target in both arms are two equal copies of the pre-shower, which is a sandwich of five repeated elements, each composed of Pb layer controlling the early development of the shower, a plastic scintillator sampling the energy deposit and an Al-plates spark chamber visualizing the tracks. The novel method[12] combined in this way visual and non-visual approaches, each providing 10^{-2} rejection to obtain an overall rejection of the pions of a few 10^{-4}. Not having observed the *HL*, possibly for the non-point-like nature of the proton, Zichichi moved the search to the highest energy e^+e^- collider available, ADONE at Frascati[13] using similar techniques. Once more, the HL was not found, and only a lower limit of 1 GeV could be established for its mass[14]. The reason is that the HL mass is 1.777 GeV, too large to be produced at ADONE, which had a maximum energy of 3 GeV. The lepton was finally discovered at the higher energy e^+e^- collider, SPEAR at SLAC, using the same technique, by M. Perl and collaborators[15] in 1975, and called tau.

The discovery of the other member of the 3rd lepton family, the tau-neutrino, had to wait another quarter of a century with the DONUT experiment at Fermilab[16]. The neutrino beam was produced using 800 GeV protons from the Tevatron. The detector must have a large mass, because neutrino cross-sections are small and micrometre level spatial resolution, provided by emulsion techniques. The specific technique, the emulsion cloud chamber (EEC), had been developed by the Nagoya group. The EEC is a sandwich of 1 mm thick Fe sheets, providing the mass, interleaved with emulsion sheet pairs, providing the images of the track segments. A short track, a kink between the mother and the daughter, is the signature of the tau lepton.

3. Neutrino change flavour

Two mechanisms observed changing neutrino flavour: oscillation in vacuum, which corresponds to the kinetic part of the Hamiltonian and transformation in matter (Mikehev-Smirnov-Wolfestein effect[17]), which is a dynamical phenomenon due to interaction potential between ν_e and electrons

Two types of experiment should be distinguished: those looking for the appearance of a flavour different from the original one, in which the far detector must be sensitive to the new flavour, and those searching for the disappearance of the initial flavour, for which the initial flux, its composition and spectrum must be accurately known.

The mentioned phenomena may happen when neutrinos of definite flavour are not stationary states (mass eigenstates). The latter, ν_1, ν_2 and ν_3, have definite masses, m_1, m_2 and m_3. The two basis are linked by an orthogonal transformation that can be expressed in terms of three rotations, through angles that we shall call θ_{12}, θ_{23} and θ_{13}, and of phase factors. If neutrinos are Dirac particles, as assumed in the Standard Model, all but one of the phase factors can be absorbed, as in the case of quarks, in the wave functions of the states. However, neutrinos might be completely neutral, namely 'Majorana particles'. In this case two more phases, which we shall call Majorana phases ϕ_1 and ϕ_2, are physically observable.

In conclusion, writing c_{ij}=cosθ_{ij}, and s_{ij}=sinθ_{ij}, the unitary transformation $\nu_l = \sum_{1=1}^{3} U_{li}\nu_i$ is

$$\begin{pmatrix} \nu_e \\ \nu_\mu \\ \nu_\tau \end{pmatrix} = \begin{pmatrix} 1 & 0 & 0 \\ 0 & c_{23} & s_{23} \\ 0 & -s_{23} & c_{23} \end{pmatrix} \begin{pmatrix} c_{13} & 0 & s_{13}e^{-i\delta} \\ 0 & 1 & 0 \\ -s_{13}e^{i\delta} & 0 & c_{13} \end{pmatrix} \begin{pmatrix} c_{12} & -s_{12} & 0 \\ s_{12} & c_{12} & 0 \\ 0 & 0 & 1 \end{pmatrix} \begin{pmatrix} 1 & 0 & 0 \\ 0 & e^{i\phi_1} & 0 \\ 0 & 0 & e^{i\phi_2} \end{pmatrix} \begin{pmatrix} \nu_1 \\ \nu_2 \\ \nu_3 \end{pmatrix} \quad (1)$$

Notice that Majorana phases are irrelevant for the oscillation and matter effects. They are observable in an extremely rare, not yet observed phenomenon, the neutrino-less double beta decay (see §5).

Experiments have measured cross sections and energy spectra relevant for the oscillation phenomena. The frequency of the oscillation in vacuum is proportional to the absolute value of the difference of the squares of two neutrino masses. The flavour conversion in matter, in the Sun in particular, depends also on the sign of the difference.

The values of two of the larger mixing angles, θ_{12} and θ_{23}, have been known with increasing accuracy since many years. Information on θ_{12} is mainly due to solar neutrinos and reactor antineutrinos, that on θ_{23} to atmospheric neutrinos, reactor antineutrinos and accelerator neutrinos. The smallest mixing angle the third angle, θ_{13}, has been measured in 2012 by disappearance experiments on $\bar{\nu}_e$ from power reactors, and ν_e appearance on a ν_μ beam from a proton accelerator, as we shall see.

A global fit to all the measurements, e.g. Fogli et al.[18] gives

$$\sin^2\theta_{12} = 0.307 \pm 0.017 \qquad \theta_{12}\ ;\ 33Þ$$
$$\sin^2\theta_{23} = 0.390 \pm 0.033 \qquad \theta_{12}\ ;\ 40Þ\text{-}50Þ. \qquad (2)$$
$$\sin^2\theta_{13} = 0.0242 \pm 0.0026 \qquad \theta_{12}\ ;\ 9Þ$$

and for the square mass differences

$$\delta m^2 = m_2^2 - m_1^2 = 75.4 \pm^{2.6}_{2.2}\ \text{meV}^2 \qquad \left|\Delta m^2\right| = \left|m_3^2 - \frac{(m_1+m_2)^2}{2}\right| = 2430^{+70}_{-90}\ \text{meV}^2 \qquad (3)$$

Fig. 1 shows schematically the neutrino square-mass spectrum. It consists of a singlet (ν_3) and a doublet (ν_1, ν_2). The approximate flavour composition of the eigenstates is also shown. We do not know either the absolute scale or whether the mass of the singlet is larger or smaller than that of the doublet.

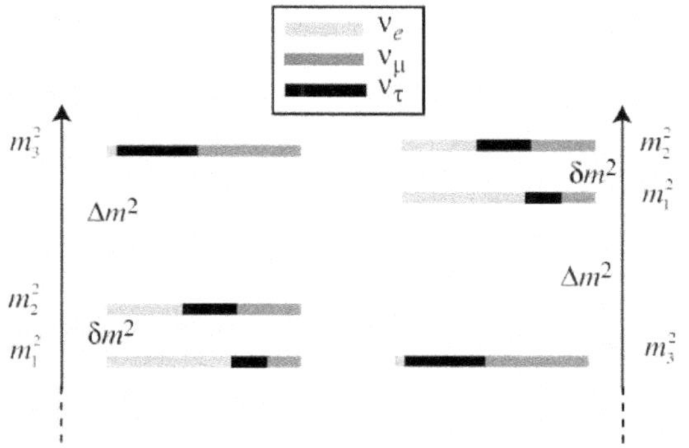

Fig. 1. Neutrino square-mass spectrum. The flavour contents of the eigenstates are also shown

4. What next

The smallest of the mixing angles θ_{13} was measured, as the last one, in 2012, in two different ways: in disappearance experiments, DayaBay[19] and, soon afterward, RENO[20], on $\bar{\nu}_e$ originated on a power nuclear reactor over 2 km base-line, and the T2K[21] appearance experiment searching for ν_e in a ν_μ beam produced by the J-Park proton accelerator, over about 300 km base line. The oscillation probability of the former is (L in km, E in GeV)

$$P(\nu_e \to \nu_x) = \sin^2(2\theta_{13})\sin^2\left(1.27\Delta\frac{L}{E}\right) \approx 4\theta_{13}^2\sin^2\left(1.27\Delta\frac{L}{E}\right). \qquad (4)$$

and that of the latter, in a first approximation

$$P(\nu_\mu \to \nu_e) = \sin^2\theta_{23}\sin^2 2\theta_{13}\sin^2\left(1.27\Delta\frac{L}{E}\right) \approx 2\theta_{13}^2\sin^2\left(1.27\Delta\frac{L}{E}\right) \qquad (5)$$

T2K reported, with an exposure corresponding to 6.4×10^{20} p.o.t. (protons on target), which is 9% of the foreseen total, to have observed 28 ν_e interactions, with an estimated number of background events of 4.6.

Neutrino oscillations and neutrino effects in matter cannot determine the absolute scale of neutrino masses, but, of the still unknown quantities, they can still give the sign of Δm^2 and the phase δ corresponding to CP violation. Both of them can be measured with a high intensity neutrino beam form an accelerator. The former is accessible also with different approaches, as we shall now summarise.

The first option is to consider the interference between the oscillations of the shorter and longer periods in a $\bar{\nu}_e$ reactor disappearance experiment, which correspond to higher order terms in the oscillation probability, which in vacuum becomes

$$P(\bar{\nu}_e \to \bar{\nu}_e);\ 1-\sin^2 2\theta_{12} \sin^2 \Delta_{21} - 4\theta_{13}^2 \cos^2 \theta_{12} \sin^2 \Delta_{31} - 4\theta_{13}^2 \sin^2 \theta_{12} \sin^2 \Delta_{32}\,. \qquad (6)$$

where $\Delta = 1.27\Delta m^2 \left(\mathrm{GeV}^2\right) \frac{L(\mathrm{km})}{E(\mathrm{GeV})}$ and $|\Delta_{31}^2| = |\Delta_{32}^2| + |\Delta_{21}^2|$ for $\Delta m^2>0$ (normal hierarchy) and $|\Delta_{31}^2| = |\Delta_{32}^2| - |\Delta_{21}^2|$ for $\Delta m^2<0$ (inverted hierarchy).

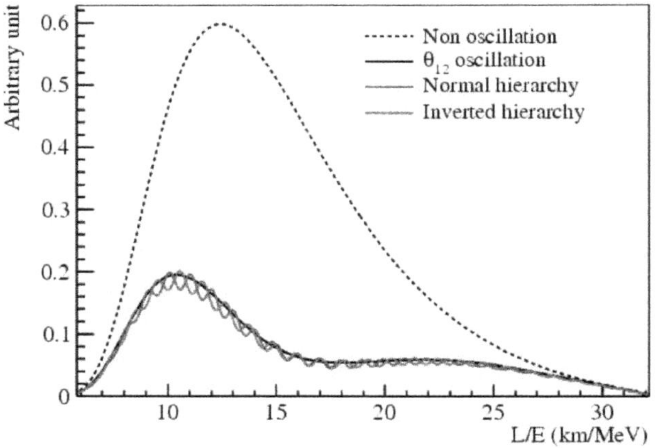

Fig. 2. Survival probability of electron antineutrinos

Fig. 2 shows the corresponding survival probability as a function of *L/E* for the two mass hierarchies. The difference between them is maximum around *L/E*=10 km/MeV, namely about 50 km for reactor neutrinos.

In the JUNO[23] experimental proposal in China the far detector will have 20 kt mass of liquid scintillator. It will be located at 53 km from the Yangjiang and Taishan nuclear power plants, both of which are under construction. The planned total thermal power of these reactors is 36 GW. The detector will be located underground under 700 m rock overburden, to be shielded from cosmic rays. Construction, including the excavation of the experimental hall and its access tunnel, is planned to start in 2013 and to be completed in 2019. The planned detector needs to be two orders of magnitudes in volume larger than those of the present generation. This looks feasible. The energy resolution must be better than 3% at 1 MeV to be able to resolve the small wavelength oscillations seen in Fig. 2. This is a real challenge, a substantial improvement on the state of the art detector, BOREXINO.

A similar proposal, RENO-50, is under study in South Korea.

Atmospheric muon neutrinos offer another possibility to determine the mass hierarchy. Again a subdominant effect must be employed. Neutrinos and antineutrinos feel a different matter potential when crossing the Earth. Consequently the survival probabilities of muon neutrinos and antineutrinos are a little different. The difference is maxim at the MSW resonance. The India based Neutrino Observatory (INO)[24] project in India is based on the MONOLITH idea originally proposed at LNGS. INO is a tracking magnetised Fe calorimeter that will measure the tracks of the muons produced by ν_μ interactions, their momentum and their charge. It will be made of RPC (Resistive Plate Chambers) tracking and timing (needed to distinguish up-going from down-going muons) elements interleaved by magnetised Fe plates (*B*=1.3 T). The oscillation probability as a function of the zenith angle, is obtained as the ratio of the fluxes at that angle from below (that have oscillated) and above (not oscillated). Different zenith angles correspond to different flight length inside the Earth and different densities crossed by the neutrino. The difference between oscillation probability of neutrinos and antineutrinos (negative and positive muons) is sensitive to the mass hierarchy. In this case the crucial issue is the accuracy in the measurement of the muon momentum.

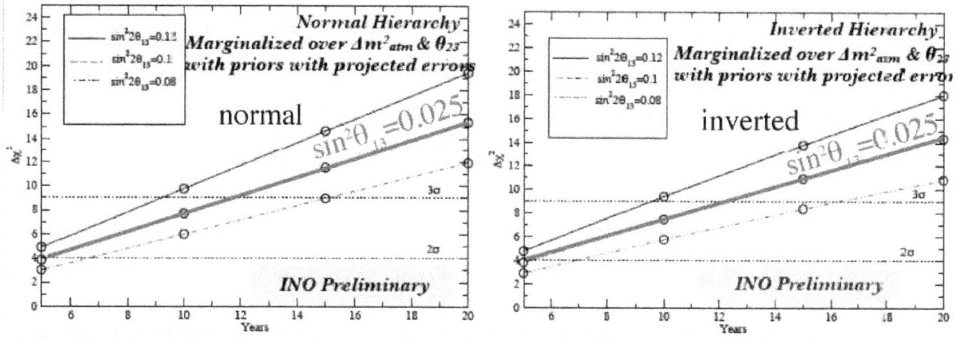

Fig.3 Sensitivity to the mass hierarchy of INO, for different values of θ_{13}. We now know that the correct one is the middle one

Fig. 3 shows the calculated sensitivity to the mass hierarchy of a 50 kt mass detector, showing that, for the now known value of $\sin^2\theta_{13}$, the hierarchy can be determined at 2 σ in 4 yr and 2 σ in eleven years. The international community might join efforts and double or triple the detector mass at a moderate cost, reducing the exposure time needed for a 3 σ result to a few years.

Consider now the ν_μ to ν_e appearance experiments at accelerators. One experiment, T2K, is running in Japan, as already mentioned, and a second larger one, NOvA[25] is starting now in the USA. It will have initially 2.5 kt sensitive mass to gradually reach 14 kt. The power on target at FNAL will start at 300 kW to ramp in time to 700 kW. The base line is 810 km, enough to have some sensitivity to matter effects.

Once more, the new neutrino physics has to be found in second order effects. Notice that disappearance experiments are not sensitive to the *CP* phase, due to *CPT* invariance. Consequently, in absence of matter effects, Eq. (4) is valid also in presence of a *CP* phase. On the contrary, terms containing δ are to be considered in the appearance probability. In vacuum we can express it as the sum of the 0th order term P_0 of Eq. (5) and three 1st order terms

$$P(\nu_\mu \to \nu_e) = P_0 + P_{\sin\delta} + P_{\cos\delta} + P_3 \,. \tag{7}$$

were

$$P_0 = \sin^2\theta_{23}\sin^2 2\theta_{13}\sin^2\left(1.27\Delta\frac{L}{E}\right); \qquad \alpha \equiv \frac{\delta m^2}{\Delta m^2};\ 0.03$$

$$P_3 = \alpha^2\cos^2\theta_{23}\sin^2 2\theta_{12}\sin^2\left(1.27\Delta\frac{L}{E}\right)$$

$$P_{\sin\delta} = \alpha\times\sin\delta\times\cos\theta_{13}\times\sin 2\theta_{12}\times\sin 2\theta_{23}\times\sin 2\theta_{13}\times\sin^3\left(1.27\Delta\frac{L}{E}\right) \tag{8}$$

$$P_{\cos\delta} = \alpha\times\cos\delta\times\cos\theta_{13}\times\sin 2\theta_{12}\times\sin 2\theta_{23}\times\sin 2\theta_{13}\times\cos\left(1.27\Delta\frac{L}{E}\right)\sin^2\left(1.27\Delta\frac{L}{E}\right)$$

Notice that $P_{\sin\delta}$ has the opposite sign for antineutrinos. Notice also that additional terms due to matter effects are to be considered for long base lines, in practice for NOvA but not for T2K.

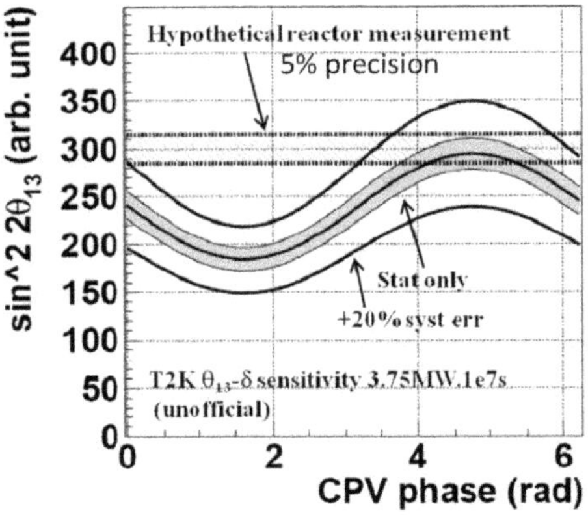

Fig. 4. Left panel

Fig. 4 shows the expected sensitivity of T2K at the end of the exposure and that of the disappearance experiments expected in the same years. Both are obviously in an arbitrary position. We shall be lucky if they are like that.

Due to the different values of *L/E* the sensitivity region of NOvA in the same diagram will be different. The experiment will so provide new independent pieces of information.

We can then conclude that the experiments running in the next few years will bring us very interesting results, possibly giving the first determination of both the mass hierarchy and δ.

Programmes of neutrino facilities over longer distances for next generation high intensity neutrino sources at accelerators are being studied in Europe (LAGUNA-LBNO), in the USA (LBNE) and in Japan (HyperK). Such projects require very large capital investments, of the order of 10^9 €. Only very strong scientific motivations can be convince the Governments to approve so expensive programmes, which can be finalised and optimised only after the results of the above mentioned present generation experiments will be available. Consequently, the maximum effort should be concentrated on them and on the development of a staged strategy.

5. Are neutrinos and antineutrinos different particles?

We have seen that flavour lepton numbers are not conserved, something that make it plausible that even the total lepton number might not be absolutely conserved too. Being it the only quantum number that distinguishes neutrinos from antineutrinos, neutrino and antineutrino may be the same particle. In this case neutrinos are completely neutral particles, as, for example, photons and π^0 are. The properties of the spin ½ completely neutral particles are somewhat more un-usual. They obey, in particular, the Majorana 1937[26] rather than Dirac, equation. Let us recall that the Dirac equation is $(\gamma_\mu p^\mu - m)\psi = (E\gamma_0 - \vec{p}\cdot\vec{\gamma} - m)\psi = 0$, where the bi-spinor ψ has four components, namely four degree of freedom

$$\psi(x) = \begin{pmatrix} \psi_1 \\ \psi_2 \\ \psi_3 \\ \psi_4 \end{pmatrix} = \begin{pmatrix} \varphi \\ \chi \end{pmatrix}; \quad \varphi = \begin{pmatrix} \varphi_1 \\ \varphi_2 \end{pmatrix}; \quad \chi = \begin{pmatrix} \chi_1 \\ \chi_2 \end{pmatrix}$$

In the representation in which γ_5 is diagonal the Dirac equation, in terms of the two spinor components, the equation is

$$(E + \vec{p}\cdot\vec{\sigma})\chi - m\varphi = 0; \qquad (E - \vec{p}\cdot\vec{\sigma})\varphi - m\chi = 0. \tag{9}$$

As it is well known, the two spinor components φ and χ correspond to the particle and the antiparticle.
Majorana showed to be possible to construct with the components of φ only, a spinor X that transforms like χ and that can take its place in the Dirac equation, and, similarly, with the components of χ only, a spinor Φ transforming like φ. They are:

$$\Phi = i\sigma_2\chi^* \qquad\qquad X = i\sigma_2\varphi^*$$

The Majorana equations for the 2-component spinors are then

$$(E+\vec{p}\cdot\vec{\sigma})\chi - im_a\sigma_2\chi^* = 0; \qquad\qquad (E-\vec{p}\cdot\vec{\sigma})\varphi - im_b\sigma_2\varphi^* = 0. \tag{10}$$

We see that the two equations are decoupled. Consequently, they describe possibly different particles, with different masses. We have taken that into account by writing m_a and m_b.
Notice that for mass less particles, as neutrinos of the Standard Model, the Dirac and Majorana equations are identical. Consequently, there is no difference between the theories in this case. However, Nature has chosen to give a mass to all the spin ½ particles and the two theories give different predictions, which we might be able to observe, as we shall see. It is easy to see that the charge conjugate of the Majorana field is the field itself.
As we already recalled, neutrino of a certain flavour is by definition the neutral particle produced together with a positive lepton of that flavour in a W^-decay, antineutrino the neutral particle produced with a the positive lepton. The *V-A* structure of charged current weak interactions implies that only left, meaning negative chirality, neutrinos and antineutrinos exist. If they are massive, they are both super-positions of positive and negative helicity, namely (see for example[27])

$$\nu_L \;;\; \frac{m}{E}\nu_L^+ + \nu_L^- \qquad\qquad \bar{\nu}_L \;;\; \nu_L^+ + \frac{m}{E}\nu_L^- \qquad . \tag{11}$$

where the upper index is the helicity, the lower one the chirality.

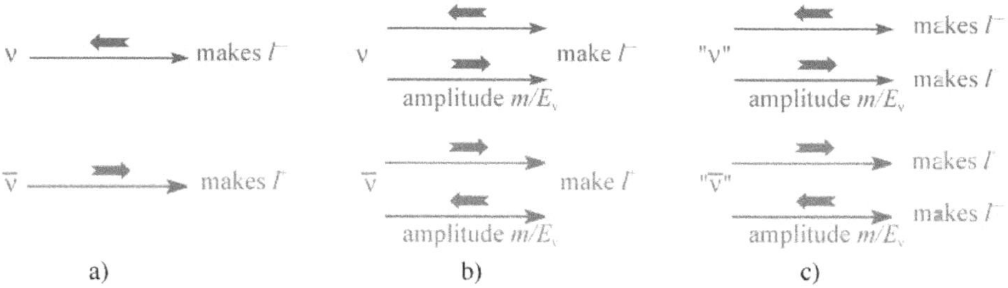

Fig. 5. Helicity components of neutrinos if a) mass less, b) massive Dirac, c) massive Majorana. Long arrows are velocities, short thick ones are the spins.

Fig. 5 shows schematically three different cases; $l=e$, μ, τ is the lepton flavour. The charged current $\bar{\nu}_{lL}\gamma^\mu l_L$ creates and absorbs negative chirality neutrinos due to its *V–A* structure in all the cases we shall consider. Fig. 5 a) represents mass less neutrinos. Left chirality neutrinos are eigenstes of both lepton number (L=+1) and helicity (h=–1), similarly for left antineutrinos, with L=–1 and h=+1. Fig. 5 b) represent Dirac massive neutrinos, with mass, say, m. Now neutrinos are in a eigenstate of the lepton number, with L=+1, but not of the helicity, rather they are superpositions of a (predominant) negative helicity component and a (small) fraction, with amplitude m/E, of positive helicity. *Vice versa* for antineutrinos. Both helicity components, when such neutrinos interact produce electrons, positrons for antineutrinos. Fig. 5 c) shows the case of Majorana neutrinos. Lepton number does not exist for them, the two states being distinguished by helicity only. A Majorana particle produced with a positron has predominantly negative helicity with a small m/E amplitude of positive one, exactly as in the Dirac case, but now when the particle interacts, its negative helicity component produces electrons, while the positive helicity one produces positrons. *Vice versa* for the particle produced with an electron.

In principle, one might investigate whether neutrinos are Majorana particles searching for reactions like $\nu_\mu + N \rightarrow \mu^+ + N'$, as first suggested by G. Racah already in 1937[28], before the structure of weak interactions was known. However, the factor $(m/E)^2$ in the cross section, for energies of the order of GeV (to be above threshold) is a small as 10^{-20}, making the detection impossible. Consequently, the *V–A* structure of the CC weak interaction and the smallness of neutrino masses are sufficient to guarantee the absence of the above mentioned reaction types, without need to invoke lepton number conservation.

There is however a possibility to look for the Majorana nature, the neutrino-less double-beta decay, as first suggested by W. H. Furry in 1939[29].

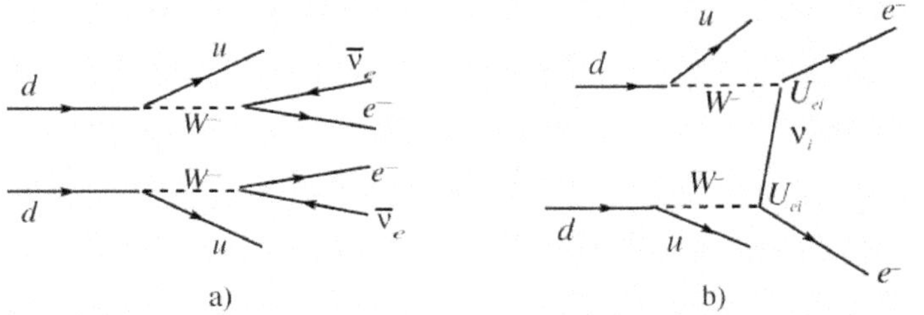

Fig. 6. Feynman diagrams for a) the $2\nu 2\beta$ decay and b) $0\nu 2\beta$ decay

As it is well known, beta decay is energetically forbidden for even-even nuclei. The open channel is a second order weak process, the two-neutrino double beta ($2\nu 2\beta$) decay, in which two nucleons experience beta decay contemporarily, $(A,Z) \rightarrow (A,Z+2)+2\bar{\nu}_e+2e^-$. The underlying process at the quark level is shown in Fig. 6 a).

The extremely long $2\nu 2\beta$ half-lives of several nuclei have been measured. For example, it is for ^{76}Ge 150±10 10^{19} yr, for ^{100}Mo, 0.71±0.04 10^{19} yr and for ^{130}Te, 90±10 10^{19} yr.

If neutrinos are Majorana particles, the neutrino-less double beta ($0\nu\beta\beta$) decay, $(A,Z) \rightarrow (A,Z+2)+2e^-$, is possible. It is shown at the quark level in Fig. 6 b). The process violates the lepton number by two units.

Notice that Majorana neutrino couples to *W* exactly as Dirac neutrino. The SM violation is in the propagator. The status created at one vertex has definite flavour, hence is a superposition of mass eigenstates. At the upper vertex in Fig. 6 b) a left (=negative chirality) "electron neutrino" is produced as

$$\nu_e^L = \sum_i U_{ei} \nu_i^L \tag{12}$$

At the second vertex the negative chirality neutrinos are absorbed. Of these, for the process considered, the positive helicity components, which act as "antineutrinos". The mass of a system of three Majorana neutrinos is a 3 x 3 matrix (non-diagonal elements correspond to flavour changing processes). The relevant matrix element in the present case is M_{ee}, called effective Majorana electron neutrino mass

$$M_{ee} = \left| \sum_i U_{ei}^2 m_i \right| = \left| c_{13}^2 c_{12}^2 m_1 + c_{13}^2 s_{12}^2 m_2 e^{i\phi_2} + s_{13}^2 m_3 e^{i\phi_3} \right| \approx \left| 0.83 m_1 + 0.55 m_2 e^{i\phi_2} + 0.15 m_3 e^{i\phi_3} \right| . \tag{13}$$

The measured quantity is the half-life, related to the Majorana mass by the equation

$$T_{1/2}^{-1} = G_Z \left(Q_{\beta\beta} \right) \left| M^{0\nu} \right|^2 M_{ee}^2 . \tag{14}$$

where G_Z is the phase space volume, depending on the nucleus, which is calculated without uncertainties, and $M^{0\nu}$ is the "nuclear matrix element". Much progress has recently been done

in understanding the nuclear physics effects and the present uncertainties have been reduced to a factor 2-3.
Notice that the differences between initial and final nuclear levels have been accurately measured. Consequently, the sum of the two electrons energies, called $Q_{\beta\beta}$, (which is typically a few MeV), is known, usually within a fraction of a keV.
Fig. 7 shows schematically the spectrum of the sum energy: the continuous part is the $2\nu 2\beta$ decay, the small peak the $0\nu 2\beta$, if it exists. Clearly, its size is not known, but its position is. The peak is very narrow in the figure as is the case of per cent (FWHM) energy resolution. Notice also that in real life other backgrounds are present. The "background index" b in the region of interest, is the number of counts per unit exposure time, per unit detector mass and unit energy interval. The lowest background indexes in the current generation of experiments is b=10^{-2}-10^{-3} $yr^{-1}kg^{-1}keV^{-1}$.

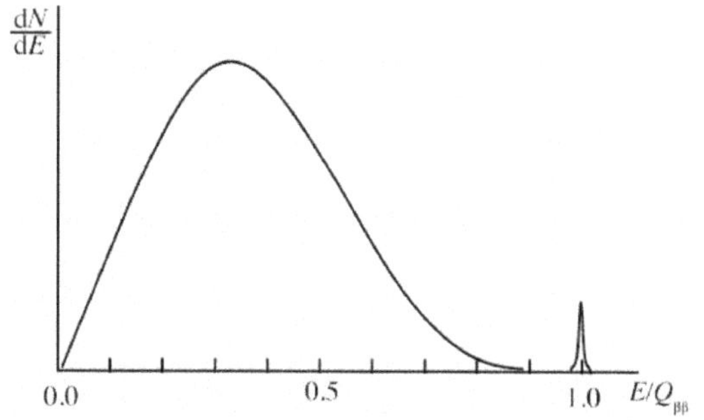

Fig. 7. Sum energy spectrum in the double beta decay

The sensitivity of an experiment with background index b, sensitive mass M, live time T and energy resolution (FWHM) ΔE to M_{ee} scales as $\left(\frac{\varepsilon a}{A}\right)^{1/2}\left(\frac{MT}{b\Delta E}\right)^{1/4}$ where a is the isotopic abundance, ε the detection efficiency and A the atomic number.
The fourth-root dependence implies that to gain, for example, an order of magnitude in the Majorana mass the sensitive mass must be increased by four orders of magnitude, if the background index is kept the same! If however the background is zero (or almost so), say ΔE, the sensitivity to M_{ee} scales as $\frac{\varepsilon a}{A}(MT)^{1/2}$.
In conclusion, the relevant quantity is the number of background events expected in the region of interest, namely $b\,\Delta E$.
Fig 8 a) is a not to scale cartoon of how the three complex addenda can add up to M_{ee}; Fig. 8 b) shows M_{ee} as a function of the smallest neutrino mass[30] for the two signs of Δm^2. Depending on the phases large cancellations may happen.
The experiments presently running and under construction will reach sensitivities able to explore down to about M_{ee}=100 meV. The corresponding half-lives lay in the range $T_{1/2}$ = 1 - 4 × 10^{26} yr, depending on the nuclear matrix element calculation and on the isotope. For these values, in one kilo-mol isotope mass, which contains 6 × 10^{26} nuclei, one expects a few events per year. Consequently, the "background free" condition demands a figure of merit $b\,\Delta E$ of a few times 10^{-3} /(kg yr).

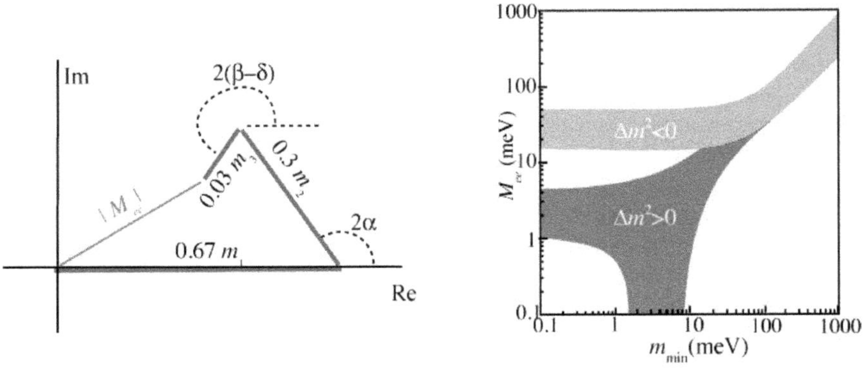

a) b)

Fig. 8. a) Cartoon showing how the three complex addenda can sum up, depending on the relative phases. b) M_{ee} vs. lightest neutrino mass[30]

Two experiments using ^{136}Xe enriched at 85% and one with enriched ^{76}Ge at 86%, GERDA (Phase 1) have recently reported results, with sensitive mass and/or background index substantially improved over the previous generation experiments.

EXO-200 is a liquid Xe two-phase (liquid and gas) TPC, in the WIPP site, employing 200 kg of enriched ^{136}Xe. About half of the isotope, in the central volume is used as fiducial, the other half being used as an active shield against radiation coming from outside. This self shielding feature avoids the presence of physical surfaces, which are always the site of background, near to the fiducial volume. A powerful tool for background rejection provided by the reconstruction of the events in a TPC is the separation of "single site events", as the signal is, from "multiple side events", as the main backgrounds are (gammas produce a series of Compton scatterings). The published data correspond to an exposure of 32.5 kg yr with an energy resolution FWHM ΔE=97 keV. The background index is unprecedented, b=1.5 x 10^{-3} cts/(kg keV yr), which is substantially the design figure. No signal is observed, corresponding to the limit $T_{1/2}^{0\nu}\left(^{136}\mathrm{Xe}\right)>1.6\times10^{25}$ yr [31] at 90% c.l. The figure of merit achieved so far by the experiment is $b\,\Delta E$=150 x 10^{-3} /(kg yr). Work is underway to improve it.

KAMLand-Zen employs, in the Kamioka Observatory, enriched ^{136}Xe dissolved in liquid scintillator. The published data correspond to an exposure of 89.5 kg yr, with an energy resolution FWHM ΔE=240 keV and a background index b=5 x 10^{-3} cts/(kg keV yr), which is an order of magnitude larger than the design figure. The upper limit on the half-life is $T_{1/2}^{0\nu}\left(^{136}\mathrm{Xe}\right)>1.9\times10^{25}$ yr [32] at 90% c.l. The figure of merit achieved so far by the experiment is $b\,\Delta E$=1200 x 10^{-3}/(kg yr). Work is underway to improve it.

GERDA-phase1, which is just concluded, employed, in the LNGS INFN Laboratory, about 18 kg of enriched (86%) ^{76}Ge diodes, which are both source and detectors, embedded naked in a liquid Ar bath. Ge counters provide a powerful tool for background rejection, because the pulse shapes of single site and multiple side events are different. The total exposure was 21.6 kg yr with an energy resolution of about ΔE= 4 keV. Both background model and background rejection algorithms were developed with a blinded interval of ±20 keV around Q_{bb} = 2039 keV. Data were later partially unblinded, but ±5 keV around Q_{bb}, to check. The number of observed events was found consistent with predictions. No line is predicted near Q_{bb}. The background index is b=10^{-2} cts/(kg keV yr) after pulse shape discrimination. This is both the design figure and one order of magnitude lower than in previous experiments. The sensitivity to the half-life before pulse shape discrimination is $T_{1/2}^{0\nu}\left(^{56}Ge\right)>1.9\times10^{25}$ yr @ 90% c.l. [33].

Results will be delivered just after this school. GERDA-phase 2 plans to double the sensitive Ge mass and to reduce the background index, profiting of the experience of phase 1, to $b=10^{-3}$ cts/(kg keV yr) and run for an exposure of 100 kg yr. The figure of merit achieved so far by the experiment is $b\,\Delta E=40\ 10^{-3}$ /(kg yr) and the expected one in phase 2 is $b\,\Delta E=4\ 10^{-3}$/(kg yr).
Another experiment using Ge diodes, in a different background reduction scheme, is MAJORANA under development in the SURF Laboratory in the USA.
An experiment in advanced construction phase at LNGS is CUORE. It is based on TeO_3 bolometers. The active ^{130}Te isotope has a natural abundance of 34%. The total isotope mass will be 200 kg. The energy resolution will be good, almost as for Ge, $\Delta E \approx 6$ keV. The background index is dominated by the radioactive contaminations of the surfaces, of the detectors, which are cubes of 5 cm side, and, mainly, of their Co holders, which are in vacuum. This even after the extremely advanced and demanding ultra-cleaning procedures developed by the experiment. It is expected, also on the basis for the data from CUORE-0, a prototype which is already taking data, to be around $b=36\text{x}10^{-3}$ cts/(kg keV yr), with reference to the ^{136}Te mass. CUORE plans to start commissioning at the beginning of 2015. The expected figure of merit of the experiment is $b\,\Delta E=200\ 10^{-3}$/(kg yr)
The SNO+ experiment is under construction too, at SNOLab. It will employ natural Te dissolved in liquid scintillator. The isotope mass will be large, 800 kg, but at the price of a very poor energy resolution, 300 keV FWHM. This feature will not allow, in particular, seeing a line at $Q_{\beta\beta}$. Rather the $0\nu\beta\beta$ would appear as a shoulder. Any signature of the signal is lost.
The NEXT[34] experiment is under construction in the Canfranc underground laboratory (LSC) under the Spanish Pyrenees. It will consist in a TPC with 100 kg of enriched ^{136}Xe. The strategic elements are the good energy resolution and the visual reconstruction of the tracks with specific ionization measurement along them. The read out does not employ charge amplification at the sense wires, as usual in the TPCs, to avoid losing energy resolution due to gain fluctuations. Instead, the electroluminescence of the secondary photons produced by the drifting charges at the wires is read out by an array of photosensors (Si photodiodes). Work on prototype detectors has shown an energy resolution extrapolated at $Q_{\beta\beta}$ (=2457.83±0.37 keV) of FWHM $\Delta E/E=0.75\%$, or $\Delta E=20$ keV.

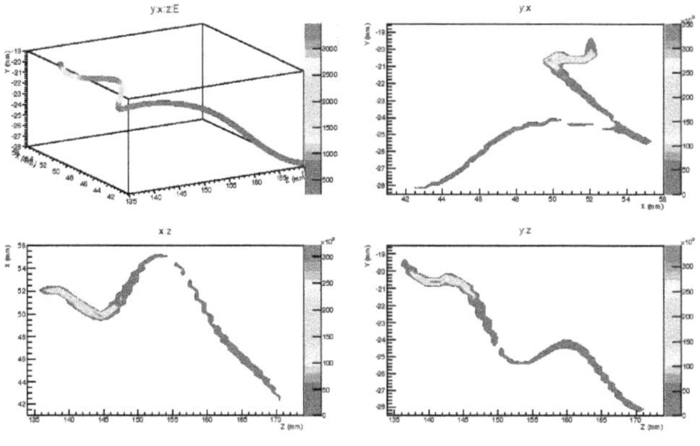

Fig. 9. A 3D reconstruction and 2D projection of an electron track from a Compton scattering of a 6oo keV gamma source. Colours code the ionisation

Fig. 9 shows the 3D reconstruction and the 2D projection of an electron track from a Compton scattering of a 600 keV gamma source. The specific ionisation is coded in the colour. The Bragg peak at the end of the track is clearly detected, providing a powerful handle to discriminate the background, which has a blob at one end only, from signal, which, being two electrons, has blobs at both ends. The $2\nu\beta\beta$ decays have obviously the same topology, but can be eliminated using the good energy resolution. Monte Carlo calculations show an expected background index better than b=5 10^{-4} cts/(keV kg yr). A 10 kg TPC, similar to the final one, is under construction to test the background model and other critical aspects. If the background index will be confirmed, the NEXT figure of merit will be $b\,\Delta E$=10 x 10^{-3} cts/(kg yr).

In conclusion we can say that the present generation of experiments, with a hard work of continuous improvement of the background control will be able to reach the 100 meV scale in several years from now. They may also see which techniques will be the best candidates for the next generation of experiments, with the ton scale mass and background index down to some 10^{-4} cts/ (keV kg yr), namely one count per keV every ten years in a ton, necessary to reach the 50 meV in Majorana mass.

6. Conclusions

Neutrino physics has opened the first window beyond the Standard Model, since the discovery of the solar neutrino puzzle almost 50 years ago.

Neutrino physics has made tremendous progress in the last 20 years. Different powerful sources both natural, the Sun and cosmic rays, and artificial, beams produced at proton accelerators and nuclear power reactors have been used, over baselines ranging from 1 km to 10 000 km. Several different detector techniques have and are being used: ultra low background measurement and control, radiochemical detection, cryogenic calorimeters, large mass Cherenkov and tacking detectors, etc.

Laboratories underground and with accelerators have been employed.

We have measured the absolute values of the mixing elements, but not their phases. The latter and the corresponding *CP* violation in the neutrino sector might be able to explain the matter-antimatter asymmetry in the Universe.

We have measured the mass differences, but not one of their signs neither the absolute mass scale.

We do not know if neutrinos and antineutrinos are different particles or not

Next phases of neutrino physics will be demanding and exciting. Being resources limited, their use should be optimized, concentrating efforts on the most cost effective alternatives, with the aim to maximize the scientific output, in a globally co-ordinated programme.

References

1. G. Mangano et al. *Phys. Lett.* **B534** (2002) 8-16
2. P. A R. Ade eta al. (PLANCK) arXiv:1303.5076 [astro-ph. CO]
3. Th. A. Mueller et al. *Phys. Rev.,* **C 83** (2011) 054615
4. G. Mention et al. *Phys. Rev.*, **D 83** (2011) 073006
5. C. Zhang et al. *Phys. Rev.* **D 87** (2013) 073018
6. C. Polly. Results from MiniBooNE experiment. Neutrino 2012 Conference. Kyoto
7. M. Antonello et al. (ICARUS) *Eur. Phys. J.* C (2013) 73:2599; arXiv:1209.0122
8. N. Agafonova et al. (OPERA), *JHEP* **1307** (2013) 004, arXiv:1303.3953.
9. A. A. Aguilar-Arevalo (MiniBooNE) ArXiv 1207.4809v2
10. Danby, G. et al.;. *Phys. Rev. Lett.* **9** (1962) 36
11. Conversi, M. et al.; *Phys. Lett.* 5 (1963) 195
12. Massam, T. *Nuovo Cimento* **39** (1965) 464. See also CERN-63-26. Nuclear Physics Division, June 27, 1963
13. Bernardini, M. et al.; A proposal to search for leptonic quarks and heavy leptons produced by ADONE. INFN/AE-67/1, 20 March 1967
14. V. Alles Borelli et al..; *Nuov. Cim. Lett.* **4** (1970) 1156
15. M. L. Perl et al.; *Phys. Rev. Lett.* **35** (1975) 1489

16. K. Kodama et al. (DONUT); *Phys. Lett.* B 504 (2001) 218
17. S. P. Mikheyev and A. Smirnov; Yad. Fiz. 42, 1441 [Sov. J. Nucl. Phys. 42, (1985) 913]
18. G. L. Fogli, et al. ArXiv: 1205.5254 v3 (2012)
19. F. P. An et al. (DayaBay) Phys. Rev. Lett. **108** (2012) 171803
20. J. K. Ahn et al. (RENO) Phys. Rev. Lett. **108** (2012) 191802
21. K. Abe et al. (T2K) arXiv:1304.0841 (2013)
22. S.T. Petcov et al., Phys. Lett. **B 533** (2002) 94; S. Choubey et al., Phys. Rev. **D 68** (2003) 113006; J. Learned et al., hep-ex/0612022; L. Zhan, Y. Wang, J. Cao, L. Wen, Phys. Rev. **D78** (2008) 111103; Phys. Rev. **D79** (2009) 073007
23. See: Miao He "Future reactor experiments" arXiv 1310.7343v1
24. Proceedings INSA (*Indian National Science Academy)* - Part A Vol. 70A No. 1. http://www.imsc.res.in/~ino/OpenReports/Insa/insa.html (2004)
25. G. Feldman et al. (NOvA); http://www-nova.fnal.gov/nova_cd2_review/tdr_oct_23/tdr.htm
26. E. Majorana; *Nuovo Cimento*, **5** (1937) 171
27. A. Bettini. Introduction to elementary particle physics. Cambridge University Press. Cambridge UK (2008).
28. G. Racah; *Nuov. Cim.* **14** (1937) 322
29. W. H. Furry; *Phys. Rev.* **56** (1939) 1184
30. S. M. Bilenki and C. Giunt.; Neutrinoless double-beta decay. A brief reviews. arXiv:1203.5250 (2012)
31. M. Auger et al., (EXO-200). *Phys. Rev. Lett.* **109**, (2012) 032505
32. A. Gando et al., (KamLANDZen) *Phys. Rev. Lett.* **110** (2013) 062502
33. M. Agostini et al. (GERDA). arXiv:1306.5084 (2013)
34. V. Álvarez et al. (NEXT). arXiv:12012.0721 (2012)

CHAIRMAN: A. BETTINI

Scientific Secretaries: R. Depalo, L. Menculini

DISCUSSION

- Vignati:

If EXO will observe a peak, would you believe it is double beta decay or not? The sensitivity of EXO is in principle very high, but it is computed assuming a flat background. However we learned from Heidelberg-Moscow and from KamLAND-Zen that there are may be also peaks in the background. Consequently, I think that the energy resolution is a critical parameter and I want to know your opinion about this. So, would you believe in the result of EXO or not?

- Bettini:

EXO is a liquid xenon TPC, in which the event position is localized. This allows defining a fiducial volume in the central part, in which the background du to gammas coming from outside is reduced (self-shielding). Of course this is very expensive, because part of the enriched xenon is used as a shield. In practice the fiducial volume is abut one third of the total. An additional handle to distinguish signal from background is as follows. A gamma-ray coming in the detector will make in general a few Compton scatterings separated by several centimetres (multiple site topology). On the contrary, the double beta decay releases all the energy in one point (single site topology). The screening of the materials done before building the experiment allows knowing, at least approximately, the radioactive traces contributing to the background. A background model is built in a simulation code producing the SS and MS events energy spectra. The spectra are compared with the data and the sizes of the different contributions fitted. One can have surprises here, but it was not the case of EXO.You are right, energy resolution is very important. However, the one of EXO about 4% FWHM is adequate for the present level of exposure. Its background index is one of the smaller, 1.5 10^{-3} cnts/(kg keV yr). The background model foresees about 4 events in the region of interest and no peak. In such conditions a statistically significant excess would point to signal.

- Vignati:

But also KamLAND-Zen didn't expect any peak...

- Bettini:

True, meaning that the background model developed by KamLAND-Zen before doing the experiment was not accurate enough. Notice that KamLAND , which followed BOREXINO in using the same technique, never reached the very low background of the latter. This can happen of course. In this case, one tries to understand which are the additional background sources and modify the experiment to correct for that, as KamLAND-Zen is now doing. Under these conditions energy resolution is even more important. KamLAND-ZEN resolution is about twice worse than EXO. The tail of the $2\nu 2\beta$ spectrum contributes to background of KL-zen, but not of EXO. The energy resolution of EXO is good enough for the present level of sensitivity; not so for a ten times larger one.

- Vignati:

So you wouldn't ask for a comparison with a non-enriched detector? Does it make sense to make a detector like EXO, with non-enriched xenon, to really understand the background?

- *Bettini:*

All the xenon contributes to background, both active and non active, so you increase the background compared to the signal if you don't enrich.

- *Vignati:*

But in GERDA it is very useful to compare enriched and non enriched detectors

- *Bettini:*

The background model of GERDA is based on the enriched detectors. Every detector has a different background, so the natural detectors are used for checking. Notice that their history of exposure to cosmic radiation is different.

- *Wever:*

On slide 12 you mentioned that there are 4.5 unknowns, but what do you mean exactly by 4.5?

- *Bettini:*

I had been joking, sorry. We know the three angles, the value and the sign of one the two square mass differences, only the absolute value of the other one (this is my 0.5).

- *Wever:*

Is there something similar to the see-saw mechanism for Dirac neutrinos?

- *Bettini:*

The see-saw mechanism is a way, a theoretical model, to understand why neutrino masses are so small. You need to have a Majorana contribution in the mass matrix for that, to my understanding.

- *Sydorenko:*

In your talk you mentioned that the *CP*-violating phase cannot be effectively measured in the current experiments. Are you aware of any future proposals or plans to make future experiments that will lead to significantly restrict the allowed area for the *CP*-violating phase?

- *Bettini:*

I did not say exactly that. I said that it's not yet measured. Consider the plane of δ vs. $\sin^2\theta_{13}$ (*shown*). The disappearance experiments at reactors give a vertical band. The appearance experiments give a wavy band, as already done by T2K. The region of overlap of the two bands gives the experimental limits for δ. The sensitivity is not enough today, but in a while, in a couple of years, the widths of the bands will be much smaller (*prediction shown*). The NOνA experiment that is starting now will contribute substantially. It is a big detector at 800 km from Fermilab; the intensity of the proton beam on target will go up from 300 kW to 700 kW in a year o two. If we are lucky, if δ is close to 90° e.g., in three years from now we will get some idea. If not, we shall need a next generation, with both a bigger detector (several hundred kilotons) and a very intense source with up to one MW proton beam on the target. This implies a considerable investment in the proton accelerators system.

- *Hartmann:*

Regarding the excess of events seen by LSND and MiniBoone, especially the MiniBoone: if true it would imply *CPT* violation, and *CPT* violation would imply violation of Lorentz invariance. There are Standard Model extensions in the literature where Lorentz invariance violations are implemented. If these experiments are not disproven, would you say that this is the way to go in model building, to start considering models violating Lorentz invariance? Or what is your opinion in this area?

- *Bettini:*

CPT invariance is one of the pillars of all the quantum field theories. It is verified with extreme precision. Certainly one cannot reasonably claim a violation with neutrino physics. In principle, if you observe any difference between the oscillation $\nu_e \rightarrow \nu_\mu$ and $\overline{\nu}_e \rightarrow \overline{\nu}_\mu$, in vacuum, you discover *CPT* violation. Or your experiment is wrong. The latter is by sure the case if the effect is so big as the one claimed by MiniBooNe. *CPT* violation cannot be at the level of percent at which neutrino experiments are sensitive.

- *'t Hooft:*

There are three phases in the neutrino mass-mixing matrix, δ, ϕ_1 and ϕ_2, while in the quark-sector there is only one, which gives *CP* violation. How would one measure ϕ_1 and ϕ_2? It should be more difficult than for quarks.

- *Bettini :*

The only available process in practice is the neutrino-less double beta decay. Here one measures one element of the Majorana mass matrix, M_{ee}. Moreover, we need to know the thee neutrino masses, having determined the mass hierarchy and the absolute neutrino mass scale. We need also to know the nuclear matrix elements with a better precision then now.

- *'t Hooft:*

But there are two phases, how would you distinguish ϕ_1 and ϕ_2? I would expect that double beta decay would only give us one number, not two.

- *Bettini :*

Yes, if you know everything else, you still have two unknowns and one measurement.

- *'t Hooft:*

Unless perhaps you have different nuclei that make different double-beta decays?

- *Bettini:*

No, because the same M_{ee} appears in the decay width of all the nuclei. The only differences are in the other factors, the nuclear matrix elements and the phase space.

- *Ferrara :*

Probably the only one thing to do is: you measure and then you make a fit, so that with a single measurement you might find the two.

- *Bettini :*

I do not see how. In principle, independent information could be provided by the other elements of the Majorana mass matrix. $M_{\mu\mu}$, $M_{e\mu}$,...have different dependence on the phases than M_{ee}. They appear in processes like an electron produced by an incident muon or a μ^+ produced by a μ^-. But unfortunately the cross section are so small that such experiments are far away to be feasible with any thinkable muon beam intensity.

- *Csanad:*

About the nuclear matrix elements, how certain they are and how much can experiments constrain their values? How much do their uncertainties affect the experiments?

- *Bettini :*

It's a rather difficult calculation because nuclei are complex to treat. There are different approaches: the shell model, the quasi random phase approximation and the interactive boson model. In principle the precision of 10-20% is reachable, but we are not yet at this level. The present accuracy is within a factor of 2 or so. In addition the axial vector coupling g_A is not well known. It is taken to be 1.25, but in simple beta decay inside the nucleus it is renormalized and becomes smaller, namely about 1. The transition probabilities depend on the 4th power g_A and consequently it represents a considerable factor of uncertainty. Measurements are being done in processes that can provide input to the theoretical calculations. There is progress.

- *Inguglia:*

About *CP* violation, suppose that you use an anti-neutrino beam, for example at T2K, could you use the same detector? Or you need to do big changes in the detector?

- *Bettini:*

No, you can use the same detector.

- *Inguglia:*

And is someone planning to use an anti-neutrino beam to measure *CP* violation?

- *Bettini:*

To measure *CP* violation you don't need to operate with neutrinos and anti-neutrinos, but is better if you use both. For that case you can use the same accelerator and the same detector. With anti-neutrino the cross section is three times less than for neutrinos, but for example the NOvA experiment plans to run with neutrinos at the beginning, and then go to anti-neutrinos, and the detector will be the same.

- *Krylov:*

If there would be a supernova explosion now, how many laboratories are ready to detect a neutrino signal from it?

- *Bettini:*

In Gran Sasso there is the LVD experiment, of Prof. Zichichi. It is 1 kt of liquid scintillator. If the supernova is in the centre of the galaxy, LVD will detect a few hundreds events. Super-Kamiokande would measure several thousand events. BREXINO is only 300 t, but still will see several events. The HALO experiment at SNOlab in Canada is based on lead and ^3He counters. A detector similar, but smaller than LVD, operates in the Baksan Laboratory in the Russian federation. IceCube at the South pole will also contribute. So there are several detectors which would be able to observe a supernova explosion, providing complementary information. We expect 3 supernova per century in our galaxy. There have been detectors operational for 30 years now, which did not see any signal. It is time for one to come! (using a wrong statistical argument).

WHAT IS THE ONTOLOGICAL STATUS OF THE HIGGS BOSON?

Notes on the Englert-Brout-Higgs Mechanism as an Unfinished Project

Tian Yu CAO

Boston University

1. Introduction

The recent observation of a massive scalar particle -- first explicitly suggested in a context of broken symmetry by Peter Higgs in 1964[1] and referred to as the Higgs boson ever since -- by the ATLAS and CMS experiments at CERN's Large Hadron Collider is a significant development in particle physics[2]. It has changed the perspective from which the electroweak theory and the future of fundamental physics are conceived. The very existence of a fundamental scalar boson has opened new vistas for, and set severe constraints on, further theoretical and experimental explorations about the true nature of the scalar sector as well as the naturalness of the hierarchy of fundamental scales, which involve many speculative notions, such as supersymmetry, technicolor, extra dimensions, inflaton, dark energy, and many more.

As a philosopher and conceptual historian of particle physics, what interests me the most is the change of perspective on this question as a result of understanding how this discovery was made and what its implications really are.

One popular assessment is that it confirms a mechanism -- first suggested by Francois Englert and Robert Brout in 1964[3] and then refined by Peter Higgs in 1966[4], so let us call it the Englert-Brout-Higgs mechanism -- which is supposed to contribute to our understanding of the origin of mass of subatomic particles[5]. To what extent the EBH mechanism is confirmed -- or what is supposed to have been and what is actually confirmed -- by the observation and in what sense the mechanism contributes to our understanding of the mass generation, however, critically hinges on the understanding of what is meant by the mechanism. In this lecture, the conceptual foundation of the mechanism will be clarified (section 3) through a historical examination of its genesis (section 2); then the ontological status of massive bosons and the EBH mechanism itself, realist or instrumental, will be discussed through an examination of the reality status of its foundation (section 4) , with a conclusion that the uncertainty in the ontological status of the EBH mechanism and thus of the Higgs boson remains before the two puzzles, the transmutation of the Goldstone modes' dynamic identity and the fixity in reorganizing the physical degrees of freedom, are properly addressed (section 5).

2. The genesis of the Englert-Brout-Higgs mechanism

Historically, the EBH mechanism emerged as a result of three lines of development: (1) broken symmetry manifested in a degenerate vacuum, (2) scalar field theory with a broken symmetry solution, and (3) mechanism for gauge bosons to be massive.

With the developments since the mid-1950's, such as PCAC, Sakurai's and Glashow's gauge theory, SU(3) flavour symmetry and current algebra, particle physicists increasingly recognized the desirability of a proper understanding of broken symmetry. More pertinent to our concern, however, was an observation, made by Abraham Pais as early as 1953, about the existence of a hierarchy of (strong, electromagnetic, and weak) interactions with regressively less symmetry[6]. This notion of a hierarchy of interactions and their associated symmetries was suggestive of Heisenberg's non-linear unified field theory proposed in 1958[7]: his unified theory would make no sense if he could not derive phenomena in various types of interaction possessing various symmetries from the equations of the underlying field that possess a higher symmetry than the phenomena themselves.

On November 21, 1958, Lev Landau wrote Heisenberg in support of his unified theory, with a novel idea which was perhaps derived from his early work on phase transitions: "the solutions of the equations will possess a lower symmetry than the equations themselves."[8] In resonance with Laudau's suggestion, Heisenberg explored the notion of the vacuum, whose properties underlie the conceptual structure of a field theory. He first invoked the idea of a degenerate vacuum to account for internal quantum numbers in 1958, and then asserted in 1959 that "it is by no means certain a priori that the theory must give a vacuum state possessing all the symmetrical properties of the starting equation", and thus "it should be considered not really a vacuum but rather a world state, forming the substrate for the existence of elementary particles. This state must then be degenerate" and "is the basis for the symmetry breaking."[9] Heisenberg's idea of a degenerate vacuum was influential. But he never reached a satisfactory understanding of the origin, mechanism and physical consequences of broken symmetry, nor did he give a convincing mathematical formulation of it.

The physical realization in field theory of Heisenberg's idea was first provided by Nambu, who applied the BCS-Bogoliubov formulation of superconductivity to a chiral invariant field theory, with the former's phonon-mediated interaction being replaced by a non-linear interaction among fermions.[10] In Bogoliubov's formulation, the elementary excitations are pseudo-particles or doubly charged Cooper pairs, whose energy gap explains superconductivity. The states of the charged pairs are not eigenstates of the charge, and thus not gauge invariant. So the vacuum as the condensate of the pairs is also not gauge invariant and thus degenerate. An important result in Nambu's work is that one of the exact solutions of the vertex part equation, namely the collective mode of the quasi-particles, which leads to the Ward Identity and thus ensures the gauge invariance of the whole theory, is the bound state of a pair with zero spin and zero energy-momentum. So the existence of the massless spinless bound states appears to be the logical consequence of gauge invariance when the vacuum is degenerate. Nambu transplanted this reasoning to a chiral invariant field theory and interpreted, with some modifications, the symmetry restoring collective modes as pions, the bound nucleon-anti-nucleon pairs, which successfully explained the successes of PCAC

Nambu's work in fact has already disproved a widely adopted misnomer, spontaneous breakdown of symmetry. Symmetry is always broken by some physical mechanism (but never spontaneously), such as the Cooper pairing in superconductors, or his own non-linear fermion interactions. It is some dynamic mechanism that brings the system into energetically favourable asymmetrical states as compared to the symmetrical ones. This dynamic origin of symmetry breaking can also be seen clearly in Goldstone's work.[11]

The theoretical context in which Goldstone explored Heisenberg's idea of broken symmetry manifested in a degenerate vacuum was quite different from Nambu's. While Nambu examined the consequences of a broken symmetry solution for the nature of the vacuum and for the inevitable appearance of massless spinless bound states in a non-renormalizable model of a self-interacting fermion field, Goldstone examined the conditions for a field theory to have a broken symmetry solution in a renormalizable model of a self-interacting boson field. Both Nambu and Goldstone observed the necessary appearance of the symmetry-restoring massless boson in a symmetry-breaking field theory. But Goldstone's boson, in contrast with Nambu's bound fernion-antifermion pair, arises from a primary scalar field. While the chiral symmetry in Nambu's case was broken by an unknown self-interaction of the fermion field, Goldstone found that, in addition to the self-coupling of the boson field, only when the boson mass squared is negative and the coupling constant satisfies a certain inequality, does the field model he examined have a broken symmetry solution, that is, the field would have a degenerate vacuum, consisting of an infinite number of lowest energy states separated by superselection rules.

A novelty of Goldstone's model is the introduction of a primary scalar field that has nothing to do with a "more fundamental" fermion field. The step taken by Goldstone appeared to be arbitrary or at best an *ad hoc* way of exploring issues accompanying symmetry-breaking with no other reasons, and thus was regarded as "a serious flaw" by Leonard Susskind[12] and by many others. Numerous efforts have been made to get rid of this fundamental scalar field, so far all of them have failed. In the five decades since its introduction, the notion of a fundamental scalar field has become a new organizing principle in the exploration of symmetry-breaking in field theoretical models.[13]

The most consequential result of Goldstone's work was his observation that whenever the original Lagrangian has a continuous symmetry, any broken symmetry solution will be invalidated by the entailed but non-existent massless boson, the so-called Goldstone boson.[14] This observation was the first revolutionary step crucial for all the subsequent developments in the exploration of broken symmetry. Goldstone's observation was generalized into a theorem[15] and thus posed a great challenge to the exploration of a broken symmetry solution to gauge theory. This prompted the enthusiasts of broken symmetry to set an agenda for evading it.[16] Although mathematically no evasion is possible[17], the theorem itself turned out to be almost irrelevant in the ensuing developments. Crucial to the proof of the theorem was the assumption of Lorentz invariance. But in the context of gauge theory, a Lorentz-invariant formulation contains nonphysical degrees of freedom, such as timelike gauge bosons with negative probability. For this reason, all discussions of the subject in this context used the Coulomb gauge, which makes manifest the menu of experimentally accessible particles of the theory, without explicit Lorentz invariance. This reasonable move was regarded as a departure from the assumptions of the theorem.[18] Still, Goldstone's original observation about the symmetry- restoring massless scalar boson remained valid, and had to be properly addressed.

This pursuit was readily possible because another innovative scheme pertinent to the issue proposed by Schwinger was already available by then.[19] One of the motivations for having a broken symmetry solution in gauge theory was to have massive gauge bosons to account for the short range nuclear forces. Schwinger explored the issue from a dynamic perspective without explicitly mentioning broken symmetry because he believed that "the general requirement of gauge invariance no longer seems to dispose of this essentially dynamic question." In a toy model of two dimensional massless QED, Schwinger demonstrated that, when the gauge field is strongly coupled with a symmetry current, it might not be massless if its vacuum polarization tensor possesses a pole at light-like momenta. The assumed field-current coupling indicates that the context in which the gauge boson was examined was an interacting theory rather than a pure gauge theory, involving another field in the current, whose interactions with the gauge field under certain conditions may contribute to the appearance of the pole. Although the notion was not invoked by Schwinger, both the pole itself, which is connected with the non-vanishing vacuum expectation value of a field interacting with the gauge field -- either a primary field or a composite -- and its consequence of giving the photon a mass, were closely related with broken symmetry. Schwinger did not specify the reason why a pole would appear in the vacuum polarization tensor; he only asserted that it was dynamically produced, most likely by a bound state of Nambu's type. But its scalar nature makes it open to the interpretation that it is produced by a primary scalar field. This interpretive flexibility was soon exploited fruitfully by Englert and Brout,[20] resulted in a mechanism which is the major subject of this note.

Schwinger's model was trivial, but his insights set a grandiose framework for many to follow. The first was Philip Anderson,[21] who used the case of the plasmon to vindicate Schwinger's insights: Nambu's massless collective mode or bound state is converted into a massive plasmon by interacting with the electromagnetic field. From this case, he suggested that "the only mechanism for giving the gauge field mass is the degenerate vacuum type of theory." He even claimed that "the Goldstone zero-mass difficulty is not a serious one, because we can probably cancel it off against an equal Yang-Mill zero mass problem."

In terms of ideas, Anderson was the first to associate Schwinger's mechanism with broken symmetry and the Goldstone boson. In terms of physics, however, the suggested association was quite tenuous. His non-relativistic example of the plasmon did established some connection between broken symmetry and Schwinger's strategy of gauge bosons acquiring mass through interacting with another field; but the physics for symmetry-breaking involved in his case came from Nambu's bound states rather than Goldstone's primary scalar field.

The first serious effort to fit Goldstone's scalar system into Schwinger's grandiose framework, in concrete physical terms rather than vague ideas, was made by Englert and Brout. In their model of scalar QED (which was extended to nonabelian symmetries without substantial structural changes) in which the symmetry-breaking scalar system was coupled to the gauge field, they directly interpreted the symmetry-restoring massless boson, in lowest order perturbative calculation of the vacuum polarization loop for the gauge field, as the physical base for Schwinger's pole, which gives mass to the gauge boson. This interpretation can be viewed as a realization of Anderson's conjecture. But it should be taken as a real breakthrough in physics, in terms of understanding the ***symbiotic*** nature of Goldstone's scalar system and its coupled gauge system: Neither the scalar system with a broken symmetry solution nor the pure gauge system could exist separately without being invalidated by the entailed but non-existent massless (scalar or vector) bosons; only jointly as two inseparable moments of a symbiont could they have broken symmetry solutions. One defect in their understanding of the nature of the symbiont was their erroneous claim that "the symmetry is broken through the gauge fields themselves;" actually, it is broken through the self-interactions of its scalar moment.

3. The conceptual foundation of the Englert-Brout-Higgs mechanism

The conceptual situation was greatly clarified once the novel notion of "induced symmetry breakdown" (ISB) was introduced, with great scientific creativity, by Higgs in 1966.[22] When the scalar system with a broken symmetry is Yukawa-coupled to a spinor system that contains no additional mechanism for symmetry breaking, Higgs asserted, the symmetry breaking in the scalar system breaks the symmetry of the spinor system to an extent that depends on the Yukawa coupling constant, and thus allows the system to have non-symmetrical states of massive fermions, whose masses are accordingly in proportion to their Yukawa coupling constants. What was suggested by Englert and Brout fits perfectly into

Higgs's notion: The gauge coupling there played the same role as the Yukawa coupling in Higgs's case, namely, for inducing symmetry breaking in the vector system by the symmetry-breaking scalar system. Thus the symmetry breaking in the vector and spinor system, manifested in the massive vectors and fermions, is not related to their own degenerate vacua, but is induced by the primary symmetry-breaking of the scalar system through the gauge coupling and the Yukawa coupling respectively. The notion of ISB has revealed a deep truth of the physical world: What renders possible a broken symmetry solution for a multi-field complex system (such as the scalar-spinor-vector system in Weinberg's unified electroweak theory[23]) may only be one degenerate vacuum constituted by the self-interaction of its scalar component rather than a set of separate vacua, each for a component field. Once the notion of ISB, which grounds the scalar-field-based EBH mechanism, was absorbed, consciously or unconsciously, into the understanding of broken symmetry physics, a clear picture, as proposed in Weinberg's unified theory, emerged. At the most fundamental level, the scalar field's self-coupling is the mechanism for constituting a degenerate vacuum (the related non-vanishing vacuum expectation value of the scalar field is a universal parameter in the electroweak domain, the weak scale). Its gauge coupling to gauge fields underlies the constitution of the massive gauge bosons and is responsible for the symbiosis of the scalar-vector complex manifested, through a redefinition of the fields involved, in a set of massive gauge bosons and a massive Higgs boson. Its Yukawa coupling to the fermion system underlies the constitution of massive fermions and contributes to CKM matrix, CP violation, and flavour physics. The gauge field's gauge coupling dictates the electroweak interactions, but contributes nothing to the constitution of the degenerate vacuum. The fermion system builds up its broken symmetry solutions, not from its own degenerate vacuum -- there is no such vacuum -- but from its Yukawa coupling to the scalar system, and thus can be viewed as conceptually less fundamental in the electroweak domain, although it itself is a primary existence that cannot be derived from anything else.

Thus the EBH mechanism can be properly understood as simply a set of scalar field couplings: its self-coupling is responsible for the broken symmetry solution, its gauge coupling and Yukawa couplings are responsible for the broken symmetry solutions for the gauge and spinor fields manifested in massive gauge bosons and massive fermions. If the EBH mechanism is understood in this way, it seems that the claim that it "contributes to our understanding of the origin of mass of subatomic particles" is justified.

4. The ontological status of massive bosons and the EBH mechanism: Realism versus Instrumentalism

But the above happy conclusion about the significance of the EBH mechanism is somewhat disturbed by a seemingly naive question about its foundation: Does the broken symmetry solution of a scalar field exist? Surely it exists, the common wisdom asserts: Its massless modes are combined with ("absorbed by") the massless gauge bosons, resulting in the desirable massive gauge bosons, whose consequences have all been confirmed without any trace of doubt. As to its massive mode, which has been elusive for a long time, now it has finally been observed by experiments at CERN's LHC. With all the consequences of its existence confirmed, one may feel that its existence is confirmed. And this also seems to suggest the confirmation of the EBH mechanism built on it. A pundit may argue that it is logically fallacious to claim the validity of the antecedent by affirming the consequent. Yet confirming the consequences at least has provided evidential support for the antecedent. In this case, however, there are more seriously disturbing problems coming from physics than from the incomplete validity of inductive logic.

The validity of the common wisdom is explicitly or implicitly based on two assumptions. First, both the broken symmetry solution of a scalar field and the symmetrical solution of gauge fields exist in reality, at least at a deep, experimentally inaccessible level. Secondly, at the experimental ("physical") level, some of them always appear in a redefined (combined) way as the massive vector bosons. That is, both the massless scalar modes and massless vector modes are real, but not observationally-empirically-experimentally-cognitively accessible in a separate way. However, if there is simply no way to have any access to their separate existence, especially to the effects derivable from the Goldstone modes, either directly through their derivative couplings or indirectly through separating the effects produced by the Goldstone modes from the massive gauge boson-mediated events, then what is the ground for believing in their separate existence? An instrumentalist may thereby deny the reality of the broken symmetry solution of the scalar field and the symmetrical solution of gauge fields, relegating them into the fictitious status of phlogiston and the ether, whose only function is to construct the observable particles (massive gauge bosons and the Higgs boson) and measurable parameters (the weak scale and masses, whose ratios are the relevant couplings), which, according to the positivist-instrumentalist philosophy, are the only realities in the physical world. That is, an instrumentalist would take Weinberg's model as physically nothing more than Glashow's model,[24] except for an additional Higgs boson.

4.1 A realist's response: a scalar-vector symbiont as a new primary entity

But there could be a realist reading of the EBH mechanism based on assumptions that differ from those taken by the common wisdom. Perhaps, rather than a set of scalar and vector fields, what exists in reality is a scalar-vector symbiont possessing broken symmetry solutions for its scalar and vector moments. Perhaps the physical foundation for the EBH mechanism is this primary symbiont -- whose internal dynamics, as we shall see shortly, explains the EBH mechanism -- but not a primary scalar field and its set of couplings. Although the symbiont, a physically primary non-decomposable (even at the deep, experimentally inaccessible level of reality) single entity is mathematically describable by analytically separable structures, no mathematical separation and manipulations would have any ontological meaning. That is, there could be no massless excitations of its scalar or vector moments arising from the symbiont: they are only the artefacts of an illegitimate separation of the non-decomposable structure. Thus the holistic structure of the symbiont is very different from, e.g. the structure of a coupled electron-photon system, whose components (electron and photon) in the gauge-less limit can have separate existence, and there is no way for its components to be recombined in different ways to produce different field configurations. To such a symbiont, an old Christian maxim applies: What God has joined together, let no man put asunder.

The ontological assumption of an inseparable symbiont may not prevent us from assuming that its two moments are dynamically separable at the deep level of reality, or even at the experimentally accessible level. That is, each moment has its own dynamical identity, its characteristic ways of coupling to other systems without being affected by those of the other moment. Thus the scalar moment has its self-coupling and Yukawa coupling without being affected by the vector moment's gauge coupling; similarly, the vector moment has its characteristic way of gauge coupling to other systems without being affected by the scalar moment's ways of coupling. This assumption is crucial for a realist reading of the EBH mechanism because otherwise, at the deep experimentally inaccessible level of reality, no self-coupling and Yukawa coupling of the scalar moment (and the degenerate vacuum and massive fermions constituted thereby) would be definable; and at the experimentally accessible level of reality, no vector moment's gauge coupling (and thus their electroweak interactions) would be definable.

4.2 A spacetime analogy: reorganisable moments of a symbiont

There is a well-known predecessor of such a holistic symbiont. In 1908, Hermann Minkowski declared: "Henceforth space by itself, and time by itself, are doomed to fade away into mere shadows, and only a kind of union of the two will preserve an independent reality."[25] Similarly, we may say that Goldstone's scalar system and Glashow's gauge system are doomed to fade away into mere shadows, and only a symbiont of the two (scalar and vector) moments exists as a holistic entity in the electroweak domain of the world, aside from the fermion system.

Just as the distinction between spatial and temporal aspects of the world remains, even though only spacetime is an independent reality, we can still meaningfully talk about the scalar moment and the vector moment of the symbiont -- and their experimentally accessible manifestations: massive scalar and vector bosons – characterized by their dynamical identity, with an understanding that they are just different moments (manifestations) of a single physical entity, having no separable existence. They are different from the scalar field and the vector field separately or joined in a coupled system because the symbiotic nature of the two moments allows the reorganization of their degrees of freedom within the symbiont. The role the reorganization plays is similar to the role the Lorentz transformation plays in the case of spacetime.

4.3 Realism of symbiont differs from realism of fields

Twenty one years ago I asked Peter Higgs at a SLAC conference: "What is the ontological status of the Higgs boson?"[26] The question had two levels of meaning: (i) a choice between a real particle and a fictitious one (a phenomenological indicator of something complicate situation); and, assuming it is real, (ii) a choice, at that time, between a composite of Nambu's type, which underlay the pursuit of dynamical breakdown of symmetry, and a fundamental scalar boson of Goldstone's type. The LHC confirmation seems to have delivered the verdict in favour of the latter, within the EW domain, of course. Beyond that domain, the space for speculative ideas is virtually infinite, but the space for real physics is almost point-like.

But if the symbiont is a primary entity in the electroweak world, then the observation of the Higgs boson (or the W and Z bosons) cannot be taken as a vindication of the excitation of the scalar field (or gauge fields), as the old realist ontology would suggest. Rather, it can only be taken as a vindication of the excitation of the scalar (or vector) moment of the symbiont. The difference between the two ontological assumptions is scientifically significant. The old view takes the recent observation of the Higgs boson as a vindication of the reality of the scalar field, which has a package of physical implications. This may tempt people to use this new observation to address open questions in fundamental physics, such as inflation or dark energy. The symbiont view, however, would take the observation of the Higgs boson as a vindication of the reality of the scalar moment of the symbiont. That is,

> ***we take the massive boson as the quantum excitation of the scalar moment of the scalar-vector symbiont, which can be observationally registered when it is in interactions with other field quanta in the EW***

part of the physical world, but is physically tightly connected with the other excitations of the whole complex, including the gauge bosons.

Thus wherever the implications of the Higgs boson lead, the whole package of implications of the existence of the symbiont, including but not restricted to the effects of the W-bosns and the Z-boson, should also be taken into serious consideration. "What God has joined together, let no man put asunder." This will put severe constraints on the pursuit of these implications. But it will also give great predictive power to the pursuit.

4.4 Two puzzles

The reorganizability of the degrees of freedom (from the inseparable moments of a symbiont) for a redefinition of fields is the characteristic feature of the symbiont, and is also crucial for grounding the EBH mechanism. In his classic 1967 paper, Weinberg indicated that the redefinition of the field (through the recombination of the degrees of freedom from two moments of the symbiont, as happened in the redefinition of gauge bosons) implied the reordering of the perturbation theory. The details of how the two are connected may be an effective technical means for addressing the subtle conceptual issues explored here. Unfortunately, Weinberg did not spell them out. But it seems reasonable to assume that what is to be recombined for redefinition of the fields, and reordered in the perturbation theory, should have their identities maintained before and after; otherwise an uncontrollable conceptual mess would ensue.

One important aspect of the identity of a degree of freedom is its dynamical identity mentioned above; namely, its characteristic ways of coupling to other systems. Thus, a scalar degree of freedom has its characteristic way of coupling to itself (self-coupling) and to fermions (Yukawa coupling). Through the redefinition, Weinberg noticed, "the Goldstone bosons (the massless mode of the scalar moment) have no physical coupling." By "physical coupling," surely he meant only experimentally accessible coupling, not all its dynamical interactions with other physical entities. The massless modes of the scalar moment do not disappear through the redefinition of the fields; they are only to be reorganized into vector fields, being their longitudinal components. Thus, their dynamical capability (manifested in their couplings to other physical entities) does not disappear, only gets to reappear in different incarnations. In fact, the dynamical interactions of the massless modes of the scalar moment with other physical entities can easily be understood as being realized through the dynamical interactions of the gauge bosons with other entities, acting as their longitudinal components. So it seems that nothing gets to disappear mysteriously through redefinition.

But there is indeed a mystery, a mystery of different kind: the mystery of the transmutation of the nature of the couplings of the massless modes of the scalar moment. In their interactions with another mode of the scalar moment (the Higgs mode) and with fermions, their original self-coupling and Yukawa coupling have been transmuted into the gauge coupling through the reorganization.

This mystery has not been addressed or even noticed by physicists. The closest approach to pondering on the subject, though not directly addressing it, was offered by Weinberg almost 20 years after his classic paper. In the section 21.1 of his book *The Quantum Theory of Fields* (*volume II: modern applications*), Weinberg was able to show that in the gaugeless limit, "the gauge boson exchange matrix element is the same as the Goldstone boson exchange matrix element" for a generic physical process.[27] The same mathematical result coming from different couplings may have indicated an intimate connection between the two, but is not an adequate explanation of the transmutation from one to the other through the relocation of the same degree of freedom. It is not clear from his mathematical manipulations what the physical mechanism for such a transmutation is.

If we want to take the EBH mechanism realistically, rather than merely as a fictitious instrument for obtaining the empirical parameters that the Glashow model needed; or, more generally, if we want to take all the terms used in a theoretical discourse as having some physically real meaning, then this mystery has to be dispelled. Yet this task remains to be done.

A closely related or even deeper issue is: What are the physical processes through which various degrees of freedom coming from two moments of the scalar-vector symbiont can be reorganized, resulting in the redefinition of the field involved? In the analogous symbiont of spacetime, the reorganization of its spatial and temporal aspects is driven by a relative velocity between two reference frames, and thus the reorganization is flexible and variable, depending on the variable velocity involved, and can result in various different configurations of spatial and temporal aspects. Is there a similar flexibility and variability in the reorganization of our scalar-vector symbiont? If so, then what is the physical ground for such flexibility? So far no such ground has been explored, and in fact no trace of flexibility and variability has been displayed in the literature, aside from the unique reorganization of the degrees of freedom from the original Lagrangian into the massive gauge and Higgs bosons. If this fixity cannot be dissolved, the EBH mechanism threatens to be an *ad hoc* device for obtaining the observable particles and measurable parameters, and thus cannot be taken realistically.

5. Conclusion: Uncertainty remains

Pending the resolution of the coupling-transmutation mystery and the dissolution of the fixity in reorganizing the fields, the uncertain status of the EBH mechanism -- a physical reality in the subatomic realm or an *ad hoc* fictitious mental device -- has not been even slightly changed by the observation of the Higgs boson. Moreover, this uncertainty also has implications for the understanding of the origin of the mass of subatomic particles. Since all the masses of subatomic particles are expressed in terms of a product of the weak scale (which is constituted by scalar moment's self-coupling) and these particles' couplings to the scalar moment, if the scalar moment and its couplings are real, then the EBH mechanism does contribute to our understanding of the dynamic origin of the mass of subatomic particles. But if the EBH mechanism itself is merely an *ad hoc* fictitious conceptual device, if the scalar moment involved in the mechanism is only a fiction, and thus the weak scale and coupling constants are merely measurable parameters; then the mechanism contributes nothing to our understanding of the origin of mass.

In order to justify the claim that the EBH mechanism contributes to our understanding of the origin of mass, the mechanism itself has to be shown to be real. But this is yet to be done. For this reason, the EBH mechanism can only be taken as an unfinished project.

Acknowledgement: I owe hearty thanks to Stephen Adler, James Bjorken, Paul Forman, Heinrich Leutwyler, John Stachel and Sam Schweber for discussions, comments, suggestions and helps.

[1] P. W. Higgs (1964): Phys. Rev. Letters, 13: 508.
[2] CERN Press Office: "New results indicate that particle discovered at CERN is a Higgs boson." 14 Mar 2013.
[3] F. Englert and R. Brout (1964): Phys. Rev.Letters, 13: 321.
[4] P. W. Higgs (1966): Phys. Rev. 145:1156.
[5] "The Nobel Prize in Physics 2013". *Nobelprize.org.* Nobel Media AB 2013. Web. <http://www.nobelprize.org/nobel_prizes/physics/laureates/2013/>
[6] A. Pais (1953): Physica, 19: 869.
[7] W. Heisenberg (1958): *Proceedings* of International Conference on High Energy Physics, CERN, Geneva, 119.
[8] L. Landau: (1958): Quoted from L. M. Brown and H. Rechenberg, Preprint MPI-PAE/Pth 42/88 (July 1988): 30.
[9] W. Heisenberg *et.al.* (1959): Zeit. F. Naturf. 14A:441; W. Heisenberg (1960): *Proceedings* of International Annual Conference on High Energy Physics, X, (Rochester).
[10] Y. Nambu (1960): *Proceedings* of International Conference on High Energy Physics, IX (Moscow), 121; *Proceedings* of Midwest Conference on theoretical Physics (Purdue). Y. Nambu and G. Jona-Lasinio (1961): Phys. Rev. 122: 345.
[11] J. Goldstone (1961): Nuovo Cimento, 19: 154.
[12] L. Susskind (1979): Phys. Rev. D, 20: 2619..
[13] Theoretically, the possibility of having Nambu-like composites pursued within the framework of dynamical breakdown of symmetry since the 1970s cannot be ruled out. But this possibility can only be pursued beyond the scope of the electroweak theory since it requires the introduction of new binding force for the primary spinor field to form a bound state, and to make the phenomenological coupling between the composite and the primary spinor field consistent with the self-coupling of the primary spinor field. Thus within the electroweak domain, the fundamentality of the scalar field is finalized, as Heisenberg would say.
[14] When the general issue of the necessary consequence of a broken symmetry solution -- the appearance of a massless scalar mode is addressed, the massless scalar mode has been called for many years Nambu-Goldstone boson. This is justifiable since Nambu definitely enjoys a priority in the discovery of this massless mode. But Nambu and Goldstone derived the consequence from different theoretical contexts. Nambu's was a non-linear spinor field model with the scalar mode as a composite while Goldstone's was a primary scalar field model in which the massless mode is also a primary rather than a derivative (composite) mode. My general position on this issue is that Goldstone's contribution was revolutionary in the sense that he broke free from Nambu's spinor field obsession so that a wider conceptual space was open to further explorations. Nambu's approach might be seen as more attractive from a Heisenberg's perspective (a fundamental unified field perspective), but in fact it was counterproductive in the particular stage of the development of particle physics from the late 1950s up till now: all efforts along his line of explorations failed, while Goldstone' approach of assuming a primary scalar field has brought rapid developments in particle physics. The whole industry of the EBH mechanism and Weinberg's model would be impossible if one started from a Nambu type of non-linear spinor field theory with a broken symmetry solution. From a "level" point of view, up to the weak scale level, Goldstone's approach is superior to Nambu's even though perhaps at a much higher energy scale, Nambu's approach may produce something more fundamental. The situation is similar to the study of quantum chemistry: is a QCD style of understanding of proton and neutron as composites of quarks more productive than simply taking them as elementary particles? So in fact I was deliberately using

Goldstone boson rather than Nambu-Goldstone boson here. As Steve Adler rightly indicated in an email exchange that here I am editorializing. This note may help to make my view on this issue explicit.

[15] J. Goldstone, A. Salam, and S. Weinberg (1962): Phys. Rev. 127: 965.

[16] See, for example, P. W. Anderson (1963): Phys. Rev. 130: 439; A. Klein and B. W. Lee (1964): Phys. Rev. Letters, 12: 266; P. W. Higgs (1964): Phys. Letters, 12: 132.

[17] W. Gilbert (1964): Phys. Rev. Letters, 12:713; G. S. Guralnik, C. R. Hagen, and T. W. B. Kibble (1964): Phys. Rev. Letters, 13: 585.

[18] G. S. Guralnik, C. R. Hagen, and T. W. B. Kibble (1964): Phys. Rev. Letters, 13: 585; R. F. Streater (1965): Phys. Rev. Letters, 15: 475.

[19] J. Schwinger (1962): Phys. Rev. 125: 39; 128: 2425.

[20] See note 3.

[21] See note 15.

[22] See note 4.

[23] S. Weinberg (1967): Phys. Rev. Letters, 19: 1264.

[24] S. L. Glashow ((1961): Nucl. Phys. 22: 579

[25] H. Minknowski (1908): *Goett. Nachr,*. 53

[26] T. Y. Cao: "Spontaneous Symmetry Breaking: Comments, in *The Rise of the Standard Model* (eds. Lillian Hoddeson, *e.t.*a, Cambridge University press, 1997), 500-5051

[27] S. Weinberg (1996): *The Quantum Theory of Fields(*Cambridge university Press), sections 19.2 and 21.1, especially pages 299-300.

Roadmap at the LHC to the Higgs Boson and Beyond

Peter Jenni
(Albert-Ludwigs-University Freiburg and CERN)

Abstract

The journey for the search of the standard model Higgs boson at the LHC started more than two decades ago. For the past three years, experiments at the Large Hadron Collider (LHC) have started exploring physics at the high-energy frontier. Thanks to the superb turn-on of the LHC already a rich harvest of initial physics results has been obtained by the general-purpose experiments ATLAS and CMS. The initial data have allowed one to test, at the highest collision energies ever reached in a laboratory, the Standard Model (SM) of elementary particles, and to make many searches Beyond the Standard Model (BSM). Clearly the most important results have been obtained with the discovery of a (or 'the') Higgs Boson, which establishes the postulated electro-weak symmetry breaking mechanism in the SM. These important, and successful, SM physics measurements are giving confidence that the experiments are in good shape for their journey into the uncharted territory of new physics anticipated at the LHC.

1. Introduction and early history

The SM has been tested by many experiments over the last few decades and has been shown to successfully describe high-energy particle interactions. Despite its incredible success, the SM is incomplete because many questions remained unanswered, the foremost being the origin of particle masses. The list of fundamental open questions can be continued with: What is the nature of the Universe Dark Matter? Why is only matter observed in the Universe as primary constituents and not anti-matter? What are the features of the primordial plasma present ~10 μs after the Big Bang? What happened in the first moments of the Universe ~10^{-11} s after the Big Bang? Are there additional forces to the known four ones? Are there additional (microscopic) space dimensions? And so on... New physics beyond the Standard Model is needed to answer these and other questions. The huge amount of precise experimental data collected so far indicates that this new physics could manifest itself at the TeV energy and mass scale being explored by the LHC.

In the 1980s nobody could foresee what the physics landscape would be like in the decades to come. Nevertheless, many physicists were excited by the prospect of making a giant leap into unchartered territory. Experimental and theoretical physicists as well as accelerator experts from all over the world came together in the 1980s in several workshops where ideas were vigorously discussed about the accelerators and detector concepts most capable of tackling the major open questions in particle physics, often paraphrased as the 'known unknowns', and to discover new physics beyond the Standard Model, the 'unknown unknowns'. Finding the Higgs boson was clearly a "dream" in the first category.

The accelerator favoured (in 1987) was a proton-proton (pp) collider operating at a centre of mass energy of around 15 TeV and at an instantaneous pp collision rate of about 800 million per second. A saying prevalent in the late 1980s and early 1990s captured the challenge that did lay ahead for the experimentalists: "we think we know how to build a high energy, high luminosity hadron collider – we don't have the technology to build a detector for it".

From the onset of the LHC project it was known that finding the Higgs would be tricky. The possibility of detection of the SM Higgs boson over the wide mass range, and its diverse manifestations or signatures, played a crucial role in the conceptual design of the ATLAS and CMS experiments, and served as a benchmark to test the physics performance of these detectors. Outstandingly versatile instruments, with diverse detection capabilities, are required to search for the Higgs boson so as to ensure that the LHC would be able to provide a definitive answer about its existence or not, regardless of its mass. It was felt that these experiments would then also be ready to explore the potentially new physics beyond the SM.

The concepts of the ATLAS and CMS experiments were born with Letters of Intent (LoI) that were submitted to the LHC Experiments Committee at CERN (LHCC) more than 20 years ago, on 1st October 1982. The LHCC approved ATLAS and CMS to proceed to a more detailed phase. The detector concepts were laid out in the Technical Proposals submitted to the LHCC at the end of 1994. These formed the basis for their formal approval in January 1996 by CERN, and subsequently by all the Funding Agencies supporting each of the Collaborations. A long process followed of refining the designs, selecting the final technologies for the component parts of the experiments, based on tests of large prototypes. These tests were conducted at CERN and elsewhere, using particle beams of known energy. The measured performance of the components or detectors guided the final choice between competing options. The optimization of the experiments, over almost 10 years, was a very complex and iterative process, and constrained not only by the technical requirements but also by the desired physics performance, tested by simulations based on extrapolations from test measurements. A crucial issue was also that of the overall cost, and most importantly, the human and financial resources available within the experiment communities.

The designs for the detectors in the individual layers were turned into detailed technical documents for construction in a series of more than 15 voluminous so-called Technical Design Reports for each experiment, spanning 1995 to 2005. These were reviewed by the LHCC and the 'green light' given to commence construction, which started for most components around 1998. During this period, each one of the ATLAS and CMS Collaborations grew from about 800 scientists in 90 Institutions signing the LoI, rather rapidly in the beginning, adding missing expertise in some areas, to their present size of over 3000 scientists from 180 Institutions in around 40 countries.

The LHC machine was proposed in 1993 to the CERN Council, and approved end of 1994 as a staged, reduced-energy first phase collider, and finally end of 1997 at its full energy of 14 TeV as a single-stage project.

An account of the early history of the LHC and the experiments can be found in [1-4], and references therein.

2. The experimental challenges

The high pp collision rate in LHC leads to a number of formidable experimental challenges. The counter-rotating LHC beams are organized in bunches comprising some 10^{11} p/bunch separated by 25ns (the accelerator currently operates at 50 ns), leading to a bunch crossing rate of 40 MHz. The event selection process ("trigger") must reduce this rate to ~ 0.5 kHz for storage and subsequent detailed analysis. New collisions occur in every crossing and it is not feasible to make a trigger decision in 25 ns, yet a trigger decision must be made for every crossing. The initial stage trigger decision takes about 3 μs. During this time the data must be stored in on-board pipelines associated with almost every one of 100 million electronic channels in each experiment.

Around 1000 charged particles, from around 20 pp interactions, emerge from the interaction region in every bunch crossing. Highly granular detectors, with fast response and good time resolution, are required to give low occupancy in each electronic channel. The resulting tens of millions of detector electronic channels require very good time synchronization to cleanly separate the different waves of particles emerging from the interaction point every 25 ns. The enormous flux of particles emerging from the interaction region leads to high radiation levels.

The event selection process proceeds in two or three steps; the first step reduces the rate to a maximum of 100 kHz, and comprises custom hardware processors that use coarse information from the detectors; the data from these events are transferred from the on-detector readout electronics through the readout network, to a processor farm which operates offline-quality reconstruction algorithms to decrease the event rate further to around 0.5 kHz before data storage. The tens of petabytes that are generated per year per experiment are distributed for offline analysis to scientists located across the globe. This data management problem motivated the development of the so-called World-wide LHC 'Computing Grid' [1].

To accomplish the physics goals new technologies had to be invented and most of the others had to be pushed to their limits [5].

3. The design of the ATLAS and CMS experiments

The mass of the SM Higgs boson was not predicted by theory and its rate of production (cross section) and natural width vary widely over the allowed mass range (100 – 1000 GeV). Once produced the Higgs boson disintegrates immediately in one of several ways (decay modes) into known SM particles. A search has to be envisaged over a large range of masses and decay modes: in pairs of photons, Z bosons, W bosons, τ leptons, and b quarks. Not only is the putative SM Higgs boson rarely produced in the proton collisions, it also rarely decays into particles that are the best messengers at the LHC of its production: photons, electrons, and muons. As these particles emerge directly from the decay they do not have any extra associated energy in the area surrounding their paths. Such particles are said to be "isolated". In one combination, of production and decay, used in the discovery, this happens only in 1 out of 10'000'000'000'000 pp collisions (one in 10 trillion!). This means that a very large number of proton-proton collisions per second have to be studied, the current number being around 500 million/second, corresponding to an instantaneous "luminosity" of 7.10^{33} cm^{-2} s^{-1}. Hence the ATLAS and CMS experiments have to operate in a very harsh environment created by these huge rates.

At masses below around 140 GeV the natural width of the SM Higgs boson is a few MeV and the width of any observed peak would be entirely dominated by instrumental mass resolution, placing most stringent performance requirements on the general-purpose detectors (GPDs), ATLAS and CMS. The Higgs decays generally accepted to be most promising are the decays into two photons, occurring a few times in a thousand, and those, occurring even less often, into a pair of Z bosons each of which in turn decays into a pair of oppositely charged leptons (electrons or muons). However, with a powerful enough GPD these particles can be detected and their energies measured precisely enough with the prospect of seeing a narrow peak, over background, in the distribution of the invariant masses of the decay particles (e.g. two photons). The invariant mass of a system of particles is a quantity obtained from combining the energy and momentum of each constituent particle.

Other signatures, geared to the discovery of the Higgs boson at higher masses, also played a crucial role in the design of the GPDs. Most of these signatures appear at a higher rate but are plagued by larger background as the signal signatures are less distinctive. For example, some of these signatures include collimated "sprays" of particles, known as "jets", resulting from the fragmentation of quarks, which represent the most likely final states from the SM Higgs boson decays but are overwhelmed by the copious production from known SM processes. Among these jets are b quark jets characterized by a finite short, sub-millimeter flight path in the detector before disintegrating. Finally we have the neutrinos, which as neutral leptons, leave the detector without direct trace. Energy balance in the transverse plane of the colliding protons "appears" to be violated, as the neutrino takes away the "missing" energy leading to the signature of missing transverse energy (E_T^{miss}). For example, when the Higgs boson decays into two Z bosons, one can decay into a charged lepton pair, and the other into a pair of neutrinos, leaving as final state an oppositely charged lepton pair and E_T^{miss}. Powerful charged particle tracking, also close to the interaction point, and extensive calorimetric coverage were therefore considered essential.

The single most important and constraining aspect of the overall detector design is the choice of the magnetic field configuration for the measurement of muons. The two basic configurations are solenoidal and toroidal, characterized by magnetic fields with direction parallel and azimuthal to the beam axis, respectively. CMS chose a superconducting high-field solenoid, with a large length-bore ratio, ATLAS chose a superconducting air-core toroid, constituting complementary choices. The CMS solenoid, in addition, provides the magnetic field for the inner tracking system whilst ATLAS has an additional smaller solenoid magnet to carry out the same function.

The ATLAS detector [6]

The ATLAS Collaboration was born out of two proto-collaborations that developed conceptual detector designs around 1990 based on a toroidal muon magnet configuration, enabling the design of a well-balanced general-purpose detector without strong geometrical and technical constraints. The challenge however was to develop a very large and novel superconducting air-core toroid magnet system for the detection of muons, containing about 80 km of superconductor in eight separate barrel coils (each 25x5 m^2 in a 'racetrack' shape) and two matching end-cap toroid systems. The toroids were complemented with a smaller solenoid (2.5 m radius, 6 m length) at the centre, providing an axial magnetic field of 2 Teslas for the measurement of the momentum of charged particles emerging from the pp collisions.

ATLAS chose to meet the physics requirements with the following main detector components:
- Highly granular and precise *electromagnetic calorimeter* (em) complemented by full-coverage *hadronic calorimeter* for accurate jet and E_T^{miss} measurements. Novel techniques were developed in substantial R&D over many years. This resulted in a fast and very "hermetic" cryogenic liquid Argon (LAr) – lead sampling calorimeter, with so-called 'accordion' geometry, for the em part and a plastic scintillator – iron sampling hadronic calorimeter with an innovative geometry for the barrel part of the experiment (LAr hadronic calorimeters are used in the regions near the beam axis). The em and hadronic calorimeters have 200'000 and 10'000 independent cells respectively, and are located in an almost field-free region between the toroids and the solenoid. They provide both fine lateral and longitudinal segmentation.

- The trajectories of *muons,* tracked over ~ 5 m or more in the air core toroid field, are unperturbed by material and their momentum can be precisely measured. About 1200 large muon chambers, of various shapes, with a total area of 5000 m^2 measure the track positions with an accuracy of better than 0.1 mm. Another set of about 4200 fast chambers are used to provide the 'trigger'. All the chambers were built in about 20 collaborating institutes in three continents that formed a "sub-system community". This organization was typical, and true also for other components of the experiment.

- The reconstruction of all *charged particles*, including that of secondary vertices, is achieved in the inner detector that combines highly granular pixel (80 M channels) and microstrip (6 M channels) silicon semiconductor sensors placed close to the beam axis, and a 'straw tube' gaseous detector (350'000 channels) which provides about 30-40 signal hits per track. The latter also helps in the identification of electrons using information from the effects of transition radiation.

Listing these main components does only partial justice to the technological challenges that had to be overcome in developing, constructing and installing the components in the large underground cavern (50 m long, 35 m diameter, and 100 m below the surface) around the collision point. The photograph (Figure 1) shows one end of the cylindrical barrel detector after about four years of installation work, 1.5 years before completion. The ends of four of the barrel toroid coils are visible indicating the 8-fold symmetry of the structure. The photograph illustrates well some of the complexity of the engineering tasks.

Figure 1: Photograph of one end of the ATLAS detector barrel with the calorimeter end-cap still retracted before its insertion into the barrel toroid magnet structure (February 2007 during the installation phase).

The CMS detector [7]

The concept of a compact detector for LHC based on a 4 Teslas solenoid, the Compact Muon Solenoid (CMS) detector, was presented for the first time in the Aachen workshop in October 1990. The world's largest superconducting solenoid magnet with a stored energy of 3 Gigajoules (GJ) reached the full field in September 2006.

In designing CMS the detection of the *muons* was optimized first, having in mind the "golden" discovery channel $H \rightarrow ZZ \rightarrow 4\mu$ that produces energetic muons in all directions. To identify these muons and measure their momenta CMS chose to surround the interaction region with enough absorber material, equivalent to about 2 m of iron, in order to stop all the particles produced except muons (and neutrinos). The muons follow spiral trajectories, in a strong magnetic field, which are reconstructed with precise drift chambers. The "muon problem" is a magnet problem and the choice made by CMS was to go for the maximum magnetic field considered feasible at the time, that of 4 Teslas. The 4 Teslas field called for a superconducting coil with a four-layer winding of a reinforced Nb-Ti superconductor, through which some 20,000 Amperes should flow. Economic and transportation constraints limited the outer radius of the coil to 3 m and the length to 13 m. The field is returned through a 1.5 m thick iron yoke, which houses four muon "stations" to ensure robustness of measurement and full geometric coverage. The iron yoke is sectioned into five barrel "wheels" and three end-cap "disks" on each side for a total weight of 12,500 tonnes. The sectioning enabled the experiment to be assembled, and tested, in a very large surface hall whilst the underground cavern was being prepared. The sections were then lowered one-by-one in spectacular operations between October 2006 and January 2008.

The second design priority of CMS was driven by the search for the Higgs boson via its decay into two photons, which was considered to be the main discovery channel for a low mass Higgs. This called for an *em calorimeter* with the best possible energy resolution.

Scintillating crystals were CMS' preferred option. A new type of crystal was selected: lead tungstate scintillating crystal ($PbWO_4$). Five years of research and development (1993-1998) were necessary to improve the transparency and the radiation hardness. It took round-the-clock production over ten years (1998-2008) to manufacture the 75848 crystals necessary to cover almost the full solid angle. The last of the crystals was delivered in March 2008, and the full "crystal calorimeter" was installed by mid-August 2008, comprising more crystals than all of the past HEP experiments put together. The crystals are arranged in a quasi-projective geometry. This "compact" crystal calorimeter fits well with the compact design of CMS leaving enough room inside the coil for powerful inner tracker and calorimeters.
The CMS solution to *charged particle tracking* was to opt for a relatively small number of measured points (~13) each with a high position resolution (~15 μm per point) leading to a very large number of cells distributed inside a cylindrical volume of 5.8 m in length and 2.5 m in diameter: 66 million 100x150 μm^2 silicon pixels and 9.3 million ~10 cm x100 μm silicon microstrips. With 198 m^2 of active silicon area the CMS tracker is by far the largest silicon tracker ever built.

Finally the *hadron calorimeter* together with the crystal calorimeter is designed to measure the energy flow in ~ 3,000 small solid angle projective towers covering almost the full solid angle. It comprises alternate plates of ~5 cm brass absorber and ~4 mm thick scintillator plates that sample the shower. The scintillation light is detected by novel photo detectors (hybrid photodiodes) that can operate in the strong magnetic field. Figure 2 shows the transverse view of the barrel part of CMS in 2007 during the installation phase in the underground cavern.

Figure 2: Transverse section of the barrel part of CMS illustrating the successive layers of detection starting from the centre where the collisions occur: the inner tracker, the crystal calorimeter, the hadron calorimeter, the superconducting coil and the iron yoke instrumented with the four muon stations. The last muon station is at a radius of 7.4 m.

The ATLAS detector was assembled directly in the huge underground cavern whilst the CMS was assembled in a large surface hall and lowered in section into the underground cavern. Connecting all the detector elements to the electronics and computers in a neighbouring service cavern was an enormous and delicate task. Access to repair any significant fault, or faulty connection, buried inside the experiment, would require many months to just open the experiments. As an example each experiment has more than 50'000 cables with a total length exceeding 3'000 km, and more than 10'000 pipes and tubes for servicing the apparatus. The nested structure of the experiments makes access for maintenance very difficult, time consuming and highly constrained. Hence, a high degree of long-term operational reliability, which is usually associated with space-bound systems, had to be attained. This was uppermost in the minds of the engineering teams of the experiments.

4. The preparation of the experiments and computing

The activity of commissioning of the component parts of the experiments, very important already during the final years of installation, became frenetic in the year before the LHC start-up (2008). Even though all components were thoroughly checked at the various construction sites around the world before delivery to CERN, they had to be tested again after their installation in the underground caverns. They were first operated and fine-tuned separately and later with more and more of the components of the experiment together. During these years, the experiments made use of the constant flow of cosmic rays impinging on the earth. Even 100 m deep underground there is still a small flux of muons: a few hundred per second traversing each of the experiments. The passage of each of the hundreds of millions of cosmic ray muons was recorded. The muon tracks were used not only to check the whole chain from the hardware to the analysis programs of the experiments, but also to adjust the huge number of parameters and constants needed to align all the detector elements and calibrate their response to the traversal of muons. This work was crucial in preparing the experiments for physics before the first LHC collisions finally occurred.

It was recognized early that the large volume of data that will be generated by ATLAS and CMS (tens of Petabytes of data per year) and the large number of physicists distributed all around the world would require a new approach to software and computing. A fully distributed computing model had to be developed. The aim was to allow any user anywhere access to any data recorded or simulated during the lifetime of the experiments. The computing system consists of a hierarchical architecture of tiered centres, with a single large Tier-0 centre at CERN, about ten large Tier-1 centres at national computing facilities, and about one hundred Tier-2 centres at various institutes. The Tier-0 receives the raw data from the experiments. It carries out prompt reconstruction, almost in real time, and exports the raw and reconstructed data to the Tier-1 centres. It must keep pace with the event rate of 0.5 kHz (~1.5 MB of raw data per event) from each experiment. The large Tier-1 centres provide long-term storage of raw data and reconstructed data outside of CERN (a second copy). They carry out second-pass reconstruction, when better calibration constants are available. Finally the Tier-2 and numerous even more local Tier-3 centres are used by the physicists to analyse the data. The also required large number of Monte Carlo simulated events is produced mainly in Tier-2 centres.

The system is based upon Grid middleware, with the common Grid services at the centres defined and managed through the Worldwide LHC Computing Grid (WLCG) project, which is a collaborative effort between LHC experiments, computing centers, and middleware providers [8].

During the three years from 2006 to 2008 a series of increasing-scale software/computing exercises called "data challenges" were performed, ending with all available components operating at full scale in as realistic a way as possible.

Writing the hundreds of thousands of lines of code in C++ necessary for the reconstruction of four-momenta of all detected particles, starting from the raw data, is another formidable challenge. The reconstruction software was developed by hundreds of physicists and software engineers from the many different participating institutes. Special tools are used to develop code in a unique software repository and to control its quality.

5. The operation of the LHC and the experiments

The operation of the LHC accelerator started on the 10th of September 2008. On the 19th of September, during the final powering tests of the main dipole circuit in the last of the eight sectors (3-4) of the LHC to be fully powered-up, an electrical fault occurred in one of the tens of thousands of connections, which resulted in large release of helium from the magnet cold mass leading to mechanical damage to around 30 magnets. These were repaired in 2009 and actions were taken to mitigate the effects of any similar incident in the future. Furthermore, it was decided to run during the first years at a reduced energy of 3.5 TeV per beam (i.e. at half the nominal energy).

The first proton-proton collisions (at an energy of 450 GeV per beam) occurred on the 20th of November 2009; the first high-energy collisions (at 3.5 TeV per beam) were recorded on 30th March 2010, and since then the machine has operated superbly, providing the two general-purpose experiments ATLAS and CMS with data samples corresponding to an integrated luminosity of close to 5 fb^{-1} during the pp running periods in 2010 and 2011, and another 23 fb^{-1}, at the slightly higher energy of 4 TeV per beam in 2012. In total these data correspond to the examination of some 3 x 10^{15} collisions. The outstandingly fast progress and excellent performance of the LHC has been a challenge for the experiments in many aspects, with typically 20 overlapping pp collisions (pile-up) in the same bunch crossing as the collision of interest. Nevertheless ATLAS and CMS have recorded collision data very efficiently, reaching a data taking efficiency of about 95% whenever the LHC machine is operating in stable conditions. Both detectors work with an impressive reliability. Some 98% of the roughly 100 million electronic read-out channels in each experiment are performing at design specification. This outstanding achievement is the result of a constant and dedicated effort by the teams of physicists, engineers and technicians responsible for the hardware, software and the maintenance of the detectors. This efficient operation of the accelerator and the experiments has led to the rapid discovery of the Higgs-like boson.

6. Standard Model physics results

Observing, and measuring accurately, at the new collision energies, the known particles from the SM, can be considered to be a necessary stepping stone towards exploring the full potential of the LHC with its many promises of possible new physics discoveries. The SM processes were often called 'standard candles' for the experiments. However there is much more value to measuring the SM processes than this: never before could the SM physics be studied at a hadron collider with such sophisticated and highly accurate detectors, ultimately allowing a test of detailed predictions of the SM with unprecedented precision and minimal instrumental systematic errors, as for example already published for many ATLAS and CMS QCD results.

In the following a few examples will be shown from a very extensive body of publications and publically released conference contributions, which are all available at [9] for ATLAS and [10] for CMS, where detailed specific references can be found.

The charged and neutral Intermediate Vector Bosons (IVB) W and Z are the major benchmark measurements at the LHC for demonstrating the excellent detector performance, as well as for testing model predictions to a high degree of accuracy. The Z decays into electron and muon pairs can be extracted almost free of any backgrounds, as shown in Fig. 3.

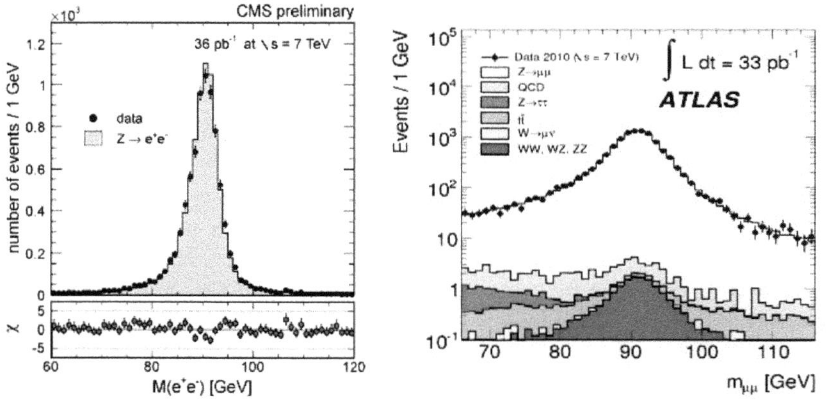

Figure 3. The CMS electron-pair mass distribution on a linear (left) and the ATLAS muon-pair mass distribution on a logarithmic (right) vertical scale, in the Z mass region. The estimated small background contributions are indicated, as well as the expected signal shape from MC simulations.

The classical W decay signatures into an electron or muon and the associated neutrino are an excellent test for the missing transverse energy (E_T^{miss}) performance of the detector due to the undetected neutrino. E_T^{miss} is inferred from the measured energy imbalance in the transverse projection of all observed signals w.r.t. the beam axis. The E_T^{miss} spectrum for events with a well-identified muon candidate is shown in Fig. 4a, and shows a clear W signal over the expected SM background sources. After applying a selection of events with $E_T^{miss} > 25$ GeV only a small residual background remains present under the W signal, as indicated in the transverse mass distribution given in Fig. 4b.

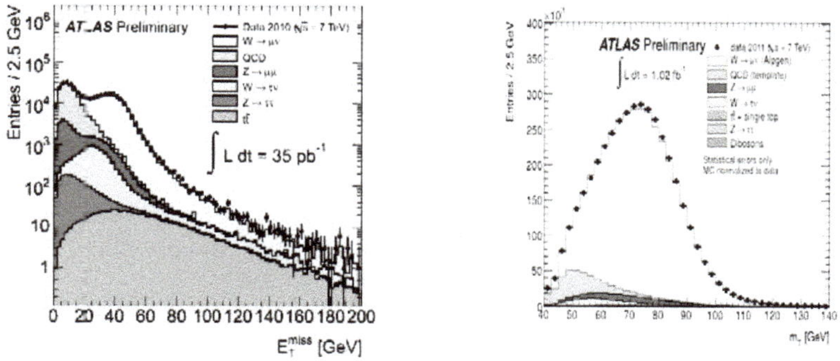

Figure 4. (a, left) Missing transverse energy distribution for events with a muon candidate. (b, right) Transverse mass distribution for W to muon decays. The expected background contributions are indicated as well (examples from ATLAS).

The good agreement between the measured and expected cross-sections times the leptonic decay branching ratios (which is the expected rate for W and Z bosons to be produced and then decay to leptons) is illustrated in Fig. 5. With the present data samples the experimental uncertainties still dominate, but with the addition of the full 2012 data, the measurements are expected to already strongly constrain the theoretical model parameters. Figure 5a shows the cross-section measurements and predictions as a function of the collision energy, whereas in (b) the W and Z cross-section results are displayed in a 2-dimensional plot including their correlated error ellipse, and compared to predictions with various parton distribution functions (describing the quark and gluon momentum distributions inside the protons). Detailed measurements of properties for IVB production and decay at the LHC have been published already, including, for example, the lepton charge asymmetry measurements for W decays, which were an important signature of the electro-weak nature of the W at the time of their discovery some 30 years ago.

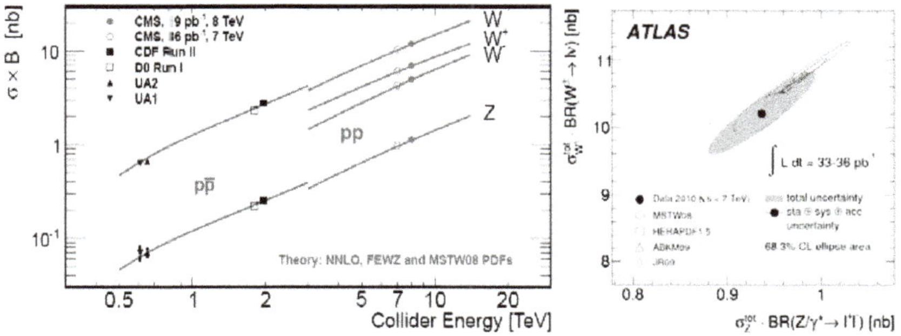

Figure 5. (a, left) CMS W and Z production cross-sections times leptonic branching ratio as a function of the collision energy, showing also previous measurements at lower energy colliders. (b, right) Correlation of the measured (solid dot) leptonic W and Z cross-section from ATLAS, compared to theoretical expectations with various choices for the parton distribution functions (open symbols).

Hard collisions (characterized by having final state particles with significant transverse energy) at the LHC are dominated by the production of high transverse momentum jets, which are the collimated sprays of particles from the hadronization of the initially scattered partons (quarks, gluons) in the colliding protons. At work is the strong interaction described by Quantum Chromo Dynamics (QCD). Most commonly two jets emerge at opposite azimuth with balanced transverse momenta, from an initial lowest order parton-parton scattering process. However, higher order QCD corrections alter this picture significantly, and detailed measurements of multi-jet configurations are very important to constrain the QCD descriptions of hadronic processes.

The most impressive results at this stage are the inclusive jet and the di-jet cross-section measurements; an example from ATLAS for them is shown in Fig. 6a. These measurements cover unprecedented kinematical ranges spanning typically over jet transverse momenta from 20 GeV to 2 TeV, in many angular (pseudorapidity) bins up to $|\eta| < 4.4$ (i.e. very close to the beam axis). The cross-sections vary over these ranges by up to 12 orders of magnitude. In general the agreement with perturbative QCD calculations, including next to leading order (NLO) corrections, is well within the systematic uncertainties. This cannot be seen in Fig. 6a directly, only in ratio plots measurement/theory for a given η-interval as shown for CMS data in Fig. 6b. The systematic uncertainties in the ratios are typically less than 20%, which is a great achievement compared to previous such measurements. The systematic uncertainties on the measurements are dominated by the jet energy scale uncertainty (calibration of the detectors for the energy of jets), which, thanks to a considerable effort, has been determined to typically better than 3%.

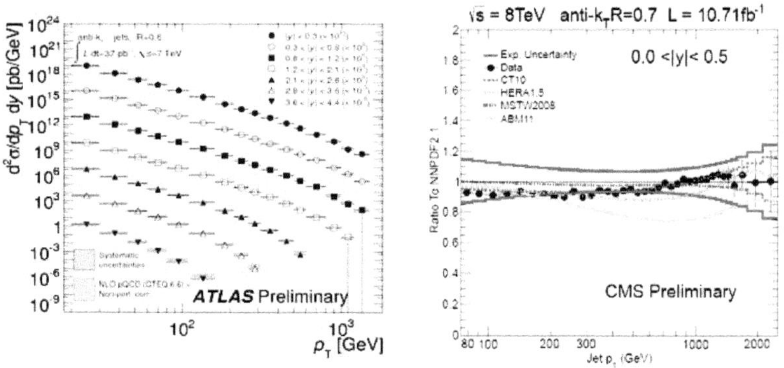

Figure 6. ATLAS measurements of inclusive jet (a, left) cross-sections, (b, right) CMS jet measurements compared to NLO perturbative QCD predictions plotted as ratio data/calculation.

Jets can also be produced together with W and Z bosons, so-called QCD corrections to the Intermediate Vector Boson production. Many results of these processes have been published. A good understanding of them is particularly important as they are, in many cases, a dominant source of backgrounds to the search for new particles, as well as to the measurements of top quark production discussed next.

The heaviest known particle in the SM is the top quark with its roughly 175 GeV mass. It decays almost exclusively into a W and a bottom quark. The measurement of top quark pair production typically requests that at least one of the W decays leptonically (also needed to trigger the events), and therefore the final states require one or two leptons (electrons or muons), E_T^{miss}, and jets, some of which, coming from the b-quarks, can be tagged by the displaced secondary vertices due to the finite life times of b-hadrons. Whilst it is beyond the scope of this report to describe the sophisticated analyses employed, the message is that there are clear top pair signals in ATLAS and CMS, both in the single and two-lepton channels, when considering the correct jet topologies. The resulting cross-sections are shown in Fig. 7 which also illustrates the expected large rise of the cross-section with the collision energy increase from 2 TeV at the TeVatron to 7 TeV and 8 TeV at the LHC. Good agreement with NLO QCD calculations is seen within the present few % measurement errors. It can be mentioned that both ATLAS and CMS have also reported first single top measurements (events with just one top quark) at a rate in good agreement with QCD expectations, as well as detailed studies of top properties like its mass.

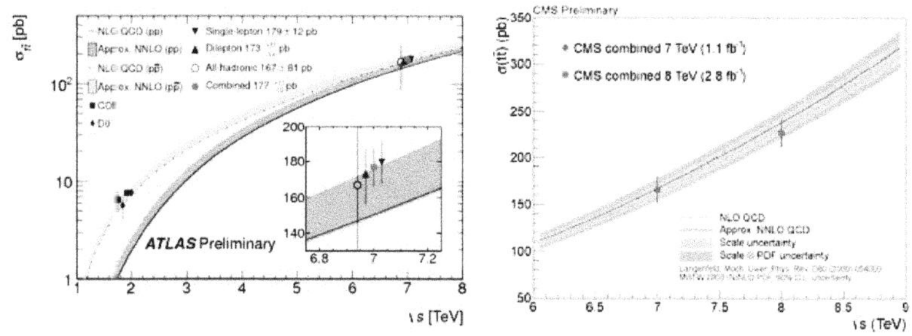

Figure 7. Top pair production cross-section as a function of the collision energy, showing the TeVatron and LHC measurements.

A summary of many measurements of interesting Standard Model production cross-sections is given in Fig. 8, which includes also the low cross-section processes of IVB pair productions.

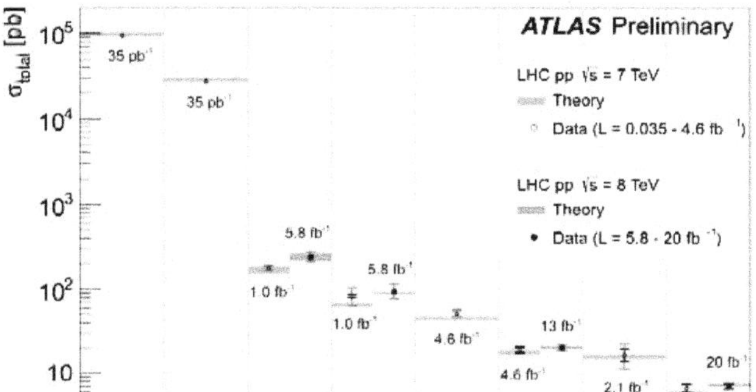

Figure 8. Summary of various Standard Model production cross-sections (similar results are available from CMS).

The excellent performance in measuring all these Standard Model physics gave confidence for the readiness of the experiments to search for new physics.

7. Higgs boson

The search for the Higgs boson H, as the decisive manifestation of the Brout, Englert, and Higgs mechanism for electro-weak symmetry breaking [11], postulated in 1964, was one of the major motivations for initiating the LHC project almost 30 years ago. The ability to detect it unambiguously over the full possible mass range from its lower experimental limit (set finally at 114 GeV at the LEP collider) up to one TeV, with very different favoured final states (decay modes) at different masses, was the major benchmark in the conception of the ATLAS and CMS detector designs.

As the mass of the Higgs bosons is not predicted by theory a search has been made over a broad mass range. But then, initial search results from the LHC and the Tevatron pointed to a range of low masses, maybe just a few tens of GeV above the lower LEP limit. Indirect indications from fits of all measured electroweak parameters also suggested a low mass. Furthermore, in December 2011, the ATLAS and CMS experiments reported a "tantalizing" excess of events near a mass of 125 GeV. The announcement of the discovery of new boson was finally made by the ATLAS and CMS Collaborations during a historic event on 4th July 2012, and published one month later [12]. A didactic description of the long journey to the Higgs boson, and details of the analyses accessible also to non-experts is available in [2].

The ATLAS and CMS searches are performed in the low-mass region in several decay modes. Favoured for making a discovery however were the two final states with the best mass resolution, namely H→γγ and H→ZZ with the Z bosons decaying to electron or muon pairs. The results including the full data samples collected in the 2011 and 2012 LHC data taking periods are shown in Figs. 9 and 10, for both channels and from both experiments [13].

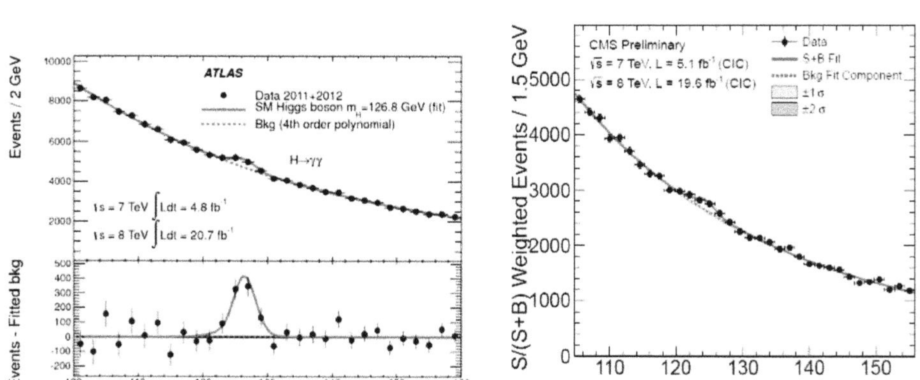

Figure 9. The H→γγ signals in the ATLAS (left) and CMS (right) two-photon invariant mass distributions.

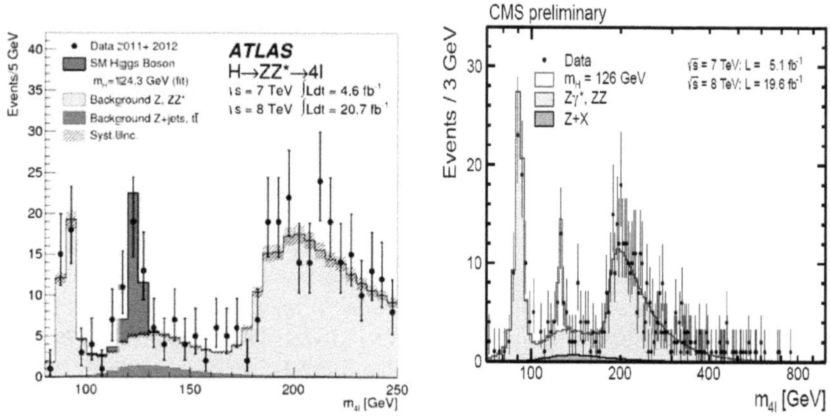

Figure 10. The H→ZZ signals with the Z bosons decaying to electrons or muons in the ATLAS (left) and CMS (right) four-lepton invariant mass distributions.

Both experiments have observed, each independently, highly significant signals in these two channels alone, at a mass very well compatible among them: 125.5±0.2±0.6 GeV for ATLAS and 125.7±0.3±0.3 GeV for CMS. Here the first error is statistical and the second represents the systematic uncertainties. The overall significance is given in Fig. 11a as an example for ATLAS, reaching 10 standard deviations away from a hypothesis of a pure background fluctuation. The compatibility of the observed signal with the expectations for the SM Higgs boson is usually expressed as so-called signal strength μ, which is defined as the ratio of the observed over expected cross section, being unity for a SM signal and zero for a background only hypothesis. The parameter μ is shown for all channels for example from CMS in Fig. 11b. The observed values for μ are 1.30±0.20 from ATLAS and 0.80±0.14 from CMS, well compatible among the two, and also with unity as expected for the SM H, given the still large errors.

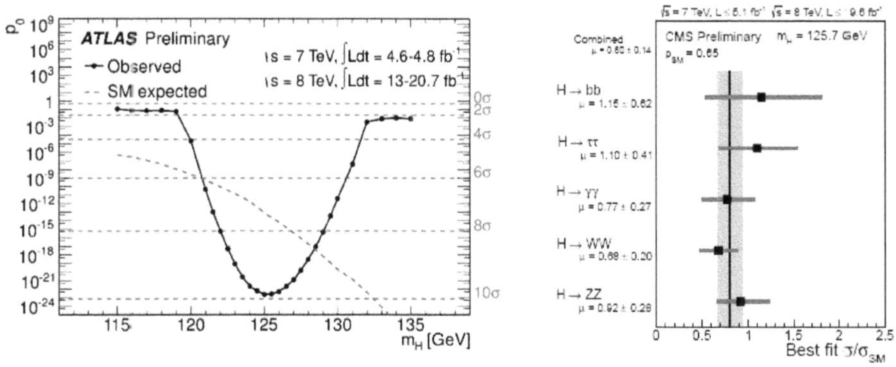

Figure 11. (a, left) the overall signal significance from ATLAS after combining all channels, (b, right) the signal strength μ for different channels from CMS.

Remarkably, the discovery of the new boson has been made at half the LHC design energy, under much more severe pileup conditions than anticipated at this stage, and only with one-third of the integrated luminosity that was originally judged necessary. This is a testament to the careful design, construction and preparation of the ATLAS and CMS experiments; their hardware, software, computing and physics analysis.

Detailed studies of the production and decay properties have started in order to characterize the new particle. It will be important to understand with best possible precision if the observed new particle is the only scalar boson of the SM Brout-Englert-Higgs mechanism postulated to break the electroweak symmetry, or if it is part of a broader physics picture, going beyond the Standard Model. These studies will be among the most central ones in the decades to come both at the LHC and its upgrades, and at possible other future colliders. The basic characteristics of a Higgs boson that have to be answered are: is its spin 0 (i.e. is it a scalar particle), is its charge and parity (CP) even, is it the neutral component of an electroweak doublet, and is its associated field the origin of mass for the W and Z bosons as well as for the quarks and charged leptons? Initial indications, albeit within large uncertainties at this early stage of data accumulation, support all these criteria at the 2 to 3 standard deviation level. But much higher precisions are needed to detect possible effects of New Physics in any of these characteristics.

8. Beyond the Standard Model

Besides the quest to elucidate the mechanism of the electro-weak symmetry breaking by searching for the Higgs boson, the major excitement for LHC comes from the great potential to explore uncharted territory of physics Beyond the Standard Model (BSM), thanks to its highest collision energy ever available in a laboratory so far. Since the beginning of the project, the search for Supersymmetry (SUSY) was a strong motivation, and besides the H boson it has been the other main benchmark physics that was guiding the detector designs. However, many other hypothetical new processes can be searched for, and indeed ATLAS and CMS have already reported in many publications a very broad spectrum of searches for BSM signatures (mass peaks for new particles or kinematical distributions with deviations from the expectations of known physics processes). No such new effect has yet been found, and all of these searches result in highly-improved, stringent exclusion limits, often well beyond the one TeV scale already. Only a few examples are mentioned in the following.

The most popular searches concern SUSY, which predicts additional fundamental particles. The search for SUSY is motivated in part by the prospect that the lightest stable neutral SUSY particle (LSP) could be an excellent candidate for explaining the Dark Matter (DM) in the Universe. The mysterious existence of DM was postulated by Zwicky, and rather convincingly evidenced by Rubin, both astronomers, in the 1930s and 1970s respectively. The SUSY searches at LHC are very complex as they must be sensitive to many (model-dependent) decay chains, implying a large variety of possible final state topologies. A common feature for most of them is the existence of significant missing transverse energy, E_T^{miss}, due to the escaping LSPs (an experimental signature similar to that of the neutrinos in the W decays). Furthermore the SUSY signatures include often high transverse momentum jets, some tagged as b-jets for third-generation squarks as particularly motivated by naturalness arguments developed in other lectures, and leptons. The expected topologies depend not only on the model parameters, but also on the mass relations between various squarks and gluinos (the SUSY partners of the SM quarks and the gluons). A summary of 95% CL mass exclusion regions from many SUSY searches is shown in Fig. 12 from ATLAS. Very similar results are available from CMS.

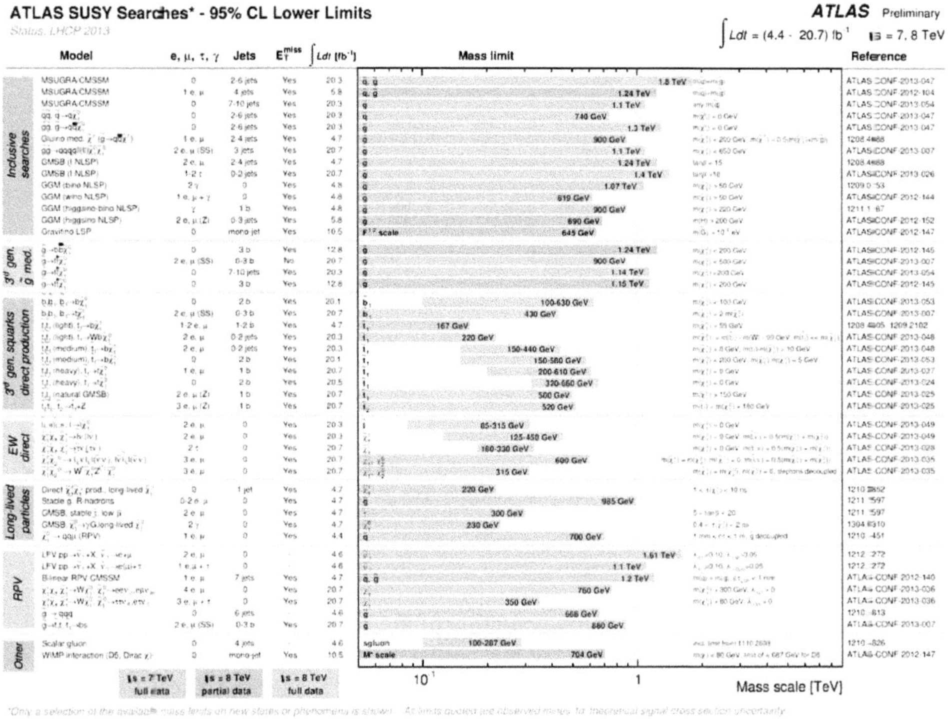

Figure 12. A summary of 95% CL mass exclusion limits from many SUSY searches as obtained by ATLAS (very similar results are available from CMS as well).

Many other searches aimed at exploring BSM physics have been conducted, at this stage all with finding no excess of observed event rates over the expected backgrounds from the known SM processes. However, much more stringent limits and constraints could be established than what were available up to now. A non-exhaustive summary of 95% CL limits is displayed in Fig. 13 from CMS, with very similar results being reported by ATLAS.

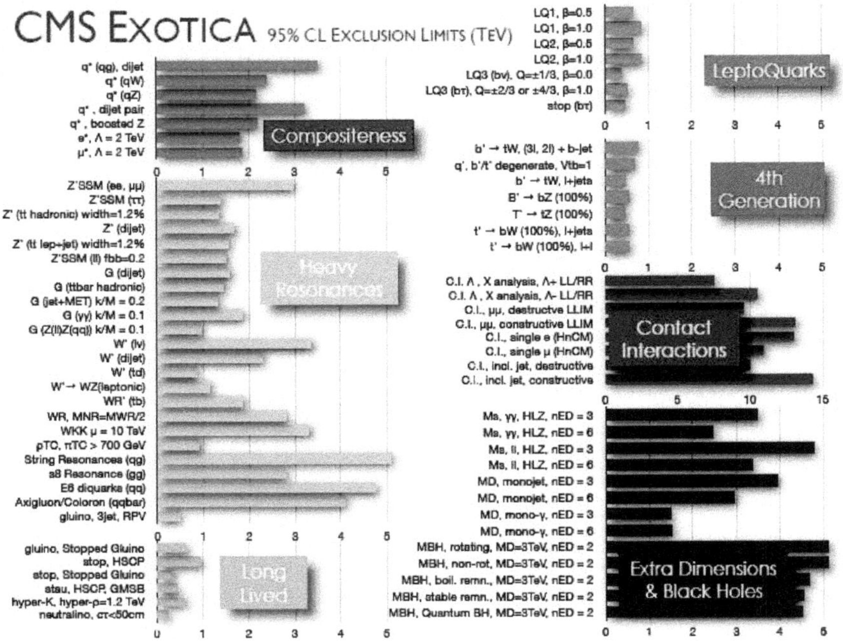

Figure 13. A summary of 95% CL exclusion limits from many BSM searches other than SUSY as obtained by CMS (very similar results are available from ATLAS as well).

9. Conclusions and outlook

After only three years of high-energy operation of the LHC, ATLAS and CMS (as well as LHCb and ALICE, not covered in these lectures) have already produced a very rich harvest of physics results. This should in the first place be seen as a success of the overall LHC project with its three major components having worked in an extraordinary performing and coherent way so early on: the collider, the experiments and the novel computing grid structure WLCG.

ATLAS and CMS have already amply demonstrated that they masters the analyses of the Standard Model physics, and have brought many SM measurements into domains of details and a degree of sophistication unprecedented at lower energy machines, setting a challenge for accurate theoretical descriptions. The SM results also demonstrate that the experiments are well understood and operate at full potential to attack the main goals of LHC, namely to make the conclusive investigations of the electro-weak symmetry breaking mechanism, i.e. establish accurately the properties of the discovered Higgs boson, and explore the physics Beyond the Standard Model. The discovery in July 2012 of *the* (or *a*) Higgs boson is a first culmination of this effort. The searches for BSM physics are exploring unchartered territory, thanks to the superb performance of the collider. One should remain aware, however, that the full physics potential will only become available in future years with the machine reaching its full design luminosity and energy. After a two-year (2013-2014) shutdown, to refurbish parts of the accelerator so as to be able to operate close or at the design energy (13-14 TeV collisions), data taking will recommence in spring 2015 allowing deeper studies of the newly discovered heavy boson and continuing the search for physics beyond the SM. There is no doubt: particle physics has exciting times ahead with the LHC project now operational.

Acknowledgements

I warmly thank Professor Antonino (Nino) Zichichi, Director of the Ettore Majorana Foundation and Centre for Scientific Culture, for having invited me to this motivating school, and for his generous hospitality in the beautiful town of Erice. I also thank all the brilliant students for their great stimulating discussions, which make me confident that our field will be in good hands for the future.

For much of the historical discussion I have greatly benefitted from pleasant common work with my CMS colleagues Michel Della Negra and Tejinder Singh Virdee.

Further reading and references

[1] L. Evans, Ed., The Large Hadron Collider, a Marvel of Technology, EPFL Press, Lausanne, Switzerland, (2009).
[2] M. Della Negra, P. Jenni, T.S. Virdee, Science 338, 1560 (2012).
[3] P. Jenni and T.S. Virdee, CERN Courier June 2013, 22 (2013).
[4] P. Grannis and P. Jenni, Phys. Today 66(6), 38 (2013).
[5] N. Ellis and T.S. Virdee, Annu. Rev. Nucl. Part. Sci. 44, 609 (1994).
[6] ATLAS Collaboration, JINST 3, S08003 (2008).
[7] CMS Collaboration, JINST 3, S08004 (2008).
[8] LHC Computing Grid, Technical Design Report CERN-LHCC-2005-024 (2005).
[9] ATLAS physics web site: https://twiki.cern.ch/twiki/bin/view/AtlasPublic
[10] CMS physics web site: http://cms.web.cern.ch/news/cms-physics-results
[11] F. Englert and R. Brout, Phys. Rev. Lett. 13, 321 (1964);
P.W. Higgs, Phys. Lett. 12, 132 (1964);
P.W. Higgs, Phys. Rev. Lett. 13, 508 (1964);
G.S. Guralnik, C.R. Hagen, T.W.B. Kibble, Phys. Rev. Lett. 13, 585 (1964).
[12] ATLAS Collaboration, Phys. Lett. B 716, 1 (2012);
CMS Collaboration, Phys. Lett. B 716, 30 (2012).
[13] ATLAS Collaboration, arXiv: 1307.1427v1[hep-ex], accepted in Phys. Lett. B;
CMS Collaboration, CMS-PAS-HIG-13-001 and CMS-PAS-HIG-13-002

CHAIRMAN: P. JENNI

Scientific Secretaries: R. Lopes de Sá and N. Ruthmann

DISCUSSION I

- Jenni:

Short comments, perhaps not answers, to some of the questions raised before. The first question was about "smarter machines". Yes, we should work on smarter machines. There is no doubt about that and R&D should go on. But we are clearly far away from any concrete possibility to provide high energy collision with smarter machines in the foreseeable future. So this is not a "yes or no" question. If you want a continuity and progress in our field, then there is no choice. As Prof. Zichichi said, there is one way to go to high energy [collisions] and I will comment more on this point.

There was also a question whether I prefer high energy or high intensity. Both frontiers are important. But if you think in terms of future accelerators, I think the high energy is the preferred way. I want to stress very much that the high intensity, and by that I mean not only colliders like the LHC, but also lower energy machines to look at very rare decays, for instance the muon to electron conversion experiment, are very important for our field. I have no doubt that these experiments need to be done. Facilities should be operated in the world with this goal. They are complementary. In the European Strategy, which takes into account the global context, we were talking more about very large facilities and less about low energy, high intensity, machines, which are and will be done more at a national level.

The point I disagree with Horst [Wenninger] a little bit is that we cannot just wait, because, as Prof. Zichichi pointed out, it takes a long time to develop the accelerator complexes and to develop detectors. In the European Strategy, we have made very clear what is the road map in the coming 15 years and that is the upgrade of the LHC to high luminosity. This is the only realistic option to progress in many of our measurements that will allow us to pin down whether what we observed is "the Higgs", "a Higgs" or something else. So, if you want to have detectors and machines ready in the years 2020 – 2030, then it is clear that the steps toward the detectors have to be taken now. And they are actually being taken. What we are not at the position to say right now is what will be the next project after the LHC. That is something for which we will have some guidance from the 14 TeV data that will come but, again, we should be careful in our community not to put us in a corner from where we can only argue for a new machine when we have a concrete prediction. I think we should keep in mind and, hopefully everyone will see, that we need to explore the high energy frontier without knowing what we will find.There was an interesting question about the time scale. I am not much younger than Horst, but I still think that if we want to have a machine in a foreseeable time scale it is not too early to think also about a large tunnel. In fact, the scenario that was considered in the European Strategy is to operate the LHC at high luminosity until 2030. In an ideal case, our Japanese colleagues will build a linear collider which will, in the most optimistic case, start operations around 2030. After that, it will take a decade to produce physics. And it will be rich physics. If CERN is to be ready with a follow up project to start sometime in 2040, that would set the time scale for such a tunnel. That is also a realistic time scale from what we know from civil engineers about building this 100km tunnel.The 100km tunnel is easier to build than the 80km, as I will show tomorrow. That is because you avoid the Salève Mountain. Civil engineers try to avoid mountains because you can have water pockets with huge pressure. The cost for such a tunnel has been estimated, at least in terms of approximate figures, around 5 billions [euros], which is a huge number, but comparable to the ILC. To that you still have to add the machine, which is in the order of another several billions, but the total cost of the VLHC is comparable to the total cost of the ILC project. If you plan for that well ahead and if the community is willing to concentrate on that, despite being a lot of money, it is possible. The last point I want to address is about muon colliders. This will be a marvelous machine if brought to maturation. There is R&D going on, unfortunately not as vigorously as needed. The European Strategy recognized that would be ideal if other regions, specially the United States, could work more on this project. The energy you could reach is approximately 5 on 5 TeV muons. So, 10 TeV in the center of mass. As a comparison, with the 100 km tunnel you would have collisions at about 100 TeV. In the visionary case, if you make it even bigger, like the ELOISATRON, then you get the number mentioned by Prof. Zichichi [300 TeV]. Clearly you must be smart, but we should also have the courage to see a road map for machines which are feasible, which can be built. Then, we could have smarter follow-up project. That's all I wanted to say, but I wanted to make these points clearer.

- *Csanad:*

I would like to ask about nucleus-nucleus and proton-nucleus collisions. How do you see the role of LHC in that area and the road map in general?

- *Jenni:*

Well, first of all, we will have a full talk about this topic by the spokesperson of the ALICE experiment [Giubellino]. I think, based on the three years experience we have now gathered, that it fits well in the LHC running scenario. It is certainly an enrichment for the LHC physics. There are very interesting results coming out from ALICE. Actually I think it helps that other experiments, in particular CMS and ATLAS, are also interested in this type of physics since they can study specific signatures which ALICE is not build for. For example, the direct observation of jet quenching was something made at ATLAS. I think it has its place and, based again on the European Strategy, when the LHC is upgraded to high intensity, at least in the first years, it will still have a place. Now, I am told by specialists that the physics of p-Pb collisions is very interesting, but I am sure that Guibellino will talk about them in detail since there are a lot of papers coming out from this run. I think it has an established place until early 2020, when we will then have the last moment of ALICE.

- *Hartmann:*

Thank you for elaborating on those early questions. My question is about the impressive numbers one can find in the LHC. I was impressed when you said that there are 10^{11} protons per bunch, 50ns between bunches and 10^9 interactions per second. It is really amazing. As a theorist I am amazed by how experimentalists can keep track of all these data. So, my question is: when you say that you sort out the data, how do you avoid being biased when choosing which data to throw away, and how not to keep only what you already expect to see?

- *Jenni:*

That is a very good question. First of all, we are trying to select signatures as inclusive as possible and, when I mentioned two or three types of trigger, this was an oversimplification. There are around 600 different triggers. We also take a sample of completely unbiased events. Of course, with those you can only find unexpected phenomena if they occur quite frequently. These are the type of things we try to do. As I said, the effort is always to be as unbiased as possible. About the impressive operation of the LHC, we have to thank our accelerator friends for having operated the machine so well. In fact, in order to make it a little bit easier for us, we are insisting very much that, when the machine comes back up, it will be running with a bunch spacing of 25ns. That is what the detectors were really designed for and, for the same luminosity, you would have half the pile-up. That is not how we say, though: we want to double the luminosity with the same pile-up because we have proved that we can handle it. One tries to be flexible and if you, as a theorist, have an idea about a signature we should be looking at, you are always welcome. First there will be discussions, as it is common in these collaborations, but the system is flexible enough to accommodate new ideas.

CHAIRMAN: P. JENNI

Scientific Secretaries: M. Schröder, M. Seidel

DISCUSSION II

- *Inguglia:*

Why cannot the $H \rightarrow \mu\mu$ decay be measured at the ILC?

- *Jenni:*

The foreseen luminosity at the ILC will not be high enough.

- *Inguglia:*

In the dijet invariant-mass spectrum m_{jj} measured in 13/fb at 8 TeV (slide 123 in the backup), there is a deviation of 2 sigma between the data and the simulation in the bin around m_{jj} = 2 TeV. This was also present in the first 5/fb of data only. Is this some systematic effect that should be checked?

- *Jenni:*

This is nothing to worry about, you would expect such statistical fluctuations given the number of bins. In fact, you might even expect more fluctuations than actually observed But the comment is acknowledged.

- *Tawfik:*

Can you elaborate on the ATLAS heavy-ion programme? What are the highlights?

- *Jenni:*

Paolo Giubellino will talk in more detail about it. But one of the highlights is certainly using the ATLAS calorimetry to directly measure jet quenching in heavy-ion collisions. In addition, both multipurpose experiments CMS and ATLAS have performed detailed measurements of the particle flow. These measurements are really complementary to the ALICE measurements because ATLAS and CMS have a much larger coverage in pseudorapidity, which is beneficial in particular for the particle flow measurements, while ALICE has the better ability to measure low-momentum particles and the better particle-identification capabilities.

- *De Palma:*

In last year's LHC publications, a small discrepancy has been observed of the $H \rightarrow \gamma\gamma$ data compared to the SM prediction. What has happened to this, is it still present?

- *Jenni:*

ATLAS observes a slightly larger $H \rightarrow \gamma\gamma$ rate than predicted (1.8 sigma) while CMS observes a lower rate. This is very much compatible with statistical fluctuations.

- *Mojaza:*

Now that a Higgs-like boson has been discovered that looks very SM-like, is it worth to continue performing the current, dedicated BSM searches or should we focus on SM precision-measurements and look for unexpected deviations there?

- *Jenni:*

I would not agree to that statement. Clearly, the SM is not sufficient and we need to find new clues.

- *Mojaza:*

For LHC physics, why is the SM not sufficient at the moment?

- *Jenni:*

That is an interesting question. But the goal of our life is not only to satisfy LHC physics but to satisfy physics in general.

- *Mojaza:*

My point is more on how searches are performed. Would it not be better, also in order not to waste our energies, to perform all measurements with the null-hypothesis that only the SM is valid. And then, if deviations from this hypothesis are observed, starting with dedicated BSM searches? Or, to state it differently, can the BSM searches see effects that SM measurements would not see?

- *Jenni:*

But this is what the BSM searches do: we look for deviations from the SM expectations. No energies are wasted. The searches with the available data are being finalised now, and then we will prepare for the new data at higher centre-of-mass energies.

- *Lopes de Sa:*

Both ATLAS and CMS measure the WW cross-section approximately 1.5σ above the SM prediction at 7 TeV (ATLAS, CMS) and at 8 TeV (CMS). The measurements are now systematically limited. What is the main limitation in the measurements, and what is being done to shed light on this slight disagreement?

- *Jenni:*

The measurements have correlated systematic uncertainties (e.g. PDF), so that a combination would be still compatible with SM expectations.

References:

arXiv:1203.6232 (ATLAS 7 TeV), arXiv: 1306.1126 (CMS 7 TeV),
arXiv:1301.4698 (CMS 8 TeV)

- *Lange:*

What are the main differences between ATLAS and CMS, and would you have done anything very differently in the design of the experiments after several years of experience?

- *Jenni:*

It is remarkable and reassuring that two quite different approaches have brought the same result. It was a main point of the LHC program to have the two general-purpose detectors with complementary technologies. The inner detector resolution of CMS is better than ATLAS, on the other hand the hadronic calorimetry is a factor of two better in ATLAS. In retrospective, only details would have been done differently.

- *Leontsinis:*

Do you think that in the next phase of LHC data taking ATLAS can compete with LHCb in precision measurements, such as $B^0_s \rightarrow \mu\mu$?

- *Jenni:*

On this particular channel, CMS and ATLAS will be competitive. At the moment LHCb has still the lead, also thanks to its very good resolution. Clearly, both general-purpose experiments are pushing these measurements as well. Though some of these analysis are not going as fast as they could because, despite 3000 collaborators, there is a lack of manpower in the general-purpose experiments.

References:

arXiv:1211.267 (LHCb), arXiv:1204.0735 (ATLAS), arXiv:1107.5834 (CMS)

- *Esposito:*

Is it true that the Higgs mass measured in the $H \rightarrow \gamma\gamma$ channel is slightly different from that in the $H \rightarrow 4l$ channel?

- *Jenni:*

In the case of CMS the masses are the same, in the case of ATLAS there is a discrepancy of the level of 2-2.5σ. But this is well within the uncertainty and nothing to worry about. Enormous effort has gone into the understanding of energy scale, and nothing wrong has been found.

References:

ATLAS-CONF-2013-014, CMS-PAS-HIG-13-005

- *Tannenbaum:*

Why does everybody think that the Standard Model is solved? There are many open questions, for instance: why is the ratio of the muon and electron masses ~207, etc.?

- *Jenni:*

This is a comment, not a question!

New Spectroscopy with Charm and Beauty Multiquark States

Luciano MAIANI
Dipartimento di Fisica, Università di Roma "La Sapienza" , Piazzale A Moro 5, Roma, I-00185, Italy;
CERN, PH-TH, 1211 Genève 23, Switzerland.

Talk given at the Erice School of Subnuclear Physic, June 30, 2013, EMFCSC, Erice, Italy.

I. INTRODUCTION

For a long time, we lived happily with the Gell-Mann Zweig paradigm:

$$\begin{aligned} \text{mesons} &= q\bar{q} \\ \text{baryons} &= qqq \end{aligned} \tag{1}$$

Paradigm rested on the absence of $I = 2$, $\pi\pi$ resonances and of $S > 0$ baryons.
The case had to be revisited, because the lowest lying, scalar mesons - the $\sigma(600)$, $f_0(980)$, and $a_0(980)$-do not fit in the picture. It was also found, much later, that *new charmonium* states seen by Belle and BaBar, do not fit the charmonium picture at all.
The discovery of Θ^+, $S = +1$ baryon (a pentaquark ?), seemed to make the case...but the first observations have not been confirmed.
The situation has considerably progressed since 2007, when I gave two lectures on tetraquarks in Erice [1]. The evidence for charmonium and bottomonium states made by four valence quarks is now established, with the discovery of *charged charmonia and bottomonia* and the wounderful experiments performed by BELLE, BaBar, CDF and D0, BES, CLEO and now CMS and LHCb. Some of the new particles have been observed at Tevatron and LHC, data on decay modes and masses of the new particles are accumulating and theoretical models explored.
We are still in a growing stage, with contrasting evidences coming and going. The situation looks quite similar to the one prevaling for meson spectroscopy in the sixties, when the Erice School started and the Standard Model was on the making, see the wonderful photo in Fig. 1. From the report of that school [2], we learn that:
Professor Hughes from the University of Glasgow, UK, presented a review of the current situation regarding the meson resonances.... Then the data ond masses and quantum numbers of the bosons were examined in detail in order to determine **what is certain, what remains to be confirmed and what should probably be forgotten** (emphasis added).
There will be, no doubt, many data and theories about the new resonances that will have to be forgotten. But non conventional hadrons are going to stay with us, giving important clues to the understanding of strong interactions dynamics.

II. LIGHT SCALAR MESONS

A light scalar meson nonet, with masses below 1 GeV, has been long suspected or claimed, but only recently established. The spectrum of these particles is completely inverted with respect to the the spectrum of a standard meson nonet, e.g. the vector meson nonet. This is exemplified in Fig. 2. The isospin one states are degenerate to an isosinglet, which is appropriate if they are made by the light quarks. However, the $a_0(980)$, $I = 1$, and $f_0(980)$, $I = 0$, are heavier than both the strange meson $\kappa(800)$ and the other singlet $f_0(600)$. One could explain the low mass of $f_0(600)$ as being pushed down by a mixing with a higher lying glueball, as assumed e.g. in [3], but this would not explain why the strange meson is lower than the isovector.
As noted by Jaffe long ago [4], antisymmetric diquarks work better. A pair of light quarks in fully antisymmetric state could have sufficent attraction to make the basic unit of a new series of hadrons (Jaffe [4], Jaffe and Wilcezck [5]). For light quarks, we denote by $[qq]$, $(q = u, d)$, the state:

$$[qq]: \quad \text{color} = \bar{3},\ \text{SU(3) flavour} = \bar{3},\ \text{orbital angular momentum } L = 0,\ \text{total spin} = 0 \tag{2}$$

FIG. 1: Making the Standard Model, Erice School in Subnulear Physics, 1967. Going around the table from left: Bruno Zumino, Sidney Coleman, Nino Zichichi, Nicola Cabibbo, Sheldon Glashow and Murray Gell-Mann. CERN Courier, Ref. [2]

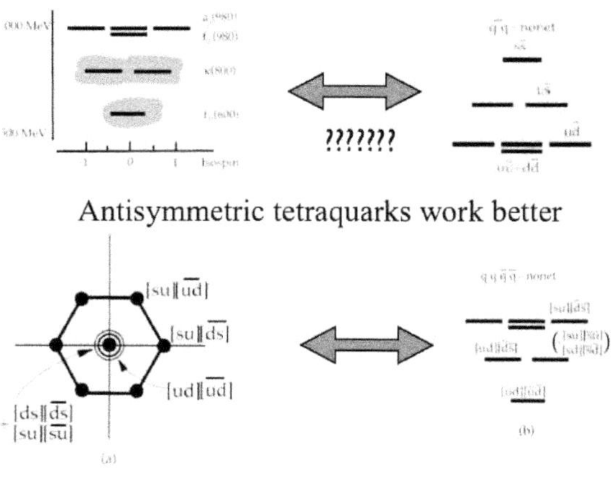

FIG. 2: .

A color singlet, diquark-antidiquark pair would make a flavor nonet with a composition such as to reproduce perfectly the ordering of the light scalars, with the $I = 1, 0$ composed as $[sq][\bar{s}\bar{q}']$ being the heaviest components, as shown in the lowest part of Fig. 2.

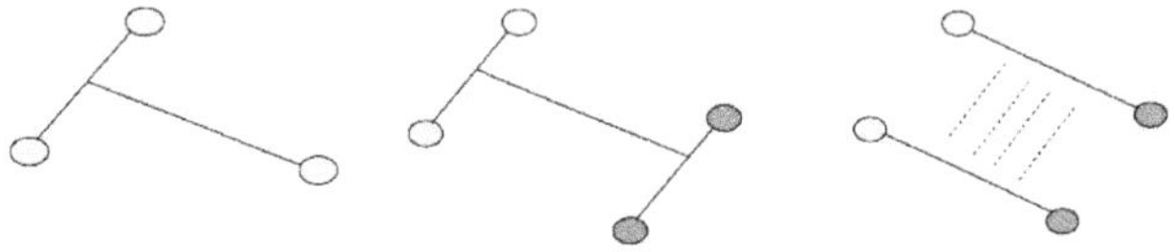

FIG. 3: Diquark binding by color forces: (a) in baryons, (b) in tetraquarks; by contrast, meson-molecules (c) are bound by residual color forces due to normal meson exchange.

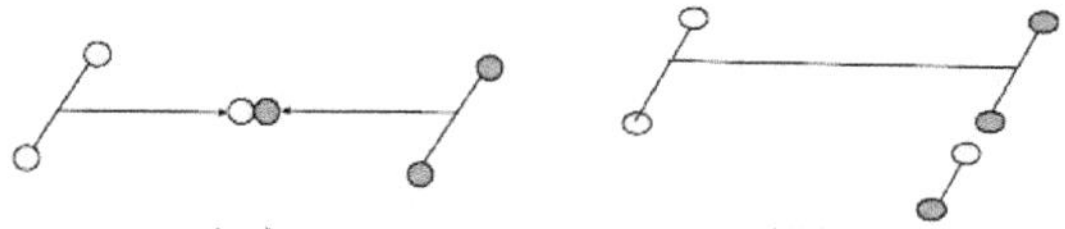

FIG. 4: (a) Tetraquark decay in Baryon-Antibaryon pair, (b) de-excitation in a lower tetraquark.

III. COLOR FORCES, DECAYS AND HEAVY DIQUARKS.

Diquarks needs to combine with other colored objects to form color neutral hadrons. Binding to another quark, with Y-shaped string, a baryon is formed, Fig. 3(a). Diquarks bound to an anti-diquark, with an H-shaped string, form a new kind of mesons, Fig. 3(b). We call these states *tetraquarks* for brief. Tetraquarks in QCD at large N have been considered recently by S. Weinberg [6].

We shall consider also diquarks containing one heavy quark and the corresponding *hidden charm or beauty tetraquarks* with composition: $[cq][\bar{c}\bar{q}']$. The spin-spin interactions of the heavy quarks are suppressed in QCD, as exemplified by the small splitting of charmed and beauty pseudoscalar and vector mesons. Thus it is reasonable to imagine that diquarks with one heavy quark may be formed irrespectively of the value of the spin of the pair:

$$S_{cq} = 0, 1 \tag{3}$$

Hidden charm/beauty tetraquarks should make full flavor multiplets corresponding to the composition of the diquark and antidiquark total spin, with radial and orbital excitations.

A different configuration of four quarks is that of a meson-meson molecules, Fig. 3(c). Such hypothetical states would be bound by the residual forces of QCD, similar to the nuclear forces which act upon color singlets due to light mesons exchange.

The question for molecules is wether these forces are at all able to bind them. The residual forces will not be flavor blind and, most likely, will give rise to few states, in each channel, possibly with incomplete SU(3) and even SU(2) multiplets

The string topology of tetraquarks is related to the baryon-antibaryon channel [7]. As shown in Fig. 4 (a), by breaking the string one gets a $B\bar{B}$ pair:

$$[qq][\bar{q}\bar{q}] \to B\bar{B} \tag{4}$$

In many cases, such decays are forbidden by energy conservation. A tetraquark may decay into a lighter one with the same composition by emission one meson, Fig. 4 (b):

$$([qq][\bar{q}\bar{q}])_A \to ([qq][\bar{q}\bar{q}])_B + \text{meson(s)} \tag{5}$$

or dissociate directly into a J/Ψ or a Υ and one or more light mesons, if diquarks contain one charm or one beauty quark.

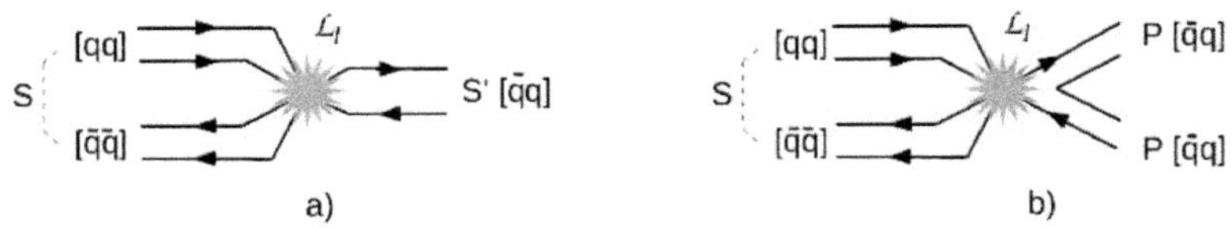

FIG. 5: (a) Mixing of the light tetraquark with a $q\bar{q}$ meson, (b) mechanism for the $f_0 \to 2\pi$ decay.

Instanton effects. Instantons produce an effective lagrangian corresponding to a six-quark vertex [8]. For tetraquarks, they are able to produce important effects , which were identified precisely in Erice in 2007 [9], see also [10]:

(a) mixing of the lightest tetraquark scalars with the lowest $q\bar{q}$ mesons, identified with the multiplet made by $a_0(1474)$ (I=1), $K_0(1412)$, (I=1/2), and three isosinglets: $f_0(1370)$, $f_0(1507)$ and $f_0(1714)$ (one could be a glueball);

(b) a mechanism for the decay $f_0(980) \to \pi^+\pi^-$.

IV. CONVENTIONAL AND LESS CONVENTIONAL QUARKONIA

The accuracy with which the spectra of $Q\bar{Q}$ states (Q=c, b) can be predicted and measured makes it possible to discover new states by difference [12, 13]. Terminology of $Q\bar{Q}$ states in S and P wave is as follows.

A state with:

$$\text{spin} = S, \;\text{orbital ang. mom.} = L, \;\text{total ang. mom.} = J, \;\text{radial excitation} = n$$

is denoted by: $n\ ^{2S+1}L_J$.

Lowest lying states of charmonium are given special notations:

$$\eta_c(1S) = 1\ ^1S_0;\; J/\Psi = 1\ ^3S_1 \tag{6}$$

$$h_c(1P) = 1\ ^1P_1,\; \chi_c(1P) = 1\ ^3P_J \tag{7}$$

The spectrum reported in Fig. 6 gives an idea of the complexity of bottomonium lines, with resonances attributed to levels by comparing the measured masses and decay branching ratios to the theoretical calculations. The situation is reminiscent of the XIX century determination of the elemental composition from the spectral lines of stars. And, similar to this example, sometimes *unanticipated lines* appear.

A recent list of unaccounted for states is given in Refs. [12, 13] and reproduced here[1], Fig. 7.

The established convention for these states is that X are neutral states which have positive parity if their decay (e.g. into $J/\Psi + \rho$) happens to be in S-wave, Y are neutral $J^{PC} = 1^{--}$ resonances seen in e^+e^- annihilation, and $Z^\pm$ are electrically charged states, again with positive parity for S-wave decay into $J/\Psi + \pi^\pm$.

Some of the new lines, e.g. $X(3872)$, are very close to some meson-meson threshold:

$$\begin{aligned} M[X(3872)] &= 3871.68 \pm 0.17 \text{ MeV} \\ M[X(3872)] - M(D^0 + D^{*0}) &= -0.12 \pm 0.30 \text{ MeV} \end{aligned} \tag{8}$$

but some are not, e.g. $Y(4260)$.

---

[1] In a recent paper by T. Friedmann [11] the existence of radial excitations is challenged and the states usually classified as such are interpreted as tetraquark states. The list of "unanticipated states" is therefore considerably increased, to include even the well known ψ', usually classified as $\psi(2\ ^3S_1)$ and now included in the class of Y states. I shall not follow this line, but note that this interpretation would put the observed decay $\psi' \to J/\Psi\ \pi^+\pi^-$ in line with the decays of $X(3872)$ or $Z_c(3900)$ into $J/\Psi+$ meson(s).

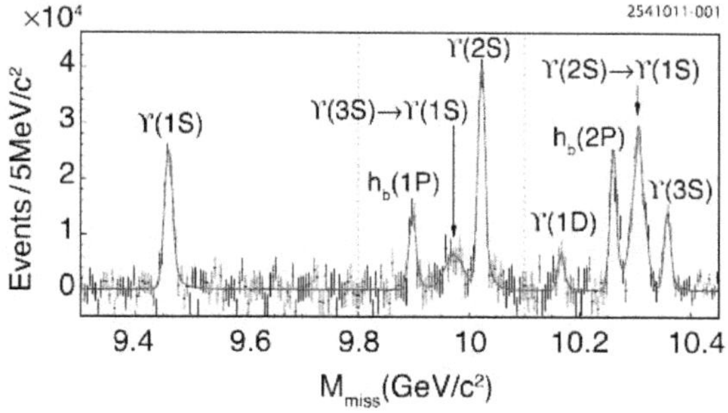

FIG. 6: Mass recoiling against $\pi^+\pi^-$ in e^+e^- collisions. Spectrum from BELLE reported in Ref. [12].

V. WHAT IS X(3872) MADE FROM?

Several hypotheses are still on the market, waiting to be supported or excluded by conclusive experimental data. Here the most frequently considered interpretations.

- The **tetraquark hypothesis** would give rise to an hadron whose components are held together by color forces, to make an hadron of normal size ($\approx$ 1 fm). Variants of the four quark picture can be envisaged with diquarks bound in color 6 or $q\bar{q}$ pairs in color 8.

- The **hybrid option**, with a $c\bar{c}$ pairs in color 8 bound to a *constituent gluon*, see e.g. [14]. This option, however, would not apply to charged states like $Z_c(4430)$ or Z_b and Z_b'.

- **Molecule**: the small value of the difference in (8) is in favor of a loosely bound state of the two mesons, held together by forces similar to the nuclear forces. Perhaps too much loosely bound, since the radius estimated form the binding energy is definitely anomalously large. From the non-relativistic Schroedinger equation one finds:

$$R \approx \frac{\hbar c}{\sqrt{2M_{red}E_{bind}}} = \frac{200 \text{ MeV} \cdot \text{fm}}{20\text{–}30 \text{ MeV}} = 7 \text{ to } 10 \text{ fm} \tag{9}$$

  where M_{red} and E_{bind} denote the reduced mass and binding energy and the numerical value follows from (8).

- **Hadro-charmonium**. A charmonium state surrounded by a light quark cloud has been recently envisaged [15]. One could have also excited charmonium in the core, which could explain why $Z(4430)$ decays into $\psi(2S)$. The wave function of hadro-charmonium should depend from the relative separation, R, of c and $\bar{c}$. For large R, light quarks/antiquarks may be attracted between the heavy particles to neutralize color forces, thereby mimicking a molecular state. Thus hadro-charmonium would have a molecular component, which however should be dominant if we wish to mantain the relation with meson masses, reproducing the puzzle of the anomalously large radius. In addition, it is not clear what would stabilize the light quark/antiquark cloud, which in principle is present also in normal charmonia, and promote some of the cloud's components into valence quarks.

Production at Hadron Colliders. Production of $X(3872)$ is seen at the Tevatron, by CDF, and LHC, by CMS and LHCb, at a lower but not so dissimilar level as charmonia, see Fig. 8.
Using Pythia to estimate the probability to find a $D\bar{D}$ pair in the relevant phase space, factors of 10^{-2} with respect to the X(3872) cross section measured by CDF ($\approx$ 30 nb) are found [16].
The observation at hadron colliders speaks against extended objects and in favor of compact hadrons, like tetraquarks. Several explanations have been advanced by the supporters of the molecular picture but, all in all, this remains to me a very weak point of this picture.

State	m (MeV)	Γ (MeV)	J^{PC}	Process (mode)	Experiment (#σ)	Year	Status
$X(3872)$	3871.68±0.17	< 1.2	$1^{++}/2^{-+}$	$B \to K(\pi^+\pi^- J/\psi)$	Belle [36,37] (12.8), BABAR [38] (8.6)	2003	OK
				$p\bar{p} \to (\pi^+\pi^- J/\psi) + ...$	CDF [39–41] (np), D0 [42] (5.2)		
				$B \to K(\omega J/\psi)$	Belle [43] (4.3), BABAR [23] (4.0)		
				$B \to K(D^{*0}\overline{D}^0)$	Belle [44,45] (6.4), BABAR [46] (4.9)		
				$B \to K(\gamma J/\psi)$	Belle [47] (4.0), BABAR [48,49] (3.6)		
				$B \to K(\gamma\psi(2S))$	BABAR [49] (3.5), Belle [47] (0.4)		
				$pp \to (\pi^+\pi^- J/\psi) + ...$	LHCb [50] (np)		
$X(3915)$	3917.4 ± 2.7	28^{+10}_{-9}	$0/2^{?+}$	$B \to K(\omega J/\psi)$	Belle [51] (8.1), BABAR [52] (19)	2004	OK
				$e^+e^- \to e^+e^-(\omega J/\psi)$	Belle [53] (7.7), BABAR [23] (np)		
$X(3940)$	3942^{+9}_{-8}	37^{+27}_{-17}	$?^{?+}$	$e^+e^- \to J/\psi(D\overline{D}^*)$	Belle [54] (6.0)	2007	NC!
				$e^+e^- \to J/\psi(...)$	Belle [20] (5.0)		
$G(3900)$	3943 ± 21	52±11	1^{--}	$e^+e^- \to \gamma(D\overline{D})$	BABAR [55] (np), Belle [56] (np)	2007	OK
$Y(4008)$	4008^{+121}_{-49}	226±97	1^{--}	$e^+e^- \to \gamma(\pi^+\pi^- J/\psi)$	Belle [57] (7.4)	2007	NC!
$Z_1(4050)^+$	4051^{+24}_{-43}	82^{+51}_{-55}	?	$B \to K(\pi^+\chi_{c1}(1P))$	Belle [58] (5.0), BABAR [59] (1.1)	2008	NC!
$Y(4140)$	4143.4 ± 3.0	15^{+11}_{-7}	$?^{?+}$	$B \to K(\phi J/\psi)$	CDF [60,61] (5.0)	2009	NC!
$X(4160)$	4156^{+29}_{-25}	139^{+113}_{-65}	$?^{?+}$	$e^+e^- \to J/\psi(D\overline{D}^*)$	Belle [54] (5.5)	2007	NC!
$Z_2(4250)^+$	4248^{+185}_{-45}	177^{+321}_{-72}	?	$B \to K(\pi^+\chi_{c1}(1P))$	Belle [58] (5.0), BABAR [59] (2.0)	2008	NC!
$Y(4260)$	4263^{+8}_{-9}	95±14	1^{--}	$e^+e^- \to \gamma(\pi^+\pi^- J/\psi)$	BABAR [62,63] (8.0)	2005	OK
					CLEO [64] (5.4), Belle [57] (15)		
				$e^+e^- \to (\pi^+\pi^- J/\psi)$	CLEO [65] (11)		
				$e^+e^- \to (\pi^0\pi^0 J/\psi)$	CLEO [65] (5.1)		
$Y(4274)$	$4274.4^{+8.4}_{-6.7}$	32^{+22}_{-15}	$?^{?+}$	$B \to K(\phi J/\psi)$	CDF [61] (3.1)	2010	NC!
$X(4350)$	$4350.6^{+4.6}_{-5.1}$	$13.3^{+18.4}_{-10.0}$	$0/2^{++}$	$e^+e^- \to e^+e^-(\phi J/\psi)$	Belle [66] (3.2)	2009	NC!
$Y(4360)$	4361 ± 13	74±18	1^{--}	$e^+e^- \to \gamma(\pi^+\pi^-\psi(2S))$	BABAR [67] (np), Belle [68] (8.0)	2007	OK
$Z(4430)^+$	4443^{+24}_{-18}	107^{+113}_{-71}	?	$B \to K(\pi^+\psi(2S))$	Belle [69,70] (6.4), BABAR [71] (2.4)	2007	NC!
$X(4630)$	4634^{+9}_{-11}	92^{+41}_{-32}	1^{--}	$e^+e^- \to \gamma(\Lambda_c^+\Lambda_c^-)$	Belle [72] (8.2)	2007	NC!
$Y(4660)$	4664±12	48±15	1^{--}	$e^+e^- \to \gamma(\pi^+\pi^-\psi(2S))$	Belle [68] (5.8)	2007	NC!
$Z_b(10610)^+$	10607.2±2.0	18.4±2.4	1^+	$\Upsilon(5S) \to \pi^-(\pi^+[b\bar{b}])$	Belle [73,74] (16)	2011	NC!
$Z_b(10650)^+$	10652.2±1.5	11.5±2.2	1^+	$\Upsilon(5S) \to \pi^-(\pi^+[b\bar{b}])$	Belle [73,74] (16)	2011	NC!
$Y_b(10888)$	10888.4±3.0	$30.7^{+8.9}_{-7.7}$	1^{--}	$e^+e^- \to (\pi^+\pi^-\Upsilon(nS))$	Belle [75,76] (2.0)	2010	NC!

FIG. 7: Unanticipated hidden charm and beauty states from Ref. [12].

Charged partners of the $X(3872)$. Negative results for the search of a narrow, charged counterparts of the $X(3872)$ have been published by BABAR [17] and BELLE [18]. The pure $I = 1$ hypothesis for $X(3872)$ is convincingly excluded, but there is still a non neglible parameter space for a mixed $I = 1, 0$ situation. The possibility that the charged X state may be wide enough to have escaped detection has been considered in Ref. [19]

VI. TETRAQUARK PICTURE OF UNEXPECTED X PARTICLES

We consider the simplest tetraquark configurations, in $S - wave$ and composition:

$$[cq][\bar{c}\bar{q}']_{L=0},\ (q, q' = u, d) \tag{10}$$

The total spin of each diquark can assume the values $S = 1, 0$ and the composition of the two spins gives rise to a supermultiplet of states with different total angular momentum, J, and positive parity, $P = +1$.

For each value of J, light quark flavours imply that there are four states with isospin $I = 1, 0$, analogous to the $f_0(980)$- $a_0(980)$ complex.

We may also analyze the states in terms of the total spin of the quark-antiquark pairs, $S_{q\bar{q}'}$ and $S_{c\bar{c}}$, which also may take the values $0, 1$. An important quantum number is the Charge Conjugation parity of the neutral

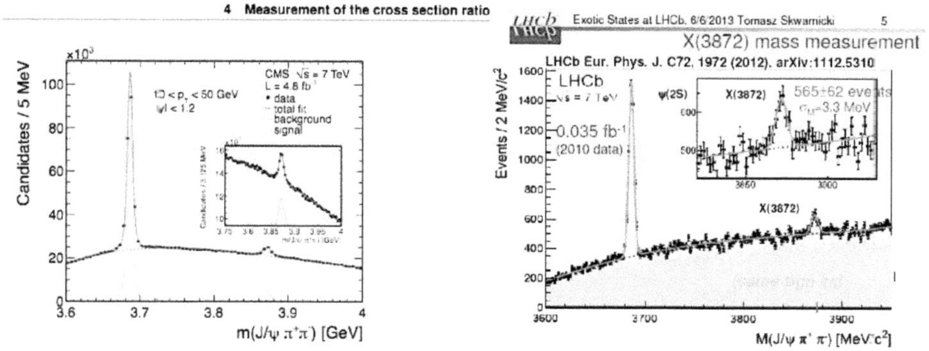

FIG. 8: Observation of X(3872 at the LHC by CMS, left, and LHCb, right. The production of ψ' gives an idea of the cross section involved.

components, C. The value of C, in a state with given values of $S_{q\bar{q}'}$ and $S_{c\bar{c}}$ is:

$$C = (-1)^{S_{q\bar{q}'}+S_{c\bar{c}}} \tag{11}$$

If isospin is exact, we may also consider the G parity of each isospin multiplet:

$$G = C(-1)^I \tag{12}$$

where C refers to the neutral component and I is the isospin of the multiplet. The two neutral mesons of a given J^{PC} can mix (like in ρ-ω mixing) with G and I violated and C conserved.

The predicted tetraquark multiplets are listed below (in bold the tentative attribution of observed states):

- $J^{PC} = 2^{++}$, (2 multiplets)
- $JPC = 1^{++}$, (1 multiplet): **X**(3872);
- $J^{PC} = 1^{+-}$, (2 multiplets): $\mathbf{Z}_c$ (3900), Z_c'; $\mathbf{Z}_b$(10610), $\mathbf{Z}_b'$(10650)
- $J^{PC} = 0^{++}$, (2 multiplets)

In analogy with the non-relativistic costituent quark model, one may try to describe the mass spectrum of the various states, starting from a common diquark mass, $M_{[cq]}$ and mass differences due to spin-spin interactions. The corresponding Hamiltonian is [20]:

$$H = 2M_{[cq]} + 2\sum_{i<j} \kappa_{ij}(\vec{s}_i \cdot \vec{s}_j) \tag{13}$$

The spin-spin interaction coefficients can be in part related to the baryon and meson spectrum (e.g. $\rho - \pi$, $\Delta - N$, etc., mass differences).

The inter-diquark coupling, $(\kappa_{cq})_{\bar{3}}$ is directly related to charmed baryon mass differences. Critical couplings are the extra-diquark couplings, $\kappa_{q\bar{q}}$, $\kappa_{c\bar{c}}$ and $\kappa_{c\bar{q}}$. From the meson spectrum we obtain the corresponding couplings for the quark pairs in color singlets, while in the case at hand these pairs are in a superposition of color singlet and octet. Moreover, the spin-spin interaction is proportional to the overlap probability of the pair. We may relate the latter couplings to the meson ones, assuming [20]:

- one-gluon exchange: this means that the interaction constant κ_{ij} for a pair in a given color state c is proportional to the value of the Casimir operator in the color state:

$$(\kappa_{ij})_c \propto <\sum_A \frac{\lambda_i^A}{2}\frac{\lambda_j^A}{2}>_c \tag{14}$$

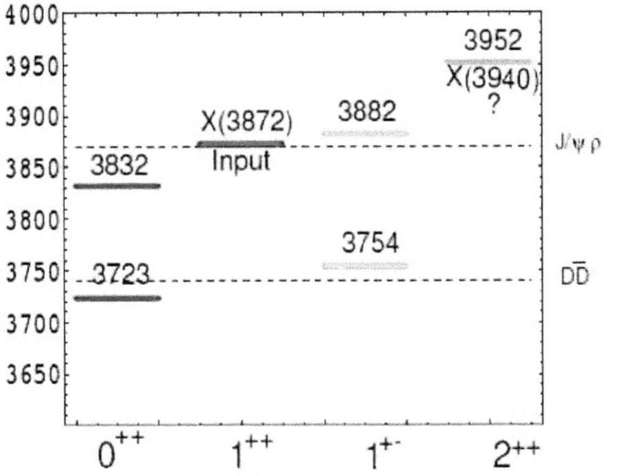

FIG. 9: .

- same $|\psi(0)|^2$ as in the mesons

The result is:

$$\kappa_{q\bar{q}} = \frac{1}{4}(\kappa_{q\bar{q}})_0 \tag{15}$$

and the numerical values [20] are given in the Table.

$(\kappa_{cq})_{\bar{3}}$	$\kappa_{q\bar{q}}$	$\kappa_{c\bar{c}}$	$\kappa_{c\bar{q}}$	$(\kappa_{bq})_{\bar{3}}$	$\kappa_{b\bar{q}}$	$\kappa_{b\bar{b}}$
22	79	15	18	6	6	9

TABLE I: Spin-spin couplings (in MeV) for tetraquarks with hidden charm and beauty derived from the meson and baryon spectrum. See text for the additional assumptions.

The constituent quark masses derived from meson and baryon spectrum, are:

$$M_q = 305, \; M_c = 1670, \; (\text{in MeV}) \tag{16}$$

while the diquark mass, fixed by the $X(3872)$ mass, is:

$$M_{[cq]} = 1933 \text{ MeV} \tag{17}$$

that is, additivity is satisfied within few tens of MeV.

The mass spectrum of hidden charm X and Z particles from spin-spin interactions thus obtained is shown in Fig. 9.

Tetraquarks with hidden beauty have been considered by Ali and collaborators, Ref. [21]. The $[bq]$ diquark mass was derived from (17) by adding the beauty-charm constituent quark mass difference:

$$\begin{aligned} \Delta_{bc} &= M_b - M_c = 3335 \\ M_{[bq]} &= M_{[cq]} + \Delta_{bc} = 5267 \end{aligned} \tag{18}$$

The mass difference of $Z_b(10610)$ and $Z_b'(10650)$ agrees with the B^*-B mass difference, as expected in the molecule model, see [22]:

$$\begin{aligned} M(Z_b) - M(Z_b') &= 45 \text{ MeV} \\ M(B^*) - M(B) &= 45.78 \pm 0.35 \text{ MeV} \end{aligned} \tag{19}$$

Tetraquark model tends to predict larger splitting, depending however on the value of the poorly known $\kappa_{q\bar{q}}$, see [21].

The $J^{PG} = 1^{+,\pm}$ charmonium tetraquarks. We consider isotriplets only. There are three states with different charge conjugation properties of the neutral component and with the indicated distribution of the diquark total spin:

$$C = +1: \frac{|1,0\rangle + |0,1\rangle}{\sqrt{2}}; \text{ (3872 MeV, input)} \tag{20}$$
$$C = -1: \frac{|1,0\rangle - |0,1\rangle}{\sqrt{2}}; \text{ (3882 MeV, computed)}$$
$$C = -1: |1,1\rangle_{J=1}; \text{ (3755 MeV, computed)} \tag{21}$$

The state 1^{++} is a mass eigenstate, while states with $C = -1$ are mixed by a 2×2 mass matrix:

$$\mathcal{M}(1^{++}) = 2M_{[cq]} + \frac{\kappa_{c\bar{c}} + \kappa_{q\bar{q}}}{2} - \kappa_{c\bar{q}} - (\kappa_{cq})_{\bar{3}} \tag{22}$$
$$\mathcal{M}(1^{+-}) = 2M_{[cq]} - \frac{\kappa_{c\bar{c}} + \kappa_{q\bar{q}}}{2} + \begin{pmatrix} \kappa_{c\bar{q}} - (\kappa_{cq})_{\bar{3}} & \kappa_{q\bar{q}} - \kappa_{c\bar{c}} \\ \kappa_{q\bar{q}} - \kappa_{c\bar{c}} & -\kappa_{c\bar{q}} + (\kappa_{cq})_{\bar{3}} \end{pmatrix} \tag{23}$$

Decays, observed and (not yet?) observed are:

$$X(3872) \to J/\Psi + \rho/\omega,\ D\bar{D}^* \tag{24}$$
$$Z_c \to J/\Psi\ \pi, J/\Psi\ \eta\ ?, \eta_c\ \rho? \tag{25}$$

Corresponding molecular states. It is interesting to note the prediction of the molecular picture [23]. In the latter case, one would associate S-wave bound states to the $D\bar{D}^*$ and $D^*\bar{D}^*$ thresholds. One finds also in this case three $J^P = 1^+$ states, to wit

$$C = +1: \frac{|D,\bar{D}^*\rangle + |\bar{D},D^*\rangle}{\sqrt{2}}; \text{ (3876 MeV)}$$
$$C = -1: \frac{|D,\bar{D}^*\rangle - |\bar{D},D^*\rangle}{\sqrt{2}}; \text{ (3875 MeV)}$$
$$C = -1: |D^*,\bar{D}^*\rangle_{J=1}; \text{ (4017 MeV)} \tag{26}$$

where for each state we have given the value of the corresponding threshold for the charged D and $\bar{D}^*$ meson pairs. Loosely bound charged molecules should differ by few MeV's. There are still two states of opposite C approximately degenerate and close to the $X(3872)$ but the third state is at mass higher than the $X(3872)$ and $Z_c(3900)$ masses.

VII. THE UNEXPECTED $J^{PC} = 1^{--}$ RESONANCE Y(4260) AND ITS UNEXPECTED DESCENDANTS

The first Y-like unexpected charmonium was discovered by Babar in 2006, in processes with Initial State Radiation [24]:

$$e^+\ e^- \to \gamma\ (ISR) + Y(4260) \to \gamma + J/\Psi\ \pi^+\pi^- \tag{27}$$

Since the photon is emitted by the initial pair, the resonance at 4260 MeV has $J^{PC} = 1^{--}$ is interpreted as a P-wave tetraquark. The Dalitz plot of the decay shows rather clearly the $f_0(980)$ band, which led us [25] to assume the quark composition:

$$Y(4260) = [cs][\bar{c}\bar{s}]_{\text{P-wave}},\ S_{[cs]} = S_{[\bar{c}\bar{s}]} = 0 \tag{28}$$

However, after discovery of the instanton effect, one realizes that the ansatz:

$$Y(4260) = [cq][\bar{c}\bar{q}],\ q = u,\ d \tag{29}$$

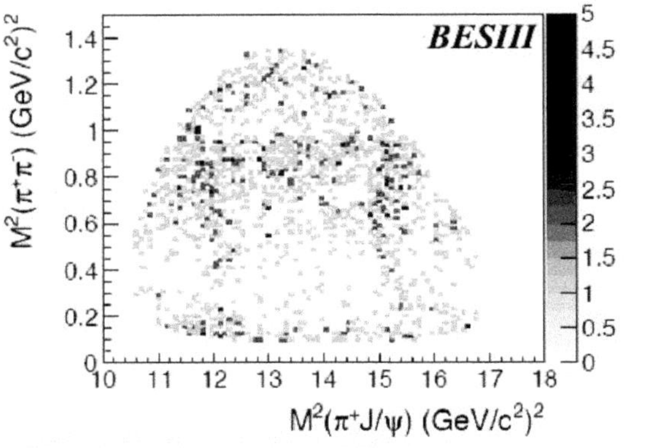

FIG. 10: Dalitz plot of the decay in (31) reported by BES III [26]. The f_0 line (horizontal, at $M^2 \approx 0.9$) and the Z_c line (vertical, at $M^2 \approx 15$), are visible.

would also give rise to the f_0 in the final state while fitting better the mass of this particle as an orbital excitation. I shall follow this intepretation.

The final state in (27) is in line with the de-excitation mode:

$$Y(4260) \to Z_c^{\pm}(3900) + \pi^{\mp} \tag{30}$$

BES III [26], BELLE [27] and CLEO [28] have observed the decays:

$$Y(4260) \to \left\{ \begin{array}{c} J/\Psi \; f_0 \to J/\Psi \; \pi^+\pi^- \\ Z_c^{\pm} \; \pi^{\mp} \to J/\Psi \; \pi^+\pi^- \end{array} \right\} \tag{31}$$

Fig. 10 shows the Dalitz plot reported by BES III, with the visible lines corresponding to the f_0 and to the $Z_c^+(3900)$. At a similar mass, CLEO reports a 3σ signal of a Z_c^0 line in the decay:

$$Y(4260) \to Z_c^0 \; \pi^0 \to J/\Psi \; \pi^0\pi^0 \tag{32}$$

The projected mass spectra are reported in Fig. 11.

As we have seen, the tetraquark model predicts also second resonance in the same channel, around 3800 MeV, whose possible presence is discussed in [29]. The molecular model, on the other hand, would correlate the second peak to the $D^*\bar{D}^*$ threshold, some 100 MeV above. So, the decay channel (31) may provide a significant opportunity to discriminate the two pictures. More precise experiments are clearly called for.

The situation can be compared with the hidden beauty situation, where two peaks, $Z_b(10610)$ and $Z_b'(10650)$, are clearly seen in the decay of $\Upsilon(5S)$: why this difference?

Heavy quark spin conservation. Spin changing interactions in QCD are inversely proportional to quark masses. Thus it is reasonable to assume that the total spin of the heavy quarks is conserved in strong decays. The value of the total $c\bar{c}$ spin in the initial state would favour decay modes into J/Ψ ($S_{c\bar{c}} = 1$) or η_c ($S_{c\bar{c}} = 0$) accordingly.

It is easy to see that the state in (20) has $S_{c\bar{c}} = 1$. In fact, to obtain $C = +1$ one has to have $S_{q\bar{q}} = S_{c\bar{c}}$, see eq. (11), and the pair $(S_{q\bar{q}}, S_{c\bar{c}}) = (0,0)$ cannot reproduce the total angular momentum $J = 1$. This goes well with the observed decays of $X(3872)$ into J/Ψ in eq. (24).

For the states (21) we may argue as follows.

The matrices representing the operators $(\vec{s}_q \cdot \vec{s}_{\bar{q}})$ and $(\vec{s}_c \cdot \vec{s}_{\bar{c}})$ in the basis (21) are easily found from eq. (23):

$$\begin{aligned} 2(\vec{s}_q \cdot \vec{s}_{\bar{q}}) &= \frac{\delta\mathcal{M}(1^{+-})}{\delta\kappa_{q\bar{q}}} = -\frac{1}{2} + \sigma_1 \\ 2(\vec{s}_c \cdot \vec{s}_{\bar{c}}) &= \frac{\delta\mathcal{M}(1^{+-})}{\delta\kappa_{c\bar{c}}} = -\frac{1}{2} - \sigma_1 \end{aligned} \tag{33}$$

These two matrices are simultaneously diagonalized by two eigenvectors:

$$v_{\pm} = \frac{1}{\sqrt{2}} \begin{pmatrix} 1 \\ \pm 1 \end{pmatrix} \tag{34}$$

The plus sign corresponds to the eigenvalue: $2(\vec{s}_q \cdot \vec{s}_{\bar{q}}) = +1/2$, i.e. $S_{q\bar{q}} = 1$, and the minus sign to: $2(\vec{s}_q \cdot \vec{s}_{\bar{q}}) = -3/2$, i.e. $S_{q\bar{q}} = 0$. In correspondence, we have for $S_{c\bar{c}}$ the complementary values: $S_{c\bar{c}} = 0$ and $S_{c\bar{c}} = 1$, as required by the negative Charge Conjugation value of the states (21), see again eq. (11).

Denoting by v_{high} the higher mass eigenvector of $\mathcal{M}$, eq. (23), the probability of its decay into J/Ψ is proportional to:

$$P(J/\Psi) = | < v_- | v_{high} > |^2 \tag{35}$$

The parameters of Tab. I lead to a higher eigenvector very close to v_+, therefore suppress the decay into J/Ψ vs. the one into η_c. However this depends critically from the assuptions used to obtain the spin coupling between the quark and the antiquarks inside different diquarks from the meson couplings. If one allows for a reduction of the overlap function $|\psi(0)|^2$, one can move v_{high} towards v_-.

To illustrate the trend, we fix $(\kappa_{cq})_{\bar{3}}$ from Tab. I and replace $\kappa_{q\bar{q}} \to r \cdot \kappa_{q\bar{q}}$, allowing for a reduction factor $r = 1.0, 0.3, 0.0$ for all quark-antiquark couplings, always fixing the diquark mass from $X(3872)$. We find the results shown in Tab. II.

r	M_{high}	M_{low}	$M_{[cq]}$	$P(J/\Psi)$
1.0	3882	3754	1933	0.0
0.3	3897	3846	1943	0.12
0.0	3916	3872	1947	0.50

TABLE II:

A measurement of the decay rate of $Z_c^{\pm} \to \eta_c \rho^{\pm}$ would be extremely informative.

Decay modes and widths of Zc(3900). The $Z_c(3900)$ resonance is much broader than $X(3872)$. In Ref. [29], on the basis of phase space and order of magnitude dimensional arguments, we estimate a total rate of 60 MeV for the tetraquark model. For molecular models, see [30, 31].

VIII. CRITICAL OPEN ISSUES

There are a few critical issues which could distinguish between systems bound by QCD forces between colored objects (tetraquark) or by residual forces between colo singlet objects (molecular models in different variations).

The first issue is that of complete isospin multiplets. In the first option, one should see neutral partners of Z and charged partners of X and Y particles. The states observed until now correspond to channels involving J/Ψ + charged pions. The others require the observation of final states with one or more π^0s, which may be more diffcult to pin down.

Another issue is isospin conservation, which would give intersting clues on the dynamics of bound states and their decay. Is isospin maximally violated with mixing between $I = 0, 1$ states ?

A third issue is the conservation of total heavy quark spin, e.g $S_{c\bar{c}}$, which is very plausible in QCD. The control of the decays into η_c, again with many neutral particles in the final state, is required and it would allow to probe into the structure of the four quark bound states, wether made by colored or by color neutral subsystems.

Finally, the existence of exotic states with different angular momentum and parity would be very illustrative of the angular momentum of the total spin of the subsystems: are there $J = 2$ or $J = 0$ exotic hidden charm or beauty states?

IX. CONCLUSIONS

Exotic mesons, with hidden charm and beauty ARE there, but their nature is still unclear. Correlation of their mass to meson meson thresholds would speak in favor of these mesons to be meson-meson molecules.

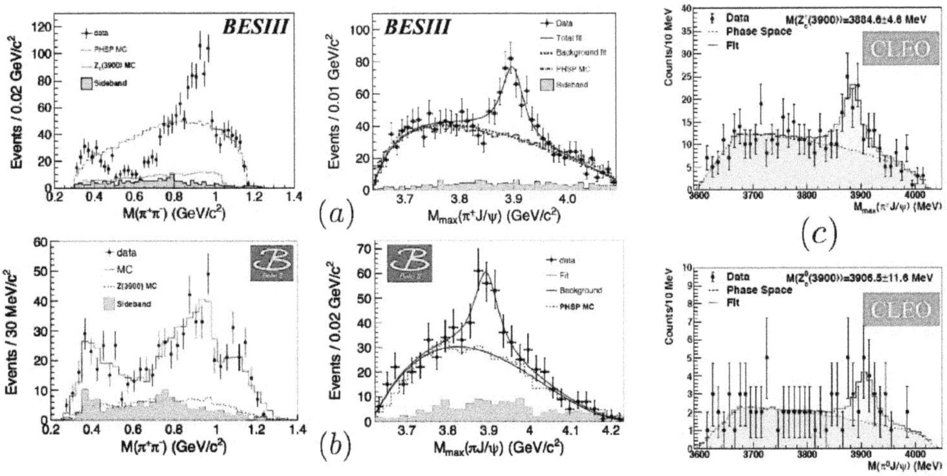

FIG. 11: Mass projections of the Dalitz plot of the decay in (31) reported by: (a) BES III [26], (b) BELLE [27], (c) CLEO [28]. The distribution for the neutral Z_c reported by CLEO is also shown in (c).

However, their sizeable production at hadron colliders disfavors extended structures vs. compact structures, with subsystems bound by color forces, of which tetraquarks are working examples.

The search for one additional peak in $Y(4260)$ decays, below or above the $Z_c(3900)$ mass may be crucial to tell molecules from tetraquarks.

The search for the missing partners with respect to electric charge and spin is called for, in all channels. This involves, in general, the control of decay modes with π^0, η, η_c, which may be quite a formidable challenge.

All this may take quite some time to be accomplished... and a lot of work at flavor factories and LHC, to keep us busy for a while.

NOTE ADDED

In fall 2013, important new informations have been added by BESIII to the issue of the charged Z_c states, which we analyze briefly here.

- A $(D\bar{D}^*)^\pm$ mass peak in $e^+e^- \to \pi D\bar{D}^*$ is observed [32], with $M = 3884$ MeV and $\Gamma = 25$ MeV, consistent within errors with being another decay mode of $Z_c(3900)$ with an estimated rate in $(D\bar{D}^*)^\pm$ about 6 times the $\pi\ J/\Psi$ rate.
- A charged $D^*\bar{D}^*$ state with $G = +1$ and presumably $J^P = 1^+$ has been observed with $M = 4026$ MeV [33], some 10 MeV above the $D^*\bar{D}^*$ threshold.
- The mass of $Z_c(4026)$ is at variance with the tetraquark model prediction discussed here in Eq. (21). It requires a change in the spin-spin interaction, suggesting a very small reduction factor, r, for the quark-antiquark, spin-spin couplings and a value of the quark-quark coupling, $(\kappa_{cq})_3$, larger than what estimated from the baryon spectrum. As seen in Tab. II, a small r was also indicated by the seemingly unsuppressed rate $Z_c(3900) \to J/\Psi\pi^\pm$.
- For $r = 0$, one finds the two first states in (21) to be degenerate and the third one at a larger mass, with $\Delta M = +2(\kappa_{cq})_3$. The spectrum in Fig. 9, in the $r = 0$ case, would comprise one $J^P = 2^+$ state and one 0^+ state at the same mass as the $D^*\bar{D}^*$ state, and one additional 0^+ state with $\Delta M = -2(\kappa_{cq})_3$ with respect to $Z_c(3900)$. Similar considerations would apply to the Z_b states.

- An additional peak in the channel $h_c(1P)\pi^\pm$ is observed at $M = 4026$ [34]. This particle would have $G = +1$, similarly to the previous ones, but to be the same particle as the $D^*\bar{D}^*$ state it should decay in P-wave. It is not clear at this moment if this is supported by data. S-wave decay would indicate $J^P = 1^-$, i.e. a charged Y state.

Acknowledgements

I am indebted to A. Polosa for numerous discussions on the subject of this lecture. Interesting discussions with F. Close, A. Pilloni, V. Riquer and X. Shen are gratefully acknowledged. I thank Prof. A. Zichichi for supporting my partecipation in the Erice School.

---

[1] L. Maiani in *Searching for the 'Totally Unespected' in the LHC Era*, A. Zichichi and G. 't-Hooft eds., 2007, Proc. of the International School of Subnuclear Physics, Erice, 29 August - 7 September 2007.
[2] CERN Courier, **8** (1997) 148.
[3] H. Frizsch, lecture at this School.
[4] R. L. Jaffe, Phys. Rev. **D15**, 281 (1977); R. L. Jaffe, Phys. Rept. **409**, 1 (2005).
[5] R. Jaffe and F. Wilcezck. Phys. Rev. Lett. **91**, 232003 (2003) [hep-ph/0307341].
[6] S. Weinberg, Phys. Rev. Lett. **110** (2013) 261601 [arXiv:1303.0342 [hep-ph]].
[7] L. Montanet, G. C. Rossi and G. Veneziano, Phys. Rept. **63** (1980) 149.
[8] see e.g. G. 't-Hooft lectures at this School.
[9] G. 't Hooft, G. Isidori, L. Maiani, A. D. Polosa and V. Riquer, Phys. Lett. B **662** (2008) 424 [arXiv 0801.2288 [hep-ph]].
[10] A. H. Fariborz, R. Jora and J. Schechter, Phys. Rev. D **77** (2008) 094004 [arXiv:0801.2552 [hep-ph]].
[11] T. Friedmann, Eur. Phys. J. C **73** (2013) 2298 [arXiv:0910.2229 [hep-ph]].
[12] Particle Data Group: S. Eidelman et al., Developments in Heavy Quarkonium Spectroscopy, 2012
[13] Quarkonium Working Group: N. Brambilla et al., Heavy quarkonium: progress, puzzles, and opportunities, 2011.
[14] F. E. Close, Int. J. Mod. Phys. A **20** (2005) 5156 [hep-ph/0411396].
[15] M. Voloshin, ArXiv:1304.0380.
[16] C. Bignamini, B. Grinstein, F. Piccinini, A. Polosa, C. Sabelli, Phys Rev Lett, **103**, 162001 (2009).
[17] B. Aubert et al.[BaBar Collaboration] Phys. Rev. **D 71** (2005) 031501.
[18] S.-K. Choi *et al.* [Belle Collaboration], Phys. Rev. **D 84** (2011) 052004.
[19] K. Terasaki, Prog.Theor.Phys. **127** (2012) 577; arXiv:1107.5868 [hep-ph].
[20] L. Maiani, F. Piccinini, A. D. Polosa and V. Riquer, Phys. Rev. D **71** (2005) 014028.
[21] Z_b , Z_b' as Tetraquark: Ali, Hambrock, Wang,Phys. Rev. D **85** (2012) 054011.
[22] Z_b , Z_b' as Molecule: A. Bondar et al., Phys. Rev. D **84** (2011) 054010.
[23] F. E. Close and P. R. Page, Phys. Lett. B **628**, 215 (2005); E. Braaten and M. Kusunoki, Phys. Rev. D **69**, 074005 (2004); F. E. Close and P. R. Page, Phys. Lett. B **578**, 119 (2004); N.A. Tornqvist, Phys. Lett. B **590**, 209 (2004); E. S. Swanson, Phys. Rep. **429**, 243 (2006); S. Fleming, M. Kusunoki, T. Mehen, and U. van Kolck, Phys. Rev. D **76**, 034006 (2007); E. Braaten and M. Lu, Phys. Rev. D **76**, 094028 (2007); D **77**, 014029 (2008).
[24] B. Aubert et al.[BaBar Collaboration] Phys. Rev. Lett. **95** (2005) 142001.
[25] L. Maiani, V. Riquer, F. Piccinini and A. D. Polosa, Phys. Rev. D **72** (2005) 031502 [hep-ph/0507062].
[26] M. Ablikim *et al.* [BESIII Collaboration], Phys. Rev. Lett. **110** (2013) 252001.
[27] Z. Q. Liu *et al.* [Belle Collaboration], Phys. Rev. Lett. **110** (2013) 252002 [arXiv:1304.0121 [hep-ex]].
[28] T. Xiao, S. Dobbs, A. Tomaradze and K. K. Seth, arXiv:1304.3036 [hep-ex].
[29] L. Maiani, V. Riquer, R. Faccini, F. Piccinini, A. Pilloni and A. D. Polosa, Phys. Rev. D **87** (2013) 111102, [arXiv:1303.6857 [hep-ph]].
[30] Q. Wang,C. Hanhart,Q. Zhao, ArXiv:1303.6355[hep-ph].
[31] T. Gutsche,V. E. Lyubovitskij, ArXiv:1306.0824[hep-ph]; Y. DongA. Faessler,T. Gutsche,V. E. Lyubovitskij,J.Phys. G **40** (2013) 015002.
[32] M. Ablikim et al. [BESIII Collaboration], arXiv:1310.1163 [hep-ex].
[33] M. Ablikim et al. [BESIII Collaboration], arXiv:1308.2760 [hep-ex].
[34] M. Ablikim et al. [BESIII Collaboration], arXiv:1309.1896 [hep-ex].

CHAIRMAN: L. MAIANI

Scientific Secretaries: P. Biancofiore, A. Esposito

DISCUSSION

- *Wang:*

What's the significance of the second resonance compared to the Belle result with just one resonance? From the Belle result the significance of the Z is 9σ. What's the significance of the second resonance?

- *Maiani:*

This is not claim to be a demonstration that there are two peaks. We are just trying to see whether a second peak could be fitted into the data. So, for instance, you can certainly make a fit without the second resonance, this is certain. Within the error. If you have a second resonance it improves the χ^2/DOF, and that's the only thing you can say. It seems more difficult to fit a second resonance on this side [the right one] where D*Dbar* threshold is than on this side [the left one]. But the difference is not this much so…It's clear that one resonance [the main one] is there. It is clear above any possible background and, in fact, the two collaborations [Belle and BESIII] agree. They give very compatible fits, as you can see from this numbers.The issue of a second resonance is completely open but the fact is that if you go to the analogous decay of the beauty you really see two peaks. So, I doubt that there are two peaks also in the charm case.

- *Wang:*

Have you tried to fit the spectrum of π^+J/ψ and π^-J/ψ separately? Have you tried that?

- *Maiani:*

Well, this is the Dalitz plot produced by BES and that's all you have. It's possible to identify here a peak, in $J/\psi\pi^+$, and in fact it is there. I do not have the projection on $J/\psi\pi^-$ axis.

- *Wang:*

We know that the glueballs and hybrids and tetraquarks have been searched for many years: there are many difficulties from the experimental side. Suppose we can measure the mass, the branching ratios and other parameters of these candidates precisely. Are there other difficulties on the theoretical side?

- *Maiani:*

As I said glueballs are out of this game because these resonances decay with a J/ψ and it is clear that the charm-anticharm pair was already there from the very beginning. So glueballs are out of the question. I think that finding a glueball is going to be very difficult because of course they can mix with singlets composed by light quarks and, as I said, the only evidence we have for glueballs it that in the region between 1 and 1.5 GeV where there are ten scalar mesons. So there is one singlet more than expected from q-qbar states. I think this is the best we can say.

- *Carlucci:*

If I have understood it correctly, as a starting point of your fitting method you take the X(3872) and interpret it as two diquarks. You fix from it the mass of the diquark and then you consider the spin-spin interaction. Is it only present the spin interaction or also some kind of radial interaction between the two diquarks?

- *Maiani:*

The model used here gives a very good reproduction of mesons and baryons, with very consistent values for the parameters that define the spin-spin interaction between one type of quark and another type of quark. For instance you can describe with that the baryon octet, the baryon decuplet, the vector mesons etc. All this is well known so it is quite reasonable to have a model like that for states that have four quarks. So, that's only that.

- *Carlucci:*

So you have only spin-spin interactions?

- *Maiani:*

Yes. The only difficulty is that the constants are known in situations where the pairs are in a 3 [color representation], like in a baryon, or in a singlet, like mesons. So you have to extrapolate to the octet. This is why I wrote explicitly the color structure of the interaction. If you assume that the color structure is this one then of course there is a fixed relation between the matrix element in an octet and in a singlet, which is given by the Casimir of $SU(3)_{color}$. And so you get a prediction on what it is [the constant in the octet configuration]. But, of course, this may be a very crude approximation.

- *Carlucci:*

But it seems to work, if I understand.

- *Maiani:*

Well, it seems to work up to a point because, for instance, in this case [the two resonances in the case of the bottom quark] you find a splitting which is essentially equal to M(B*)-M(B). Now, of course, inside B and B* there are also spin-spin interactions but, if you assume that this is the molecule, what is not there is the spin-spin interaction of, say, one quark here [one of the mesons composing the molecule] with an antiquark there [the other meson], while in the tetraquark they are all there and, presumably, this makes the splitting predicted in the tetraquark picture larger than the observed mass difference. It is about twice that. But of course this depends on the assumption that you can extrapolate the spin-spin interaction from one place to another.

- *Carlucci:*

And, as far as the mass part of your effective formula is concerned, why have you fixed the mass of the diquark from the X and not from the baryons?

- *Maiani:*

You could say that this $M_{diquark}$ is just the mass of a charm quark and a light quark. This is not so different from using it as a parameter and fix it later. In fact these numbers here are not so different from those you would get from the two quarks.

- *Carlucci:*

Yes, I was talking just about a baryon. At the beginning, in one picture, you showed a baryon as a diquark plus a quark. So I take away the mass of the quark, some interactions and I get the mass of the diquark.

- *Maiani:*

It is possible. I didn't try but if you consider that the constituent mass of the charm is 1.8 GeV and the constituent light quark mass is about 150 MeV…so, they are very close.

- *P. Minkowski:*

I would expect here other perfect isospin multiplets and I don't see it very well. You set the charged [tetraquark] to isospin 1, so there must be a neutral one and the neutral one in the singlet. I don't see evidence for these other states nearby.

- *Maiani:*

That's the point. Well, here you could find the neutral one. But as you can see here [the plot relative to the data from CLEO] the neutral is much more confused than it was in the evidence for the charged. The reason is clear: if you want to look at $J/\psi\pi^+$ is one thing, if you want to look at $J/\psi\pi^0$ it is much more difficult: there is more confusion and background. There is another point about that which also may have some relevance and it is the fact that the X(3872) decays into two modes, and they are both seen: two pions and three pions. If you make the mass distribution the two pions it looks like a ρ and the three pions one looks like an ω. That, of course, would be a signal of a violation of isospin as if there were two partners [of the X] mixed. This is possible, like ω^0-ρ^0 mixing. However, the second neutral has not been found yet, and this is another problem. It could be an incomplete multiplet with an isospin violation because the second neutral state does not exist, or it could be that it has simply not been found yet. The problem of complete or incomplete multiplets is vital to distinguish between the bound state of quarks [the tetraquark] or the bound state of hadrons [the molecule]. It is the most crucial point. So, from this point of view, the search for the neutral partner of this would be very important.

- *Close:*

Can I just add a comment? I think it is convincing what you demonstrated: there are states beyond the ordinary q-qbar that are made of boson-boson. The question is whether those bosons are color-triplet diquarks of whether they are conventional singlet hadrons making molecules. The latter picture is clearly producing less states and one of the problems I have always felt for the diquark are the many states which are not seen. What I think it would be an interesting program that someone may already have done for diquark is: if you have a state that you identify as a particular diquark state, whose production rate is known experimentally, to predict what the production rate of the other state would be in order to have some benchmark to know whether they are not showing in a significant way or not.

Present Status of the EMC effect

Klaus Rith

Friedrich-Alexander-Universität Erlangen-Nürnberg
Physics Department
Erwin-Rommel-Str. 1
D-91058 Erlangen, Germany
E-mail: klaus.rith@physik.uni-erlangen.de

Abstract

The present status of the EMC effect, the modification of the per nucleon cross section in deep-inelastic lepton nucleus scattering by the nuclear environment, is reviewed.

1 Introduction

The 30th anniversary of the publication by the European Muon Collaboration (EMC) [1] on the modification of the per-nucleon cross section of nucleons bound in nuclei, named *EMC effect*, is a good opportunity to review the status of the effect and to summarize the available experimental and theoretical information. The EMC effect can be interpreted as a modification of quark and gluon distributions in bound nucleons by the nuclear environment. The dependences of these nuclear modifications on kinematics and various nuclear properties like mass, density or radius are meanwhile rather well known, nevertheless the origin of the effect is still not fully understood. Recent data, discussed in section 6, provide new important input.

In my opinion further measurements of the effect in electron-nucleus, neutrino-nucleus and proton-nucleus scattering are needed to determine nuclear quark and gluon distributions (nPDFs) down to very low parton momentum fractions x and to constrain the initial state for the AA program at RHIC and LHC, or for the correct interpretation of, e.g., (future) νA experiments.

2 Deep-inelastic scattering, structure functions and quark distributions

Nuclear modifications of parton distributions have been studied mainly by measurements of cross sections in deep-inelastic lepton-nucleus scattering. For electromagnetic interactions of charged leptons with nuclear targets and in the approximation of one-photon exchange the cross section reads

$$\frac{d^2\sigma}{dQ^2dx} = \frac{4\pi\alpha^2}{Q^4}\frac{\mathrm{F}_2(x,Q^2)}{x}\left[1-y-\frac{Q^2}{4\mathrm{E}^2}+\frac{y^2+Q^2/\mathrm{E}^2}{2\left[1+\mathrm{R}(x,Q^2)\right]}\right]. \tag{1}$$

Here $-Q^2$ represents the squared four-momentum of the virtual photon that mediates the interaction with coupling strength α and $x = Q^2/2\mathrm{M}\nu$ can be interpreted as the fraction of the longitudinal nucleon momentum carried by the struck quark, in a frame where the nucleon moves with infinite momentum in the direction opposite to that of the virtual photon. The variable y denotes, in the target rest frame , the virtual-photon energy ν with respect to the lepton-beam energy E.

At leading order in QCD the structure function F_2 is defined as the sum of the momentum distributions $\mathrm{q}(x, Q^2)$ and $\bar{\mathrm{q}}(x, Q^2)$ of quarks and anti-quarks of flavor $\mathrm{q} = \mathrm{u}, \mathrm{d}, \mathrm{s}, ...$ weighted by x and z_q^2, where z_q is the quark charge (in units of the elementary charge $|e|$):

$$\mathrm{F}_2(x, Q^2) = \sum_{\mathrm{q=u,d,s..}} x z_\mathrm{q}^2 \left[\mathrm{q}(x, Q^2) + \bar{\mathrm{q}}(x, Q^2)\right] \ . \tag{2}$$

The quantity

$$\mathrm{R} = \frac{\sigma_\mathrm{L}}{\sigma_\mathrm{T}} = \frac{\mathrm{F}_2}{2x\mathrm{F}_1}\left[1 + \frac{4\mathrm{M}^2 x^2}{Q^2}\right] - 1 = \frac{\mathrm{F}_\mathrm{L}}{2x\mathrm{F}_1} \tag{3}$$

is the ratio of the longitudinal to transverse virtual-photon cross sections. In the quark-parton model, $\mathrm{R} = 0$ for the interaction of the virtual photon with a point-like spin-1/2 particle. Quark transverse momenta, quark masses and gluon radiation cause R to deviate from zero. If R is independent of the nuclear mass number A (see the discussion in section 4.4), then the ratio of cross sections for two different nuclei is equal to the ratio of their structure functions F_2.

Subsequently, we will always discuss the ratio of structure functions (cross sections) per nucleon for a nucleus with mass number A (i.e., A nucleons) and the deuteron D. The latter is, to a good approximation, equal to the proton-neutron averaged structure function $\mathrm{F}_2^\mathrm{D} \approx (\mathrm{F}_2^\mathrm{p}+\mathrm{F}_2^\mathrm{n})/2$. The x dependence of the structure functions F_2^p and F_2^n is different (for free nucleons they are approximately related by $\mathrm{F}_2^\mathrm{n}/\mathrm{F}_2^\mathrm{p} \approx 1 - 0.8x$). Results for the nuclear structure function F_2^A (cross section σ^A) for nuclei with Z protons and N neutrons will always be corrected for neutron excess by

$$\mathrm{F}_2^\mathrm{A} = (\frac{\mathrm{F}_2^\mathrm{p} + \mathrm{F}_2^\mathrm{n}}{2})^\mathrm{A} \cdot \left[1 - \frac{\mathrm{N} - \mathrm{Z}}{\mathrm{N} + \mathrm{Z}} \cdot \frac{1 - \mathrm{F}_2^\mathrm{n}/\mathrm{F}_2^\mathrm{p}}{1 + \mathrm{F}_2^\mathrm{n}/\mathrm{F}_2^\mathrm{p}}\right] , \tag{4}$$

where it is assumed that proton and neutron structure functions are modified equally by the nuclear environment. Thus, F_2^A is the structure function per nucleon for a hypothetical isoscalar nucleus with an equal number ($\frac{\mathrm{A}}{2}$) of protons and neutrons.

3 The discovery

The historical result of the EMC effect [1] (updated results were published in [11]) is presented in the left panel of Fig. 1. It shows the ratio of the structure function F_2 per nucleon for iron and deuterium, both uncorrected for Fermi motion, as a function of x. The shaded area indicates the range for the errors on the slope of a linear fit to the data, the point-to-point systematic uncertainties are somewhat larger. In addition there is an overall uncertainty of $\pm 7\%$.

The ratio is seen to be different from unity. It falls from ~ 1.15 at $x = 0.05$ to a value of ~ 0.89 at $x = 0.65$ and doesn't follow the expectations from Fermi-motion calculations. This

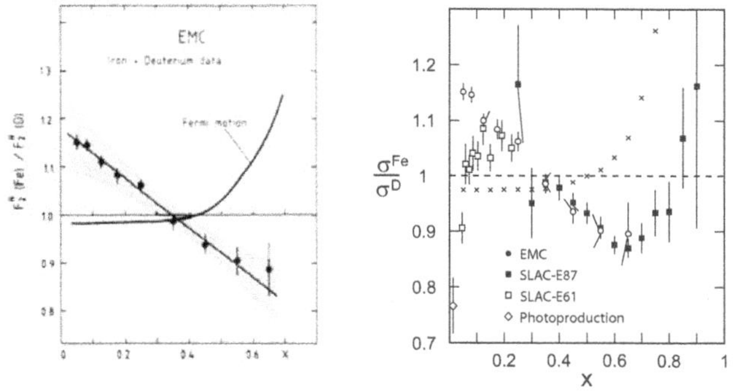

Figure 1: The ratio F_2^{Fe}/F_2^{D} (σ^{Fe}/σ^{D}) as a function of x. Left panel: the original EMC result [1]; right panel: The SLAC result [3] together with low-Q^2 SLAC data for Cu/D [6] and the photoproduction result at $Q^2 = 0$ GeV^2 [8]. Also shown is one expectation for the effect of Fermi motion on F_2^{Fe} in absence of other nuclear effects [5].

result demonstrated for the first time that the structure function F_2 is modified when nucleons are embedded in a nucleus. At the time when these data were presented the effect appeared quite astounding. Especially members of the high-energy physics community could hardly believe that at momentum transfers several orders of magnitude larger than typical nuclear binding energies quark distributions should be affected by the nuclear environment. However, 'quarks in nuclei' and phenomena that could possibly modify the quark distributions in bound nucleons were already discussed by the nuclear-physics community for some time before this discovery. (For a first review of such ideas see Ref. [2]). The EMC result was quickly confirmed by the SLAC-MIT-Rochester group that recovered and reanalyzed the data stemming from the aluminum and steel cell walls of the liquid hydrogen and deuterium targets from the experiments E49B and E87. The result for σ^{Fe}/σ^{D} [3] is shown in the right panel of Fig. 1 together with earlier data from SLAC-E61 at lower Q^2 for Cu/D in the range $0.04 < x < 0.25$ [6] and a data point at $Q^2 = 0$ GeV^2 from an experiment that had investigated shadowing in photoproduction [8]. (Similar data were published for aluminum [4]). At high x, these data were in good agreement with the EMC data, but the rather large value of F_2^{Fe}/F_2^{D} for the two low-x EMC points was in disagreement with the low-Q^2 SLAC data which indicated that for x below ~ 0.15 the ratio decreases again with decreasing x. (Indeed, it was found out later that the low-x EMC points suffered from correlated tracking inefficiencies affecting the deuterium but not the iron data).

This exciting result caused enormous activities in both experiment and theory, resulting at present in about 1000 citations of the EMC publication [1]. The experimental data (with the exception of the recent JLAB data presented in Sect. 6) and many of the theory papers have been discussed in great detail some time ago in the excellent review by P. Norton [9]. Therefore, I will only summarize some of the key results and the main theoretical ideas for the interpretation of the effect.

4 The experimental data

After the discovery, nuclear effects have been studied experimentally in charged lepton-nucleus scattering by the muon experiments BCDMS [10], EMC-NA38 [12], EMC [13] and NMC [14, 15, 16, 17, 19, 20, 21] at CERN and E665 [22, 23] at FNAL, in electron scattering at SLAC [24, 25, 26, 27, 28], DESY [29] and JLAB [30, 31, 32, 33], in neutrino-nucleus scattering [34, 35] and in the Drell-Yan process [38, 39]. I will only discuss the most important results.

4.1 Data from SLAC-E139

At large x the most precise data are those from the SLAC experiment E139 [24] that measured the cross section ratio σ^A/σ^D for 8 nuclei ranging from ^{4}He to ^{197}Au. The results of an updated analysis with an improved treatment of radiative corrections [25] are shown in Fig. 2.

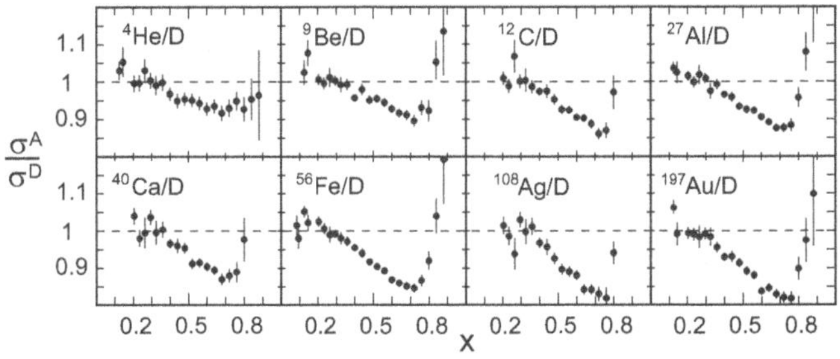

Figure 2: The ratio σ^A/σ^D as a function of x for various nuclei measured by SLAC-E139 [24, 25].

For all nuclei one observes, in the region $0.3 < x < 0.8$, a reduction of the per-nucleon cross section σ^A compared to the 'free nucleon' one, σ^D. The x dependence of this reduction has a very characteristic universal shape with a minimum near $x \approx 0.7$. The effect is already present for helium, its magnitude increases with the atomic mass number A. The A dependence will be discussed in more detail in section 4.8.

4.2 The universal x dependence

The E139 data provide precise information for $x > 0.2$. The region of lower x is covered by the data from SLAC-E61[6], from the HERMES experiment [29], where the 27.6 GeV electron beam of HERA was scattered from internal gas targets of various nuclear species, and the muon experiments. As an example, the x dependence of the cross section ratio $\sigma^{C(N)}/\sigma^D$ measured in electron scattering by E139 [25] and HERMES [29] is presented in Fig. 3. Also shown are data from JLAB-E03103 [32] taken with a beam energy of 5.8 GeV.

This figure nicely summarizes the universal x dependence of the nuclear effects. It can be subdivided into four x regions (plus a fifth one at $x > 1$ which will be discussed separately in section 6):

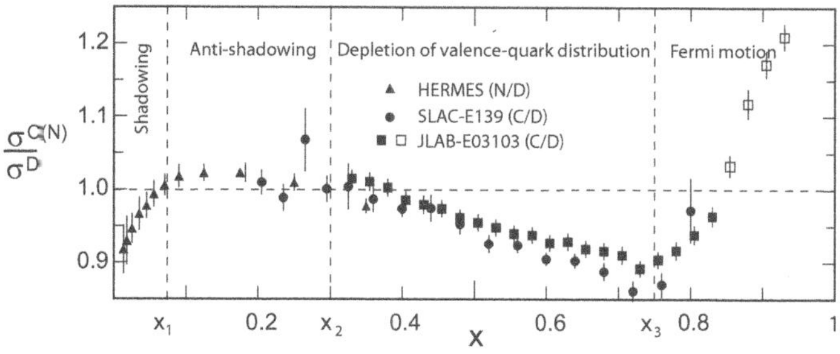

Figure 3: The ratio $\sigma^{C(N)}/\sigma^{D}$ as a function of x from HERMES [29], SLAC-E139 [25], and JLAB-E03103 [32]. Open squares denote W^2 below 2 GeV^2, where W is the invariant mass of the photon-nucleon system.

- the 'shadowing' region ($0 < x < x_1 \simeq 0.06$), where the structure function ratio is smaller than unity and decreases with decreasing x down to the value measured in photoproduction. Here, the dominant contribution to the cross section is due to sea quarks. The essential longitudinal distances Δz probed in the deep-inelastic interaction (see section 5.1) are $\Delta z > 3$ fm, much bigger than the size of a nucleon;
- the 'anti-shadowing' region ($x_1 < x < x_2 \simeq 0.3$, 3 fm $> \Delta z > 0.7$ fm), where the ratio shows a small increase of a few percent over unity;
- the region ($x_2 < x < x_3 \simeq 0.8$, $\Delta z < 0.7$ fm), where the ratio is smaller than unity with a minimum near $x \approx 0.7$. Here, the sea-quark distribution is essentially negligible and the ratio reflects the behavior of the valence-quark distributions;
- the region ($x_3 < x < 1$), where the ratio increases rapidly with increasing x. This behavior is dominantly a kinematic effect since the free-nucleon cross section vanishes for $x \to 1$. It is partly also due to the Fermi motion of the bound nucleons in the nucleus.

4.3 Low-x data

In deep-inelastic scattering from stationary targets, the kinematic region of very low x can only be accessed with muon beams, since those can be produced with much higher energies than electron beams. The first of such measurements was performed by EMC-NA28, using a muon detection system at small scattering angles down to 2 mrad and nuclear targets of C and Ca [12]. This experiment demonstrated that shadowing persist also at high values of Q^2. The low-x region was then explored in detail by NMC with nominal incident muon energies of 90–200 GeV [14, 15, 16, 17, 19, 20, 21] and at even lower values of x by E665 at FNAL with a mean incident muon energy of 470 GeV.

NMC had the main objective to study the nuclear modification of the structure function F_2 with high precision. Cross section ratios were measured for nine nuclear species. In one set of measurements [14, 16, 18, 19] ^{4}He, ^{6}Li, ^{12}C and ^{40}Ca were compared to deuterium. All these nuclei are isoscalar and no correction for neutron excess (Equ. (4)) is necessary. In another set of measurements [20], ^{9}Be, ^{27}Al, ^{40}Ca, ^{56}Fe, ^{119}Sn and ^{207}Pb were compared to carbon. One important peculiarity of the NMC experiment was the multiple target arrangement. Targets of different materials were placed in a row at longitudinally well separated locations along the spectrometer axis and exposed simultaneously to the beam. Two such rows, differing in the ordering of materials, were placed on a common platform. The rows of targets were positioned in the beam in turn by lateral displacement of the platform at approximately 30–60 min intervals. With this arrangement beam flux and spectrometer acceptance corrections canceled in the determination of cross section ratios. These high-precision measurements are the low-x counterpart of the E139 data at large x.

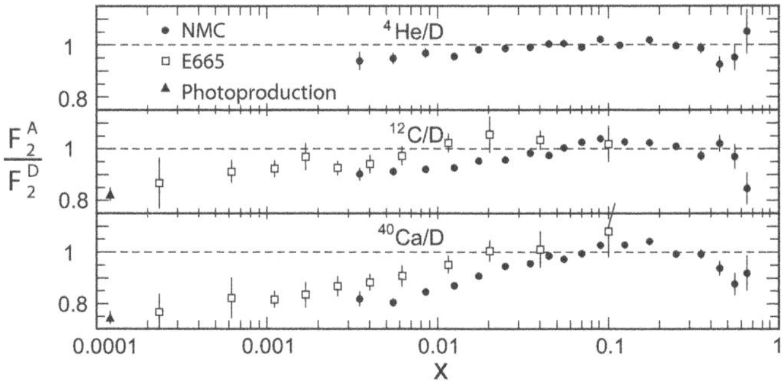

Figure 4: The ratio F_2^A/F_2^D for He, C and Ca measured at low x by NMC [19] and E665 [23]. Note that for both experiments $Q^2 < 1$ GeV^2 for x values below ≈ 0.005.

As an example the structure function ratios of ^{4}He, ^{12}C and ^{40}Ca to deuterium measured by NMC [19] and E665 [23] are shown in Fig. 4. The E665 data nicely extrapolate to the shadowing results measured in photoproduction [8]. The NMC and E665 data do not agree very well, but the E665 data move downwards by a few percent, when another method of radiative corrections is applied [23]. As already seen in Fig. 3, the ratio is smaller than unity below $x_1 \approx 0.06$ and decreases with decreasing x. The effect is already visible for ^{4}He, it increase with A.

4.4 Nuclear dependence of R

As stated in section 2, σ^A/σ^D is only equal to F_2^A/F_2^D, if the quantity R is independent of the nuclear mass number A. The nuclear dependence of R has been studied by SLAC-E140 [26, 27], NMC [15, 21] and HERMES [29]. All measurements are consistent with R being independent of A. A reanalysis of all SLAC data with an improved radiative-corrections calculation procedure [27] resulted in $R^{Fe} - R^D = 0.001 \pm 0.018(stat.) \pm 0.016(sys.)$, and the authors conclude that possible contributions to R from nuclear higher-twist effects and possible spin-0 constituents in nuclei are not different from those in free nucleons. This conclusion is supported by the more recent HERMES measurement [29]. Averaging over all measurements of R^A/R^D for light and medium heavy nuclei (^{3}He, ^{4}He, ^{12}C, ^{14}N) HERMES obtained an average value for R^A/R^D of 0.99 ± 0.03. The result is unchanged if also the data on the heavier nuclei are included in the average.

4.5 Q^2 dependence of the nuclear effects

Already from the good agreement between the muon and the electron data one can conclude that the Q^2 dependence of the nuclear effects is very small, since, at the same value of x, their average Q^2 typically differs by more than an order of magnitude.

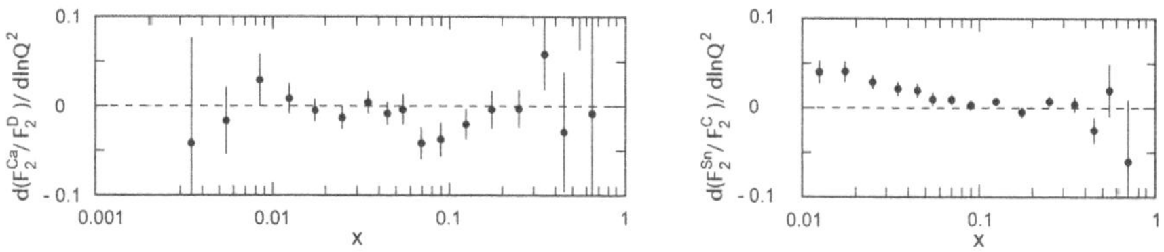

Figure 5: NMC results for the x dependence of the logarithmic slope $d(F_2^A/F_2^D)/d\ln Q^2$. Left panel: Ca/D [19], right panel: Sn/C [21].

The Q^2 dependence was studied in some detail by NMC [18, 21], FNAL-E665 [23] and HERMES [29]. As an example, the left panel of Fig. 5 shows the x dependence of the logarithmic slope $d(F_2^{Ca}/F_2^D)/d\ln Q^2$ from fits of the form $F_2^A/F_2^D = a + b \cdot \ln Q^2$ to the NMC data [19]. In the covered Q^2 range there is little indication for a Q^2 dependence of the ratio. The NMC Sn/C data [21], shown in the right panel of Fig. 5, are the only exception where one observes an indication of a small Q^2 dependence at small values of x.

4.6 Neutrino data

The available measurements in neutrino/antineutrino-nucleus scattering [34, 35] suffer from their large statistical and systematic uncertainties and do not provide additional information. Apart from [34] all these measurements have been performed with bubble chambers. Precise measurements of neutrino and antineutrino scattering from deuterium and heavy nuclear targets would be very helpful for a separation of nuclear effects in sea-quark and valence-quark distributions and a determination of nuclear parton distribution functions (nPDFs). But such measurements have to wait for the future realization of a very-high-luminosity neutrino factory.

4.7 Drell-Yan data

The nuclear modification of anti-quark distributions can also be investigated by the Drell-Yan process [37] in proton-nucleus scattering: $pA \to (l^+ l^-)X$. In this process a quark (anti-quark) with the four-momentum fraction x_1 from the beam proton and an anti-quark (quark) of the target nucleon with x_2 annihilate electromagnetically into a virtual photon, which immediately decays into a charged-lepton pair: $q(x_1)\bar{q}(x_2) \rightarrow \gamma^* \rightarrow l^+l^-$.

The longitudinal momentum of the l^+l^- pair in the proton-nucleon center-of-mass system is approximately given by $p_L^{l^+l^-} \approx (x_1 - x_2)\sqrt{s}/2$ and its invariant mass by $(M_{l^+l^-})^2 = x_1 x_2 s$. Here, $\sqrt{s}$ is the nucleon-nucleon center-of-mass energy. The Drell-Yan cross section reads

$$\frac{d^2\sigma}{dx_1 dx_2} = K\frac{4\pi\alpha^2}{9x_1x_2}\frac{1}{s}\sum_{q=u,d,s...} z_q^2\left[q(x_1)\bar{q}(x_2) + \bar{q}(x_1)q(x_2)\right] , \qquad (5)$$

where $K \approx 2$ is a factor representing the deviation from the simple parton model due to QCD corrections. By a suitable choice of the kinematics of the $\ell^+\ell^-$ pair the quark and anti-quark distributions in the target can be determined separately. If one requires for instance $x_1 - x_2 > 0.3$, then the second term in (5) can be neglected and the cross section ratio for two nuclei $A1$ and $A2$ is to a good approximation equal to the ratio of anti-quark distributions $\bar{q}^{A1}/\bar{q}^{A2}$ in the target.

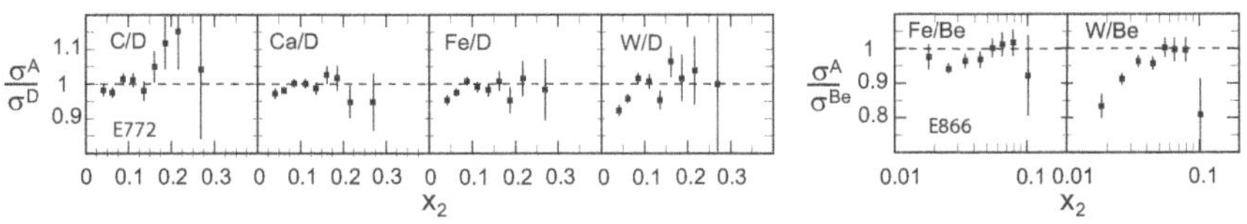

Figure 6: Drell-Yan results for the cross-section ratio σ^A/σ^D [38] (left) and σ^A/σ^{Be} [39] (right)as a function of x_2.

In the left panel of Fig. 6, σ^A/σ^D for four nuclei (C, Ca, Fe, W) measured by FNAL-E772 [38] is shown as a function of x_2, whereas the right panel shows σ^A/σ^{Be} for (Fe, W) measured by FNAL-E866 [39], both with a proton-beam energy of 800 GeV. At the smallest x_2 values the data show an indication of shadowing, while above $x_2 \approx 0.06$ the ratios are consistent with unity (with the exception of the C/D data) leading to the conclusion that anti-shadowing is very likely not a sea-quark effect or caused by nuclear pions. This aspect will be studied in detail by the experiment E906/Seaquest at the 120 GeV beam of the FNAL Main Injector [40].

Unfortunately the Drell-Yan cross section is very small and it is very difficult to study the process with colliding ion beams at much higher center-of-mass energies. Nevertheless such measurements in proton-deuteron and proton-nucleus collisions at RHIC or the LHC would be very desirable for the determination of nPDFs at low values of x.

4.8 Dependence on nuclear properties

Dependence on nuclear mass A. From Figs. 2 and 4 it is obvious that the nuclear effects increase continuously with nuclear mass number A. This aspect has been studied in detail by E139 [25] and NMC [20].

The left panel of Fig. 7 shows the NMC results [20] for the dependence of F_2^A/F_2^C on A for the two bins $x = 0.0125$ and $x = 0.175$, and the upper plot in the right panel shows the E139

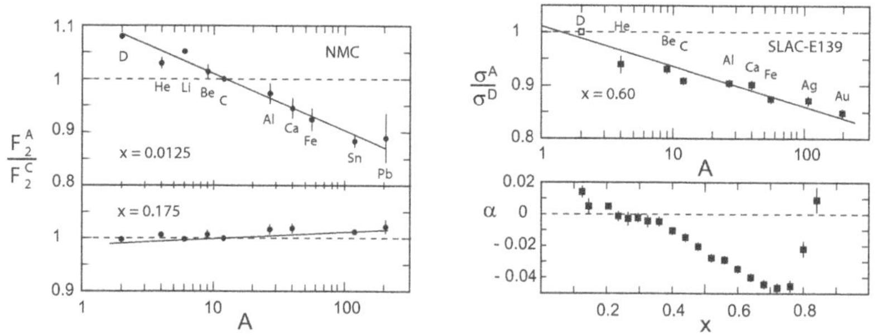

Figure 7: F_2^A/F_2^C as a function of nuclear mass A at low x from NMC [20] (left) and σ^A/σ^D at high x from SLAC-E139 [25](right). The lower right panel shows the coefficient $\alpha(x)$ from a fit of the form $\sigma^A/\sigma^D(x) = c(x)A^{\alpha(x)}$.

results [25] for σ^A/σ^D in the bin $x = 0.60$. Obviously the nuclear effects increase to a good approximation linearly with logA. Small deviations from this linear dependence are observed in the left panel for He and Li and in the right panel for He and C. Obviously there are other nuclear properties than A affecting the nuclear dependence. The lines are results of fits of the form: Ratio = $c(x)A^{\alpha(x)}$. The coefficient $\alpha(x)$ determined from the E139 data is shown in the lower right panel as a function of x. If we rewrite $\alpha(x)$ as $\alpha(x) = \alpha'(x) - 1$, then $A^{\alpha'(x)}$ can be interpreted as the effective number of nucleons participating in the interaction.

Dependence on nuclear density ρ. One important aspect for the understanding of the nuclear medium effects is their dependence on the nuclear density ρ. In Fig. 8 the data presented in Fig. 7 are shown as a function of ρ in four bins of x.

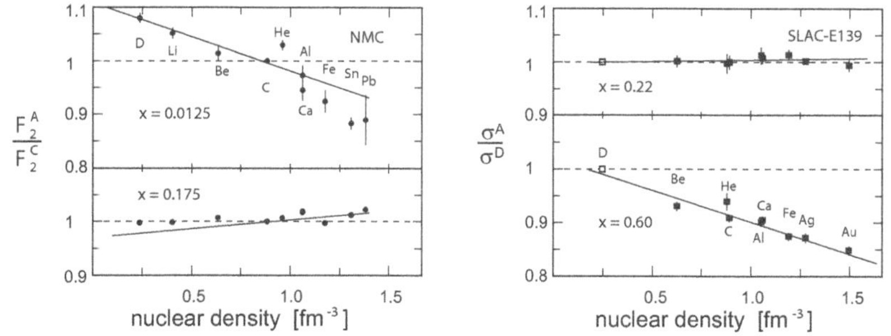

Figure 8: F_2^A/F_2^C as a function of nuclear density ρ at low x from NMC [20] and σ^A/σ^D at high x from SLAC-E139 [25] .

Here $\rho(\mathrm{A})$ is given by $\rho(\mathrm{A}) = 3\mathrm{A}/4\pi \mathrm{R}_e^3$, with $\mathrm{R}_e^2 = 5 < r^2 > /3$, and $< r^2 >$ is taken from Ref. [41]. At large x the cross section ratio approximately scales with ρ, with the exception of the rather special nuclei ^{4}He and ^{9}Be. The deviation of ^{9}Be from this linear behavior is, however, much smaller than in the analysis of JLAB-E03103 [32], where a density based on ab initio few-body calculations [42], scaled by $(\mathrm{A}-1)/\mathrm{A}$ is used. The small-x data are not well described by a linear function of nuclear density.

Dependence on nuclear radius r. The deviations from a linear behavior observed in Fig. 8 indicate that besides the nuclear density other parameters like the nuclear radius or the nuclear surface may play a role. The nuclei ^{4}He, ^{6}Li, ^{12}C and ^{40}Ca have been used by NMC [14, 15, 18, 19] to possibly differentiate between effects originating from the nuclear density or from the nuclear radius. These nuclei differ primarily either in radius (r) or density (ρ). In particular, ^{6}Li ($r = 2.6$ fm, $\rho = 0.04$ fm^{-3}) and ^{12}C ($r = 2.5$ fm, $\rho = 0.09$ fm^{-3}) have nearly equal radii but different densities, whereas ^{4}He ($r = 1.7$ fm, $\rho = 0.09$ fm^{-3}) and ^{12}C have nearly equal densities but different radii. ^{12}C and ^{40}Ca ($r = 3.5$ fm, $\rho = 0.11$ fm^{-3}) differ more in radius than in density. The analysis presented in [15] demonstrates that there is a rather complicated interplay between the dependences on radius and density: the depletion at low x is larger in ^{12}C than in ^{6}Li, showing that at the same radius the effect increases with ρ; the comparison of the pairs (^{4}He,^{12}C) and (^{40}Ca,^{12}C) indicates that at similar densities the effect increases with radius; the depletion in ^{40}Ca is twice as large than in ^{6}Li, implying that the depletion increases with both radius and density; and the Li/He ratio is consistent with unity over the common x range, indicating that the opposing dependencies on radius and density tend to cancel. Furthermore NMC has shown that at low x the data are best described by a fit of the form $\mathrm{F}_2^{\mathrm{A}}/\mathrm{F}_2^{\mathrm{C}} = a + b\mathrm{A}^{-1/3} + c\mathrm{A}^{-2/3}$ [20].

5 Interpretations

It is beyond the scope of this lecture note to address the multitude of possible explanations for the observed nuclear medium effects (for a rather detailed discussion see Ref. [9]). Instead, only the main ideas of some classes of models will be summarized without a discussion why and where they fail to reproduce the data correctly.

5.1 Shadowing

The first class of models deals with the shadowing region. The term 'shadowing' has been introduced to explain the reduction of the nuclear cross sections in photoproduction, but has then also been used in the discussion of the low-x modification of the nuclear cross section in inelastic lepton-nucleon scattering. There are two possibilities to explain this phenomenon. In the first approach the parton distributions in bound nucleons remain unchanged compared to those in the free nucleon. The interaction is viewed in the rest system of the nucleus. The nuclear effects are attributed to a modification of the interaction of the virtual photon with the atomic nucleus by fluctuations of the virtual photon into quark-antiquark pairs. Such a pair then interacts with the nucleus via the *strong* interaction. Since the strength of the latter is much larger than the electromagnetic one, the interaction does no longer happen incoherently with all the nucleons in the nucleus but preferentially with those at the front surface. The nucleons being in the 'shadow' of the nucleons at the front surface then do not or much less contribute to the cross section. Quantitatively this happens, when the fluctuation length $\Delta d \approx 1/Mx \approx 0.2$ fm$/x$ is larger than the mean free path L $\cong 2.5$ fm of the quark-antiquark pair, i.e., for x below ≈ 0.08.

In the second approach the effect is attributed to a modification of the quark and gluon distributions in the nucleus. The interaction of the virtual photon with the nucleus is viewed in a fast moving system, where the nucleon with diameter D is Lorentz contracted to a disc of thickness $\mathrm{D}' \approx \mathrm{D} \cdot \mathrm{M}/|\vec{\mathrm{P}}|$ and the mean nucleon distance $\mathrm{d} \approx 2$ fm to $\mathrm{d}' = \mathrm{d} \cdot M/|\vec{P}|$. The longitudinal position of its constituents, however, has an uncertainty of $\Delta z = 1/x|\vec{P}|$. At small values of x this uncertainty Δz can be much larger than D$'$ and in a nucleus much larger than d$'$. For $x < 1/d \cdot 1/\mathrm{M} \approx 0.1$ there will be a spatial overlap of sea quarks and gluons of different nucleons. The smaller x, the larger is the number of nucleons sharing their contents of sea-quarks and gluons. The effect increases with increasing mass number A and saturates for $x \approx 1/\mathrm{D_A} \cdot 1/\mathrm{M}$, where $\mathrm{D_A}$ is the diameter of the nucleus in its rest frame. Thus, the density of gluons and sea quarks at the position of a nucleon in a nucleus can be much larger than for a free nucleon. Due to this 'overcrowding' the probability for an interaction between sea quarks and gluons is increased and by pair annihilation their density is reduced again, resulting in the observed reduction of the nuclear structure function. Momentum conservation requires that this reduction of the number of partons at low values of x gets compensated by an enhancement at larger x, i. e., anti-shadowing.

Both approaches for shadowing are equivalent. They describe the same phenomenon but viewed in a different reference frame. More details can be found, e. g., in Refs. [43, 44].

5.2 Convolution Models

In the second class of models the structure function of a nucleus A is described as the incoherent sum over contributions of all kind of clusters c with structure functions $\mathrm{F}_2^{\mathrm{c}}(x/y)$ convoluted with the probability $f_{\mathrm{c}}^{\mathrm{A}}(y)$ to find a certain cluster c of momentum y in the nucleus:

$$F_2^{\mathrm{A}}(x, Q^2) = \sum_{\mathrm{c}} \int_x^A dy f_{\mathrm{c}}^{\mathrm{A}}(y) F_2^{\mathrm{c}}(x/y). \tag{6}$$

Examples for such clusters are the nucleon itself, undisturbed or with a reduced mass due to nuclear binding or with an increased size due to different boundary conditions in the nuclear environment, extra pions being responsible for nuclear binding, Δ-isobars, multi-quark clusters like bags of 6 quarks, 9 quarks or 12 quarks (i.e., α particles) or the whole nucleus as one big bag with free quark and color flow throughout the whole nuclear volume.

There is a lot of freedom in these approaches, concerning as well the choice of $f_{\mathrm{c}}^{\mathrm{A}}(y)$ as the parameterization of $F_2^{\mathrm{c}}(x/y)$ that both are badly known. Consequently, it is not very surprising that they succeed to reproduce a portion of the data rather well, at least in the medium x range.

5.3 Rescaling models

In the third class of approaches the EMC effect is explained by a change of either the Q^2 scale or the x scale for the nuclear structure function compared to the free nucleon's one.

Q^2 rescaling. In Q^2 rescaling models, first proposed in Refs. [45] and [46], the EMC effect is related to a change of confinement size inside the nucleus. The A *dependence* of quark and gluon distributions for bound nucleons and the Q^2 *evolution* of these distributions for free nucleons both have the same origin. They are caused by the color forces between quarks and gluons which ensure confinement and are the origin of scaling violations, i.e., the increase of the structure function with Q^2 at low values of x and the decrease with increasing Q^2 at large values of x. The qualitative argument is as follows: The strength of the strong force between quarks is not only determined by the transverse resolution $1/\sqrt{Q^2}$ at which they are probed, but also by the radial extension r_{A} of the volume in which they are confined. Therefore, the relevant parameter for the strength of the strong coupling constant α_{s} is not just Q^2 but $(Q \cdot r_{\mathrm{A}})^2$. (This is similar to the situation in nuclear physics where the form factors for spherical nuclei have the identical oscillating pattern when plotted against $(Q \cdot r_{\mathrm{A}})$). If the confinement size is modified inside the nucleus, either due to a 'swelling' of nucleons, the formation of multi-quark bags, short-range nucleon-nucleon correlations or free quark and color flow throughout the whole nucleus, then, as a consequence, quark and gluon distributions obtained for nuclei A and B are related by

$$q_{\mathrm{A}}(x, Q^2) = q_{\mathrm{B}}(x, \xi \cdot Q^2), \; g_{\mathrm{A}}(x, Q^2) = g_{\mathrm{B}}(x, \xi \cdot Q^2). \tag{7}$$

ξ is a rescaling parameter determined by the two confinement scales r_{A} and r_{B}. In the so-called 'dynamical rescaling' models it is given by

$$\xi = (r_{\mathrm{A}}/r_{\mathrm{B}})^{2\alpha_s(\mu_{\mathrm{A}}^2)/\alpha_s(Q^2)}, \tag{8}$$

where μ_A is a low-momentum cut-off for radiating gluons. Consequently, at the same value of Q^2, $F_2^A/F_2^B > 1$ for small values of x and $F_2^A/F_2^B < 1$ for large values of x, if $r_A > r_B$.

x rescaling. In the x-rescaling models, first proposed in [47, 48] and then refined by numerous authors, the depletion of the nuclear structure function at medium x is explained by conventional nuclear binding and Fermi-motion corrections. The x dependence can be reasonably well reproduced if, for a nucleus, the scaling variable x is replaced by a modified one $x^* > x$. For a nucleon i moving with momentum $\vec{\mathrm{p}}_i$ in a nucleus the variable $x = Q^2/2\mathrm{M}\nu$ has to be replaced by

$$x_i = Q^2/2\mathrm{p}_i q = Q^2/\left[2(\mathrm{M} + \mathrm{E}_i)\nu - 2\vec{\mathrm{p}}_i\vec{q}\right], \tag{9}$$

where E_i is the removal energy of the nucleon ($< \mathrm{E}_i > \cong -25$ MeV) and $\vec{q}$ is the momentum of the virtual photon. Consequently, at the same kinematics of the scattered lepton, the effective x^* at which the structure function is probed in a nucleus, is larger than x for a free nucleon. At large x the structure function F_2 is steeply falling with x and a depletion of the structure function ratios is naturally explained.

6 The new ingredient: $x > 1$

Recently their has been renewed interest in the EMC effect and its possible origin by JLAB measurements of the cross section ratios in inclusive electron scattering $A(e, e')$ in the kinematic region $x > 1$ where the cross section vanishes for scattering from free nucleons.

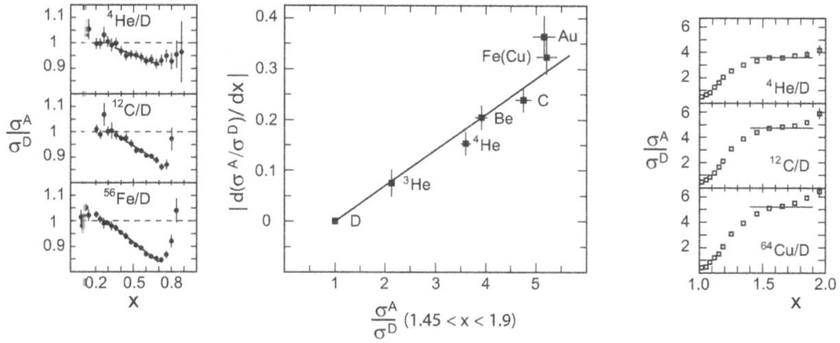

Figure 9: The cross section ratio σ^A/σ^D measured by SLAC-E139 at $x < 1$ (left panel) and by JLAB-E02019 in the region $1 < x < 2$ (right panel). The middle panel shows the correlation between the slope $|d(\sigma^A/\sigma^D)/dx|$ in the region $0.35 \leq x \leq 0.7$ and the height of the plateau of σ_A/σ_D in the region $1.45 < x < 1.9$ (after [36]).

The CLAS experiment [30, 31] has measured $\sigma^A/\sigma^{^3He}$ for ^{4}He, ^{12}C and ^{56}Fe and JLAB-E02019 [33] measured σ^A/σ^D for ^{3}He, ^{4}He, Be, C, Cu and Au. In the right panel of Fig. 9 the cross section ratios ^{4}He/D, C/D and Cu/D are shown as a function of x. They rise with x until they reach a plateau at $x \approx 1.4 - 1.5$. The height of this plateau increases with A. Such a behavior has already been observed with less accuracy by an experiment at SLAC [28]. A similar pattern is seen in the CLAS comparison of the nuclear cross sections to ^{3}He. Here the measurements extend up to $x = 3$ and an indication of a second plateau is observed for $x > 2.25$. Such a behavior is being interpreted as the manifestation of short-range nucleon-nucleon (mostly p–n) correlations (three-nucleon correlations for x beyond 2) or, in another approach, as the ratio of the probabilities to find 6-quark or 9-quark clusters in nuclei compared to the reference nucleus.

Fig. 9 shows a very interesting observation [36]. In the left panel the E139 cross section ratios ^{4}He/D, C/D and Fe/D are shown for $x < 1$. The lines correspond to the slopes $d(\sigma^A/\sigma^D)/dx$ in the region $0.35 \leq x \leq 0.7$, that characterize the strength of the EMC effect in this region and are unaffected by overall normalization uncertainties. In the middle panel the slopes of the E139 data (tabulated in [36] and corrected by $x_A = x_p \cdot AM_p/M_A$ [49]) are plotted against the height of the plateaus [33]. Obviously there is a strong correlation between these quantities as indicated by the straight line. It is rather unlikely that this correlation is purely accidental and one can therefore rather safely assume that a large fraction of the strength of the EMC effect in the valence quark region is due to short-range nucleon-nucleon correlations.

7 Summary and Outlook

The EMC effect is with us now for 30 years. It has stimulated huge experimental and theoretical efforts, but its origin is still not fully understood. Recent data shed some new light on its possible origin, i.e., short-range nucleon-nucleon correlations may play an important role for the observed nuclear modifications. Still more precise data are needed, especially on the nuclear gluon and antiquark distributions at very low x to constrain the initial state for the AA program at RHIC and LHC. These hopefully will come from future measurements at JLAB12, RHIC and LHC and eventually also from the proposed projects EIC and LHeC.

References

[1] J. J. Aubert et al. (EMC), Phys. Lett. **B 123** (1983) 275.

[2] K. Rith, Proceedings of the Eightteenth Rencontre de Moriond, Vol. 2, J. Tran Thanh Van ed., Editions Frontieres, 1983, pp. 207-222.

[3] A. Bodek et al. (SLAC), Phys. Rev. Lett. **50** (1983) 1431.

[4] A. Bodek et al. (SLAC), Phys. Rev. Lett. **51** (1983) 534.

[5] A. Bodek and J. L. Ritchie, Phys. Rev. **D 23** (1981) 1070; **D 24** (1981) 1400.

[6] S. Stein et al. (SLAC), Phys. Rev. **D 12** (1975) 1884.

[7] S. Rock and P. Bosted, Phys. Lett. **B 518** (2001) 34.

[8] D. O. Caldwell et al., Phys. Rev. Lett. **42** (1979) 553.

[9] P. R. Norton, Rept. Prog. Phys. **66** (2003) 1253.

[10] G. Bari et al. (BCDMS), Phys. Lett. **B 163** (1985) 282; A. C. Benvenuti et al. (BCDMS), Phys. Lett. **B 189** (1987) 483.

[11] J. J. Aubert et al. (EMC), Nucl. Phys. **B 293** (1987) 740.

[12] M. Arneodo et al. (EMC-NA38), Phys. Lett. **B 211** (1988) 493; Nucl. Phys. **B 333** (1990) 1.

[13] J. Ashman et al. (EMC), Phys. Lett. **B 202** (1988) 603; Z. Phys. **C 57** (1993) 211.

[14] P. Amaudruz et al. (NMC), Z. Phys. **C 51** (1991) 387.

[15] P. Amaudruz et al. (NMC), Phys. Lett. **B 294** (1992) 120.

[16] P. Amaudruz et al. (NMC), Z. Phys. **C 53** (1992) 73.

[17] P. Amaudruz et al. (NMC), Nucl. Phys. **B 371** (1992) 553.

[18] P. Amaudruz et al. (NMC), Nucl. Phys. **B 441** (1995) 3.

[19] M. Arneodo et al. (NMC), Nucl. Phys. **B 441** (1995) 12.

[20] M. Arneodo et al. (NMC), Nucl. Phys. **B 481** (1996) 3.

[21] M. Arneodo et al. (NMC), Nucl. Phys. **B 481** (1996) 23.

[22] M. R. Adams et al. (FNAL-E665), Phys. Rev. Lett. **68** (1992) 3266; Phys. Lett. **B 287** (1992) 375.

[23] M. R. Adams et al. (FNAL-E665), Z. Phys. **C 67** (1995) 403.

[24] R. G. Arnold et al. (SLAC-E139), Phys. Rev. Lett. **52** (1984) 727.

[25] J. Gomez et al. (SLAC-E139), Phys. Rev. **D 49** (1994) 4348.

[26] S. Dasu et al. (SLAC-E140), Phys. Rev. Lett. **60** (1988) 2591.

[27] S. Dasu et al. (SLAC-E140), Phys. Rev. **D 49** (1994) 5641.

[28] L. Frankfurt et al., Phys. Rev. **C 48** (1993) 2451.

[29] A. Airapetian et al. (HERMES), Phys. Lett. **B 567** (2003) 339.

[30] K. Egyan et al. (JLAB-CLAS), Phys. Rev. **C 68** (2003) 014313.

[31] K. Egyan et al. (JLAB-CLAS), Phys. Rev. Lett. **96** (2006) 082501.

[32] J. Seely et al. (JLAB-E03103), Phys. Rev. Lett. **103** (2009) 202301.

[33] N. Fomin et al. (JLAB-E02019), Phys. Rev. Lett. **108** (2012) 092502.

[34] H. Abramowicz et al. (CDHS), Z. Phys. **C 25** (1984) 29.

[35] A. M. Cooper et al., Phys. Lett. **B 141** (1984) 133; M. A. Parker et al., Nucl. Phys. **B 232** (1984) 1; V. V. Ammosov et al., JETP Lett. **39** (1984) 393; J. Hanlon et al., Phys. Rev. **D 32** (1985) 2441; A. E. Asratyan et al., Sov. J. Nucl. Phys. **41** (1985) 763; Sov. J. Nucl. Phys. **43** (1986) 380; M. Aderholz et al., Phys. Lett. **B 173** (1986) 211; J. Guy et al., Z. Phys. **C 36** (1987) 337; J. Guy et al. , Phys. Lett. **B 229** (1989) 421; T. Kitagaki et al., Phys. Lett. **B 214** (1988) 281.

[36] L. B. Weinstein et al., Phys. Rev. Lett. **106** (2011) 052301.

[37] S. Drell and T.-M. Yan, Phys. Rev. Lett. **25** (1970) 316; Erratum-ibid. **25** (1970) 902.

[38] D. M. Alde et al. (FNAL-E772), Phys. Rev. Lett. **64** (1990) 2479.

[39] M. A. Vasiliev et al. (FNAL-E866), Phys. Rev. Lett. **83** (1999) 2304.

[40] P. Reimer, Proceedings of Spin 2010, Jülich, Germany; J. Phys.: Conf. Ser. **295** (2011) 012011.

[41] H. De Vries et al., Atomic Data and Nuclear Tables **36** (1987) 495.

[42] S. C. Pieper and R. B. Wiringa, Ann. Rev. Nucl. Part. Sci. **51** (2001) 53.

[43] B. Z. Kopeliovich, J. Morfin and I. Schmidt, Prog. Part. Nucl. Phys. **68** (2013) 314.

[44] L. Frankfurt, V. Guzey and M. Strikman, Phys. Rept. **512** (2012) 255.

[45] F. E. Close et al., Phys. Lett. **B 129** (1983) 346.

[46] O. Nachtmann and H. J. Pirner, Z. Phys. **B 21** (1984) 277.

[47] C. H. Garcia Canal et al., Phys. Rev. Lett. **53** (1984) 1430.

[48] M. Staszel et al., Phys. Rev. **D 29** (1984) 2638.

[49] L. Frankfurt and M. Strikman, Int. J. Mod. Phys. **E 21** (2012) 1230002.

CHAIRMAN: K. RITH

Scientific Secretaries: Á. Lukács, J. Seele

DISCUSSION

- Csanád:

People are trying to put in the EMC effect in their nuclear modification calculations in RdA and RAA. What do you think about this, and can measurements of the nuclear modification ever be discriminative somehow to the EMC effect?

- Rith:

My naïve opinion is that these effects should be included, when you look at effects that are different in AA collisions compared to pA collisions. The problem is how to get that information. There are now several calculations of nuclear PDFs on the market, but I showed in one transparency that all the information you get below x=10^{-2} is just speculation. Somehow the behavior just reflects the parameterization people put into their code. The systematic uncertainties blow up, and, in fact, you don't have any idea how nPDFs behave at very low x. What one really would need is pd collisions, not just p-heavy ions, and also electron-deuteron collisions. These need to be compared to pA or eA, to really get out the nuclear effects. You need to understand the initial state precisely. It will, of course take some time to get the experimental input, and then also to study what happens when one puts the results into the codes.

- Backhouse :

Can you comment on the application of the EMC effect to neutrino cross sections? There are experiments recently that are talking about nucleon-nucleon correlations and they say they have evidence for that. But if you apply the EMC slope naively in the Monte Carlo it doesn't approve the agreement.

- Rith

On slide 45 you can see all the trustable information we have on the EMC effect with neutrino beams. Essentially all this information is coming from bubble chambers, there has never been a fixed target experiment with a deuteron target. Most of the data in this figure is coming from two experiments and you could say that within the experimental uncertainties, neutrino experiments do not see any EMC effect. But if you put the data on top of the universal curve of the EMC effect, you can see that the neutrino data from the bubble chambers are in agreement with it. Now comes the problem: since there are no experimental deuteron data one has to construct hypothetical neutrino-deuteron data to compare them with the nuclear data. People tried to put all of the known effects into their calculations to understand the neutrino-nucleus data. An example is mass effects: when a neutrino interacts with a strange-quark this is converted to a charm-quark. There are subtle mass effects which influence the data.They have to be taken properly into account for the description of the deuteron data. The only thing I can say at the moment is that for the global fitters there is still quite some work to be done. On slide 46 you see on the left hand side F_2 and on the right hand side xF_3, the red curves are the nuclear PDFs from de Florian et al. They agree reasonably well with the data. The authors have found some tensions between the data but they also conclude that there is nothing very problematic. The situation is different when you look at the results from a subgroup of CTEQ shown in slide 47. They tried to extract the EMC effect from neutrino data, using their method to calculate the hypothetical neutrino-deuteron data. You observe a kind of anti-shadowing down to the lowest values of x, i.e., an anti-EMC effect. De Florian et al. argue that their assumptions are of better quality and that their hypothetical neutrino-deuteron data are in agreement with their nuclear PDFs. For me it is not possible to decide which group uses the best information and assumptions. As an experimentalist I can only say that if there will be future high intensity neutrino beams, then also measurements with a long deuterium target should be performed with high priority. CDHS once performed such measurements using a 6 m long hydrogen target with 35 m^3 of liquid hydrogen. So it should also be possible with a liquid deuteron target. There is another very crucial point: most the information about the x dependence of the distributions of strange-quarks and antistrange-quarks comes from neutrino-nucleus scattering. If the fitters don't succeed to reconstruct a hypothetical neutrino-deuteron measurement why should I really trust their strange-quark distributions?

- *Haidt:*

a) I just wanted to make a comment. It is not even known whether there is a difference between neutrino–s-quark and neutrino–anti-s-quark. b) To my knowledge, there exists data on neutrino–deuteron. I remember a plot, where the comparison shows essentially no effect within error, but I should check. I think by BEBC.

- *Rith:*

Yes, I have shown BEBC data (slide 42). There is a comparison of neon-deuteron bubble-chamber data, but this is unfortunately of insufficient statistics. If you want to answer with high significance whether you see the same effect in neutrino data as in charged-lepton data and whether there is the same kind of shadowing in neutrino data, then you need to do this kind of measurements: a comparison of nuclear data with those from a long deuterium target with really high luminosity.

- *'t Hooft:*

There is a long dispute today about the next generation of accelerators, and these are p-p-machines, p-ion, ion-ion and electron-electron, electron-positron machines. Could you make a strong case for any, supposing high or very high energy electron-proton or electron-nucleus machines.

- *Rith:*

Yes, I tried to demonstrate that there are many reasons to argue for an electron-ion or electron-proton, electron-nucleus collider. The field of deep-inelastic scattering, investigating the nucleon structure, has evolved enormously. Now we want to study, for example, generalized parton distributions, which allow to map out the spatial distributions of quarks in the nucleon and not just their longitudinal momentum distribution. There is a long list of arguments for such a machine. The left hand side of slide 42 shows the kinematic reach of the American proposal, the EIC, for two different values of sqrt(s). In the mid 90'ies there was a European proposal to build such an electron-ion collider at the GSI. But at a 'shoot-out' meeting it was concluded that such measurements could better be done at HERA. But for several sad reasons such measurements never materialized. Now it is the American community that tries to get such a project realized.

- *'t Hooft:*

But now, at CERN, there is the same discussion about what to do in the LHC tunnel perhaps, and hopefully it will be used with the next generation machine, and all the time the question comes up, if we should collide electrons against protons or heavy ions perhaps, and it usually comes out negative as you say.

- *Rith:*

Yes, but I briefly mentioned in my lecture one of the many arguments: the gluon distribution shoots up tremendously at low values of *x*. At some point, one must see there some saturation effects, and then one has to go from linear equations to the nonlinear ones, for the description of the evolution of parton distributions. In the picture I showed about shadowing in the infinite momentum frame one gets an increase in the gluon and quark distributions already due to the non-localization effect in a nucleus. Consequently one needs much smaller values of Q^2 to see the onset of the saturation effects. At HERA there are some indications for an onset of this behavior, it would be important to look for it in electron-ion collisions at the LHeC.

- *Tannenbaum:*

A comment: one thing that EIC will be able do at RHIC, that is unique, is we can polarize protons, and electrons will be polarized too, but HERA also had that, so we can measure spin structure of the proton. No other machine will be able to do that, neither HERA nor the CERN machine; it is very difficult, because they are not level and/or too many spin preserving Siberian snakes would be required.

- *Alam:*

In your fifth transparency you mention that the ratio of the cross sections is equal to the ratio of the F_2 structure functions. Why have the other structure functions been neglected?

- *Rith:*

That was due to the lack of time. There were detailed measurements done at SLAC and NMC to study the influence of the second structure function. As you correctly said and can see on slide 29, the assumption that the cross section ratio is equal to the structure function ratio is only valid if the other structure function F_1 is affected equally by nuclear effects, i.e., if R is the same in the two nuclei. There is a kinematic factor in the cross section formula that allows to determine the two structure functions separately by measurements at different beam energies. The difference of R or the ratio for different nuclei has been measured in detail. I show this in the next transparency for medium or large values of Q^2. There is not so much information, but the assumption of no nuclear modification of R is very well justified within the uncertainties of the experiments. The only exceptions are these HERMES points. They are at the lowest value of Q^2~1GeV^2. It may well be that there is a small effect. After the 12 GeV upgrade of JLAB there is quite a big program to measure the nuclear effects as a function of the beam energy and to measure a possible nuclear modification of R with high precision. There are also theoretical studies being done, e.g., by Guzey et al. You can just take the cross section measurements and assume that R(nucleus) differs by a certain amount from R(nucleon) and study the effect on F_1 and F_2. The effect on F_2 comes out very small, at large values of x the effect on F_1 and F_2 is similar, but it turns out that anti-shadowing disappears for F_1 while it remains for F_2. Since F_1 is purely transverse with respect to the virtual photon polarization, this means that anti-shadowing appears only for longitudinally polarized virtual photons.

«ETTORE MAJORANA» FOUNDATION AND CENTRE FOR SCIENTIFIC CULTURE
INTERNATIONAL SCHOOL OF SUBNUCLEAR PHYSICS
50th Anniversary Celebrations 2011-2012-2013

Theme of the School:
REFLECTIONS ON THE NEXT STEP FOR LHC

Technology and Computing in relation to "Reflections on the Next Step for LHC"

Talk by Horst Wenninger / CERN

References:
CERN Photo Library (copy rights), CERN programs and achievements, CERN current long-term plan, "The European Strategy for Particle Physics", ISSP2012 (Erice), CAS School 2013 (Erice), EPS Technology & Innovation Workshop (Erice 2012) private communications (L. Rossi, S. Myers L. Evans, et al.)

Before discussing next steps let us look back to CERN of the years 1975 -1980.
Prior to 1980 CERN used to built accelerators and detectors and also operate most of these facilities. Around 1980 the physics community proposed to enlarge CERN's experimental program - limited so far to proton beams (PS, SPS) and p-p collisions (ISR) and just starting a new proton – antiproton collider SPS-pp_{bar} - by proposing experiments with a collider of electrons on positrons (LEP, Large Electron Positron collider)

Indeed, a "LEP Working Group" of ECFA - the European Committee for Future Accelerators - under Chairman A. Zichichi - presented a Progress Report on April 15, 1980, which paved the way to build a 27km tunnel for LEP (Large Electron Positron collider) & future LHC (Large Hadron Collider).

The construction of this new CERN complex (27 km tunnel infrastructure and accelerator plus detectors) within a constant budget forced CERN from 1980 to 1989 to concentrate its resources mainly to the SPSpp_{bar} program and the LEP project. Not much was left for other activities. In addition the ISR and bubble chambers had to be closed to free resources for LEP construction.

Fortunately, during this period Prof. A. Zichichi initiated the LAA Project, implemented at CERN in 1986 as „another CERN programme of activities". This project allowed the recruitment of some 40 LAA staff. In addition over 80 unpaid scientists worked for the project with focus on R&D in preparation for the LHC.

Among all the different components of the R&D project, it allowed to boost microelectronics and silicon-pixel technology at CERN. Detailed examples are reported in Ref (1).

The technology R&D related to the LEP accelerator consisted mainly of the development, industrial fabrication, and implementation of

- Low field „ concrete magnets „
- Control systems
- Getter pumping for LEP vacuum NEG = Non Evaporable Getter
- Sputtered superconducting RF cavities

The R&D on superconducting separators started in the early 70's by Karlsruhe and CERN for SPS physics, super-conducting-cavities were used in beams of the CERN Omega Spectrometer to separate kaons and antiprotons (energy range 10 to 40 GeV/c) (W. Bauer/H. Lengeler). LEP needed RF power to accelerate electrons and positrons to high energy in compensation of increasing synchrotron light losses of the LEP beams. For this reason an extensive development of superconducting sputtered cavities was launched at CERN and the technology thereafter transferred to industry for the fabrication. The technology was also transferred to DESY for the TESLA project and is now used for the light source of the European XFEL project. In future, linear colliders are planned with super-conducting cavities.

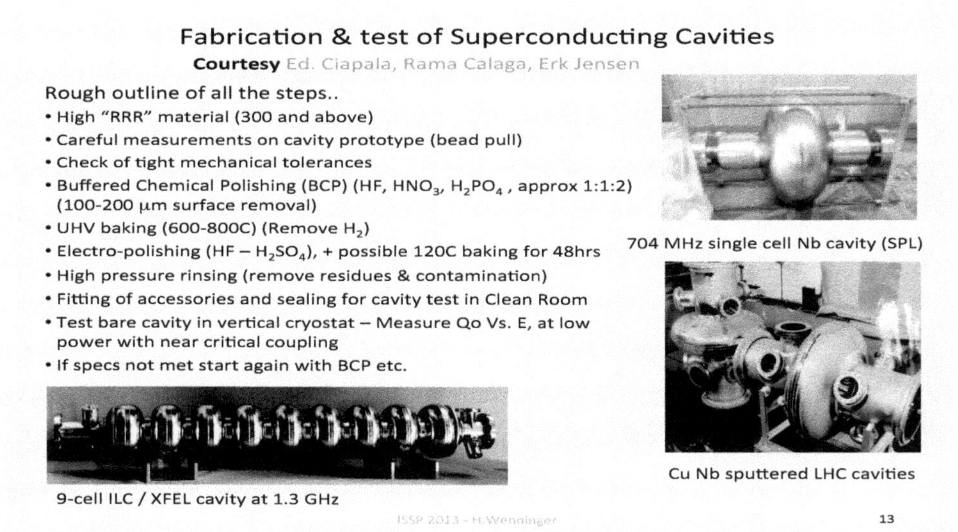

Figure 1: SPL - LHC - ILC superconducting cavities fabrication and test

The technology R&D related to the LEP experiments concerned superconducting solenoids (ALEPH, DELPHI), read-out electronics and new detectors, e.g.:

- Si - pixel detectors
- BGO photon detectors (L3)
- Time-Projection-Chambers (TPC)
- Calorimeters

Some highlights of the LEP program are the precision measurements at the Z resonance (3 families), W+W-production, ZZ production, indirect limits on the Higgs mass, contributions to the exploration of the "Cabibbo-Kobayashi-Maskawa " quark mixing matrix and on the LEP measurements of the coupling constant α_s ……

The LHC preparation for the accelerators started also in the early 80th, in parallel to the LEP construction, as outlined in the long-term plan below:

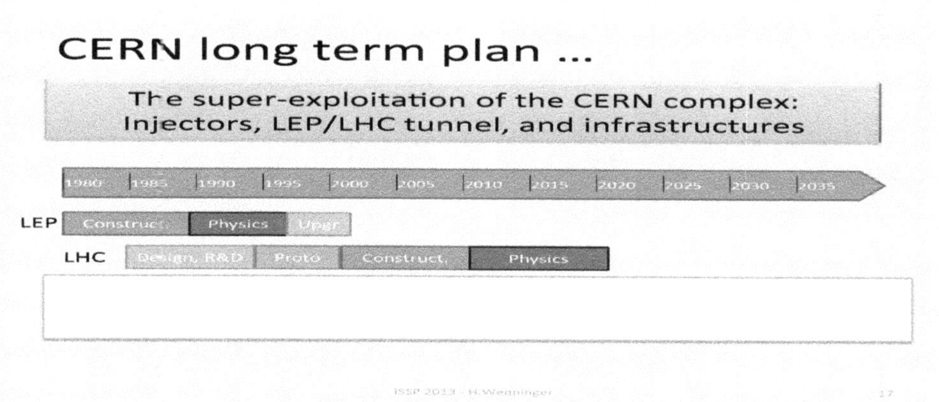

Figure 2: CERN long term plan for LEP and LHC

A challenge for LHC was to develop high field super-conducting dipole magnets to reach highest possible bending power allowing for highest energy proton collisions in the given 27 km tunnel. Niobium-Titanium cables were used to wind coils which had to be assembled to dipole-, quadrupole-, and corrector-magnets, operated at 1.9 K superfluid helium temperature.

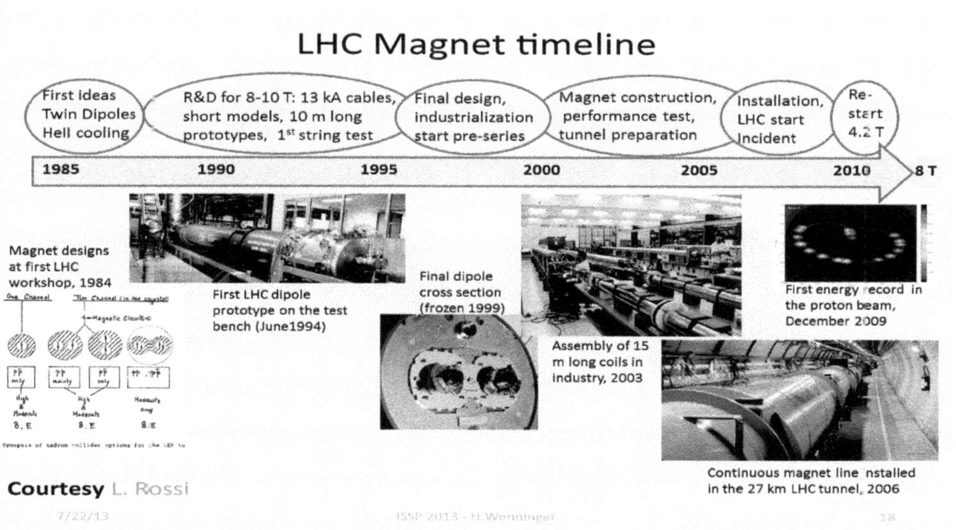

Figure 3: LHC magnet development timeline

Accelerators, such as the LHC, cannot be realized without extensive use of superconductivity and high quality magnets – but also the detectors require the highest possible fields in their solenoid - (CMS) or their toroid magnets (ATLAS). The design and construction of the major 4 LHC detectors was the result of an outstanding achievement of many thousands of people around the globe who collaborate in the LHC experiments.

Physics experiments started in 2011 and by 4th of July 2012 the observation of a new Higgs like particle at 126 GeV energy was reported, a great discovery!

Figure 4: four major LHC experiments

Following this first running period of LHC at reduced collision energy and luminosity, a long shutdown has started in 2013 to go up to design energy and luminosity

Shut-down LS1 to fix interconnects (LHC incident of Sept. 2008)

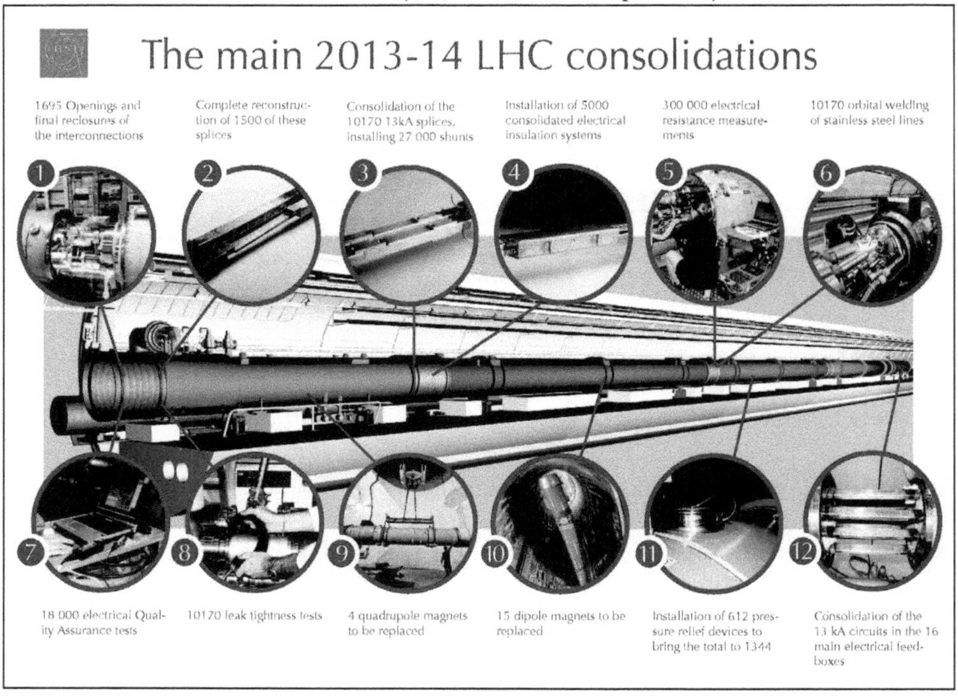

Figure 5 : Shutdown LS1 to fix interconnects (LHC incident of Sept. 2008)

All works to be performed during the shutdown is described in the Figure above. After the shutdown LS1 the LHC will resume physics runs for at least two years.
Long Shutdown 1 (2013-2014) Restart conditions:

- E=6.5TeV
- 25ns
- $\beta^* = 0.5$m (maybe lower)
- All other conditions as in 2012 i.e. LHC availability same etc.

A second shut down LS2 should then fix remaining limitations of LHC, i.e.:

- Injector and LHC Phase I upgrades (Dispersion Suppressor area collimation, New Cryo P4,…)
- go to design- luminosity of present LHC

To summarize the actual long-term plan for LHC, which extends until 2030:

2013 -2014 Shutdown to fix interconnects and overcome energy limitation
2017-2018 Shutdown to overcome intensity Limitation (collimation)

2021 Shutdown LS3 prepares LHC Phase II upgrades: HL-LHC with new focusing magnets, CRAB cavities aiming at very high luminosity. This longer term plan up to 2030 focusses on an increase of the Large Hadron Collider luminosity from **1-2×10^{34}** to **5×10^{34}** cm^{-2} s^{-1} peak luminosity by reducing the beam size at the interaction points and doubling the triplet aperture.

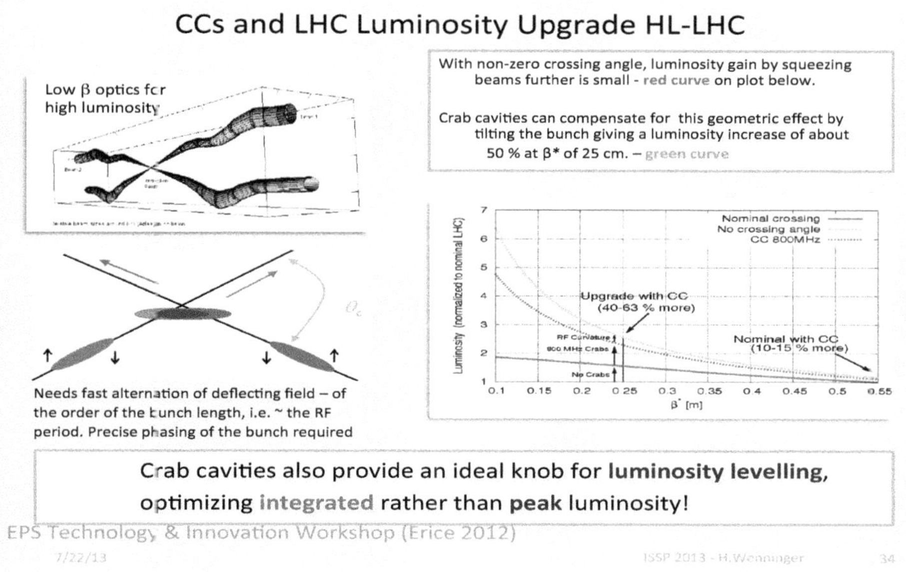

Figure 7: High Luminosity LHC – requires Crab Cavities (to be developed)

A QUESTION remains: will an increased integrated luminosity or rather a higher energy allow to find physics beyond Higgs?

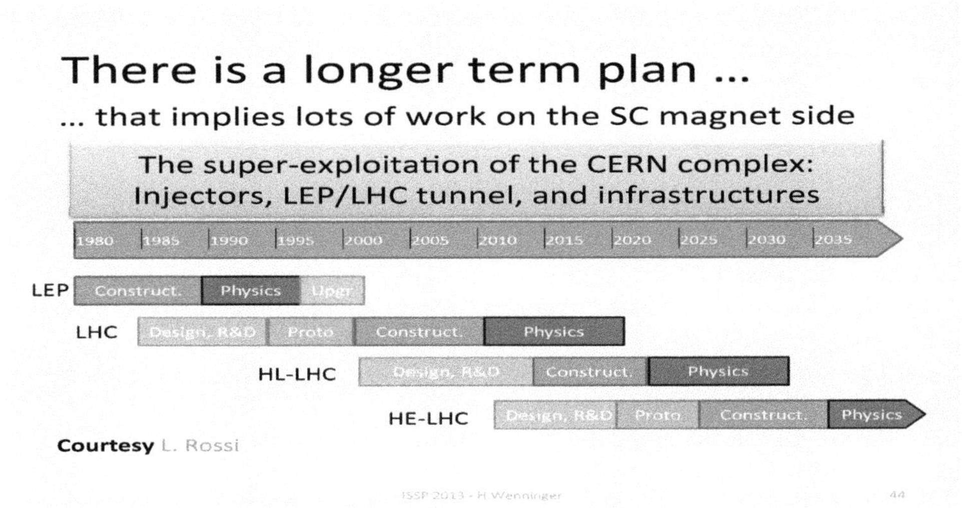

Figure 8: CERN longer term plan

To prepare the High Luminosity LHC (HL-LHC) an R&D program was already started in 2000 for conductors and magnets for collimation (11 Tesla twin LHC dipoles) and for Intersection Region Magnets (Low ß quadruplos and large aperture separation dipoles)
In parallel also studies started in collaboration with other laboratories to develop prototypes of 15 to 20 Tesla dipoles for a High Energy LHC (HE-LHC) and fast pulsing magnets for the SPS injector upgrade to 1000 GeV.

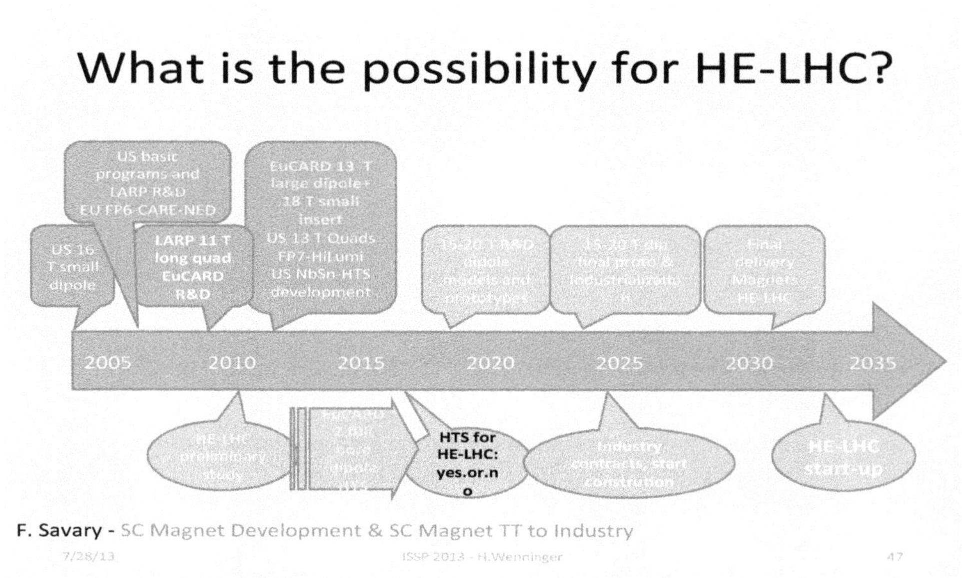

Figure 9: sc-magnet development schedule for high energy HE-LHC.

Discussions started to study a VHE-LHC, with 80 -100 TeV cms-energy in an 80 to 100 km long ring tunnel, using the existing LHC machine as an injector. We also recall the study of the Eloisatron, 200 TEV cms-energy, 200 km tunnel. (Kjiel Johnson in 1990, initiated by Professor Zichichi)

To complete the European Strategiy long term plan of particle physics, there are options such as LHeC, electrons colliding protons in LHC, the ILC (International Linear Collider), with the design study presented in June in 2013, and the CLIC (Compact Linear Collider) studied at CERN, and a Muon Collider, not to forget the Neutrino beams and experiments: long baseline neutrino experiments, and the neutrino experiment at CERN, addressing the question of sterile Neutrinos.

With all these strategic options in mind, I refer to a report from the European Physics Society – Technology and Innovation Group (EPS-TIG) Workshop 2012 in Erice, which included a review of the accelerator magnet R&D activities in preparation of LHC upgrade scenarios.

Starting a review from material to final magnets the workshop covered aspects of the superconduction magnet technology, see Ref (2)

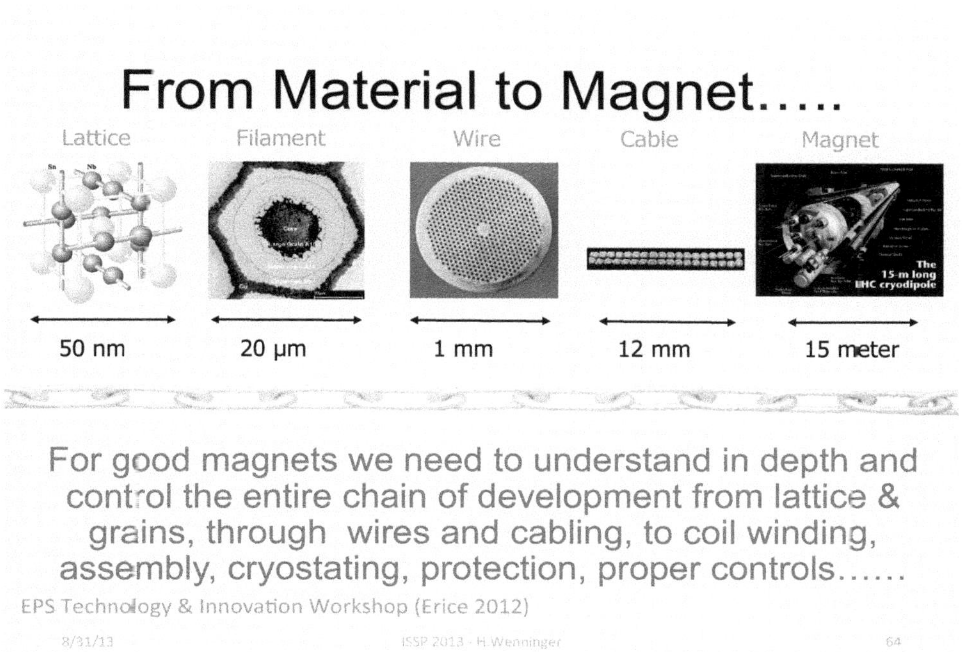

Figure 10: overview of developments of sc-magnets by F. Savary / CERN

Included in the workshop slides of Ref (2) is a review of all actual projects on high power accelerators (SNS, SPIRAL ; CERN LINAC 4, J-PARC IFMIF, FAIR, PSI, MYRRHA)

In addition LHC experiment and computing upgrades are covered which illustrate the status of Technology and Computing in relation to the Theme of the School, namely “Reflections on the Next Step for LHC”.

Ref (1): Progress Report 1988-1989, Development of integrated CMOS circuits and Silicon Pixel Detectors in the CERN-LAA Project, CERN/LAA/ SD 89-11

Ref (2): Accelerating innovation: CERN Courier April 2013
Workshop on technology of particle accelerators and detectors marked the revival of an EPS group focusing on innovative areas and potential spin offs.
http://cerncourier.com/cws/article/cern/52747
https://indico.cern.ch/conferenceDisplay.py?ovw=True&confId=215087

CHAIRMAN: H. WENNINGER

Scientific Secretaries: V. Alba, A. Morozov

DISCUSSION

- Inguglia:

You have shown some future plans for possibilities for LHC to be upgraded. And you have shown that there is a possibility of 80 km tunnel... 100km, even worse. In order to have a 100km accelerator at 2040 means that first of all we need a lot of money and second of all, that is more important, we did not make a new discovery of particles, we did not make progress in acceleration technology. I think that we should take care about new way of acceleration of the particles. In my idea of the future we do not need larger machine necessarily, we need smarter machines. It was announced last week at SLAC that they accelerated electrons to 100 GeV in 1 inch. I think that in the future we can have different kind of machine that are not so huge.

- Wenninger:

You are completely right. We have discussed for a long time a new method of accelerations. And we haven't yet put enough effort to progress. If people at SLAC showed something, you should be sure that you can accelerate electrons as well as positrons. It must be possible to build machines with new techniques of acceleration. I agree it is much more smart to make smaller machines but with as high an energy as needed. On the other hand, what I presented is the view of the European Strategic group. We discuss all the idea in CERN. I present this inErice to see if you agree that building larger and larger machines is the smartest way to progress in physics or if it would be better to invest more into new acceleration techniques. There are a couple projects that work on new technology, but now it is impossible to replace tunnels with magnets and collisions as we have with LHC. I presented the large tunnel project, because it is an extrapolation beyond LHC when we do not have 20 Tesla magnets. What will we do next: electron linear colliders ?. I agree with you that we cannot ask for more money before we know what really is the outcome of present physics running at LHC and what we are looking for. Peter Jenni will present new results beyond the Higgs and then we can go and see what is the next step. You are completely right: new technology in acceleration is welcome.

- Mojaza:

What should we do if LHC does not discover NP?

-Wenninger:

We wait to see what LHC shows before deciding.

- Hartmann:

What would be the cost of building a 80-100km ring around Geneva and how long would it take? Taking these aspects into consideration, will it be realistic to have it built? Maybe most of us will be old enough to be not interesting in it?

- Wenninger:

You are young enough to ask such question. I will be over 100 years old before the new tunnel will be built. I think that the new generation has to pick-up the idea. We should be very careful.

- Hartmann:

How much would you estimate the cost and the time?

- *Wenninger:*

Peter Jenni do you have some numbers in your slides? *(He announced later a cost of order 5 BCHF)*

- *Jenni:*

You can address all these questions to me. I disagree with some statements. I will give comments during my discussion session.

- *Wenninger:*

He disagrees because he was involved in the hot discussion at CERN and he came up with the ideas as they are presented here. But as I said I will be over 100 years old and I need you to help make a decision.

-*Shroeder:*

Could you comment on the feudality of a high energy LHC as ccompared to a muon-collider?

-*Wenninger:*

Muon-collider technologies have been studied for a long time. I don't know at the moment what is the status of such studies. But it is very worth to study.

- *Trofymenko:*

In your presentation you talked about the necessity of the development of new magnetic systems for deflecting of the beams. However, nowadays there exist a new direction of the investigation which is associated with use of small bent crystals for deflecting of beams using the channeling effect. Can you comment upon the prospects of usage of the bent crystals instead of large magnet systems for deflecting the beams.

- *Wenninger:*

There is a group working on this at CERN. And again, probably there is not enough effort put in this study to really go fast. When I saw the people they explained what they get results but they do not have enough support in CERN. On the other hand, will this replace high magnetic field magnets in the existing tunnel of LHC? So, the 20T magnets is the main line of the development of the magnet group. Some small physicist group develop ideas like using of the bent crystal and they hope to get the acceleration in this way as well. Tom Taylor is probably closer to these studies.

- *Taylor:*

For the crystal bending. The problem is the beam intensity. You can indeed bend particle beam using a crystal, but if we want to have a beam of any significant intensity this is a different matter. And, indeed, work has been done on it. But it is not for nearest future, unfortunately.

-*Wenninger:*

Bent crystals work, but the problem is to get an intense beam, focused, to a sufficient degree for acceleration.

«ETTORE MAJORANA» FOUNDATION AND CENTRE FOR SCIENTIFIC CULTURE
TO PAY A PERMANENT TRIBUTE TO GALILEO GALILEI, FOUNDER OF MODERN SCIENCE AND TO ENRICO FERMI, THE "ITALIAN NAVIGATOR", FATHER OF THE WEAK FORCES

INTERNATIONAL SCHOOL OF SUBNUCLEAR PHYSICS
50TH ANNIVERSARY CELEBRATIONS 2011-2012-2013
51st Course: REFLECTIONS ON THE NEXT STEP FOR LHC
ERICE-SICILY: 24 June – 3 July 2013

THE PROBLEM OF (CPT) INVARIANCE IN EXPERIMENTAL PHYSICS AND THE TIME OF FLIGHT (TOF) WORLD RECORD

Antonino Zichichi
INFN and University of Bologna, Italy
CERN, Geneva, Switzerland
Enrico Fermi Centre, Rome, Italy
World Federation of Scientists, Beijing, Geneva, Moscow, New York

Paul A.M. Dirac Lecture Hall – Patrick M.S. Blackett Institute
Erice, 28 June 2013

THE PROBLEM OF (CPT) INVARIANCE IN EXPERIMENTAL PHYSICS AND THE TIME OF FLIGHT (TOF) WORLD RECORD

Antonino Zichichi

INDEX

I – The Problem of Matter and Antimatter. Dirac: The Antihydrogen is not Antimatter

The word “matter” should be used to specify the elements of the Mendeleev Table. If matter was made only of hydrogen, we would not need the nuclei. Hydrogen is the only element whose nucleus is a single particle, the proton, and the atom is obtained by adding to it a fundamental fermion, the electron. Matter consists of nuclei surrounded by electron clouds. The word “matter” should not be used to indicate “particles” (mesons and baryons) or fundamental fermions: quarks and leptons.

In order to have an atom of iron we need the leptonic cloud (made of electrons) and the nucleus (made of baryons glued by mesons).

To have anti-iron we need the fundamental antifermions (positrons), the antiparticles (antimesons and antibaryons) and the property that these antiparticles must interact like the particles in order to have the antinucleus of the iron. We cannot define antimatter the fundamental antifermions and the antiparticles. We need to specify the property which allows all antielements of the Mendeleev Table to exist. This property (in addition to the electromagnetic binding) is the nuclear binding which allows the antinucleons (antiprotons and antineutrons) to form the antinuclei. Thus it seems appropriate to use the word antimatter when we refer to the specific need to build the antielements of the Mendeleev Table. The nucleus of the antihydrogen atom is an antiparticle (the antiproton). The existence of the antielectron guarantees the existence of the antihydrogen atom, thanks to QED. No nuclear forces are needed in order to product the existence of the antihydrogen.

The simplest element where an antinucleus is needed is the antideuterium. This is the reason why the antideuteron ($\bar{p}\,\bar{n}$) is referred to as being the first example of “nuclear antimatter” (“antinuclear matter” would be more appropriate).

And here comes Dirac statement in 1965.

Dirac – 1965*: The existence of antiparticles does not imply the existence of antimatter.*

«Those who say that antihydrogen is antimatter should realize that we are not made of hydrogen and we drink water, not liquid hydrogen». These are Dirac's own words to a group of physicists (Figure 1) gathered around him, who, with a single equation [1], opened new horizons to human knowledge.

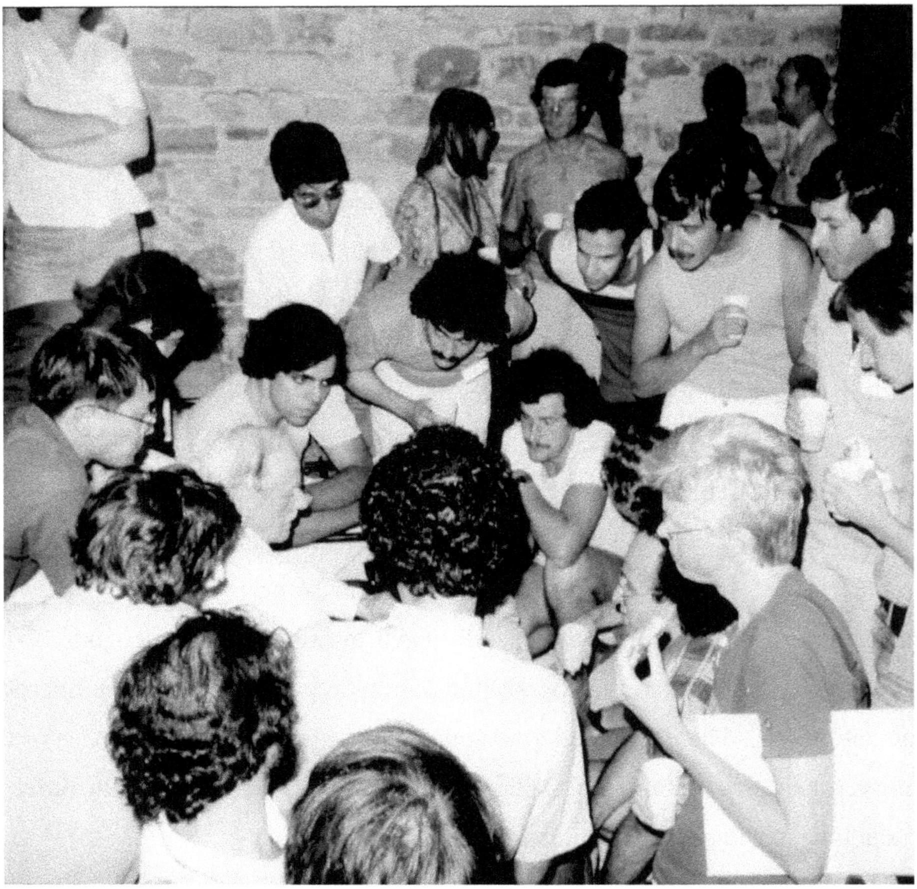

Figure 1: Dirac surrounded by young physicists in Erice, after a lecture when he explained the difference between antiparticles and antimatter. It is on this occasion that he made the statement reported at the beginning of this article.

II – THE IMPORTANCE OF NUCLEAR ANTIMATTER: WHY $\Delta m(d - \bar{d})$ IS TO BE ENCOURAGED

A meson is made with a pair of quark-antiquark confined by QCD coloured forces which act on the $(q\bar{q})$ pair. A baryon is made with three quarks confined by QCD coloured forces. We will see the reasons why it is necessary to distinguish the mesons from the baryons. The best example is the use of K-mesons in order to claim how small is the violation of CPT invariance. The extremely small value of $\Delta m_{K\bar{K}} \leq 4 \times 10^{-10}$ eV/c^2 is used in order to get the ratio

$$\frac{\Delta m_{K\bar{K}}}{m_K} \leq 10^{-18}$$

and consider it as the best check of CPT invariance in the field of mass-measurements.

This conclusion is not as straight-forward as generally believed and could even be wrong. In fact the K^0-meson consists of a quark-antiquark pair $(d\bar{s})$ confined by QCD colour forces which act on the $(d\bar{s})$ pair. Thus the mass of a K-meson receives contributions from two different sources. One is the QCD-Bag, which is a QCD non-perturbative confinement effect and, since QCD is a RQFT, it could be CPT invariant despite the problems created by the convergence of the gauge couplings at the Planck scale.

The other contribution is coming from the intrinsic quark masses, whose origin is totally unknown and certainly not due to QCD or to any other well established CPT invariant mechanism. It is therefore crucial to check if CPT invariance holds true for the source of the intrinsic quark masses. Here $\Delta m_{K\bar{K}}$ fails to be the best check. In fact, if CPT invariance is broken in the process giving rise to the intrinsic quark masses (more precisely m_d and m_s), and if QCD respects CPT invariance, $\Delta m_{K\bar{K}}$ could be exactly zero, despite a strong CPT breaking in the intrinsic quark masses. There is no theory describing CPT breaking effects but $\Delta m_{K\bar{K}} = 0$ is the result of the simplest possible flavour invariant model of CPT breaking. The reason for $\Delta m_{K\bar{K}} = 0$, despite a strong CPT breaking in the intrinsic quark masses, is due to the basic property of a meson: its structure in terms of a (quark-antiquark) pair.

Experiments in baryon-antibaryon physics are free from this danger. In fact, it is thanks to a recent experiment by Gabrielse et al. on the measurement of the charge over mass ratio $\left(\frac{Q}{M}\right)$ for protons and antiprotons that a CPT check on the intrinsic quark masses can be well established. This limit allows to conclude that, if QCD confinement effects are CPT invariant, the CPT breaking in the intrinsic quark masses (more precisely m_u, m_d) cannot be larger than

$$\frac{\Delta m_{q\bar{q}}}{\Delta m_q} \lesssim 3 \times 10^{-11} \; . \tag{1}$$

This limit is seven orders of magnitude larger than the one existing in K-meson physics but it is free from any washing effect, since a baryon has no quark-antiquark structure but only quarks. Thus, if the final result of a Gedanken experiment would be

$$\Delta m_{K\bar{K}} = \text{zero} \; ,$$

the CPT Symmetry in the intrinsic quark masses could be broken at a level as large as (1) and have no effect on the "Gedanken" result.

III – SEVEN DECADES FROM ANTIELECTRON TO ANTIMATTER AND THE UNIFICATION OF ALL GAUGE FORCES

Let me know try to give a telegraphic summary of what did happen from the discovery of the first Symmetry operator **C** (charge conjugation) to the Unification of all Gauge Couplings ($\alpha_1\ \alpha_2\ \alpha_3$).

It is the unification which re-opens the problem of matter-antimatter symmetry.

- **The validity of C invariance from 1927 to 1957.**

After the discovery by Thomson in 1897 of the first example of an elementary particle, the Electron, it took the genius of Dirac to theoretically discover the Antielectron thirty years after Thomson.

1927 → Dirac equation [1]; the existence of the antielectron is, soon after, theoretically predicted. Only a few years were needed, after Dirac's theoretical discovery, to experimentally confirm (Anderson, Blackett and Occhialini [2]) the existence of the Dirac antielectron.

1930-1957 → **Discovery of the C** (charge conjugation) Symmetry operator [H. Weyl and P.A.M. Dirac [3]]; discovery of the **P** (Parity) Symmetry operator [E.P. Wigner, G.C. Wick and A.S. Wightman [4, 5]]; discovery of the **T** (time reversal) Symmetry operator [E.P. Wigner, J. Schwinger and J.S. Bell [6, 7, 8, 9]]; discovery of the **CPT** Symmetry operator from **RQFT** (1955-57) [10].

1927-1957 → Validity of C invariance: e^+ [2]; $\bar{p}$ [11]; $\bar{n}$ [12]; $K_2^0 \to 3\pi$ [13] but see LOY [14].

- **A new era starts: C ≠ ; P ≠ ; CP ≠** (*) .

1956 → Lee & Yang P ≠ ; C ≠ [15].

1957 → Before the experimental discovery of P ≠ & C ≠, Lee, Oehme, Yang (LOY) [14] point out that the existence of the second neutral K-meson, $K_2^0 \to 3\pi$, is proof neither of C invariance nor of CP invariance. Flavour antiflavour mixing does not imply CP invariance.

1957 → C.S. Wu et al. P ≠ ; C ≠ [16]; CP ok [17].

1964 → $K_2^0 \to 2\pi \equiv K_L$: CP ≠ [18].

1947-1967 → QED divergences & Landau poles.

1950-1970 → The crisis of RQFT & the triumph of S-matrix theory (i.e. the negation of RQFT).

1965 → Nuclear antimatter is (experimentally) discovered [19]. See also [20].

1968 → The discovery [21] at SLAC of Scaling (free quarks inside a nucleon at very high q^2) but in violent (pp) collisions no free quarks at the ISR are experimentally found [22] and no free quarks in high energy weak interactions [22]. Theorists consider Scaling as being evidence for RQFT not to be able to describe the Physics of Strong Interactions. The only exception is G. 't Hooft who discovers in 1971 that the β-function has negative sign for non-Abelian theories [23].

1971-1973 → β = – ; 't Hooft, Politzer, Gross & Wilczek. The discovery of **non-Abelian** gauge theories. Asymptotic freedom in the interaction between quarks and gluons [23 bis].

(*) The symbol ≠ stands for “Symmetry Breakdown”.

- **CPT Invariance and the Unification of the three Gauge Forces thanks to Super-Supersymmetry. The problem of Matter-Antimatter Symmetry or Asymmetry.**

1974 → All gauge couplings α_1 α_2 α_3 run with q^2 but they do not converge towards a unique point.

1979 → A.P. & A.Z. point out that the new degree of freedom due to SUSY allows the three couplings α_1 α_2 α_3, **to converge towards a unique point** [24].

1980 → QCD has a "hidden" side: the multitude of final states for each pair of interacting particles: (e^+e^- ; $p\bar{p}$; πp; Kp; νp; pp; etc.). The introduction of the **Effective Energy** [25] allows to discover the Universality properties in the multihadronic final states.

1992 → All gauge couplings converge towards a unique point at the gauge unification energy: $E_{GU} \cong 10^{16}$ GeV with $\alpha_{GU} \cong \frac{1}{24}$ [26, 27].

1994 → The problem of the Gap [28] between E_{GU} & the String Unification Energy: $E_{SU} \cong E_{Planck}$.

1995 → **CPT loses its foundations at the Planck scale (T.D. Lee)** [29].

1995-1999 → **No CPT theorem from M-theory (B. Greene)** [30].

1995-2000 → A Z. points out the need for new experiments to establish if matter-antimatter symmetry or asymmetry are at work.

IV – ANTIMATTER IN COSMOLOGY AND IN SPACE

Modern cosmological theories claim to be able to account for the nine orders of magnitude which exist between the predictions of the simplest baryon symmetric big-bang model and real life. We exist because the ratio of baryons to photons is of the order of 10^{-9}. On the other hand, according to the simplest baryon symmetric big-bang model, when the universe is expanded and matter is "frozen-out", only one residual baryon or antibaryon survives out of 10^{18} and therefore the ratio of baryons to photons should be 10^{-18}.

Before the discovery of CP violation [18], the universe was "theoretically" predicted to be with an equal number of galaxies and antigalaxies, well separated in distinct volumes, to avoid annihilation.

It is the discovery of CP violation [18] which prompted A. Sakharov [31] to imagine a mechanism in order to have a matter-asymmetric universe. For this to happen, it is necessary that a violation of both C and CP takes place, in addition to baryon non-conserving interactions and a departure from thermal equilibrium.

These three Sakharov conditions do not necessarily end up with a matter-dominated universe. This is one option. The other option is that the CP symmetry violation occurs with opposite signs, thus producing well-separated microscopic domains, fully dominated by baryons or antibaryons. The inflation mechanism would later expand these microdomains into astronomical dimensions.

The "theoretical" predictions for the existence of galaxies and antigalaxies cannot be compared in rigour with the "theoretical" predictions based on the validity of the CPT theorem as it was the case for the problem of the existence of matter and antimatter. The crisis of CPT was a fundamental one, during the triumph of the S-matrix theory. The new CPT crisis is again of fundamental origin and it arises from the convergence of the gauge couplings at the Planck scale.

The crisis of the cosmological theories predicting a matter-dominated universe is by far less rigorous, as we have seen with the Sakharov model. Only experiments can establish on firm bases if our universe is matter-antimatter symmetric or not. This is the reason why the experiment, called AMS (Alpha Magnetic Spectrometer) has been designed [32] in order to search for antimatter in space.

If the CERN group [20] had proved in 1965 that antimatter did not exist, it would not have been possible to think of the AMS Project [32]. Half a century ago to establish the existence of antimatter had to do with the CPT theorem. Now, the existence of antimatter in Space has to do with the beginning of the universe and its evolution. Despite the nine orders of magnitude quoted above, the fashionable theoretical predictions are for the non-existence of antimatter in Space. Having in mind the nine orders of magnitude, it is obvious that these predictions, not based on a theorem but on a model, should be experimentally tested as precisely as possible.

V – The Recent Past: The Triumph of S-Matrix and the Difficulties of RQFT

It is interesting to recall the recent past years when Relativistic Quantum Field Theory (RQFT) was encountering big difficulties.

This apparent "**new era**" had no need for the description of the fundamental forces in terms of a RQFT. What about matter-antimatter symmetry?

The breakdown of the Symmetry Operators (C, P, CP) **and the apparent triumph of the S-matrix** were coupled with the lack of evidence for the first example of nuclear antimatter (the antideuteron) found not to be there at the 10^{-7} level: i.e. no antideuteron observed despite the 10^{7} pions, produced in the same interactions.

The discovery of CP violation was announced at the first High Energy Conference, held in the USSR, where I was reporting on the results obtained at CERN with the most intense beam of negative particles implemented in order to study the time-like structure of the proton [33, 34], to search **for a new Heavy Lepton (HL)** [35] in addition to the **muon**, and to study the existence of **nuclear antimatter at a level ten times better than previous searches**. The purpose of this negative beam was to **compete** with the physics of "**Bubble-chamber**" which dominated nearly all the physics research activities of the time.

Question. If C invariance is broken, what is expected to happen to the existence of antimatter? The answer — as said before — is not that antimatter should not be there, but that its existence is granted by the validity of CPT. And what about the well established equality between τ_{μ^+} and τ_{μ^-} ? The answer is the same. The equality

$$\tau_{\mu^+} = \tau_{\mu^-}$$

is granted by the validity of CPT invariance.

Another question. Since CP invariance does not hold in K-physics, what happens to matter-antimatter symmetry? Is it broken? The answer is no, since — as mentioned earlier — matter-antimatter symmetry, despite the breaking of CP invariance, is granted by the validity of CPT. In other words it does not matter if the Symmetry Operators, C, P, and CP are broken. The existence of antimatter depends on the validity of CPT. But the validity of CPT was (and is) deeply rooted in RQFT. Here comes the key point: the theoretical description of the strong interactions. No one knew how to describe the nuclear forces in terms of a mathematical formalism of the RQFT type. We have said that those were years of triumph for the negation of RQFT: i.e. the S-matrix theory. Since this was the status of the strong interactions, the existence of the antideuteron could not be "theoretically" predicted. In fact, the strong interactions seemed to be dominated by the S-matrix theory and the Symmetry Operators C, P and CP were all experimentally found to be broken with the T Symmetry Operator imagined to have a privileged role. All this conspired towards a breakdown of CPT invariance. Why should we believe in CPT.

Chapter III illustrates (pages 467–469) the basic steps, during seven decades, from the antielectron and the properties of an antiparticle to the properties of antimatter and the unification of all gauge forces (QED, QFD, QCD). It is the unification of all forces at $E_{GU} \cong 10^{16}$ GeV, and eventually at the Planck scale, which creates problems to the validity of CPT [29].

Let me elaborate a bit on these developments, starting **with the discovery of the first example of nuclear antimatter** (or, better, "**antinuclear matter**": see later). For this to happen, two basic technical developments were needed. One was the most intensive beam of negative particles we built at CERN. The other was the most precise **Time-Of-Flight** (**TOF**) device we had implemented [20] to reach a time accuracy of ± 100 psec (psec ≡ 10^{-12} sec). These technical developments allowed us to improve, by an order of magnitude, the search for nuclear antimatter. And this is how — **contrary to the general theoretical trend of the time and to the experimental failure to observe the first example of nuclear antimatter** — we were able to prove that the antideuteron was produced **at the level of 10^{-8}**, thus providing a firm basis for the validity of the **CPT symmetry** in the theory of the nuclear forces, despite the fact **that no one knew how to theoretically describe these forces**.

This was in 1965. The following years were even more dramatic for the **validity** of a mathematical description of the strong interactions in terms of a **RQFT**. Not only the **S-matrix theory was dominating** but an experimental discovery at SLAC came in. It was called "scaling" [21]: its physical implication was (and is) that the proton had to have elementary constituents (called partons: quarks and gluons) and that these, at high energy, had to be **free inside a proton** (this property being called asymptotic freedom). On the other hand, it was experimentally found at CERN [22] that, in violent (pp) collisions, there were

no free quarks produced. The general trend in the theoretical interpretation of these experimental results was that **RQFT was not able to describe the physics of the strong interactions**. The proof that the first example of nuclear antimatter exists **was — during those years — the only source for confidence in a RQFT description of the strong interactions**. There were **exceptions** in the theoretical front. In the volume "The Discovery of Nuclear Antimatter" [20] Luciano Maiani says: «*The discovery of the antideuteron gave more confidence to the search of a field theoretical basis for the strong interactions, that today seems so obvious to us*». The **most** interesting of these searches was the one by **G. 't Hooft** who discovered in **1971** that the β-function has negative sign for non-Abelian RQFT. By now — after many decades — our understanding of the **nuclear forces is not in terms** of a fundamental RQFT **but** in terms of an "**effective**" **theory** which should be derived from the fundamental one - QCD: i.e. the non-Abelian force acting between quarks and gluons. This is a **RQFT**; however, the discovery of the renormalization group equations (RGEs) generates the running with energy of the three couplings ($\alpha_1\ \alpha_2\ \alpha_3$) of the gauge forces (QCD, QFD, QED) towards a unique point at $E_{GU} \cong 10^{16}$ GeV where $\alpha_{GU} \cong \frac{1}{24}$ [27, 26, 28, 23] and possibly at the Planck scale. **Here the CPT theorem loses its foundation** [29], therefore all gauge forces, despite being examples of Relativistic Quantum Field Theories (RQFTs), have no fundamental reasons to obey CPT invariance, if their common origin is **at the Planck scale**. How the loss of such a theoretical fundamental reason affects the low energy domain, where the colour-neutral-QCD-binding nuclear forces act, is not known. **Only high precision experiments on the mass difference Δm**, between **nuclei and antinuclei**, the best example being, **deuteron-antideuteron**

$$\Delta m_{D\bar{D}}$$

can determine these consequences. At present all we know is very simple: **antinuclear matter** exists despite the theoretical foundations, whose existence it was based upon, **have vanished**.

VI – The Three Types of Masses

To sum up, our understanding of the physical quantity called mass can be described in terms of three sources:

i) the "**intrinsic**" mass of the fundamental fermions (quarks and leptons); the origin of these "intrinsic" masses is not known but is certainly not due to Quantum ChromoDynamics (QCD).

ii) the "**confinement**" masses : m^{Bag} , due to QCD confining colour forces; these effects produce positive masses.

iii) the "**binding**" masses which can be either of **electromagnetic** or of **nuclear** origin. These effects produce negative masses.

All these sources of masses have different origins. Two of them, the QCD "confinement" masses and the electromagnetic "binding" masses, are associated with well known forces, QCD and QED. The other two ("intrinsic" and "nuclear binding") have some peculiar features. We briefly review all of them.

"Intrinsic mass". It has still not been established how the masses of the fundamental fermions (quarks and leptons) are generated and what is the energy scale where this mechanism becomes operative. It could be the so-called "Higgs" mechanism and it could also be a new force of nature operating at the Planck scale. If it is at the Planck scale, here the CPT theorem loses its foundations [29] and it cannot be excluded that a CPT breaking occurs. Since no one knows either the energy scale where the masses of the fundamental fermions are generated, or if quark and lepton masses have the same origin, it cannot be excluded that at the energy scale where these masses are generated — for example the SUSY breaking scale — the CPT invariance breaks down. It is therefore worthwhile to investigate if the "intrinsic" masses obey CPT invariance.

"Confinement mass". We will assume for simplicity that the "confinement" masses respect CPT invariance. Since QCD is a Relativistic-Local-Quantum-Field-Theory (RQFT), it could appear reasonable to expect that QCD respects CPT invariance [36]. But the convergence of the gauge forces at the Planck scale creates problems with the foundations of CPT [29]. On the other hand, the "confinement" mass is a QCD non-perturbative phenomenon and no one knows how to deal with these divergent effects.

The "binding" masses have two origins.

Electromagnetic "binding" masses. All atoms exist because of the electromagnetic binding effects, which produce negative masses. Their origin is in Quantum ElectroDynamics (QED). If QED were a RQFT without the problems created by the convergence of the gauge forces at the Planck scale, here again it would be reasonable to expect no CPT breaking effects.

Nuclear "binding" masses. All nuclei exist because of the nuclear "binding" forces. These are residual QCD effects, acting between neutral colour quark-triplets (qqq), the nucleons, and neutral colour quark-antiquark pairs ($q\bar{q}$), the mesons. Nuclear physics is based on the interaction between nucleons mediated by mesons. Before the discovery of QCD [25], nuclear physics was thought to be an example of a RQFT; by now we know that the origin of nuclear forces is the non-Abelian theory describing the interaction between quarks and gluons, Quantum Chromodynamics, QCD; nevertheless no one has been able so far to derive the nuclear forces from QCD. What is clear is that what we call nuclear processes represent the transition from the QCD world (coloured quark and gluons) to the QCD colour neutral world of baryons and mesons. Since no one is able to describe this transition, it is probably wise to perform some experimental checks on the validity of CPT invariance in the nuclear "binding" masses, which are negative.

The conclusion is that, out of the four types of masses, two of them appear to be associated with problems still to be understood. These two are: i) the exact origin of the intrinsic quark masses and ii) the nuclear binding (negative) masses which are associated with the transition from the QCD colour-full world of quarks and gluons to the colour neutral world of mesons and baryons. This is why it is important to check CPT invariance in the field of "intrinsic" and of nuclear "binding" masses.

Let us start focusing our attention on the "intrinsic" quark masses and on their CPT invariance. The first point to clarify is the basic difference between a meson and a baryon.

VII – From the Fundamental Fermions and Antifermions to Matter and Antimatter

The synthesis of all problems from the fundamental fermions and antifermions to matter and antimatter is in Table 1.

If the Hydrogen atom were the only element in the Mendeleev Table, then the only form of antimatter would be the antihydrogen atom. There are, by now, 118 elements in the Mendeleev Table and the first element, Hydrogen, is the only example of atom made of two elementary particles: the proton and the electron. In order for (what we call) matter to exist, it is necessary that protons and neutrons can attract each other to form nuclei. For the existence of antimatter 118 elements of the new Mendeleev Table for the antielements must exist. For this to happen the fundamental forces have to be exactly like they are for the elements of the Mendeleev Table.

FROM THE FUNDAMENTAL FERMIONS AND ANTIFERMIONS TO MATTER AND ANTIMATTER

HOW MATTER IS BUILT		MASSES	HOW ANTIMATTER IS BUILT	
FUNDAMENTAL FERMIONS	NEEDED FOR ORDINARY MATTER		NEEDED FOR ORDINARY ANTIMATTER	FUNDAMENTAL ANTIFERMIONS
$\begin{pmatrix}\nu_e\\e^-\end{pmatrix}\begin{pmatrix}\nu_\mu\\\mu^-\end{pmatrix}\begin{pmatrix}\nu_{HL}\\HL^-\end{pmatrix}$ **Leptons** ($\uparrow\tau^-$)	$\Rightarrow e^-$	*Intrinsic*	$e^+ \Leftarrow$	$\begin{pmatrix}\bar\nu_e\\e^+\end{pmatrix}\begin{pmatrix}\bar\nu_\mu\\\mu^+\end{pmatrix}\begin{pmatrix}\bar\nu_{HL}\\HL^+\end{pmatrix}$ **Antileptons** ($\uparrow\tau^+$)
$\begin{pmatrix}u\\d\end{pmatrix}\begin{pmatrix}c\\s\end{pmatrix}\begin{pmatrix}t\\b\end{pmatrix}$ **Quarks**	$\Rightarrow\begin{pmatrix}u\\d\end{pmatrix}$		$\begin{pmatrix}\bar u\\\bar d\end{pmatrix} \Leftarrow$	$\begin{pmatrix}\bar u\\\bar d\end{pmatrix}\begin{pmatrix}\bar c\\\bar s\end{pmatrix}\begin{pmatrix}\bar t\\\bar b\end{pmatrix}$ **Antiquarks**
PARTICLES Mesons $(q\bar q)$ Baryons (qqq)	$\Rightarrow \pi_0^\pm$ $\Rightarrow\begin{pmatrix}p\\n\end{pmatrix}$	*Intrinsic* & *Confinement*	$\pi_0^\pm \Leftarrow$ $\begin{pmatrix}\bar p\\\bar n\end{pmatrix} \Leftarrow$	ANTIPARTICLES Antimesons $(\bar q q)$ Antibaryons $(\bar q\bar q\bar q)$
NUCLEI $(pn) \equiv D$ $(pnn) \equiv {}^3H$ $(ppn) \equiv {}^3H_e$ $(ppnn) \equiv H_e$	All **Nuclei** up to the element with 118 protons and 175 neutrons contribute to fill the Mendeleev table	*Intrinsic* & *Confinement* & *Nuclear Binding*	All **Antinuclei** up to the element with 118 antiprotons and 175 antineutrons contribute to fill the Mendeleev table	ANTINUCLEI $(\bar p\bar n) \equiv \bar D$ $(\bar p\bar n\bar n) \equiv {}^3\bar H$ $(\bar p\bar p\bar n) \equiv {}^3\bar H_e$ $(\bar p\bar p\bar n\bar n) \equiv {}^4\bar H_e$
ATOMS The simplest Atom	Hydrogen $\equiv (p\,e^-)$	*Intrinsic* & *Confinement* & *EM Binding*	Antihydrogen $\equiv (p^- e^+)$	ANTIATOMS The simplest AntiAtom
ALL OTHER ATOMS	Nuclei & Electrons Deuterium: $[(pn)\,e^-]$ Helium: $[(ppnn)\,e^-e^-]$ etc.	*Intrinsic* & *Confinement* & *EM Binding* & ***Nuclear Binding***	Antinuclei & Positrons Antideuterium: $[(\bar p\bar n)\,e^+]$ Antihelium: $[(\bar p\bar p\bar n\bar n)\,e^+e^+]$ etc. ...	ALL OTHER ANTIATOMS

Table 1

VIII – THE TOF WORLD RECORD

During many decades, high precision technology in measuring the Time Of Flight (TOF) of elementary particles has played a vital role in Subnuclear Physics [23]. Let me quote few examples:

i) the discovery of nuclear antimatter [19, 20, 29];

ii) the discovery of the time-like structure of the proton [33, 34];

iii) the search for the third lepton HL at CERN [34, 35] and at Frascati [37] many years before the Kobayashi and Maskawa [38] proposal for the existence of three families, in order to explain the origin of CP violation. Also many years before the SLAC discovery [39] which did not use the HL symbol but the Greek letter τ despite it's use during many years for the (θ–τ) puzzle. This is why the Heavy Lepton (HL) is now called τ ;

iv) the high precision search for fractionally charged particles in high energy strong and weak interactions [22];

v) the discovery of the Effective Energy in QCD [25].

The impressive series of technological developments has been going on during the last decade.

I will limit myself to give the results (Figure 2) on the time resolution obtained using two MRPCs (Multigap Resistive Plate Chamber) [40, 41] in a 5 GeV/c π–beam at CERN.

The data reported in Figure 2 give for the time resolution 16ps: a world record. The efficiency is also reported and is nearly 100% all along the range of the voltage applied. For details see the paper of my group presented at this School [42].

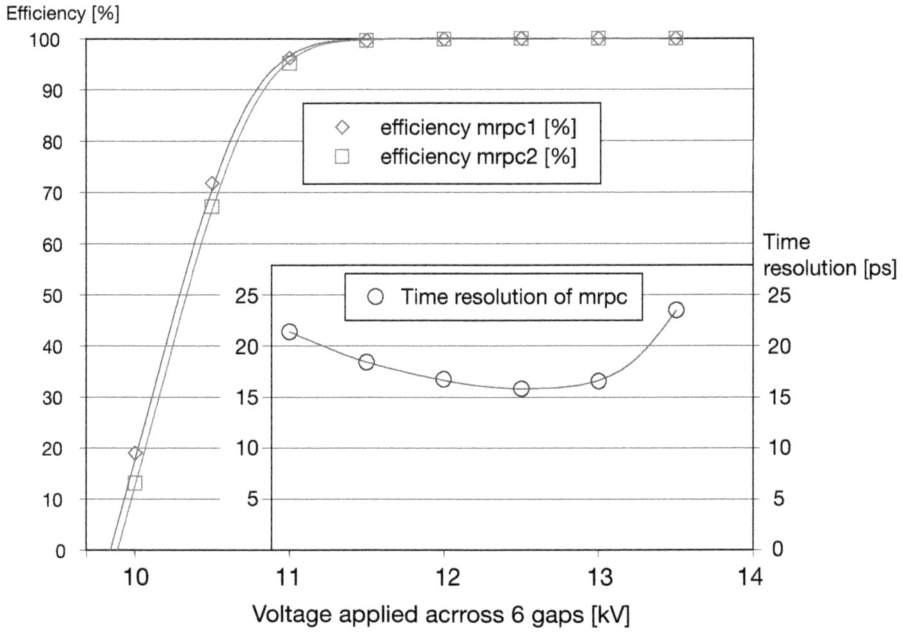

Figure 2: Efficiency and time resolution of two 24 gap MRPC tested in a 5 GeV/c pion beam.

REFERENCES

[1] P.A.M. Dirac, "*The Quantum Theory of the Electron*", *Proc. Roy. Soc.* (London) A117, 610 (1928); "*The Quantum Theory of the Electron, Part II*", *Proc. Roy. Soc.* (London) A118, 351 (1928).

[2] *The Positive Electron*
C.D. Anderson, *Phys. Rev.* 43, 491 (1933);
Some Photographs of the Tracks of Penetrating Radiation
P.M.S. Blackett and G.P.S. Occhialini, *Proc. Roy. Soc.* A139, 699 (1933).

[3] H. Weyl, "*Gruppentheorie und Quantenmechanik*", 2nd ed., 234 (1931).

[4] E.P. Wigner, "*Unitary Representations of the Inhomogeneous Lorentz Group*", *Ann. Math.*, 40, 149 (1939).

[5] G.C. Wick, E.P. Wigner, and A.S. Wightman, "*Intrinsic Parity of Elementary Particles*", *Phys. Rev.* 88, 101 (1952).

[6] E.P. Wigner, "*Über die Operation der Zeitumkehr in der Quanten-mechanik*", *Gött. Nach.* 546-559 (1931). Here for the first time an anti-unitary symmetry appears.

[7] E.P. Wigner, *Ann. Math.* 40, 149 (1939).

[8] J. Schwinger. *Phys. Rev.* 82, 914 (1951).

[9] J.S. Bell, "*Time Reversal in Field Theory*", *Proc. Roy. Soc.* (London) A231, 479-495 (1955).

[10] To the best of my knowledge, the CPT theorem was first proved by W. Pauli in his article "*Exclusion Principle, Lorentz Group and Reflection of Space-Time and Charge*", in "Niels Bohr and the Development of Physics" [Pergamon Press, London, page 30 (1955)], which in turn is an extension of the work of J. Schwinger [*Phys. Rev.* 82, 914 (1951); "*The Theory of Quantized Fields. II.*", *Phys. Rev.* 91, 713 (1953); "*The Theory of Quantized Fields. III.*", *Phys. Rev.* 91, 728 (1953); "*The Theory of Quantized Fields. VI.*", *Phys. Rev.* 94, 1362 (1954)] and G. Lüders, "*On the Equivalence of Invariance under Time Reversal and under Particle-Anti-particle Conjugation for Relativistic Field Theories*" [*Dansk. Mat. Fys. Medd.* 28, 5 (1954)], which referred to an unpublished remark by B. Zumino. The final contribution to the CPT theorem was given by R. Jost, in "*Eine Bemerkung zum CPT Theorem*" [*Helv. Phys. Acta* 30, 409 (1957)], who showed that a weaker condition, called "weak local commutativity" was sufficient for the validity of the CPT theorem.

[11] *Observation of Antiprotons*
O. Chamberlain, E. Segrè, C. Wiegand, and T. Ypsilantis, *Physical Review* 100, 947 (1955).

[12] *Anti-Neutrons Produced from Anti-Protons in Charge Exchange Collisions*
B. Cork, G.R. Lambertson, O. Piccioni, W.A. Wenzel, *Physical Review* 104, 1193 (1957).

[13] *Observation of Long-Lived Neutral V Particles*
K. Lande, E.T. Booth, J. Impeduglia, L.M. Lederman, and W. Chinowski, *Physical Review* 103, 1901 (1956).

[14] *Remarks on Possible Noninvariance under Time Reversal and Charge Conjugation*
T.D. Lee, R. Oehme, and C.N. Yang, *Physical Review* 106, 340 (1957).

[15] *Question of Parity Conservation in Weak Interactions*
T.D. Lee and C.N. Yang, *Phys. Rev.* 104, 254 (1956).

[16] *Experimental Test of Parity Conservation in Beta Decay*
C.S. Wu, E. Ambler, R.W. Hayward, D.D. Hoppes, *Phys. Rev.* 105, 1413 (1957);
Observation of the Failure of Conservation of Parity and Charge Conjugation in Meson Decays: The Magnetic Moment of the Free Muon
R. Garwin, L. Lederman, and M. Weinrich, *Phys. Rev.* 105, 1415 (1957);
Nuclear Emulsion Evidence for Parity Non-Conservation in the Decay Chain $\pi^+\mu^+e^+$
J.J. Friedman and V.L. Telegdi, *Phys. Rev.* 105, 1681 (1957).

[17] *On the Conservation Laws for Weak Interactions*
L.D. Landau, *Zh. Éksp. Teor. Fiz.* 32, 405 (1957).

[18] *Evidence for the 2π Decay of the K_2^0 Meson*
J. Christenson, J.W. Cronin, V.L. Fitch, and R. Turlay, *Physical Review Letters* 113, 138 (1964).

[19] *Experimental Observation of Antideuteron Production*
T. Massam, Th. Muller, B. Righini, M. Schneegans, and A. Zichichi, *Nuovo Cimento* 39, 10 (1965).

[20] *The Discovery of Nuclear Antimatter*
L. Maiani and R.A. Ricci (eds), Conference Proceedings 53, Italian Physical Society, Bologna, Italy (1995); see also A. Zichichi in "*Subnuclear Physics - The first fifty years*", O. Barnabei, P. Pupillo and F. Roversi Monaco (eds), a joint publication by University and Academy of Sciences of Bologna, Italy (1998); World Scientific Series in 20th Century Physics, Vol. 24 (2000).

[21] The first report on "scaling" was presented by J.I. Friedman at the 14th International Conference on *High Energy Physics* in Vienna, 28 August-5 September 1968. The report was presented as paper n. 563 but not published in the Conference Proceedings. It was published as a SLAC preprint. The SLAC data on scaling were included in the Panofsky general report to the Conference where he says «... the apparent success of the parametrization of the cross-sections in the variable ν / q^2 in addition to the large cross-section itself is at least indicative that point-like interactions are becoming involved». "*Low q^2 Electrodynamics, Elastic and Inelastic Electron (and Muon) Scattering*", W.K.H. Panofsky in Proceedings of 14th International Conference on *High Energy Physics* in Vienna 1968, J. Prentki and J. Steinberger (eds), page 23, published by CERN (1968). The following physicists participated in the inelastic electron scattering experiments: W.B. Atwood, E. Bloom, A. Bodek, M. Breidenbach, G. Buschhorn, R. Cottrell, D. Coward, H. DeStaebler, R. Ditzler, J. Drees, J. Elias, G. Hartmann, C. Jordan, M. Mestayer, G. Miller, L. Mo, H. Piel, J. Poucher, C. Prescott, M. Riordan, L. Rochester, D. Sherden, M. Sogard, S. Stein, D. Trines, and R. Verdier. For additional acknowledgements see J.I. Friedman, H.W. Kendall and R.E. Taylor, "*Deep Inelastic Scattering: Acknowledgements*", *Les Prix Nobel 1990*, (Almqvist and Wiksell, Stockholm/Uppsala 1991), also *Rev. Mod. Phys.* 63, 629 (1991). For a detailed reconstruction of the events see J.I. Friedman "*Deep Inelastic Scattering Evidence for the Reality of Quarks*" in "*History of Original Ideas and Basic Discoveries in Particle Physics*", H.B. Newman and T. Ypsilantis (eds), Plenum Press, New York and London, 725 (1994).

[22] *Quark Search at the ISR*
T. Massam and A. Zichichi, *CERN preprint*, June 1968;
Search for Fractionally Charged Particles Produced in Proton-Proton Collisions at the Highest ISR Energy
M. Basile, G. Cara Romeo, L. Cifarelli, P. Giusti, T. Massam, F. Palmonari, G. Valenti and A. Zichichi, *Nuovo Cimento* 40A, 41 (1977); and
A Search for quarks in the CERN SPS Neutrino Beam
M. Basile, G. Cara Romeo, L. Cifarelli, A. Contin, G. D'Alì, P. Giusti, T. Massam, F. Palmonari, G. Sartorelli, G. Valenti and A. Zichichi, *Nuovo Cimento* 45A, 281 (1978).

[23] A. Zichichi in "*Subnuclear Physics - The first fifty years*", O. Barnabei, P. Pupillo and F. Roversi Monaco (eds), a joint publication by University and Academy of Sciences of Bologna, Italy (1998); World Scientific Series in 20th Century Physics, Vol. 24 (2000).

[23bis] A. Zichichi in "*Subnuclear Physics - The first fifty years*", O. Barnabei, P. Pupillo and F. Roversi Monaco (eds), a joint publication by University and Academy of Sciences of Bologna, Italy (1998); World Scientific Series in 20th Century Physics, page 27, Vol. 24 (2000).

[24] *New Developments in Elementary Particle Physics*
A. Zichichi *Rivista del Nuovo Cimento* 2, n. 14, 1 (1979). The statement on page 2 of this paper, «*Unification of all forces needs first a Supersymmetry. This can be broken later, thus generating the sequence of the various forces of nature as we observe them*», was based on a work by A. Petermann and A. Zichichi in which the renormalization group running of the couplings using supersymmetry was studied with the result that the convergence of the three couplings improved. This work was not published, but perhaps known to a few. The statement quoted is the first instance in which it was pointed out that supersymmetry might play an important role in the convergence of the gauge couplings. In fact, the convergence of three straight lines ($\alpha_1^{-1}\alpha_2^{-1}\alpha_3^{-1}$) with a change in slope is guaranteed by the Euclidean geometry, as long as the point where the slope changes is tuned appropriately. What is incorrect about the convergence of the couplings is that, with the initial conditions given by the LEP results, the change in slope needs to be at $M_{SUSY} \sim 1$ TeV as claimed by some authors not aware in 1991 of what was known in 1979 to A. Petermann and A. Zichichi.

[25] V.N. Gribov, G. 't Hooft, G. Veneziano and V.F. Weisskopf "*The Creation of Quantum ChromoDynamics and the Effective Energy*", L.N. Lipatov (ed), a joint publication by the University and the Academy of Sciences of Bologna, Italy (1998); World Scientific Series in 20th Century Physics, Vol. 25 (2000).

[26] *The Effective Experimental Constraints on M_{SUSY} and M_{GUT}*
F. Anselmo, L. Cifarelli, A. Petermann and A. Zichichi, *Nuovo Cimento* 104A, 1817 (1991).

[27] *The Simultaneous Evolution of Masses and Couplings: Consequences on Supersymmetry Spectra and Thresholds*
F. Anselmo, L. Cifarelli, A. Petermann and A. Zichichi, *Nuovo Cimento* 105A, 1179 (1992).

[28] *A Study of the Various Approaches to M_{GUT} and α_{GUT}*
F. Anselmo, L. Cifarelli and A. Zichichi, *Nuovo Cimento* 105A, 1335 (1992).

[29] *Are Matter and Antimatter Symmetric?*
T.D. Lee, in Proceedings of the "*Symposium to celebrate the 30th anniversary of the Discovery of Nuclear Antimatter*", L. Maiani and R.A. Ricci (eds), Conference Proceedings 53, page 1, Italian Physical Society, Bologna, Italy (1995).

[30] *String Theory: The Basic Ideas*
B. Greene, Erice Lectures - Discussion 1999 in "*Basics and Highlights in Fundamental Physics*", A. Zichichi (ed), World Scientific (2001).

[31] A. Sakharov, *JEPT Lett.*, 5, 24 (1967).

[32] *The Discovery of Nuclear Antimatter and the Origin of the AMS Experiment*
S.C.C. Ting, in Proceedings of the *Symposium to celebrate the 30th anniversary of the Discovery of Nuclear Antimatter*, L. Maiani and R.A. Ricci (eds), Conference Proceedings 53, 21, Italian Physical Society, Bologna, Italy (1995).

[33] *Search for the Time-Like Structure of the Proton*
M. Conversi, T. Massam, Th. Muller and A. Zichichi, *Phys. Lett.* 5, 195 (1963).

[34] *The Leptonic Annihilation Modes of the Proton-Antiproton System at 6.8 (GeV/c)^2 Timelike Four-Momentum Transfer*
M. Conversi, T. Massam, Th. Muller and A. Zichichi, *Nuovo Cimento* 40, 690 (1965).

[35] C.S. Wu, T.D. Lee, N. Cabibbo, V.F. Weisskopf, S.C.C. Ting, C. Villi, M. Conversi, A. Petermann, B.H. Wiik and G. Wolf
The Origin of the Third Family, O. Barnabei, L. Maiani, R.A. Ricci and F. Roversi Monaco (eds), Academy of Sciences, Bologna University, INFN, SIF, Rome (1997); and World Scientific Series in 20th Century Physics, Vol. 20 (1998).

[36] No matter what the physics processes are, if their mathematical formulation can be expressed in terms of a relativistic, local, quantized, field theory (RQFT), these processes have to obey CPT invariance. Thus to violate this fundamental invariance of nature corresponds to break the basic conceptual structure of a RQFT. It took many decades to discover CPT invariance, as discussed in Reference 10. And many other decades were needed to discover the convergence of the gauge forces at the Planck scale.

[37] *A Proposal to Search for Leptonic Quarks and Heavy Leptons produced by ADONE*
M. Bernardini, D. Bollini, E. Fiorentino, F. Mainardi, T. Massam, L. Monari, F. Palmonari and A. Zichichi, INFN/AE-67/3, 20 March 1967;

First Search for Sequential Heavy Leptons at ADONE
A. Zichichi, CERN-PPE/93-58 and CERN/LAA/93-18, 2 April 1993. Presented at the Symposium on "*The τ particle*", in honour of Martin Perl's 65th birthday, SLAC, Stanford, CA, USA, 24 July 1992.

[38] *CP-Violation in the Renormalizable Theory of Weak Interaction*
M. Kobayashi and T. Maskawa, *Prog. Theor. Phys.* 49, 652 (1973).

[39] *Evidence for Anomalous lepton production in e^+-e^- Annihilation*
M.L. Perl et al., *Physical Review Letters* 35, 1489 (1975).

[40] *A New Type of Resistive Plate Chamber: The Multigap RPC*
E. Cerron Zeballos, I. Crotty, D. Hatzifotiadou, J. Lamas Valverde, S. Neupane, M.C.S. Williams and A. Zichichi, *Nuclear Instruments and Methods* A374, 132 (1996).

[41] *A 20ps Timing Device – A Multigap Resistive Plate Chamber with 24 Gas Gaps*
S. An, Y.K. Jo, J.S. Kim, M.M. Kim, D. Hatzifotiadou, M.C.S. Williams, A. Zichichi and R. Zuyeuski, *Nuclear Instruments and Methods* A594, 39-43 (2008).

[42] K. Doroud, M.C.S. Williams and A. Zichichi, in the present volume at page 487.

CHAIRMAN: A. ZICHICHI

Scientific Secretaries: G. Karananas, A.Tokareva

DISCUSSION

- *Leontsinis:*

What is the spatial resolution of MRPCs? Do you observe any degradation in time and spatial resolution due to ageing effects?

- *Zichichi:*

The Time resolution is 16ps: a world record; we do not observe any degradation. For more details let me give the floor to Dr.Williams.

- *Williams:*

We can go down to hundreds of microns spatial resolution. For the timing resolution the results after the irradiation were better than before. Still having the best timing resolution compared to any other detector. No degradation effects.

- *Csanad:*

What is the momentum range in which you can differentiate pions from kaons and kaons from protons with the MRPCs? You said the resolution is 16ps. What kind of PID makes that possible and how does that relate to HMPID capabilities of ALICE?

- *Zichichi:*

I have shown the spectra in my lecture. It is the highest resolution in all momentum spectra.

- *Williams:*

It is an easy calculation based on the 16ps resolution time. However, other effects (path length etc.) will play a dominant role at this level. It will increase PID MOM range.

- *Antonenko:*

Could you please explain with more details why CPT invariance should or might be broken at the Planck Scale?

- *Zichichi:*

It is broken because the Planck scale corresponds to a black hole. Three gauge couplings α_1 $\alpha_2\alpha_3$ meet at about 10^{16} GeV energy which is more than two orders of magnitude below the Planck scale. Nevertheless, taking into account all uncertainties, the gap can disappear. The point is that at the Planck scale the CPT theorem does not work due to the black hole condition. The effect of this CPT breaking at energies below the Planck scale is model dependent. In string theory for example, there is no CPT theorem.

- *'t Hooft:*

I don't think all string theorists agree with you. They say they believe in CPT, especially for black holes. Black holes can be made CPT invariant. We understand very little about Nature at the Planck scale. If you use as your starting points unitarity, causality and Lorentz invariance, CPT is conserved, even at the Planck scale. Obviously, if one cannot use Poincaré invariance to prove CPT, then everything is open again. People believe there is a cosmological constant, which would replace the back ground by anti-de Sitter space, and that does not have Poincaré invariance at infinity. That might break CPT, however I'm not convinced at all about that. We don't know for sure, we don't have a good theory.

- *Zichichi:*

The basic point is to know what experiments should we do to prove this.

- *Antonenko:*

What about causality? As far as I know, it is connected with CPT theorem.

- *Zichichi:*

I strongly believe in causality, it is a fundamental principle. Violation of CPT does not imply violation of causality.

- *Tannenbaum:*

It seems to me that with the two large proton-antiproton colliders storage rings it should be possible to measure the $p^{+}-p^{-}$mass difference very precisely. Do such measurements exist?

- *Zichichi:*

What is needed is a high precision determination of the $(d, \bar{d})$ mass difference. Here we have, in addition to the intrinsic quark and antiquark masses the gluon and antigluon. I have discussed all these points in my lecture.

- *Minkovski:*

To prove or disprove CPT in the realm including quantum gravity the very formulation is not clear to me. Why did Antonino Zichichi present the breaking of CPT in the above realm as given or at least as very probable?

- *'t Hooft:*

I always thought that the arguments from which CPT is derived (Poincaré invariance and all that) do not care what underlying theory you have. If you have any unitary, causal S-matrix with these symmetries you have CPT. And string theory does generate an S-matrix. I don't see why this argument doesn't apply to string theory unless there is a cosmological constant, which ruins Poincaré invariance at infinity. Then we have to rethink what this proof really means. In flat space we have particles still colliding in full agreement with string theory so they should obey CPT invariance.

- *Minkovski:*

I just have a comment. Classical gravity is known to be causal. It is a proven fact. But quantum gravity...

- *Zichichi:*

A theory which violates causality should be abandoned.

- *Inguglia:*

We had in the past C as a symmetry of Nature, until it was violated. The same for CP. I think we should not exclude the possibility that CPT will be violated before Planck scales are reached.

- *Zichichi:*

I'm sorry that Bruno Zumino,one of the fathers of CPT, is not here. Many years ago, when we were starting to be involved in new technology for high precision TOF, we had on this topic a very interesting discussion. I am pleased to realize that the 16ps TOF record generates now a discussion on the validity of CPT in black hole physics.

Advances in Fast Timing up to 16 ps

K. Doroud [a,c], M.C.S. Williams [a,b] and A. Zichichi [a,b,c]

[a] *CERN Geneva, Switzerland*

[b] *INFN and Dipartimento di Fisica e Astronomia, Università di Bologna, Bologna, Italy*

[c] *Museo Storico della Fisica e Centro Studi e Ricerche E. Fermi, Roma, Italy*

Abstract

During the last decade , 'fast timing' has played a vital role in the field of particle physics with an impressive series of technological developments in medical physics, especially with the TOF-PET camera using differential readout.

In this paper we examine the key issues that have led to the exceptional time resolution of 16 ps and review what we may expect to achieve in the near future.

Key words: TOF, PET

1 Introduction

If we look back in the past, we can find experiments such as the search for Anti-deuteron in 1965[1] that used scintillators readout by phototubes to select with precise timing heavy particles: and thus identify the production of negatively-charged particles with a mass of 1867 MeV/c^2 that provided the proof for the existence of anti-deuteron[2]. Forty years later one can find more modern experiments, such as Belle[3] at KEK and AMS[4] at the space station; both these experiments use scintillators readout by phototubes. During these forty years, the number of channels has increased from three [1] in 1965 to 128 (Belle[3]) in 2002 and 68 (AMS[4]) in 2010. These latter experiments use more modern phototube designs with better electronics and readout, but the time resolution remains at $\sim$ 100 ps(sigma).

The ALICE experiment in 2000 required an exceptional Time of Flight detector to identify as many as possible of the particles produced in the multitude expected from the interactions of heavy ions. In fig. 1 we show a typical heavy ion interaction. Clearly it is impossible to build a time-of-flight array with the segmentation offered by scintillators coupled to some hundreds of phototubes.

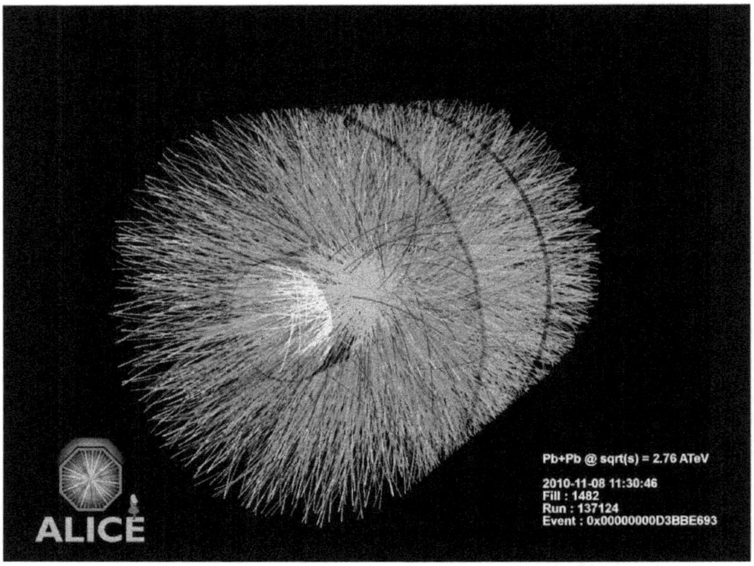

Fig. 1. A typical Pb-Pb interaction observed in the ALICE detector. Clearly scintillators read out with phototubes does not offer enough segmentation.

The Multigap Resistive Plate Chamber (MRPC) was invented in 1996 [5]. This device has been discussed in many papers and a review of its performance as a time-of-flight (TOF) detector can be found in ref [6]. The key points of this device will be discussed further in this paper.

During the R&D phase of developing the ALICE TOF, we found that only by employing differential readout a time resolution below 100 ps (sigma) could be achieved. Thus the MRPCs designed for ALICE employed differential readout. A new ASIC, the NINO [7,8], was designed, using differential architecture for all stages, in order to optimise time resolution. The importance of differential readout is the focus of this paper and will be discussed more below.

A device known as the Silicon Photomultiplier (SiPM) has been invented in Russia by Z. Sadygov [9], V. Golovin and Savedier[10]. This is a solid-state photo-detector that can be used to replace the photomultiplier tube. It operates in a magnetic field and can be highly segmented. This device is also known as a Multi-Pixel Photon Counter (MPPC) by Hamamatsu. In order to obtain enhanced time resolution we apply the same techniques as developed for the MRPC, i.e. differential readout with the NINO ASIC.

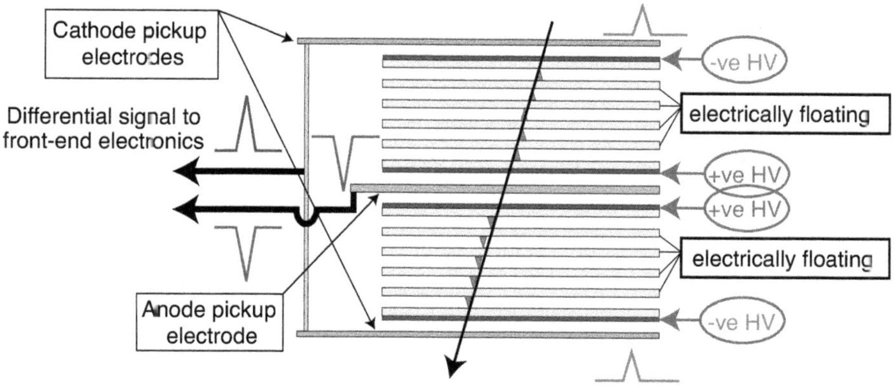

Fig. 2. A schematic cross-section of the multigap resistive plate chamber.

2 The multigap Resistive Plate Chamber (MRPC)

A schematic cross-section of the MRPC device used for ALICE time-of-flight, is shown in fig. 2. Since this device has been widely studied and discussed, we do not go into detail here, but just make some key observations.

This particular device is a double stack: each stack consists of 5 gas gaps of 250 μm in size. There is a central anode pickup electrode and two outer cathode electrodes so that the signal from the detector is fully differential. All the intermediate plates are electrically floating, but are kept at the correct voltage by the flow of electrons and positive ions into the surfaces of these plates. Any deviation from the correct voltage will affect the gain in the gas gaps and alter the ratio of positive ions and electrons. The stable state is equal gain in each gas gap. Since there are no connections to these internal plates, the chamber is relatively simple to build: one can just stack up the plates separating them with suitable spacers (fishing line in our case).

The time resolution of the MRPC depends on the size of the gas gap; smaller gaps give better timing. However the probability of a through-going particle creating an electron/positive ion cluster that then avalanches to create an observable signal obviously decreases as the gas gaps are made smaller. Thus for small gas gaps (needed to give good timing) the number of gaps must be increased to obtain high detection efficiency. A 24 gap MRPC with gaps of 160 micron has been built and tested[11]. On subsequent testing the published 20 ps time resolution was further improved to 16 ps. This is shown in fig. 3.

In addition to the excellent timing, ease of building and segmentation, the MRPC has very low dark noise and current (typically 10 nA/m^2 and a DCR of 0.2 Hz/cm^2 [12]) so that it works as an excellent trigger device.

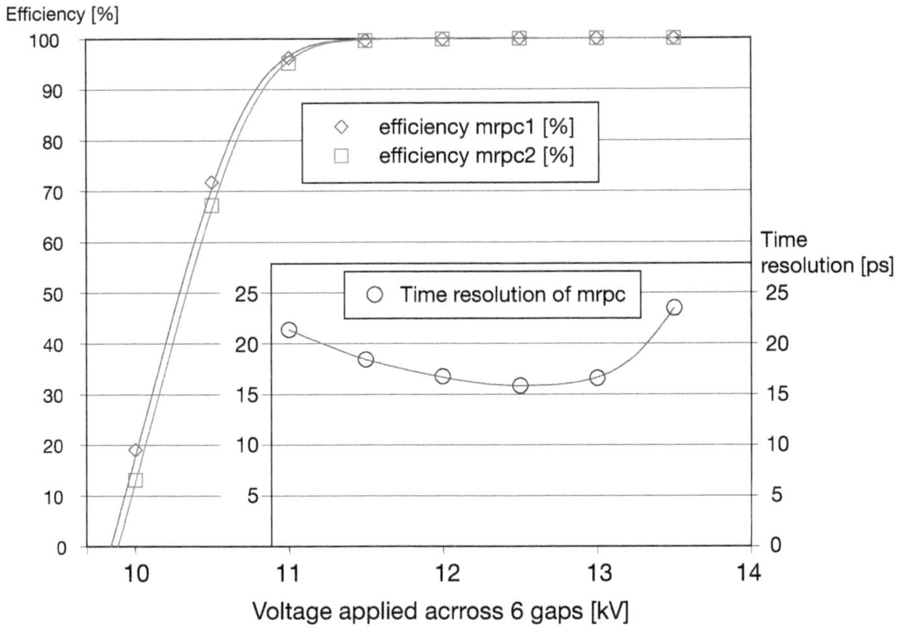

Fig. 3. Efficiency and time resolution of two 24 gap MRPC tested in a 5 GeV/c pion beam.

3 Differential readout and the NINO asic

When precise timing is required, differential readout is needed, as noted above. This concept is illustrated in fig. 4. All signals in all detectors are due to the movement of charge between two pickup electrodes. On the left we illustrate the differential readout; charge is induced on the two pickup electrodes and a current pulse flows through the front end of the electronics from one electrode to the other; thus each channel is truly independent. The case of the single-ended readout is shown on the right; here the current pulse flows through the front-end electronics into the common ground shared by all the electronic channels. For devices designed for precise timing, the signal is a current pulse that has a very short rise-time and this is coupled to ultra-fast and sensitive front-end electronics. For the single-ended case, the ground is the reference level: thus all these fast current pulses introduce noise in the reference level creating timing jitter. This is particularly important for photo-detectors reading out crystals. For precise timing one needs to measure the time of arrival of the first photon, so one is obliged to set the threshold at the 1 photoelectron level. However the signal from an energetic gamma (511 keV in the case of PET) generates tens of thousands of photons, so the signal is enormous and cross-talk will result (and thus extra noise is introduced into the common

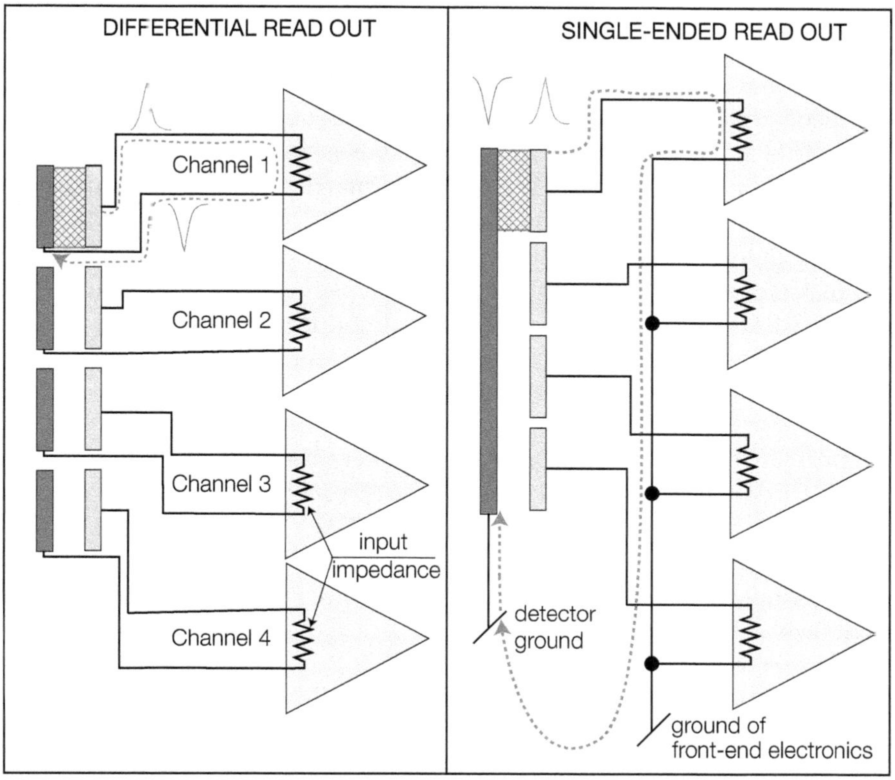

Fig. 4. The difference between the single-ended and differential readout. All signals are due to the movement of charge between two pickup electrodes, with an equal and opposite induced charge on each electrode. For the differential readout shown on the left hand side, the induced charge on one electrode flows through the front-end electronics to the other electrode. In the case of the single ended readout, fast current signals are fed into the electronics ground that is the reference level: this generates noise and thus time jitter.

ground).

For the ALICE TOF detector we observed the importance of differential read-out during the early stages of the R&D. Thus the MRPC detector was divided into strips (thus allowing access to both anode and cathode readout electrode) and the NINO ASIC was designed. The NINO[7,8] is an ultra-fast amplifier/discriminator that has differential architecture throughout. It is built in 0.25 μm CMOS technology.

4 Silicon photomultipliers read out with the NINO ASIC

The silicon photomultiplier (SiPM) is a collection of individual avalanche photodiodes run at a high gain such that the avalanche initiates a Geiger discharge. Let us consider attaching a SiPM to a crystal matrix. This could be an element of a calorimeter for a LHC experiment, or a detector for a 511 keV gamma for a PET detector. As noted above, for the best timing differential readout is needed. Unfortunately, it is difficult to implement differential readout for an array of SiPMs. The substrate is often common for a 4x4 cell array, thus all 16 cells will have a common cathode (or anode); for such a configuration only single-ended readout is possible.

For optimum timing, a fast front-end amplifier/discriminator has to be used and the threshold set to detect one photoelectron(10 to 50 fC). As discussed above the full light pulse produced by the 511 keV gamma or by a through-going minimum-ionising particle in the crystal is very large. Thus with single-ended readout and electronics, there is extra noise (increasing time jitter) and a high level of cross-talk.

The solution is to have differential readout and differential electronics; the only way to do this with the SiPM array is to create an array with individual SiPM cells and mounted as close together on a substrate. However, the electrical connections from the outer edges of these SiPM matrices to the individual cells creates a dead space. Hamamatsu has recently brought out a new range of SiPM/MPPC that allows differential readout without creating dead space. These matrices have both anode and cathode connections on the back side of the device with Through Silicon Vias (TSV) [13]. We fully expect that the timing performance of these SiPM devices to improve now that differential readout can be exploited.

5 Conclusion

The Multigap Resistive Plate Chamber has allowed large time-of-flight arrays to be built since the time resolution is excellent and the device can be easily segmented and can operate in a magnetic field. The NINO asic was designed and built to exploit the differential signal from the MRPC and underlined the importance of differential readout for highly segmented detectors.

The Silicon Photomultipliers have been developed during the last decade. They have excellent time resolution; however, as we have already learned from the R&D for the MRPC, the only possibility to exploit this time resolution is to derive a differential signal from each cell making up the matrix. Currently, the best time resolution has been obtained from the SiPM only when a differential signal has been coupled to a NINO ASIC [14].

References

[1] *'Experimental Observation of Antideuteron Production'*, T. Massam, Th. Muller, B. Righini, M. Schneegans, A. Zichichi, Il Nuovo Cimento Vol. XXXIX, (1965)10

[2] *'Symposium to celebrate the 30th anniversary of the Discovery of Nuclear Antimatter'*, L. Maiani and R.A. Ricci (eds), Conference Proceedings 53, Italian Physical Society, Bologna, Italy (1996)

[3] *'The Belle detector'*, A. Abashian, K. Gotow, N. Morgan, L. Piilonen S. Schrenk, K Abe, I. Adachi, J.P. Alexander,1, K. Aoki, S. Behari, Y. Doi, R. Enomoto,2, H. Fujii, Y. Fujita, Y. Funahashi, J. Haba, H. Hamasaki, T. Haruyama, K. Hayashi, Y. Higashi, N. Hitomi, S. Igarashi, Y. Igarashi, T. Iijima, Hirokazu Ikeda, Hitomi Ikeda, R. Itoh, M. Iwai, H. Iwasaki, Y. Iwasakib, K.K. Joo, K. Kasami, N. Katayama, M. Kawai, H. Kichimi, T. Kobayashi, S. Koike, Y. Kondo, M.H. Lee, Y. Makida, A. Manabe, T. Matsuda, T. Murakami, S. Nagayama, M. Nakao, T. Nozaki, K. Ogawa, R. Ohkubo, Y. Ohnishi, H. Ozaki, H. Sagawa, M. Saito, Y. Sakai, T. Sasaki, N. Sato, T. Sumiyoshi J. Suzuki, J.I. Suzuki, S. Suzuki, F. Takasaki, K. Tamai, M. Tanaka, T. Tatomi, T. Tsuboyama, K. Tsukada, T. Tsukamoto, S. Uehara, N. Ujiie, S. Uno, B. Yabsley, Y. Yamada, H. Yamaguchi, H. Yamaoka, Y. Yamaoka, M. Yamauchi, Y. Yoshimura, H. Zhao, R. Abe, G. Iwai, T. Kawasaki, H. Miyata, K. Shimada, S. Takahashi, N. Tamura, K. Abed, H. Hanada, T. Nagamine, M. Nakajima, T. Nakajima, S. Narita, M. Sanpei, T. Takayama, M. Ueki, M. Yamaga, A. Yamaguchi, B.S. Ahn, J.S. Kang, Hyunwoo Kim, C.W. Park, H. Park, H.S. Ahn, H.K. Jang, C.H. Kim, S.K. Kim, S.H. Lee, C.S. Park, E. Won, H. Aihara, T. Higuchi, H. Kawai, T. Matsubara, T. Nakadaira, H. Tajima, J. Tanaka, T. Tomura, M. Yokoyama, M. Akatsu, K. Fujimoto, M. Hirose, K. Inami, A. Ishikawa, S. Itami, T. Kani, T. Matsumoto, I. Nagai, T. Okabe, T. Oshima, K. Senyo, A. Sugi, A. Sugiyama, S. Suitoh, S. Suzuki, M. Tomoto, K. Yoshida, R. Akhmetshin, P. Chang, Y. Chao, Y.Q. Chen, W.S. Hou, S C. Hsu, H.C. Huang, T.J. Huang, M.C. Lee, R.S. Lu, J.C. Peng, K.C. Peng, S. Sahu, H.F. Sung, K.L. Tsai, K. Ueno, C.C. Wang, M.Z. Wang, G. Alimontj, T.E. Browder, B.C.K. Casey, F. Fang, H. Guler, M. Jones, Y. Li, S.L. Olsen, M. Peters, J.L. Rodriguez, M. Rosen, S. Swain, K. Trabelsi, G. Varner, H. Yamamoto, Y.H. Zheng, Q. An, H.F. Chen, Y.F. Wang, Z.Z. Xu, S.W. Ye, Z.P. Zhang, M. Asai, Y. Asano, S. Mori, S. Stanic, Y. Tsujita, J. Zhang, D. Zontar, T. Aso, V. Aulchenko, D. Beiline, A. Bondar, L. Dneprovsky, S. Eidelman, A. Garmash, A. Kuzmin, L. Romanov, N. Root, B. Shwartz, A. Sidorov, V. Sidorov, Y. Usov, V. Zhilich, A.M. Bakich, L.S. Peak, K.E. Varvell, E. Banas, A. Bozek, P. Jalocha, P. Kapusta, Z. Natkaniec, W. Ostrowicz, H. Palka, M. Rozanka, K. Rybicki, P.K. Behera, A. Mohapatra, M. Satapathy, Y.H. Chang,

H.S. Chen, L.Y. Dong, J. Li, H.M. Liu, Z.P. Mao, C.X. Yu, C.C. Zhang, S.Q. Zhang, Z.G. Zhao, Z.P. Zheng, B.G. Cheon, Y. Choi, D.W. Kim, J.W. Nam, S. Chidzik, K. Korotuschenko, C. Leonidopoulos, T. Liu, D. Marlow, C. Mindas, E. Prebys, R. Rabberman, W. Sands, R. Wixted, S. Choi, J. Dragic, C.W. Everton, A. Gordon, N.C. Hastings, E.M. Heenan, L.C. Moffitt, G.R. Moloney, G.F. Moorhead, M.E. Sevior, G.N. Taylor, S.N. Tovey, A. Drutskoy, R. Kagan, P. Pakhlov, S. Semenov, C. Fukunaga, R. Suda, M. Fukushima, V.I. Goriletsky, B.V. Grinyov, V.R. Lyubinsky, A.I. Panova, K.V. Shakhova, L.I. Shpilinskaya, E.L. Vinograd, B.G. Zaslavsky, R.S. Guo, F. Haitani, Y. Hoshi, K. Neichi, K. Hara, T. Hara, M. Hazumi, T. Hojo, D. Jackson, H. Miyake, Y. Nagashima, J. Ryuko, K. Sumisawa, M. Takita, T. Yamanaka, H. Hayashii, K. Miyabayashi, S. Noguchi, S. Hikita, H. Hirano, K. Hoshina, H. Mamada, O. Nitoh, N. Okazaki, T. Yokoyama, H. Ishino, S. Ichizawa, T. Hirai, H. Kakuno, J. Kaneko, T. Nakamura, Y. Ohshima, Y. Watanabe, S. Yanaka, Y. Inoue, E. Nakano, T. Takahashi, Y. Teramoto, J.H. Kang, H.J. Kim, Heejong Kim, Y.-J. Kwon, H. Kawai, E. Kurihara, T. Ooba, K. Suzuki, Y. Unno, N. Kawamura, H. Yuta, K. Kinoshita, A. Satpathy, S. Kobayashi, T. Kuniya, A. Murakami, T. Tsukamoto, S. Kumar, J. Singh, J. Lange, R. Stock, S. Matsumoto, M. Watanabe, H. Matsuo, S. Nishida, T. Nomura, H. Sakamoto, N. Sasao, Y. Ushiroda, Y. Nagasaka, Y. Tanaka, S. Ogawa, H. Shibuya, K. Hanagaki, S. Okuno, D.Z. Shen, D.S. Yan, Z.W. Yin, N. Tan, C.H. Wang, T. Yamaki, Y. Yamashita, Nucl. Instrum. Meth A 479 (2002)117.

[4] *'The Alpha Magnetic Spectrometer (AMS) experiment'*, S.C.C. Ting, International Symposium on *"Subnuclear Physics: Past, Present and Future"*, 30 October-2 November 2011, Vatican City, The Pontifical Academy of Sciences, Scripta Varia 119, pages 564-604, PAS, Vatican City (2014).

[5] *'A New type of resistive plate chamber: The Multigap RPC'*, E. Cerron Zeballos, I. Crotty, D. Hatzifotiadou, J. Lamas Valverde, S. Neupane, M.C.S. Williams and A. Zichichi; Nucl. Instrum. Meth. A374(1996)132.

[6] *'Particle identification using time of flight'*, M C S Williams, J. Phys. G: Nucl. Part. Phys. 39(2012)123001 doi:10.1088/0954-3899/39/12/123001.

[7] *'NINO : An ultra-fast and low-power front-end amplifier/discriminator ASIC designed for the multigap resistive plate chamber'*, F. Anghinolfi, P. Jarron, A.N. Martemyanov, E. Usenko, H. Wenninger, M.C.S. Williams, A. Zichichi, Nucl. Instr. Meth. A 452 (2004)183.

[8] *'NINO: An Ultrafast Low-Power Front-End Amplifier Discriminator for the Time-of-Flight Detector in the ALICE Experiment'*, F. Anghinolfi, P. Jarron, F. Krummenacher, E. Usenko, and M. C. S. Williams, Proc. IEEE Nucl. Sci. Symp., Portland, OR, Oct. 2003, vol. 1, pp. 375-379.

[9] Z. Sadygov, *'Avalanche detector'*, Russian patent RU2102820, Application from 10/10/1996, published in 01/20/1998.

[10] *'Novel type of avalanche photodetector with Geiger mode operation'*, V. Saveliev, V. Golovin, Nucl. Instr. and Meth. A 442 (2000) 223.

[11] *'A 20 ps timing device: A Multigap Resistive Plate Chamber with 24 gas gaps'*, S. An, Y.K. Jo, J.S. Kim, M.M. Kim, D. Hatzifotiadou, M.C.S. Williams, A. Zichichi, R. Zuyeuski, Nucl. Instr. Meth. A 594(2008)39

[12] *'Performance of the ALICE Time-Of-Flight detector at the LHC'*, AA. Akindinov, A. Alici, A. Agostinelli, P. Antonioli, S. Arcelli, M. Basile, F. Bellini, G. Cara Romeo, L. Cifarelli, F. Cindolo, M. Colocci, A. De Caro, D. De Gruttola, S. De Pasquale, K. Doroud, M. Fusco Girard, B. Guerzoni, D. Hatzifotiadou, D. W. Kim, J. S. Kim, S. Kiselev, S. C. Lee, D. Malkevich, A. Margotti, R. Nania, A. Nedosekin, F. Noferini, P. Pagano, A. Pesci, O. Pinazza, R. Preghenella, E. Scapparone, G. Scioli, K. Voloshin, M. C. S. Williams, C. Zampolli, A. Zichichi, Eur. Phys. J. Plus (2013) 128: 44 DOI 10.1140/epjp/i2013-13044-x

[13] *'Assembly technology of 4-side buttable MPPC'*, poster presented at Vienna Conference on Instrumentation by K. Yamamoto; VCI2013 11-15 February 2013

[14] *'Systematic study of new types of Hamamatsu MPPCs read out with the NINO ASIC'*, K. Doroud, A. Rodriguez, M.C.S. Williams, K. Yamamoto, A. Zichichi, R. Zuyeuski, Nucl. Instr. Meth. A 753 (2014)149.

DIPLOMAS

Twenty six Diplomas were open for competition among the participants. They have been awarded as follows:

- John S. BELL Diploma to:
  Leron BORSTEN
  Imperial College of London, UK

- Patrick M.S. BLACKETT Diploma to:
  Silvia NAGY
  Imperial College of London, UK

- Nicola CABIBBO Diploma to:
  Nils RUTHMANN
  Albert-Ludwigs Universitat Freiburg, Germany

- James CHADWICK Diploma to:
  Daniel BRITZGER
  DESY, Hamburg, Germany

- Sidney COLEMAN Diploma to:
  Josh NOHLE
  UCLA, Los Angeles, CA, USA

- Richard H. DALITZ Diploma to:
  Alfonso SASTRE BRUNO
  Centre de Physique Théorique, Marseille, France

- Paul A.M. DIRAC Diploma to:
  Christine HARTMANN
  CERN, Geneva, Switzerland
  Niels Bohr Institute, University of Copenhagen, Denmark

- Bruno FERRETTI Diploma to:
  Joseph SEELE
  Riken BNL, Upton, NY, USA

- Richard P. FEYNMAN Diploma to:
  Andrey PIKELNER
  Joint Institute for Nuclear Research, Dubna, Russian Federation

- Vladimir N. GRIBOV Diploma to:
  Oleksii IVANYTSKYI
  Bogolyubov Institute for Theoretical Physics
  National Academy of Sciences of Ukraine, Kiev, Ukraine

- Robert HOFSTADTER Diploma to:
  Markus SEIDEL
  Universität Hamburg, Germany

- Gunnar KÄLLEN Diploma to:
  Moritz KIEHN
  Physikalisches Institut, Heidelberg, Germany

- Seymour J. LINDENBAUM Diploma to:
  Joseph ZENNAMO
  Fermilab, Batavia, IL, USA

- Yuval NE'EMAN Diploma to:
  Stefanos LEONTSINIS
  National Technical University of Athens, Greece
  CERN, Geneva, Switzerland

- Giuseppe P.S. OCCHIALINI Diploma to:
  Marco VIGNATI
  INFN, Sapienza University of Rome, Italy

- Oreste PICCIONI Diploma to:
  Mohammad Rafi ALAM
  Department of Physics
  Aligarh Muslim University, Aligarh, India

- Bruno PONTECORVO Diploma to:
  Alexander Karl ROTHKOPF
  Albert Einstein Center for Fundamental Physics
  University of Bern, Switzerland

- Giampietro PUPPI Diploma to:
  Joern LANGE
  Institute of Experimental Physics, Hamburg, Germany

- Isidor I. RABI Diploma to:
  Chris WYMANT
  Institute for Particle Physics Phenomenology, University of Durham, UK

- Giulio RACAH Diploma to:
  Màtè CSANÀD
  Eötvös Loránd University, Budapest, Hungary

- Bruno ROSSI Diploma to:
  Matthias SCHROEDER
  Universität Hamburg, Germany

- Julian S. SCHWINGER Diploma to:
  Alessio MARRANI
  Instituut Voor Theoretische Fysica, Leuven, Belgium

- Victor F. WEISSKOPF Diploma to:
  Sergii TROFYMENKO
  Kharkov Institute of Physics and Technology, Kharkov, Ukraine

- Eugene P. WIGNER Diploma to:
  Alla SEMENOVA
  Saint-Petersburg, Russian Federation

- Bjorn H. WIIK Diploma to:
  Chiun-Tien YU
  FNAL, Batavia, IL, USA

- Chien Shiung WU Diploma to:
  Rosanna DE PALO
  University of Padova, Italy

Diploma to Celebrate 90th Birthday of Bruno ZUMINO to:
Scott DAVIES
UCLA, Los Angeles, CA, USA

Award for the BEST STUDENT to:
Christine HARTMANN
CERN, Geneva, Switzerland
Niels Bohr Institute, University of Copenhagen, Denmark

PARTICIPANTS

Ahmadjon ABDUJABBAROV **Uzbekistan**	Institute of Nuclear Physics, Ulughbek Tashkent, Uzbekistan
Robin AGGLETON **UK**	University of Bristol, UK
Mohammad Rafi ALAM **India**	Aligarh Muslim University, Aligarh, India
Vasyl ALBA **Ukraine**	ITEP, Moscow, Russian Federation; and Princeton University, Princeton, NJ, USA
Daniil ANTONENKO **Russia**	Moscow Institute of Physics and Technology Dolgoprudniy, Russian Federation
Ignatios ANTONIADIS **France**	Centre de Physique Théorique Ecole Polytechnique, Palaiseau, France
Richard ARNOWITT **USA**	Texas A&M University, College Station, TX, USA
Konstantin ASTAPOV **Russia**	Institute for Nuclear Research of RAS Moscow, Russian Federation
Christopher BACKHOUSE **USA**	CALTECH, Pasadena, CA, USA
Hend Magdi I. M. BAZA **Egypt**	Egyptian Center for Theoretical Physics MTI University, Cairo, Egypt
Viktor BEGUN **Ukraine**	Bogolyubov Institute Theoretical Physics, Ukraine
Zvi BERN **USA**	UCLA, Los Angeles, CA, USA
Sergio BERTOLUCCI **Italy**	CERN, Geneva, Switzerland
Alessandro BETTINI **Italy**	INFN & Padoa University, Italy; and Canfranc Underground Lab, Spain
Pietro BIANCOFIORE **Italy**	University of Bari, INFN Bari, Italy
Kyrylo BONDARENKO **Ukraine**	Kiev National Taras Shevchenko University, Ukraine

Leron BORSTEN **Holland**	Imperial College of London, London, UK
Daniel BRITZGER **Germany**	DESY, Hamburg, Germany
Tian Yu CAO **USA**	Boston University, MA, USA
Maria Valentina CARLUCCI **Italy**	Technische Universitaet Muenchen TUM Institute of Advanced Studies, Garching, Germany
Bianca Letizia CERCHIAI **Italy**	University of Milano, Italy
Luisa CIFARELLI **Italy**	INFN & University of Bologna, Italy
Frank CLOSE **UK**	University of Oxford, Oxford, UK
Manuel COLOCCI **Italy**	University of Bologna and INFN Bologna, Italy
Màtè CSANÀD **Hungary**	Eötvös Loránd University, Budapest, Hungary
Ivan DANILENKO **Russia**	Moscow Institute of Physics and Technology Dolgoprudniy, Russian Federation
Scott DAVIES **USA**	UCLA, Los Angeles, CA, USA
Giacomo DE PALMA **Italy**	Scuola Normale di Pisa, Italy
Rosanna DE PALO **Italy**	INFN & University of Padova, Padova, Italy
Katayoun DOROUD **Iran**	Centro Studi e Ricerche Enrico Fermi, Roma, Italy and CERN, Geneva, Switzerland
Sean DOWNES **USA**	Texas A&M University, College Station, TX, USA
Michael J. DUFF **UK**	Imperial College London, London, UK
Petr DUNIN-BARKOWSKI **Russia**	ITEP, Moscow, Russian Federation

Angelo ESPOSITO **Italy**	La Sapienza University, Rome, Italy
Sergio FERRARA **Italy**	CERN, Geneva, Switzerland and LNF–INFN, Frascati, Italy
Fernando FERRONI **Italy**	INFN, Rome, Italy
Harald FRITZSCH **Germany**	Ludwig-Maximilians-Universitaet, Muenchen, Germany
Kenji FUKUSHIMA **Japan**	Keio University, Yokohama-shi, Kanahawa, Japan
Paolo GIUBELLINO **Italy**	University of Turin, Italy
Elena GIUBEGA **Romania**	National Institute of Nuclear Physics and Engineering Bucharest-Magurele, Romania
Ivan GNESI **Italy**	Centro Studi e Ricerche Enrico Fermi, Roma, Italy
Olga GOGOTA **Ukraine**	Kiev National Taras Shevchenko University, Ukraine
Vladimir GOY **Russia**	Far Eastern Federal University Vladivostok, Russian Federation
Barbara GUERZONI **Italy**	INFN - University of Bologna, Bologna, Italy
Dieter HAIDT **Germany**	EPJC–DESY, Hamburg, Germany
Christine HARTMANN **Ukraine**	CERN, Geneva, Switzerland and Niels Bohr Institute, University of Copenhagen, Denmark
Thomas HERMANN **Germany**	Karlsruhe Institute fuer Technologie (KIT), Germany
Peter Ware HIGGS **UK**	The University of Edinburgh, UK
Tahir HUSSAIN **India**	Aligarh Muslim University, Aligarh, India
Peter JENNI **Switzerland**	CERN, Geneva, Switzerland

Gianluca INGUGLIA **Italy**	Queen Mary University of London, UK
Oleksii IVANYTSKYI **Ukraine**	Bogolyubov Institute for Theoretical Physics National Academy of Sciences of Ukraine, Ukraine
Georgios KARANANAS **Greece**	Ecole Polytechnique Federale de Lausanne, Switzerland
Moritz KIEHN **Germany**	Physikalisches Institut, Heidelberg, Germany
Iakov KONONOV **Russia**	Moscow Institute of Physics and Technology Dolgoprudniy, Russian Federation
Christian P. KORTHALS-ALTES **The Netherlands**	CNRS-Luminy, Marseille, France
Andrey KOTOV **Russia**	ITEP, Moscow, Russian Federation
Olga KRUGLINSKAYA **Russia**	Lebedev Physics Institute of RAS Moscow, Russian Federation
Nadiia KRUPINA **Ukraine**	Universitatet Mainz, Germany
Evgeny KRYLOV **Russia**	Moscow State University, Moscow, Russian Federation
Maxim KURKOV **Russia**	University of Napoli Federico II, Napoli, Italy
Maxim LALETIN **Russia**	National Research Nuclear University MEPhI Moscow, Russian Federation
Joern LANGE **Germany**	Institute of Experimental Physics, Hamburg, Germany
Andràs LÀSZLO **Hungary**	HAS Wigner Research Centre for Physics Budapest, Hungary
Emanuele LATINI **Italy**	Universitat Zürich, Switzerland
Alexander LEITHES **UK**	Queen Mary University of London, UK
Oleg LEMETS **Ukraine**	Kharkov Institute of Physics and Technology, Ukraine

Stefanos LEONTSINIS **Greece**	National Technical University of Athens, Greece and CERN, Geneva, Switzerland
Lev N. LIPATOV **Russia**	PNPI, St. Petersburg, Russian Federation
Rafael LOPES DE SÀ **Brazil**	Stony Brook University, NY, USA
Árpád László LUKÁCS **Hungary**	HAS Wigner Research Centre fur Physics Budapest, Hungary
Luciano MAIANI **Italy**	La Sapienza University, Rome, Italy
Sergii MAMEDOV **Ukraine**	G.V. Kurdyumov Institute for Metal Physics, Ukraine
Marco MARITI **Italy**	University of Pisa, Italy
Alessio MARRANI **Italy**	Instituut Voor Theoretische Fysica, Leuven, Belgium
Karina MARTIN **Russia**	Budker Institute of Nuclear Physics Novosibirsk, Russian Federation
Thomas Edward MELIA **UK**	CERN, Geneva, Switzerland
Lorenzo MENCULINI **Italy**	University of Perugia, Italy
Paul Emil MEREUTÀ **Romania**	National Institute of Nuclear Physics and Engineering Bucharest-Magurele, Romania
Emanuele MESSINA **Italy**	INFN University of Catania, Catania, Italy
Peter MINKOWSKI **Switzerland**	University of Bern, Switzerland
Tariq MOHAMMAD **India**	Aligarh Muslim University, Aligarh, India
Matin MOJAZA **Denmark**	Syddansk Universitet, Odense, Denmark
Oleg MOROZ **Ukraine**	Bogolyubov Institute for Theoretical Physics, Ukraine
Andrey MOROZOV **Russia**	ITEP, Moscow, Russian Federation

Anton MOROZOV
Russia
ITEP, Moscow,
Russian Federation

Viatcheslav MUKHANOV
Russia
Ludwig Maximilian University of Munich,
Germany

Silvia NAGY
Romania
Imperial College,
London, UK

Dimitri V. NANOPOULOS
Greece
Texas A&M University, College Station,
TX, USA

Alexander NESTERENKO
Russia
Joint Institute for Nuclear Research
Dubna, Russian Federation

Vladimir NIKITIN
Russia
Moscow State University, Moscow,
Russian Federation

Ramil NIYAZOV
Russia
Saint-Petersburg State University,
Russian Federation

Francesco NOFERINI
Italy
Enrico Fermi Centre, Rome and
University of Bologna, Italy

Josh NOHLE
USA
UCLA, Los Angeles,
CA, USA

Oleg Olegovich NOVIKOV
Russia
Saint-Petersburg State University,
Russian Federation

Maxim PERFILOV
Russia
Moscow State University,
Russian Federation

Andrei PETROVSKII
Russia
Saint-Petersburg Academie University,
Russian Federation

Piotr PIETRULEWICZ
Germany
University of Vienna, Austria

Andrey PIKELNER
Russia
Joint Institute for Nuclear Research
Dubna, Russian Federation

Fedor POPOV
Russia
ITEP, Moscow,
Russian Federation

Semyon POZDNYAKOV
Russia
Saint-Petersburg State University,
Russian Federation

Roberto PREGHENELLA
Italy
Enrico Fermi Centre, Rome, Italy and
INFN-University of Bologna, Italy

Olga PTITSYNA
Russia
Moscow State University,
Russian Federation

Ksenia PTITSYNA
Russia

Moscow State University,
Russian Federation

Ilia PUSENKOV
Russia

Saint-Petersburg State University,
Russian Federation

Anastasiia REBESH
Ukraine

Bogolyubov Laboratory of theoretical Physics,
Ukraine

Edilson Alfonso REYES ROJAS
Colombia

Universidad Nacional de Colombia, Bogotà,
Colombia

Alain RIAZUELO
France

Institut d'Astrophysique de Paris,
France

Veronica RIQUER RAMIREZ
Mexico

CERN, Geneva,
Switzerland

Klaus RITH
Germany

DESY-HERMES, Hamburg, Germany; and
FAU Erlangen-Nuremberg,
Germany

Alexander Karl ROTHKOPF
Germany

Albert Einstein Center for Fundamental Physics
University of Bern, Switzerland

Nils RUTHMANN
Germany

Albert-Ludwigs Universitat Freiburg,
Germany

Alfonso SASTRE BRUNO
Spain

Centre de Physique Théorique, Marseille,
France

Petr SATUNIN
Russia

Moscow State University, Moscow,
Russian Federation

Matthias SCHROEDER
Germany

Universität Hamburg,
Germany

Joseph SEELE
USA

Riken BNL, Upton,
NY, USA

Markus SEIDEL
Germany

Universität Hamburg,
Germany

Alla SEMENOVA
Russia

B.P. Konstantinov Petersburg Nuclear Physics
Institute
Saint-Petersburg, Russian Federation

Eva SICKING
Germany

CERN, Geneva,
Switzerland

Anna SINELNIKOVA
Russia

Moscow State University,
Russian Federation

Valentin SLEPUKHIN **Russia**	ITEP, Moscow, Russian Federation
Emeri SOKATCHEV **France**	Laboratoire d'Annecy-Le-Vieux de Physique Theorique Annecy-le-Vieux Cedex, France
Nikita SOPENKO **Russia**	Moscow Institute of Physics and Technology Dolgoprudniy, Russian Federation
Yang SUN **China**	University of North Carolina, Chapel Hill, NC, USA
Maksym SYDORENKO **Ukraine**	T. Shevchenko National University, Kiev, Ukraine
Michael J. TANNENBAUM **USA**	Brookhaven National Laboratory, Upton, NY, USA
Daniel Jesus TAPIA TAKAKI **Mexico**	IPNO – Université de Paris-Sud, Orsay, France
Abdel Naser TAWFIK **Germany**	ECPT, MTI University, Cairo, Egypt
Tom TAYLOR **Switzerland**	CERN, Geneva, Switzerland
Maksym TESLYK **Ukraine**	T. Shevchenko National University, Kiev, Ukraine
Gerardus 't HOOFT **the Netherlands**	Utrecht University, the Netherlands
Inar TIMIRYASOV **Russia**	Institute for Nuclear Research of RAS Moscow, Russian Federation
Arsenii TITOV **Russia**	Saint-Petersburg State University, Russian Federation
Dmytro TIUTIUNNYK **Ukraine**	Kharkov Institute of Physics and Technology, Ukraine
Anna TOKAREVA **Russia**	Institute for Nuclear Research of RAS Moscow, Russian Federation
William Javier TORRES BOBADILLA **Colombia**	Universidad Nacional de Colombia, Bogotà, Colombia
Sergii TROFYMENKO **Ukraine**	Kharkov Institute of Physics and Technology, Ukraine

Aleksandr TUMANOV
Russia
Saint-Petersburg Academic University,
Russian Federation

Maksim ULYBYSHEV
Russia
Moscow State University,
Russian Federation

Semen VALGUSHEV
Russia
Moscow Institute of Physics and Technology
Dolgoprudniy, Russian Federation

Marco VIGNATI
Italy
INFN, La Sapienza University of Rome,
Italy

Alan WALKER
UK
School of Physics and Astronomy
The University of Edinburgh, UK

Yadi WANG
China
Laboratori Nazionali INFN
Frascati, Italy

Horst WENNINGER
Germany
CERN, Geneva,
Switzerland

Christopher WEVER
Netherlands
Utrecht University,
The Netherlands

Crispin WILLIAMS
UK
CERN, Geneva,
Switzerland

Chris WYMANT
UK
Institute for Particle Physics Phenomenology
University of Durham, UK

Danylo YEROKHIN
Ukraine
Kharkov Institute of Physics and Technology,
Ukraine

Chiun-Tien YU
Japan/USA
FNAL, Batavia, IL,
USA

Joseph ZENNAMO
USA
Fermilab, Batavia,
ILL, USA

Antonino ZICHICHI
Italy
CERN, Geneva, Switzerland and
INFN & University of Bologna, Italy

Michele ZOCCALI
Italy
Queen Mary University,
London, UK

Roman ZUBOV
Russia
Saint-Petersburg State University,
Russian Federation

www.ingramcontent.com/pod-product-compliance
Ingram Content Group UK Ltd.
Pitfield, Milton Keynes, MK11 3LW, UK
UKHW052114170926
8624IPUK00026B/393